BIRDS *of the* SALTON SEA

BIRDS *of the* SALTON SEA

Status, Biogeography, and Ecology

Michael A. Patten, Guy McCaskie, and Philip Unitt

UNIVERSITY OF CALIFORNIA PRESS
Berkeley   Los Angeles   London

University of California Press
Berkeley and Los Angeles, California

University of California Press, Ltd.
London, England

Library of Congress Cataloging-in-Publication Data

Patten, Michael A.
  Birds of the Salton Sea : status, biogeography, and
ecology / Michael A. Patten, Guy McCaskie, and
Philip Unitt.
     p.   cm.
  Includes bibliographical references (p.   ).
  ISBN 0-520-23593-2 (cloth : alk. paper)
  1. Birds—California—Salton Sea. I. McCaskie,
Guy II. Unitt, Philip. III. Title.
  QL684.C2 P38   2003
  598′.09794′99—dc21   2002013312

Manufactured in the United States of America
13 12 11 10 09 08 07 06 05 04
10 9 8 7 6 5 4 3 2 1

CONTENTS

FOREWORD

Anybody who has been to California's Salton Sea will have lasting memories, and perhaps a recurring nightmare or two. The sea assaults one's senses with a potpourri of richly organic odors, eye-popping sunsets, the biting cold of a winter morning, or the sauna ambience of a summer afternoon. And there are the birds—the spectacular flocks in the air or rafting on the sea, the yap-yapping of Black-necked Stilts punctuating the summer heat, the whistling of the wings of winter waterfowl. Birders, hunters, and tourists from far and wide have visited the Salton Sea to enjoy the spectacle of thousands of geese in the agricultural fields and wildlife refuges at the south end, along with winter flocks of ducks, egrets, ibises, gulls, and curlews. But many birders, and most notably this book's three authors, have learned that the Salton Sea's steamier, seedier "other side"—the sweltering heat and stifling humidity that grip the region from May through September—offers even greater rewards: significant yet fragile breeding colonies of terns, skimmers, herons, ibises, and cormorants; northward-wandering subtropical waterbirds; mudflats teeming with southbound shorebirds; even surprise visits by seafaring birds such as petrels and albatrosses.

I first birded the Salton Sea in January 1968. By this time Guy McCaskie was already well known for the wizardry of his precise and systematic approach to finding new and unusual birds and birding places. Phil Unitt was just developing an interest in ornithology that would lead to a career and to the authorship of numerous important works, including the monumental *Birds of San Diego County*. And the primary author, Michael Patten, had yet to start kindergarten. All trained in different decades, each of these co-authors approaches field ornithology with a different perspective. They have collaborated to produce a timely and masterful analysis of the birdlife of one of the most important bird areas in the Western Hemisphere.

Some ornithological exploration of the Salton Trough predates the formation of the present-day sea, and in many ways true study of the sea's avifauna began with Joseph Grinnell's voyage on the *Vinegaroon* in April 1908. But not until now has there been a complete analysis of the odd biogeographic juxtapositions that constitute the area's avifauna. The inclusion of subspecies treatments continues an important tradition that unfortunately has been lost from much of the recent literature.

This book documents the past and present avifauna of the Salton Sea at a crossroads in its history. Threats to the sea are ultimately derived from the decisions by governments, corporations, and special interests to accommodate a rapidly expanding human population in the water-starved southwestern United States at the expense of the natural integrity and unique and diverse wildlife of the region. The theft of this water—whether to lavish it upon the lawns of San Diego, to wash the cars of Los Angeles, or to spout it obscenely from the garish fountains of Las Vegas—has robbed the Colorado River delta of its lifeblood and now threatens the delta's de facto replacement, the Salton Sea. The future of the Salton Sea is uncertain. We do not know if this schol-arly and thorough effort by Michael Patten, Guy McCaskie, and Phil Unitt will serve as documentation of what is lost or a catalyst for ongoing research in a region of great diversity. In either or both of these roles, however, *Birds of the Salton Sea* will quickly become and long remain a classic.

KIMBALL L. GARRETT
*Natural History Museum
of Los Angeles County*

PREFACE AND ACKNOWLEDGMENTS

It is at best a rather cheerless object, beautiful in a pale, placid way, but the beauty is like that of a mirage, the placidity that of stagnation and death. Charm of color it has, but none of sentiment; mystery, but not romance. Loneliness has its own attraction, and it is a deep one; but this is not so much loneliness as abandonment, not a solitude sacred but a solitude shunned. Even the gulls that drift and flicker over it seem to have a spectral air, like bird-ghosts banished from the wholesome ocean.

> *"E'en the weariest river*
> *Winds somewhere safe to sea";*

but for the Salton the appointed end is but a slow sinking of its bitter, useless waters, a gradual baring of slimy shores, until it comes once more, and probably for the last time, to extinction in dead, hopeless desert.

J. SMEATON CHASE (1919),
DESCRIBING THE SALTON SEA ON 30 JULY 1911

The Salton Sea evokes myriad images, many in stark contradiction. It is at once unbearably hot, foul of odor, and seemingly uninhabitable, yet it is a haven to one of the largest populations of waterbirds in western North America. Indeed, its unique biogeographic setting supports more than four hundred native species, exceeding the total for many states. Some one hundred species, including the Brown Pelican, the Gull-billed Tern, and the Black Skimmer, oddities in the landlocked Southwest, see fit to breed in the region.

Yet all is not well in the Salton Sea. The water quality has deteriorated, and much native habitat has been lost, particularly thickets of mesquite and riparian woodland of cottonwood and willow. More alarming, massive die-offs of fish and birds have been commonplace, initially sparking concern, later yielding a fervent push to preserve this important ecosystem before it is lost. But basic conservation efforts require basic data. Before we can hope to preserve the ecosystem, we need to know more about the species that constitute it—their relative abundances, their habitats, their places of origin, the last being especially important for migratory taxa.

Birds are the most conspicuous animals in the ecosystem, and we have been studying them for many years. We pooled our expertise in an attempt to produce a treatise that would both form the foundation for informed conservation decisions and introduce scientists and laypersons to the birdlife of this fascinating region.

This book was inspired by many others, but it differs from them all. Particularly notable influences were *The Distribution of the Birds of California*, by Joseph Grinnell and Alden H. Miller (1944), *The Birds of Arizona*, by Allan R. Phillips, Joseph T. Marshall, and Gale Monson (1964), *The Birds of Southern California: Status and Distribution* by Kimball L. Garrett and Jon L. Dunn (1981), *Once a River*, by Amadeo M. Rea (1983), *The Birds of San Diego County*, by Philip Unitt (1984), and *Birds of the Lower Colorado River Valley*, by Kenneth V. Rosenberg, Robert D. Ohmart, William C. Hunter, and Bertin W. Anderson (1991). To varying degrees these works included detailed information on status, distribution, habitat, and subspecies. In the present volume, we have strived to incorporate the best elements of each of these books to construct exhaustive accounts of each species' current and historical status. We have included information on ecology in the Salton Sink, the biogeography of the sink and neighboring regions and its effect on distribution and occurrence, and taxonomy at the species and subspecies levels. Together these aspects paint a detailed portrait of the avifauna of the Salton Sea.

This project could not have been accomplished without the assistance of many field ornithologists. Several individuals deserve special recognition for their input and their willingness to share their expertise. To that end, we express our deepest gratitude to Brian E. Daniels, Kimball L. Garrett, Roger Higson, Kenneth Z. Kurland, Robert L. McKernan, Kathy C. Molina, Amadeo M. Rea, W. David Shuford, and Sherilee von Werlhof.

We thank the following individuals for supplying information on specimens in their care and other museum-related information, allowing access to their collections, or lending specimens: William G. Alther of the Denver Museum of Natural History, George F. Barrowclough, Christine Blake, Emanuel Levine, and Paul Sweet of the American Museum of Natural History in New York, Louis R. Bevier and Nate Rice of the Academy of Natural Sciences, Philadelphia, Kevin J. Burns of San Diego State University, Steven W. Cardiff and J. V. Remsen Jr. of the Louisiana State University Museum of Natural Science in Baton Rouge, Carla Cicero and Ned K. Johnson of the Museum of Vertebrate Zoology in Berkeley, Charles T. Collins of California State University, Long Beach, Paul W. Collins and Krista Fahy of the Santa Barbara Museum of Natural History, René Corado of the Western Foundation of Vertebrate Zoology in Camarillo, California, Tamar Danufsky of Humboldt State University in Arcata, California, Charles M. Dardia of Cornell University, James P. Dean and Craig Ludwig of the National Museum of Natural History in Washington, D.C., Kimball L. Garrett of the Natural History Museum of Los Angeles County, John C. Hafner and James R. Northern of the Moore Laboratory of Zoology at Occidental College, Mary Hennen of the Chicago Academy of Sciences, Fritz Hertel of the Dickey Collection at the University of California, Los Angeles, Gene K. Hess of the Delaware Museum of Natural History, Janet Hinshaw of the University of Michigan Museum of Zoology, Mark A. Holmgren of the University of California, Santa Barbara, Thomas R. Huels of the University of Arizona, Robert L. McKernan of the San Bernardino County Museum, Alison Pirie and Douglas Siegel-Causey of the Museum of Comparative Zoology at Harvard, Mark B. Robbins of the University of Kansas, Gary W. Shugart of the Slater Museum at the University of Puget Sound, Paul F. Whitehead of the Peabody Museum at Yale, David Willard of the Field Museum in Chicago, Chris Wood of the Burke Museum at the University of Washington, and Robert M. Zink of the Bell Museum at the University of Minnesota.

Many other individuals supplied unpublished field notes or other important information without which this project could not have been completed: Daniel W. Anderson, James C. Bednarz, Louis R. Bevier, Jean Brandt, Aaron Brees, M. Ralph Browning, P. A. Buckley, Jutta C. Burger, Eldon R. Caldwell, Eugene A. Cardiff, Steven W. Cardiff, Carla Cicero, Luke W. Cole, Daniel S. Cooper, René Corado, James P. Dean, Jon L. Dunn, Richard A. Erickson, Shawneen E.

Finnegan, Daniel D. Gibson, John Green, Sue Guers, Robert A. Hamilton, Loren R. Hays, Gjon Hazard, Matthew T. Heindel, Tom and Jo Heindel, Steve N. G. Howell, Stuart H. Hulbert, Joseph R. Jehl Jr., Ned K. Johnson, Paul D. Jorgensen, Lloyd F. Kiff, Howard King, John R. King, Sandy Koonce, Jim Kuhn, Paul E. Lehman, Tony Leukering, Linette Lina, Margaret McIntosh, Curtis A. Marantz, Chet McGaugh, Douglas B. McNair, Eric Mellink, Bob Miller, Jason A. Mobley, William J. Moramarco, Brennan Mulrooney, Richard J. Norton, Kenneth C. Parkes, Bruce G. Peterjohn, Stacy J. Peterson, the late Allan R. Phillips, Molly Pollock, Peter Pyle, Kurt A. Radamaker, Carol A. Roberts, Michael M. Rogers, Gary H. Rosenberg, Paul Saraceni, Jack W. Schlotte, N. John Schmitt, Jay M. Sheppard, Brenda D. Smith-Patten, Mary Beth Stowe, Emily Strauss, Brett Walker, Richard E. Webster, Walter Wehtje, the late Claudia P. Wilds, Douglas R. Willick, and Thomas E. Wurster.

We thank Karen Klitz and the Archives at the University of California, Berkeley, for supplying historical photographs of the Salton Sea. Kenneth Z. Kurland graciously provided many photographs of recent birds, as did Jack W. Schlotte of recent habitat. Additional photographs of birds and habitats were supplied by Barbara A. Carlson, Deborah L. Davidson, Greg W. Lasley, Brennan Mulrooney, Brian G. Prescott, Lawrence Sansone, Brenda D. Smith, and Richard E. Webster.

Part of Michael Patten's specimen research was funded by a collection study grant from the American Museum of Natural History. Financial support for the completion of this book was generously provided by the Western Ecological Research Center of the United States Geological Survey in San Diego, through a contract administered by Douglas A. Barnum and William I. Boarman. We thank Daniel W. Anderson, Jeff N. Davis, Joseph R. Jehl Jr., Paul F. Springer, Rich Stallcup, and Nils Warnock for reading all or portions of the text. Finally, we thank Doris Kretschmer and Nicole Stephenson of the University of California Press for their invaluable guidance and advice.

MICHAEL A. PATTEN
GUY MCCASKIE
PHILIP UNITT

A History of the Salton Sink

WE BEGIN THE STORY of the Salton Sea with the Salton Trough, the rift extending from the Coachella Valley to the central Gulf of California. The Salton Sink, which lies between the southern Coachella Valley and the northern Mexicali Valley, occupies but a small portion of the trough, but the histories of the two are intertwined. From a geological standpoint, the long history of the Salton Trough ranges from its connection with the Gulf of California during the Tertiary Period (Blake 1914; Durham and Allison 1960) to the maximum spread of Lake Cahuilla beginning some 40,000 years ago, during the pluvial times of the Pleistocene (Setmire et al. 1990). During the Pliocene, a few million years ago, the entire Salton Trough was, by most accounts, merely the head of the Gulf of California (Fig. 1; Durham and Allison 1960). Massive deposits of silt from the Colorado River eventually accumulated along the southern edge of Gravel Mesa to form a barrier between the former head of the gulf and the current one (Fig. 2; Blake 1914; Kennan 1917; Loeltz et al. 1975; cf. Free 1914). The enclosed sea subsequently dried up, but reminders of the saline environment in the form of "oyster-shells and other forms of marine life" are strewn across the base of the Santa Rosa and San Jacinto Mountains at an elevation of about 100 m (Blake 1914).

The depression resulting from the desiccation of this "trapped" sea was further honed by the uplifting of the surrounding mountains (Blake 1914) and became a deep basin generally called the Salton Trough (Sykes 1914). In rough terms, this trough includes the Coachella Valley, the Salton Sink (the geographical region covered in this book), the Imperial Valley, the Mexicali Valley, and the southern portion of the Río Colorado delta. (We refer to the U.S. portion of the river as the Colorado River and to the Mexican portion as the Río Colorado.) The Peninsular Ranges, the Anza-Borrego Desert, the Sierra Juárez, and various other ranges in northeastern Baja California border it to the west. The Orocopia and Chocolate Mountains, the Algodones Dunes, and Gravel Mesa form the eastern boundary. The northern terminus lies at the base of the great rift known as the San Gorgonio Pass.

Much of this trough lies well below sea level, an area known as the Salton Sink or, sometimes, the Cahuilla Basin (Setmire et al. 1993). The sink lies in an actively spreading rift valley (Setmire

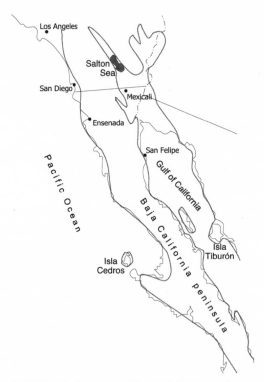

FIGURE 1. Northern extent of the Gulf of California during the Pliocene (*dotted line*), showing area now silted and the Salton Sea. Adapted from Durham and Allison 1960.

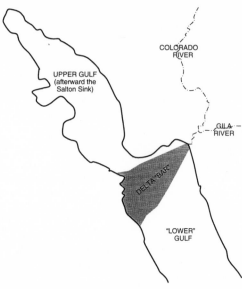

FIGURE 2. Formation of a sandy bench dividing the Salton Sink from the present Colorado River floodplain and the Gulf of California. Adapted from Kennan 1917:8.

et al. 1993). Low-lying alkaline flats typify this region, although there are a few rocky outcrops and hills, mostly volcanic in origin, such as Travertine Rock, Mullet Island, Red Hill, Obsidian Butte, Mount Signal, and Cerro Prieto (Loeltz et al. 1975). Tremendous geothermic activity not far below the earth's surface sprouts mud pots and "mud volcanoes" (Blake 1914; Nelson 1922) that further shape the terrain near the southeastern shoreline of the Salton Sea and the former Volcano Lake (now dry and occupied by Campo Geotérmico Cerro Prieto). Even so, the sink is practically featureless in comparison with most of the Sonoran Desert.

LAKE CAHUILLA

The existence of Lake Cahuilla was first recorded for posterity by William Phipps Blake during surveys for potential rail routes through the Southwest (Blake 1858). For this reason early literature often referred to the ancient lake as the Blake Sea. The moniker Lake Le Conte, after Joseph Le Conte, a well-known professor of California geography, was also widely used until 1907, when Blake himself insisted on his own priority (Gunther 1984). The name Lake Cahuilla, after the local Native Americans, has been used for the prehistoric lake ever since. The shoreline can still be seen along the eastern flank of the Santa Rosa Mountains (see, e.g., Schoenherr 1992: 421), especially around Travertine Rock, and there is evidence of the "heavy assaults of surf for some considerable period" along the northeastern shoreline, south of Mecca (Sykes 1914). Furthermore, the Algodones Dunes and presumably dunes around the Superstition Mountains are the remnants of beaches from this forerunner of the Salton Sea (Norris and Norris 1961), and it is clear that the central Coachella Valley was once an area of playas and shallow lakes similar to but commonly larger than the Salton Sea (California Department of Water Resources 1964).

Early incarnations of Lake Cahuilla had a shoreline peaking at 50 m above sea level (Setmire et al. 1990). More recent versions had a lower surface elevation, reaching a mere 12 m

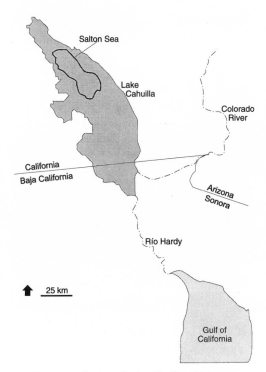

FIGURE 3. Typical extent of Lake Cahuilla at high water, with the present Salton Sea superimposed for comparison.

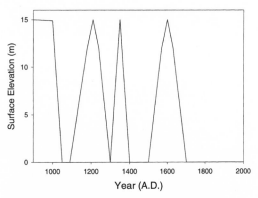

FIGURE 4. Approximate timing of high water at Lake Cahuilla, A.D. 800 to A.D. 2000. Adapted from Laylander 1997 and Smith 1999.

above sea level. In either form this mighty lake—160 km long and 56 km wide, covering an area of about 5,400 km² and reaching a maximum depth of nearly 100 m—dwarfed the present Salton Sea, which measures only 72 km long by 27 km wide, covers an area of about 1,150 km², and reaches 25 m at its deepest point (Fig. 3; Blake 1914; Woerner 1989). Indeed, the entire Salton Trough is scarcely larger than was Lake Cahuilla, which measured 208 km by 112 km (Loeltz et al. 1975).

After it dried up, Lake Cahuilla was periodically filled by floodwaters from the Colorado River; it was thus a largely freshwater lake (Walker 1961), although it may have been somewhat brackish at times, especially during early inundations (Blake 1914). The river deposited copious sediments and left the rich soil that supports flourishing agriculture in the Coachella, Imperial, and Mexicali Valleys today (Blake 1914; Cory 1915; Setmire et al. 1990). At least four major floods occurred during the past four millennia, each reforming this vast lake (Fig. 4). The

earliest flood was between 6,670 and 1,000 years ago (Waters 1983; Gurrola and Rockwell 1996). The other three were much more recent: the second flood was between 720 and 1,180 years ago, the third about 650 years ago (Wilke 1978; Waters 1983; Gurrola and Rockwell 1996), and Lake Cahuilla was last filled to capacity only about 400 years ago, meaning that it was present during the late 1500s and early 1600s (Norris and Norris 1961; Gurrola and Rockwell 1996; Laylander 1997). The appearance and evaporation of this lake little by little was well known to the Cahuilla Indians (Blake 1914), further evidence of its recency in the Salton Trough.

Since Lake Cahuilla was last filled, periodic small floods have inundated the Salton Sink, creating smaller lakes or seas in 1840, 1842, 1852, 1859, 1862, and 1867 (Sykes 1914). Most flooding has been through the New River, "so named for its unexpected appearance flowing into the desert in the year 1849" (Blake 1915, contra Schoenherr 1992). The last major flood occurred in 1891 (Sykes 1914, 1937), when a lake with a surface area roughly half that of the current Salton Sea was formed (Chase 1919). Sykes (1914) noted that some water had "found its ways down the channel of the New River toward the Salton every year since the inundation of 1891," demonstrating the capacity of natural Colorado River floods to fill the Salton Sink, a seemingly esoteric point that should be remembered when pundits decry the Salton Sea as nonnatural. While the sink may not have flooded again after the turn of the twentieth century, this result would have been achieved only through the continuous damming and diking of the great river, not because its capacity to flood was lost. But for human domination of the Colorado, who knows how full the sink would have become when tropical storm Kathleen hit in 1976 or when the river flooded during El Niño in 1983? People's taking from and giving to this ecosystem were punctuated by a fortuitous engineering blunder that brought back a vestige of the former Lake Cahuilla.

FORMATION OF THE SALTON SEA

The story of the Salton Sea begins with William Phipps Blake. Blake (1858) observed that routine flooding by the Colorado River accumulated extremely rich fluvial and silty clay soils in the Salton Sink that would produce bountiful crops if sufficiently irrigated. This seemingly wild notion was ignored for decades. Quite independently of Blake, and a half-decade earlier, another visionary, Oliver Wozencraft, saw that the rich soils in the Imperial Valley might be irrigated by diverting water from the Colorado River through the Alamo River. Blake warned that irrigation attempts might flood the Salton Sink since it was some 100 m lower in elevation than the Colorado River at Yuma. Wozencraft was less concerned about such potential problems and was undeterred by skeptics who noted that the sand dunes between the Colorado and the Salton Sink, rising a few hundred meters, would impede construction of canals. Instead, he pursued his vision with vigor for several decades, though ultimately he failed to gain the needed support.

The dream might have died had it not been for Charles Rockwood and George Chaffey, whom Woerner (1989) calls the "main players . . . in 'The Great Imperial Valley–Colorado River' play." Rockwood and Chaffey resurrected Wozencraft's dream, designing an elaborate scheme to bring irrigation water to the Imperial Valley (a name coined by Chaffey). They too chose the Alamo River as the most reasonable channel by which to convey water. A short canal and an inflow site with a headgate were needed to convey water from the Colorado to the Alamo. The headgate was placed near Yuma, and construction began on it and the canal in August 1900 (Cory 1915; Kennan 1917; Woerner 1989). By June 1901 water was being diverted into the Alamo, allowing irrigation and agriculture to begin. The canal worked smoothly for several years, until it was clogged by a deposit of heavy silt, which disrupted inflow. (The Colorado River, which had carved the Grand Canyon and other glorious rock formations in the Southwest, was known for carrying a massive load of silt.) Thus, dredging to unclog the canal was a ceaseless task. In spite of this problem, about 150,000 acres (ca. 620 km^2) of the Imperial Valley had been brought under cultivation by 1904.

The rapid establishment of an important agricultural economy increased pressure for adequate water delivery. In September 1904 the fateful decision was made to dig a temporary intake in hopes of improving water delivery to the Alamo River (Kennan 1917; Woerner 1989). Finished quickly, it amounted to little more than a ditch 15 m wide and 2 m deep dug at river level, without a headgate, some 6 km south of the main headgate. The ditch soon silted, requiring constant dredging, but the water level could not be controlled because there was no headgate. Floods on the Colorado River in the winter of 1904–5 rushed down the ditch, bringing to the Imperial Valley far more water than was needed for irrigating crops. The excess water was carried by the Alamo and New Rivers to the Salton Sink, which slowly began to fill, just as Blake had warned decades before.

Water in the sink was rising at a rate of about 1 cm per day by October 1905 as water gushed through the ditch, widened by floods to a gaping 200 m and deepened to 8 m! Almost all the river water was now flowing to the Salton Sink, with little to none continuing on to the Gulf of California (Cory 1915; Kennan 1917). Panic set in as rising waters claimed not only farms and ranches but also the saltworks and the Southern Pacific Railroad tracks. Nearly every solution imaginable, from pilings to sandbags to dyna-

4 A HISTORY OF THE SALTON SINK

mite, was thrown at the problem, but each failed as the raging torrent continued to widen the channel. By 1906 hope was nearly lost, spawning sarcastic news features and editorials publicizing Indio as a new seaport for the gulf (Woerner 1989). In August of that year the flow was nearly halted when 200 railcar loads of rock, gravel, and clay were dumped into the channel, plugging the break by November. But the solution was short lived. By December floodwaters again rose, destroying the newly constructed dam and carving a channel nearly a kilometer wide that dumped ever more water into the Salton Sea, now a vast lake. A successful dam was completed by February 1907. The level of the Colorado River rose some 4 m, and water levels in the Alamo and New Rivers decreased drastically.

Once the 1907 dam blocked inflow, the Salton Sea had a surface elevation of 60 m below sea level and was thus larger than it is today. Sur-

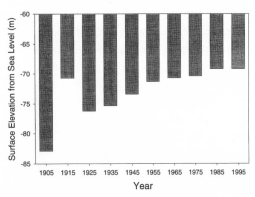

FIGURE 5. Surface level of the Salton Sea, 1905–1995. Data from the Imperial Irrigation District, 1996.

face elevations fluctuated dramatically during the first two decades after the sea's rebirth (Fig. 5), reaching a low of 75 m below sea level in 1925 (Holbrook 1928). Irrigation-effluent ditches constructed in 1922 and after restored the level of the sea to about 68 m below sea level (Holbrook

FIGURE 6. The Salton Sea on 9 April 1908, shortly after its birth, looking south from the shoreline near Mecca. Protruding from the surface are a multitude of dead shrubs killed by flooding of the sink. Photograph by Joseph Grinnell, courtesy of the Archives of the Museum of Vertebrate Zoology, University of California, Berkeley.

1928); indeed, had it not been for the vast repository that is the Salton Sea, the rich agriculture of the Imperial, Coachella, and Mexicali Valleys would not be possible. The water level is now much more stable, although it is slightly higher in spring and lower in fall (Setmire et al. 1990).

And thus the Salton Sea was born (Fig. 6). For the reason why the current lake is called the Salton Sea rather than Lake Cahuilla we once again turn to Blake (1914). Blake noted that because the Salton Sea was but a vestige of the much vaster ancient body—much as the Great Salt Lake is a vestige of Lake Bonneville (Blake 1914; Jehl 1994)—each deserved its own name.

Because the engineering accident that led to the inundation of the Salton Sink and thus formed the Salton Sea has been chronicled in detail by others, we present only a general summary. Readers interested in greater detail should refer to the lively account by Kennan (1917), which set the standard for subsequent efforts, and that of Woodbury (1941), who took a broader geographic view. The more technically inclined are referred to Cory's (1915) account, which discusses every engineering aspect of the initial irrigation project and the various efforts to stop the flooding of the Salton Sink.

Most accounts fail to mention that 1905 was a major flood year and that had it not been for the extensive channelization and draining of the Colorado River, the Salton Sink likely would have been inundated anyway. Copious inflow of irrigation water has kept the Salton Sea alive, sparing it the desiccation of Lake Cahuilla. As a result of very high evaporation rates, from 1.8 m (Tetra Tech 2000) to 2.4 m (Blake 1914) per annum, Blake (1914) asserted that the 25 m "of water now covering the Desert, and known as the Salton Sea, will require ten and a half years for its complete evaporation." Thus, aqua-engineering of the region has been a decidedly mixed blessing. Because of the engineering debacle that formed it and the agricultural runoff that maintains it, the largest inland body of water in California has been dubbed "in essence manmade [with] a manmade ecosystem" (Setmire et al. 1993). Yet given the long history of flooding in the Salton Sink and its long history of waterbird use, it may be more proper to view the Salton Sea as the latest in a long series of lakes that have occupied the basin. With this view, at the least the sea is partly manmade and partly a natural phenomenon, not wholly one or the other.

Conservation and Management Issues

I N THE 1940S AND 1950S the Salton Sea attracted tourists and fun-seekers from nearby metropolitan areas. The sea became so popular that the Salton Sea State Recreation Area was developed along the northeastern shoreline. After the introduction of marine fishes the sea also became a major sport fishery. The robust economy of the 1950s brought real estate speculators, who marketed the area as a thriving resort. As the Salton Sea aged, however, its appeal waned. The water became brown and turbid, fish carcasses littered the shore, and distasteful odors emanated from mud and backwaters. In the 1980s high water levels encroached on prime shoreline, and much of the shoreline now is dotted with dilapidated buildings and abandoned marinas, built only a half-century ago. Negative publicity about the New River, the most polluted waterway in the United States, contributed to the economic decline of the area. Pollution in this river poses a serious threat to the Salton Sea. At the international boundary, water in the New River carries raw sewage, agricultural drainage water, and power plant effluent, with the attendant detergents, pesticides, and other industrial and agricultural chemicals. Despite natural

cleansing during the river's 80-km journey, many of these harmful elements reach the Salton Sea. The Alamo and Whitewater Rivers fare better: they carry mainly agricultural runoff, but not raw sewage or industrial effluent.

The Salton Sea has thus become a cause célèbre for conservation biology. Its plight features commonly in newspapers, popular magazines, and journals (e.g., Boyle 1996; Dunlap 1999; Morrison and Cohen 1999; and Cohn 2000). Small-scale die-offs of birds have been known at the Salton Sea since its early years, with nearly 200 individuals of various species found along the shore about December 1917 (Gilman 1918). Recent massive die-offs of fish and birds (see, e.g., Saiki 1990; Jehl 1996; and Bruehler and de Peyster 1999) have caused alarm, particularly as they have occurred with increasing frequency and severity (Fig. 7), perhaps heralding a collapse of the ecosystem. Similar die-offs in the Gulf of California (Vidal and Gallo-Reynoso 1996) suggest that problems exist on a broader scale, perhaps throughout the Salton Trough. Annual fish kills at the Salton Sea are thought to result from high concentrations of sulfide and ammonia at the bottom of

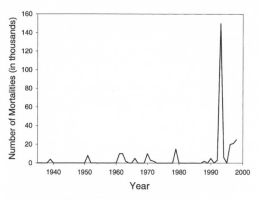

FIGURE 7. Bird die-offs at the Salton Sea since the mid-1930s. Note that the severity and frequency of mass mortalities has increased since the 1970s.

the sea mixing into surface water during the summer (Setmire et al. 1993). The causes of mass mortalities of birds are largely a matter of speculation, ranging from botulism to cholera to avian Newcastle disease (Kuiken 1999; Friend 2002), although botulism clearly is a principal culprit. Unfortunately, the environment of the Salton Sea favors botulism outbreaks, which are most likely to occur when low oxygen concentrations coincide with high water temperature and high salinity (Rocke and Samuel 1999).

Recently five major ornithological societies—the American Ornithologists' Union, the Association of Field Ornithologists, the Cooper Ornithological Society, the Western Field Ornithologists, and the Wilson Ornithological Society—adopted resolutions "in support of the Salton Sea as significant wildlife habitat" (see *Condor* 100:782–84; and Garrett 1998).

CURRENT THREATS

Because the level of the Salton Sea is maintained by a balance between runoff from agricultural irrigation and evaporation, from the time of the sea's birth various chemicals and elements have become increasingly concentrated in its waters (Table 1). High concentrations of chemicals such as calcium and potassium may prove of little consequence to the long-term health of the Salton Sea ecosystem, whereas high concentrations of boron and sodium have been a cause for concern

even if there have been few demonstrable negative effects thus far (Setmire et al. 1993). Even so, because high levels of both chlorine and boron can have detrimental effects on breeding birds (Setmire et al. 1990, 1993), increased concentrations of either are a potential problem.

Increased contamination with pesticides and heavy metals may have serious long-term effects. Levels of DDE, a metabolite of DDT, are higher in birds that frequent the Salton Sea than in those frequenting areas in nearby Mexico, including the Mexicali Valley (Mora et al. 1987; Mora 1991; Mora and Anderson 1995). Enormous amounts of pesticides are sprayed on crops in the Coachella and Imperial Valleys. The pesticides are largely washed away with irrigation runoff and carried to the Salton Sea either directly or via channels that empty into the Alamo and New Rivers. Preliminary studies of the endangered Yuma Clapper Rail (*Rallus longirostris yumanensis*) have shown that organochloride contamination, derived mostly from pesticides, may negatively affect its reproductive success at the Salton Sea (C. A. Roberts pers. comm.). The grim conclusions were summarized by Setmire et al. (1993), who noted that "waterfowl and fish-eating birds in the Imperial Valley may be experiencing reproductive impairment as a result of DDE contamination of food sources," with the highest concentrations in "birds feeding in agricultural fields on invertebrates and other food items." Worse, DDE contamination of birds wintering in the Imperial Valley has been linked to reproductive failures on breeding grounds as far away as the western Great Basin (Henny and Blus 1986; Henny 1997; Shuford et al. 2000).

Poisoning by heavy metals, especially selenium, also has been implicated as a potential threat to marine invertebrates, fish, and birds of the Salton Sea (Saiki 1990; Setmire et al. 1990, 1993; Fialkowski and Newman 1998; Setmire 1998; Bruehler and de Peyster 1999). In the late 1980s selenium levels were high enough to "cause physiological harm to fish and wildlife" (Setmire et al. 1990). By the early 1990s selenium, boron, and DDE were "accumulating in tissues of migratory and resident birds that use

TABLE 1

Increase in Concentration of Certain Constituents of Salton Sea Water (Parts per Thousand)
during the First Six Years of the Sea's Existence

	1907	1908	1909	1910	1911	1912
Sodium	1.11	1.34	1.60	1.89	2.28	2.71
Calcium	0.010	0.012	0.013	0.014	0.016	0.017
Potassium	0.023	0.028	0.032	0.035	0.038	0.038
Chlorine	1.70	2.04	2.41	2.81	3.39	3.95
Total soluble solids	3.64	4.37	5.19	6.04	7.18	8.47

Source: MacDougal 1914.
Note: Concentration of each chemical has increased substantially since 1912 (Carpelan 1958; Hely et al. 1966).

food sources in the Imperial Valley and Salton Sea," with selenium concentrations in fish-eating birds, shorebirds, and the endangered Yuma Clapper Rail "at levels that could affect reproduction" (Setmire et al. 1993). Most water discharge contains low concentrations of selenium and dissolved solids, but about one-fourth is drain water with extremely high concentrations of selenium (Setmire et al. 1993). This heavy metal poses a particular threat because of intense bioaccumulation (Dubowy 1989).

Perhaps worst of all, further energy development (there are numerous geothermal plants along the southern shoreline) is expected to change the water budget of the sea, leading to increased salinity (Dritschilo and Vander Pluym 1984). The amount of dissolved salts in the Salton Sea is currently as high as 44 parts per thousand, a salinity level 25–30 percent greater than that of the Pacific Ocean (Setmire et al. 1993; Stephens 1997b; Setmire 1998), and it generally has ranged between 32 and 43 parts per thousand since the 1920s (Grant 1982). Even a modest increase in salinity could render the sea uninhabitable for the few fish species that remain, leading to a potentially catastrophic collapse of the current ecosystem. Of course, other large saline lakes in western North America, for example, Mono Lake and the Great Salt Lake, are

far saltier than the Salton Sea is today and no longer support fish, and yet these lakes remain important to birds. But the ecosystem of the Salton Sea would likely undergo a drastic decline before reaching a new, unknown equilibrium. The combination of poisons, trace heavy metals, and high salinity have already caused severe problems in fish at the Salton Sea (Matsui et al. 1992), as have periodic anoxic conditions and high numbers of parasites. Also, high levels of salinity greatly alter the growth form and increase mortality rates of the Acorn Barnacle (Simpson and Hurlbert 1998), a species whose empty shells contribute substantially to habitat along the shoreline, affecting the whole ecosystem. High rates of evaporation, generally about 2.2 m in surface elevation of water per year (Blake 1914:6), coupled with excessive leaching of salts from the soil (Grismer and Bali 1997), contribute to this alarming problem. The Salton Sea needs an infusion of fresh, clean water, but the best means of achieving this lofty goal is debatable, and many proffered "solutions" seem unrealistic.

PROPOSED SOLUTIONS

Since the mid-1960s various ways to reduce salinity, stabilize surface elevation, and maintain agricultural, environmental, and recreational values have been suggested. However, for years the lack of political clout and money relegated the Salton Sea to the periphery. It was not until the mid-1990s, through the efforts of the late Sonny Bono, the Coachella Valley Audubon Society, and other concerned parties, that restoring the health of the Salton Sea ecosystem finally became a priority. In August 1994 the Salton Sea Authority, the Bureau of Reclamation, and the California Department of Water Resources embarked upon a cooperative effort to identify and compile potential ways to reduce the sea's salinity and the rate of evaporation. Proposed solutions were gleaned from past studies, new ideas were developed, and media announcements and public meetings invited submission of alternatives. Some 54 solutions were put forth, including 16 for pumping out highly saline water, 8 for impoundment by a system of dikes, 4 that combined impoundment and pump-out schemes, 5 for salt removal, and 2 for water importation. We do not necessarily advocate any of the solutions proposed; instead, we simply summarize each of the "finalists."

PUMPING OUT WATER

Most solutions propose salinity reduction via pumping water out of the sea and into the Gulf of California, Laguna Salada, the Pacific Ocean, or a deep aquifer. Salts would be removed with water pumped from the sea and thus could not accumulate there. Because inflow water sources (e.g., the Whitewater, New, and Alamo Rivers) carry largely fresh water, the salt load in the Salton Sea would steadily decline. However, pumping out water would decrease the total volume of the sea, leading to higher concentrations of salt and a declining surface elevation, at least in the short term. Proposed mitigation for these undesirable effects includes balancing outflow with the pumping of fresher water back into the sea. This "two-way pumping" would lessen salinity in the sea (until it reached an equilibrium) but would increase salinity in the receiving body of water. An increased salt load would not be a problem for the Gulf of California or the Pacific

Ocean, but it could cause problems for Laguna Salada or other repositories that do not have a natural outflow.

DIKE IMPOUNDMENTS

Like pump-out solutions, managing salinity by means of diked impoundments is based on the concept of providing the Salton Sea with an artificial outlet. Earthen dikes would separate the sea from large impoundments, which would act as expansive evaporation ponds. Water flowing into these impoundments through gated inlets would carry a heavy salt load. Most salt would remain in the impoundments as water evaporated from them. Occasional removal of accumulated salt would be necessary to ensure that the impoundments could still hold a sufficient amount of water. Freshwater inflow from rivers and ungauged drains would lead to a decrease in salinity in the sea. The sea's surface elevation would necessarily be reduced, but the additional water surface in the impoundment would probably compensate. A major advantage to this scheme is that, in principle, impoundments could be placed anywhere along the edge of the sea. A major disadvantage might be the potential concentration of heavy metals, as happened with selenium at Kesterson National Wildlife Refuge in California's Central Valley (Ohlendorf et al. 1986).

COMBINING OPTIONS

A number of proposed solutions combine impoundments, pumping out water, enhanced evaporation, and solar-power generation in an effort to exploit advantages of each scheme. Because controlling surface elevation and controlling salinity are conflicting objectives, combining methods may be the best solution. A balance between evaporation and inflow is the only means of stabilizing surface elevation, but lack of outflow leads to increased salinity. If a small area were enclosed, increasing salt concentration, pumping costs would be lower because less liquid would have to be pumped out to remove the same amount of salt. Thus, combining impoundments and pumping in fresh water could solve both the surface elevation and salinity problems. Furthermore, revenue from any salt mined from the evaporation ponds might offset the costs of pumping in fresh water.

REMOVING SALT INFLOWS

A major threat to the Salton Sea ecosystem is increased salinity. Removing salts from the water before it entered the sea would reduce this concern. Unfortunately, moving tons of salt over mountain ranges or long distances would be expensive, and disposal could have dire consequences for the environment. Removing salts from tributaries would help, but removal would only slow the rate of salinity increase, not solve the problem per se.

IMPORTING WATER

A few solutions involve reducing salinity by importing fresh water. However, the scarcity of high-quality water is prohibitive. Because jurisdictions have overlapping claims, more than 100 percent of the Colorado River's water is already committed for human use. Also, importing water could cause seasonal flooding, hardly contributing to stabilization of the surface elevation.

Some biologists advocate a more passive option, namely, to let the Salton Sea evaporate, forcing birds to switch to the "safer" environment of the northern Gulf of California (see Kaiser 1999). In any event, the Salton Sea Authority, the Bureau of Reclamation, and the California Department of Water Resources evaluated each proposal for feasibility and cost-effectiveness. Although the decision is not final, one of the dike-impoundment solutions appears to be the most promising for managing salinity in the Salton Sea. It would also have the least effect on overall surface area and water levels. However, detailed studies are under way to ensure

that reducing salinity and stabilizing the surface elevation will not exacerbate wildlife mortality and will maintain a healthy ecosystem. In addition, other proposals, such as a combination of impoundments and pumping water out into Laguna Salada, have not been explored fully.

In the meantime, the fate of the Salton Sea is in limbo. Despite the obvious conclusion that any solution is better than none at all, the extent to which any of the proposed solutions will restore the Salton Sea ecosystem is unclear. Maintaining the water level and reducing salinity clearly are important goals. Indeed, we argue that if salt concentrations are not reduced (or at least stabilized at current levels) and the surface elevation of the sea is not stabilized, the ecosystem as we currently know it will be altered drastically. However, these threats are not the only ones confronting the sea. Beyond ensuring continued freshwater inflow, perhaps the most pressing need is to decrease the concentration of soluble solids, from organochlorides such as DDE to heavy metals such as selenium, particularly given that the threat they pose may become magnified if a system of evaporation ponds is established (see Bradford et al. 1991). Thus, cleaning water in the rivers and irrigation channels could be the most important goal, yet this notion is not addressed in any of the proposed solutions. Furthermore, determining the cause of massive bird and fish die-offs is of utmost importance; after all, solutions to problems cannot be proposed until the causes of those problems are understood fully. The Salton Sea is a unique lake that supports one of the most diverse populations of birds in the world. We hope that a satisfactory solution to its plight will be implemented before we are faced with the horrid specter of mass die-offs and eventually a dead sea.

Biogeography of the Salton Sea

BIOGEOGRAPHY ENCOMPASSES knowledge and inferences of both past and present. It is challenging to deduce how present conditions affect distributions and abundances of a suite of organisms, to say nothing of how past conditions might be related to present ones. Here we compare what is known about the fauna of Lake Cahuilla with what is known about the fauna of the Salton Sea in order to put current distributions into perspective. But geography can only provide part of the story, so we describe current vegetation and habitats to clarify many patterns that might otherwise remain mysterious.

THE FAUNA OF LAKE CAHUILLA

That Lake Cahuilla was born from Colorado River floodwaters is attested by its similar fish fauna. Unlike the current Salton Sea, which supports only game fish (Walker et al. 1961; Riedel et al. 2002), Lake Cahuilla supported a fish community that was clearly a subset of the community on the lower Colorado River (Gobalet 1992). It included such species as the Razorback Sucker (*Xyrauchen texanus*), the Colorado Squawfish (*Ptychocheilus lucius*), the Striped Mullet (*Mugil*

cephalus), the Machete (*Elops affinis*), and the Bonytail (*Gila elegans;* Gobalet 1992). The floodwaters that filled the Salton Sea also brought other species, such as the trout *Salmo pleuriticus* and the nonnative carp *Cyprinus carpio* (Evermann 1916), but none persist today (Walker et al. 1961; Saiki 1990). Aquatic mammals typical of the Colorado River valley are also known to have been present in Lake Cahuilla. For example, the Muskrat (*Ondatra zibethicus*) occurred at Lake Cahuilla (Yohe 1998), and it occurs today in freshwater marshes flanking the Salton Sea and in lakes in the Imperial Valley.

Little is known about the avifauna of Lake Cahuilla, although there is every reason to suppose that it was very similar to that of the present Salton Sea (Patten and Smith-Patten 2003). Indeed, given the ancient lake's vast size, it was perhaps even more heavily visited by species now generally confined to the Gulf of California. Evidence from archaeological middens provides a partial picture of the bird life of Lake Cahuilla. As expected, present-day species such as the Eared and Pied-billed Grebes, the *Aechmophorus* grebes, the American White Pelican, the Black-crowned Night-Heron, the *Anas* and

Aythya ducks, and the American Coot were all in evidence at Lake Cahuilla (Wilke 1978; Beezley 1995; Patten and Smith-Patten 2003). Given the preponderance of immature American Coots captured, it is obvious that the species bred commonly at Lake Cahuilla. Furthermore, there is evidence of a heron rookery at Bat Cave Buttes (Wilke 1978:102), a former island to the southeast of North Shore, a few kilometers east of the railroad stop at Durmid. Of course, presence in a midden is only an indication of what was hunted and thus does not provide a complete portrait of the birds on the lake. For example, shorebirds were "apparently not sought by inhabitants of the Myoma Dunes" (Wilke 1978: 92), even though they were likely abundant at Lake Cahuilla; thus, there is scant evidence of their occurrence. Some species collected far from Lake Cahuilla were transported to settlements along the shore; such transport presumably explains the two Band-tailed Pigeons in the Myoma Dunes midden (Wilke 1978:97). Nonetheless, midden evidence supports the notion that prehistoric species composition was similar to that of today (Table 2; Patten and Smith-Patten 2003).

CURRENT CONDITIONS

Situated in the heart of the western Sonoran Desert, the Salton Sink typically has hot, dry weather. Despite occasional thunderstorms from the south, summers are particularly hot. For example, the maximum temperature exceeds 100°F on 110 days a year in the Imperial Valley (Hely and Peck 1964). Temperatures average slightly lower in the Coachella Valley. Conversely, rainfall is slightly higher in the Coachella Valley, although the entire sink gets little annual precipitation, with a mere 7.6 cm a year at Mecca (Hely and Peck 1964). Most rain falls from December through mid-March; it has rained only once in June since 1914 (Layton and Ermak 1976). Even so, it is often humid because of the extreme evaporation from the sea and the surrounding irrigated agricultural lands. Finally, the deep rift of the San Gorgonio Pass forms a natural funnel. Gusty winds and sandstorms are frequent in the northern Coachella Valley, but the Salton Sink is less affected.

The Coachella and Imperial Valleys remain among the most important agricultural areas in the United States, with crops ranging from grapes, dates, and sugar beets to Alfalfa, cotton, and wheat (Steer 1952; Putnam and Kallenbach 1997; Brewster et al. 1999). The Mexicali Valley is likewise important to Mexico (Stephens 1997a). This omnipresent agriculture drains an enormous amount of water from the Colorado River (Stephens 1997a), mainly via the All-American Canal, the Coachella Canal, and various canals off the Río Hardy. Irrigation runoff results in perennial water (not the naturally seasonal water) in the Alamo, New, and Whitewater Rivers, with most water reaching the sea rather than recharging the water table (Loeltz et al. 1975). Most runoff is contributed by growers in the Imperial Valley. For example, in the mid-1980s the New River discharged 512,000 acre-feet per year into the Salton Sea, but only 262,640 acre-feet per year entered the United States at the international boundary (Fogelman et al. 1986). What is even more striking, the Alamo River discharged 600,200 acre-feet per year, but a mere 1,842 entered the United States; inflow is now at or near nil (S. H. Hurlbert pers. comm.). Ungauged drains contribute another 131,000 acre-feet per year of inflow to the south end (Hely et al. 1966). All told, some 1,328,800 acre-feet of water empties into the Salton Sea each year, but evaporation maintains its surface at a fairly stable elevation, highest in the spring and lowest in the fall (Setmire et al. 1990). The current surface elevation averages −68 m, with the greatest depth about 15 m. Because the Salton Sea is less saline at the mouths of the rivers and major drains, it supports greater species diversity in those areas (Walker 1961).

The capacious waters of the Salton Sea are home to only four species of fish, none of them native (Walker et al. 1961; Saiki 1990). A hybrid tilapia (*Oreochromis mossambica* × *O. urolepis*) is the most common (Riedel et al. 2002). It is the major victim in annual mass mortalities, carcasses being strewn across many of the beaches.

TABLE 2
Bird Species Recorded from Midden Sites and Coprolite Samples at Sites around Lake Cahuilla and Their Present-Day Status at the Salton Sea

SPECIES	MNI[a]	PRINCIPAL STATUS AT SALTON SEA
Pied-billed Grebe (*Podilymbus podiceps*)	2	Fairly common breeding resident
Eared Grebe (*Podiceps nigricollis*)	8	Abundant winter visitor, transient
Aechmophorus sp.	13	Fairly common breeding resident
American White Pelican (*Pelecanus erythrorhynchos*)	2	Common winter visitor, transient
Double-crested Cormorant (*Phalacrocorax auritus*)	4	Common breeding resident
American Bittern (*Botaurus lentiginosus*)	1	Uncommon winter visitor
Great Blue Heron (*Ardea herodias*)	12	Common breeding resident
Black-crowned Night-Heron (*Nycticorax nycticorax*)	4	Fairly common breeding resident
Wood Stork (*Mycteria americana*)	1	Rare summer or fall visitor
Turkey Vulture (*Cathartes aura*)	3	Common transient
Canada Goose (*Branta canadensis*)	2	Fairly common winter visitor
Tundra Swan (*Cygnus columbianus*)	1	Casual winter visitor
Mallard (*Anas platyrhynchos*)	3	Fairly common winter visitor
Cinnamon Teal (*A. cyanoptera*)	7	Common transient
Northern Pintail (*A. acuta*)	4	Common winter visitor
Green-winged Teal (*A. crecca*)	6	Common winter visitor
Anas sp.	4	Generally common winter visitors
Canvasback (*Aythya valisineria*)	16	Uncommon winter visitor
Redhead (*A. americana*)	11	Fairly common breeding resident
Greater Scaup (*A. marila*)	7	Uncommon or rare winter visitor
Lesser Scaup (*A. affinis*)	17	Fairly common winter visitor
Aythya sp.	3	Generally fairly common winter visitors
Bufflehead (*Bucephala albeola*)	1	Uncommon winter visitor
Common Goldeneye (*B. clangula*)	1	Uncommon winter visitor
Ruddy Duck (*Oxyura jamaicensis*)	5	Common breeding resident
Red-tailed Hawk (*Buteo jamaicensis*)	1	Common winter visitor
American Kestrel (*Falco sparverius*)	1	Common breeding resident
Gambel's Quail (*Callipepla gambelii*)	2	Common breeding resident
American Coot (*Fulica americana*)	182	Common breeding resident
Sandhill Crane (*Grus canadensis*)	1	Uncommon winter visitor
Willet (*Catoptrophorus semipalmatus*)	1	Common transient and winter visitor

TABLE 2 *(continued)*

SPECIES	MNI[a]	PRINCIPAL STATUS AT SALTON SEA
Long-billed Dowitcher (*Limnodromus scolopaceus*)	1	Abundant transient and winter visitor
Band-tailed Pigeon (*Columba fasciata*)[b]	2	Casual visitor
Mourning Dove (*Zenaida macroura*)	3	Abundant breeding resident
Barn Owl (*Tyto alba*)	1	Fairly common breeding resident
Great Horned Owl (*Bubo virginianus*)	1	Uncommon perennial visitor; breeds
Burrowing Owl (*Athene cunicularia*)	1	Fairly common breeding resident
Common Raven (*Corvus corax*)	2	Fairly common breeding resident
Red-winged Blackbird (*Agelaius phoeniceus*)	1	Abundant breeding resident

Note: Historical data are bones recovered from the Myoma Dunes and the vicinity of La Quinta in the southern Coachella Valley (Wilke 1978; Patten and Smith-Patten 2003), Wadi Beadmaker and Bat Caves Buttes on the eastern shore near the mouth of Salt Creek (Wilke 1978), and the Elmore site north of the mouth of the New River (Beezley 1995; Laylander 1997).
[a]MNI = minimum number of individuals collected.
[b]The pigeons were probably collected away from Lake Cahuilla and transported there (see Patten and Smith-Patten 2003).

Less common are the Orangemouth Corvina (*Cynoscion xanthulus*), the Bairdiella (*Bairdiella icistia*), and the Sargo (*Aniostremus davidsoni*). Freshwater irrigation drains support the Bairdiella and the goby *Gillichthys mirabilis,* the mosquitofish *Gambusia affinis,* and the guppy *Poecilia latipinna* (Setmire et al. 1993). The rare Desert Pupfish (*Cyprinodon macularis*) is the only native in the region. It was reintroduced into Salt Creek and San Felipe Creek (mainly near San Sebastian Marsh) shortly after the Salton Sea formed (Schoenherr 1992).

Among aquatic invertebrates a particularly visible and abundant organism is the Acorn Barnacle (*Balanus amphitrite*). It was first recorded at the sea in 1944 (Walker 1961), supposedly introduced on the landing gear of seaplanes. Its shells now form substantial beaches, spits, and islets at Salton City and various other locales, especially at the north end. The Salton Sea was regularly stocked with both clams and shrimp at least through the 1940s (Steere 1952), and populations of the latter probably persist, providing food for numerous shorebirds and other waterbirds.

Long before the Salton Sea was formed, lagoons, playas, and marshes along the New and Alamo Rivers hosted numerous waterbirds, mainly herons, ducks, and shorebirds (Mearns 1907). Even so, the presence of the sea increased waterbird and seabird use of the area substantially, so much so that in 1930 the U.S. Fish and Wildlife Service established the Salton Sea National Wildlife Refuge at the south end. Shortly thereafter the California Department of Fish and Game established similar refuges at Wister and at Finney and Ramer Lakes. Additional land received de facto protection by virtue of the Salton Sea Naval Test Base and the Torres-Martinez Indian Reservation. Even so, most of the Salton Sea's shoreline remains privately owned.

THE SALTON SEA AND THE GULF OF CALIFORNIA

The Salton Sea is vastly different from all other inland bodies of water in North America. Its salinity certainly does not make it unique, as the Great Salt Lake and Mono Lake, to name but two examples, are also saline. Indeed, many saline lakes in the Great Basin have similar avifaunas (Jehl 1994). What make the Salton Sea unique are its location and the geography of the area (see Sykes 1937). As noted above, the sea and its adjoining valleys lie below sea level in a vast trough continuing southward to the Gulf of California (Durham and Allison 1960), but with montane barriers on the other three sides. Thus, the Salton Sink lies within a basin jutting northward from the gulf. As a result, migratory seabirds, shorebirds, and waterfowl normally associated with coastal environments (e.g., the Brant, the scoters, the Ruddy Turnstone, the Red Knot) are found regularly at the Salton Sea despite their being extremely rare inland anywhere else in the West (Patten and McCaskie 2003).

Migratory birds are not confronted with geographic barriers, such as mountain ranges, when they head north out of the Gulf of California (see Patten and McCaskie 2003). The highest points between the head of the gulf and the south end of the sea are no more than 50 m above sea level. The jagged Peninsular Ranges to the west and higher ground to the east (e.g., Mesa Andrade, the Chocolate Mountains) naturally funnel into the Salton Sink birds that were not deterred when they reached land at the head of the gulf. Furthermore, an abundance of water along the Río Hardy and the Río Colorado and in large bodies, such as at Campo Geotérmico Cerro Prieto (in the bed of the former Volcano Lake) or Laguna Salada (when flooded), is conducive to "leapfrogging" through the Río Colorado delta into the Mexicali and Imperial Valleys. Lastly, the "thermal-rich environment of the coastline in the Sonoran Desert" enables birds to soar at great heights (Anderson et al. 1977). Birds reaching the northern end of the Salton Sink will inevitably find the Salton Sea, not only because the sea covers a vast area but also because the geography of the area further pinches inward, tightly narrowing the funnel. The rugged Santa Rosa Mountains sweep eastward, skirting the northwestern edge of the Salton Sink. Similarly, the Orocopia Mountains sweep westward, nearly merging with the Little San Bernardino Mountains (Fig. 8).

In the northern part of the Salton Sink the north end of the Peninsular Ranges (i.e., the San Jacinto Mountains) meets the central Transverse Ranges (i.e., the San Bernardino Mountains) to form the San Gorgonio Pass, which rises from 50 m below sea level to more than 650 m above sea level in a stretch of only about 50 km. This great, narrow pass is one of the deepest rifts in the Americas, as it separates Mount San Jacinto (3,225 m) and Mount San Gorgonio (3,464 m), whose apexes are a mere 34 km apart. Most waterbirds moving northward through the Salton Sink in the spring are funneled through this pass, making it one of the most important migratory corridors in the West.

Winds exert a strong influence on bird movement (Gauthreaux and Able 1970; Butler et al. 1997), and wind patterns have a significant effect on such movements at the Salton Sea, again establishing a tight link between the sea and the gulf (Anderson et al. 1977; Patten and Minnich 1997). The wind flow in the summer months (May through September) is primarily from the south, with the prevailing flow stopping at the San Gorgonio Pass (prevailing winds north of the pass are always from the north). During other months the monsoon flow breaks down and winds are primarily from the north, sweeping southward from the San Gorgonio Pass through the Salton Sink (Fig. 9; Blake 1923). These wind patterns are strongly associated with the dispersal of numerous seabirds from the Gulf of California to the Salton Sea. Species include regular visitors such as the Brown Pelican and the Yellow-footed Gull, which occur in the thousands and have been increasing (Anderson et al. 1977; Patten 1996), and unusual strays such as eight species of highly pelagic Procellariiformes (i.e., albatrosses, petrels, and shearwaters), for which there are more than 30 records (Patten and Minnich 1997).

Concomitant with the southerly winds in summer is an increase in sea-surface temperatures off western mainland Mexico. During the winter months the Gulf of California is much warmer than oceanic waters off western Mexico, effectively isolating the gulf from the surround-

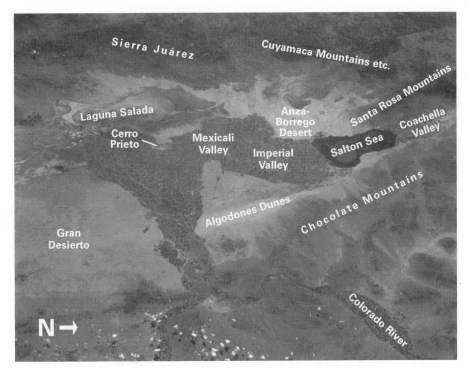

FIGURE 8. Topography of the Salton Sink, showing major landforms. Note the broad, low-lying funnel opening southward toward the Gulf of California. Base photograph courtesy of NASA (STS51F-038-005, August 1985).

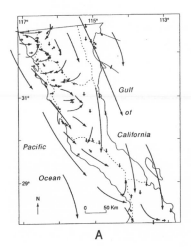

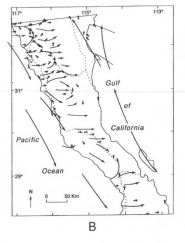

FIGURE 9. Prevailing wind flow through the Salton Sink and northern Gulf of California during January (A) and July (B). From Patten and Minnich 1997.

ing seas. The warming of the waters during the summer brings temperatures off western Mexico to nearly the same as those in the gulf, creating a more uniform environment. The breakdown of the temperature gradient barrier, aided by the monsoonal winds, facilitates dispersal of oceanic birds into the gulf (Patten and Minnich 1997).

In sum, a tight link is forged between the Salton Sea and the Gulf of California by a combination of factors, most notably their close proximity, the lack of geographic barriers between them, and the favorable northward winds in summer (Patten and McCaskie 2003), such that the regions function as a single grand ecosystem.

VEGETATION AND HABITAT

Two aspects of biogeography and macroecology are especially important to a consideration of the distribution and abundance of organisms. The actual geography and topography of the region in question constitute the first. As noted above, the strong relationship between the Salton Sink and the Gulf of California exerts a powerful influence on bird distribution in the region. Vegetation and habitat, which are directly related to the geography and climate of a region but shape animal distributions in an entirely different way, constitute the other major aspect. Obviously, if a habitat is not favorable to the survival and reproduction of a species, that species will not occur regularly in the habitat.

The dry, hot environment of the western Sonoran Desert is undeniably harsh, exacting a heavy toll on plants that grow there. A particularly harsh climate prevails in the Salton Trough, which is one of the hottest locales in all of North America, the others being Death Valley and the head of the Gulf of California (Schmidt 1989). Furthermore, even though the soils in the Salton Sink—which are a residual of Lake Cahuilla—are mainly alluvial deposits from the Colorado River, with some gravel and sand along the western and eastern shorelines and a fine, sandy loam in portions of the Coachella Valley, most are highly alkaline. Finally, the annual rainfall is exceedingly low, averaging less than 4 cm throughout the region. Vegetation is therefore clumped around water sources, such as the Whitewater River (draining the south slope of the San Bernardino Mountains), Salt Creek (draining the Orocopia Mountains), the Alamo and New Rivers (diffluents of the Colorado River), dozens of washes that carry occasional flash floods, and various lakes, ponds, marshes, and irrigation ditches. The searing heat, low annual rainfall, and highly alkaline soils exclude many plant species that are otherwise common in the deserts of western North America. Visitors who expect a multitude of cacti and other succulents will be disappointed.

Not surprisingly, extensive cultivation has drastically altered the natural vegetation of the region. Nevertheless, biogeographic affinities of native plant species in the Salton Sink are predominantly transmontane (the other side of the mountains relative to the Pacific Ocean, i.e., desert), not cismontane (the same side of the mountains as the ocean, i.e., coastal scrubs). As aptly noted by Parish (1914), "A student of the flora of the Colorado Desert speedily discovers evidences of migratory movement from the south and east." The Colorado Desert flora represents the western fringe of the arid-land flora of Sonora, Arizona, and New Mexico (Turner et al. 1995), but the xerophytic (drought-tolerant) vegetation of the Salton Sink is differentiated mainly by the preponderance of saltbush (*Atriplex* spp.) and other halophytes (salt-tolerant plants) (Parish 1914). With regard to purely native plants, one finds in a transect of the Salton Sink from west to east pinyon-junipers woodland on the dry eastern slopes of the Peninsular Ranges, rich *Yucca*–cholla (*Opuntia* spp.)–Ocotillo (*Fouquieria splendens*) vegetation in the Anza-Borrego Desert and at the edge of the Pattie Basin, Creosote (*Larrea tridentata*) scrub at lower elevations, mesquite (*Prosopis* spp.) thickets on lowlying sandy soils, and saltbush flats characterizing the central Salton Sink (Fig. 10; Parish 1914; Wilke 1978). An eastward climb out of the sink leads through a similar transition until one reaches the once extensive mesquite–willow (*Salix* spp.)–Fremont Cottonwood (*Populus fremontii*) association of the lower Colorado River valley.

Vegetation associations in the Salton Sink can be divided into eight principal formations or types, five of them native. Such divisions are somewhat artificial, as many vegetation communities grade into one another. Nevertheless, naming these formations provides a useful shorthand for discussing avian habitats in the region. The five native formations—hydrophytic, heliophytic, mesophytic, halophytic, and xerophytic—were recognized by Parish (1914). Our descriptions of them are largely summaries of his early report but are updated with current nomenclature

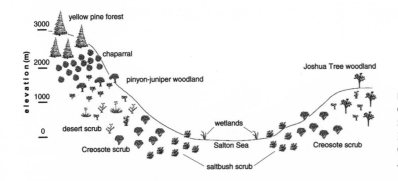

FIGURE 10. Floral zones of the Colorado Desert around the Salton Sink, from the Santa Rosa Mountains (*left*) to the Orocopia Mountains (*right*). Adapted from Wilke 1978.

and taxonomy (see Hickman 1993; and Turner et al. 1995) and modifications to accommodate recent knowledge. The three other vegetation "formations" are habitats modified by humans: suburbia, parklands, and ranch yards; orchards; and agricultural fields. Obviously such associations are artificial, so speaking of them as communities is meaningless. Nonetheless, each provides habitat for numerous species, as does the open water of the Salton Sea itself.

HYDROPHYTIC FORMATION

Truly aquatic plants are poorly represented in the Salton Sink. The ditch grass *Ruppia cirrhosa* is the only prevalent representative. It occurs sparingly in freshwater marshes and in some ditches that have sufficiently clear water to allow growth.

HELIOPHYTIC FORMATION

In contrast to aquatic plants, sun-tolerant, shallow-water plants are fairly well represented in the Salton Sink, mainly at the edges of marshes, ponds, and lakes and along rivers and ditches (Fig. 11). Dominant species are two cattails, *Typha domingensis* and *T. latifolia,* various bulrushes, especially *Scirpus americanus* and *S. maritimus,* the reed *Phragmites australis,* rushes such as *Juncus cooperi,* and the salt grass *Distichlis spicata.* The nonnative Saltcedar (*Tamarix ramosissima*) has become a significant component of this community in the past half-century, a common trend in riparian habitats of

the desert Southwest (Askins 2000:196). *Typha* is especially common in marshes and around river mouths, but Saltcedar and *Phragmites* dominate along rivers, ditches, and lake edges.

MESOPHYTIC FORMATION

Riparian forest and similar mesophytic formations are generally only moderately represented in the Salton Sink. They are found exclusively along the rivers, around seeps, and near lakes and large ponds. Although the extent of riparian habitat has decreased considerably over the past century, historical accounts (e.g., Mearns 1907) suggest that it was never abundant in the region. Saltcedars (*Tamarix ramosissima* and *T. aphylla*) now dominate this habitat. Portions of the Whitewater, New, and Alamo Rivers support the Fremont Cottonwood and Goodding's Black Willow (*Salix gooddingii*) (Fig. 12), and stands of these trees occur around some lakes and marshes. In the Imperial Valley, cottonwoods are failing to reproduce, perhaps because of high soil salinity. With the death of each old tree we witness the continued diminution of the habitat most attractive to landbirds. Aside from Saltcedar, the most numerous shrub element is the Arrowweed (*Pluchea sericea*). In stark contrast to the Mojave Desert, the western Sonoran Desert (i.e., the Colorado Desert) and especially the Salton Sink have few heavily wooded oases. A prominent tree of oases in the Colorado Desert is the fan palm *Washingtonia filifera,* but it occurs commonly only at Dos Palmas and is otherwise rare along the rivers.

FIGURE 11. Heliophytic vegetation dominated by Southern Cattail, Saltcedar, and Common Reed at the south end. Stands of cattail edging rivers, marshes, and lakes are ideal for breeding Least Bitterns, Clapper Rails, Marsh Wrens, Common Yellowthroats, Red-winged and Yellow-headed Blackbirds, and other wetland species. Photograph by Jack W. Schlotte and Philip Unitt.

FIGURE 12. Riparian forest dominated by Saltcedar (*foreground*), Fremont Cottonwood, and Goodding's Black Willow near Brawley. Such mesic habitats are crucial for breeding species like the Yellow-breasted Chat, the Blue Grosbeak, and the Song Sparrow. Photograph by Jack W. Schlotte and Philip Unitt.

FIGURE 13. Halophytic scrub dominated by Iodine Bush growing along the edge of Obsidian Butte. The Large-billed Savannah Sparrow favors such habitats at the south end of the Salton Sea. Photograph by Jack W. Schlotte and Philip Unitt.

HALOPHYTIC FORMATION

Given the highly alkaline soils of the Salton Sink, it should come as no surprise that halophytic plants are well represented. The Saltcedar is the dominant plant species along the fringe of the Salton Sea and in river bottoms and other wetlands. Understory vegetation is often nothing more than a mat of the salt grass *Distichlis spicata,* and where Saltcedar does not occur at the edge of the sea the dominant plant is generally the Iodine Bush (*Allenrolfea occidentalis*) (Fig. 13). Away from the edge of the sea various recessions of Lake Cahuilla and the initial recession of the Salton Sea exposed alkaline flats that were readily occupied by halophytes (MacDougal 1914), forming the saltbush scrub that characterized much of the preirrigation Salton Sink (Fig. 14). Dominant plants are the saltbushes *Altriplex lentiformis, A. polycarpa,* and *A. canescens.* Other common shrubs and subshrubs in this formation are the Bush Seepweed (*Suaeda moquinii*), the goldenbush *Isocoma menziesii,* Western Sea-Purslane (*Sesuvium verrucosum*), and Alkali-Mallow (*Malvella leprosa*). In more mesic areas the heliotrope *Heliotropium curassavicum* and salt grass are common elements of the ground cover.

XEROPHYTIC FORMATION

Because of its desert environment, one expects xerophytic plants to be well represented in the Salton Sink. Exclusive of cacti and true succulents, xerophytic plants are indeed common. However, true desert scrub, though well developed, is now confined to the fringes of the sink. It is especially common along the eastern edge, on the gentle slope that climbs to the Orocopia and Chocolate Mountains and the Algodones Dunes, and on the western edge as the halophytic scrub of the sink merges with the relatively lush desert scrub of the Anza-Borrego region. On sandier soils the common plant is *Larrea tridentata,* the Creosote. It is especially common along the eastern edge of the cultivated portion of the Imperial Valley. Desert scrub in the sink is otherwise characterized by lower shrubs and subshrubs, such as the Brittlebush (*Encelia farinosa*), the Turtleback (*Psathyrotes ramosissima*), the dyeweed *Tiquilia plicata,* the locoweed *Astragalus limatus,* and Burroweed (*Ambrosia dumosa*)

FIGURE 14. Desert scrub dominated by saltbush near Niland. Few birds occur in this habitat, but it is important for breeding Lesser Nighthawks and for wintering sparrows, especially the Sage and Brewer's. Photograph by Jack W. Schlotte and Philip Unitt.

(Fig. 15). Patches of both Honey and Screwbean Mesquites, *Prosopis glandulosa* and *P. pubescens*, form dense stands, especially in parts of the southern Coachella Valley, the eastern Imperial Valley, and the western Mexicali Valley and along San Felipe Creek (Fig. 16). Unfortunately, mesquite is now much rarer in the Salton Sink than it once was. Mearns's (1907) descriptions of vegetation along the New and Alamo Rivers in 1894 reveal that mesquite was once the dominant tree. It dominates in few locations today. The Desert Mistletoe (*Phoradendron californicum*), which grows mainly on mesquite in the region but also uses the Fremont Cottonwood as a host, provides important habitat and food for frugivores. In addition to mesquite, washes draining into the sink, particularly along the western edge, support stands of the Blue Palo Verde (*Cercidium floridum*), the Ironwood (*Olneya tesota*), and the Smoke Tree (*Psorothamnus spinosus*). Although characteristic of drainages in many parts of the Sonoran Desert, the Desert Willow (*Chilopsis linearis*) is scarce in the Salton Sink.

SUBURBIA, PARKLANDS, AND RANCH YARDS

In the desert Southwest, wooded suburbia, parks, and ranch yards now act as surrogates for many erstwhile riparian species (Rosenberg et al. 1987). Large broadleaf trees, ranging from the native Fremont Cottonwood to nonnative *Eucalyptus* spp., were heavily planted throughout the settled portions of the valleys. Planted pines (*Pinus* spp.) are prevalent in the larger towns, as are a multitude of deciduous trees, such as elms (*Ulmus* spp.), mulberries (*Morus* spp.), the myoporum *Myoporum laetum*, and sycamores (*Platanus* spp.), and numerous shrubs. Extensive planting has created a lush artificial habitat that is now important for many species, whether residents (e.g., the Inca Dove, the Gila Woodpecker, and the Northern Mockingbird), breeding visitors (e.g., the Black-chinned Hummingbird and the Hooded Oriole), or migrants.

ORCHARDS

Cultivated orchards constitute another important artificial habitat in the Salton Sink. The southern

FIGURE 15. Xerophytic desert scrub at the western edge of the Salton Sink. This habitat tends to support few species of birds, but it is important to the Horned Lark and formerly supported Le Conte's Thrasher. Photograph by Jack W. Schlotte and Philip Unitt.

FIGURE 16. Stands of mesquite, formerly common in the Salton Sink but now occurring only in isolated patches, are important for many nesting and foraging landbirds, including the Crissal Thrasher and the Phainopepla. Photograph by Jack W. Schlotte and Philip Unitt.

Coachella Valley supports numerous vineyards and Date Palm and citrus orchards. These habitats in turn support numerous bird species that are generally associated with wooded habitats elsewhere (e.g., the Common Ground-Dove and the Lark Sparrow). Orchards are less common in the Imperial and Mexicali Valleys (especially in the former) but nonetheless provide habitat for birds in the region.

AGRICULTURAL FIELDS

As noted above, the Coachella and Imperial Valleys are among the most important agricultural regions in the United States, and the Mexicali Valley is one of the most important in Mexico (Fall 1922; Steere 1952; Putnam and Kallenbach 1997). The Coachella Valley largely supports orchards and vineyards; open agricultural land generally is dominated by sod farms, with little devoted to vegetable or textile crops. Far fewer orchards exist in the Imperial and Mexicali Valleys, where cultivated lands are devoted to grains, Bermuda Grass (*Cynodon dactylon*), Alfalfa (*Medicago sativa*), sugar beets, vegetables, cotton, and many other crops. Crop rotation yields an ever changing matrix of fallow or weedy fields. Fields not farmed in a particular year provide habitat for numerous wintering birds, ranging from Northern Harriers and Short-eared Owls to Western Meadowlarks and various sparrows. Burning of harvested fields, especially of Asparagus, is a common practice in both the southerly valleys; such burns may provide critical habitat for wintering Mountain Plovers and frequently support large numbers of wintering Horned Larks and American Pipits. Nearly 95 percent of birds feeding or roosting in agricultural fields during a recent study (Shuford et al. 2000) occurred in fields with three broad cover types. Fields supporting grasses host more birds (39%) than do fields supporting Alfalfa (31%), which in turn support more birds than do fields of bare dirt (24%). However, about 70 percent of the birds detected during this study were of three species—in decreasing order of abundance, the Ring-billed Gull, the Cattle Egret, and the Red-winged Blackbird—so generalizations about microhabitat use cannot be made. Flooded agricultural fields host a variety of shorebirds and thousands of Cattle Egrets, White-faced Ibises, and Ring-billed Gulls (Shuford et al. 2003; Warnock et al. 2003).

THE SALTON SEA

Even though the Salton Sea does not support the growth of vegetation, remnants of vegetation are crucial to breeding waterbirds. Drowned shrubs evident when the sea first flooded (see Fig. 6 and MacDougal 1914) have long since disappeared. Rising water levels in the 1950s and 1960s drowned many mesquites, Saltcedars, and cottonwoods that formerly grew near the shoreline, especially around the river mouths and various other locales with freshwater inflow. These drowned trees remain as snags that are used by numerous breeding cormorants, herons, and egrets. Thousands of these snags are scattered along the shoreline in the vicinity of the Whitewater River delta, Wister, Morton Bay, and Bruchard Bay (Fig. 17). Similar snags at Finney and Ramer Lakes and Fig Lagoon also serve as nest and roosting sites.

Snags are the closest to vegetative habitat that the Salton Sea offers, but they are far from the only habitat it offers. The open water may be used by tens of millions of waterbirds in some years, including 2–3 million Eared Grebes. Hordes of pelicans, cormorants, ducks, gulls, and terns make ready use of the abundant habitat and its resources. The shoreline varies from mudflats around river mouths and bays to barnacle beaches at Salton City and many parts of the sea that are lapped by waves. Breakwaters, jetties, marinas, pilings, and embankments offer roosting and foraging opportunities to pelicans, shorebirds, and larids. Some even provide nest sites for tern colonies. Shallow impoundments of fresh or brackish water that border the sea along Morton Bay, near Oasis, and at various other locales provide abundant foraging habitat for shorebirds (Fig. 18; Shuford et al. 2002, 2003; Warnock et al. 2003). Foraging herons and ibises heavily use similar ponds with slightly deeper water and a marshy fringe.

FIGURE 17. Snags protruding from the Salton Sea, like these off Red Hill, provide important nesting and roosting sites for Double-crested Cormorants and various species of herons and egrets. Photograph by Jack W. Schlotte and Philip Unitt.

FIGURE 18. Mudflats and shallows in large freshwater and brackish impoundments, like these evaporation ponds along Morton Bay south of Wister, provide an abundance of habitat for migrant and wintering plovers, sandpipers, and other shorebirds. Mullet Island is visible in the background. Photograph by Jack W. Schlotte and Philip Unitt.

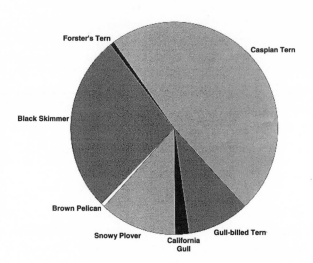

FIGURE 19. Relative abundance of seabirds breeding at the Salton Sea in the late 1990s. Displaying pairs of Heermann's Gull, the Yellow-footed Gull, and the Least Tern were noted on multiple occasions during the 1990s, but none have nested. Furthermore, a few pairs of the Laughing Gull have nested on an irregular basis. Data adapted from Shuford et al. 1999, 2002; and the species accounts.

THE AVIFAUNA

BREEDING SEABIRDS

The saline water of the Salton Sea and the sea's tight relationship with the Gulf of California combine to create a landlocked haven for seabirds, those taxa typical of coastal or saline habitats (e.g., Pelecaniformes, Charadriiformes). The Salton Sea hosts the largest breeding population of the Gull-billed Tern in the western United States and substantial breeding populations of the Black Skimmer and the Caspian Tern (Fig. 19). Small numbers of Forster's Tern also breed there. In 1997 California Gulls began to nest off Obsidian Butte. The nearest established breeding population is at Mono Lake, in the western Great Basin. In 1996 the Brown Pelican began to nest at various locales around the south end. Aside from locales in central Florida (e.g., Lake Okeechobee), the Salton Sea is the only inland site where this species breeds. The Laughing Gull, formerly a regular nesting species, still breeds there on occasion. The Salton Sea is also a breeding stronghold in the West for the principally coastal Snowy Plover.

Apart from the California Gull, each of these species breeds commonly in the Gulf of California (Anderson et al. 1976; Wilbur 1987). Several other seabirds that breed in the gulf occur at the Salton Sea with some regularity, chief among them Heermann's Gull, the Yellow-footed Gull, and the Least Tern. Displaying pairs of each

species have been observed at the Salton Sea since 1995, probably portending future breeding. Other seabirds that breed in the gulf, such as the Elegant Tern, occurred at the Salton Sea with increasing frequency in the 1990s, but without evidence of breeding activity until 2002.

BREEDING WATERBIRDS

Numerous waterbirds breed in the Salton Sea region. The breeding avifauna shows no clear affinity with that of any adjacent region, although it is most like that of the lower Colorado River valley (see Rosenberg et al. 1991). A principal difference is the sheer magnitude of breeding waterbirds in the Salton Sink, where many species have their largest population in the interior Southwest (Patten and McCaskie 2003). The relative compositions of the avifaunas differ, yet both regions share many breeding species. Notable absentees from the Colorado River region are the Cattle Egret, which nests in the Salton Sink in the tens of thousands, and the White-faced Ibis, a regular but less numerous breeder.

Chief among breeding waterbirds in the Salton Sink are the Double-crested Cormorant and various species of herons and egrets, each with thousands of pairs nesting annually. The Salton Sink is the principal breeding locale in the interior Southwest for the cormorant (Carter et al. 1995). Nesting birds occur mainly at the Salton Sea itself, with a few pairs at Finney and Ramer Lakes (Warnock et al. 2003). Birds at the sea are concentrated on Mullet Island and along the shoreline near the Whitewater River delta and between the mouths of the Alamo and New Rivers. The Salton Sea and the Imperial and Mexicali Valleys are important areas for breeding herons and ibises. Tens of thousands nest in rookeries dotting the northern and southern shorelines, and just as many breed in the massive rookery at Finney and Ramer Lakes just south of Calipatria. Breeding species include the Least Bittern, the Great Blue and Green Herons, the Great, Snowy, and Cattle Egrets, the Black-crowned Night-Heron, and the White-faced Ibis. There is even strong circumstantial evidence that California rarities like the Little Blue Heron and the Tricolored Heron have bred there, the former thrice and the latter once.

Breeding grebes occur in lesser numbers. The Salton Sink supports many dozens of the Pied-billed Grebe and both of the *Aechmophorus* grebes, with Clark's Grebe predominating around the Whitewater River delta and the Western Grebe being more numerous elsewhere in the region. Although various species of ducks have bred in the Salton Sink, only the Cinnamon Teal, the Redhead, and the Ruddy Duck do so annually in significant numbers. Sadly, the elegant Fulvous Whistling-Duck has gone from being a fairly common breeder to being nearly extirpated in the past century. Fewer than five pairs nest annually (but only one pair nested in 1998, and none in 2002), the only remaining breeding population remaining in the western United States. The American White Pelican has not bred in the Salton Sink since water levels rose in the 1950s.

Breeding rails are well represented. The American Coot and the Common Moorhen breed in large numbers, the former in the hundreds. Smaller numbers of the endangered Yuma Clapper Rail breed here, mainly in expansive marshes around the south end, such as at the Wister Unit of the Imperial Wildlife Area and Unit 1 of the Salton Sea National Wildlife Refuge. Importantly, the Salton Sink supports approximately one-third of the world population of this subspecies (Setmire et al. 1990). The Virginia

Rail is also an uncommon breeding resident, with populations mainly around Wister and along the three rivers. An unknown but perilously small number of Black Rails breed in the Salton Sink; they are much more numerous in the lower Colorado River valley.

NONBREEDING SEABIRDS AND WATERBIRDS

The extensive habitat in the Salton Sink provides a home to a plethora of seabirds and waterbirds during the winter and during migration. The region is an extremely important wintering locale for numerous species of waterfowl, all of which are typical of the Pacific, the Intermountain, and even the Central Flyway (Bellrose 1976). Snow and Ross's Geese total in the tens of thousands in the Imperial Valley in winter, and the number of wintering ducks, from dabblers to divers, is usually in the hundreds of thousands, if not millions (Heitmeyer et al. 1989). With the exception of the Greater White-fronted Goose (Ely and Takekawa 1996), Brant, and seagoing ducks, the vast majority of waterfowl migrating through and wintering in the Salton Sink breed in the prairies of central Canada and the northern Great Plains, not in Alaska and westernmost Canada (Fig. 20; Bellrose 1976; Rienecker 1976).

Among waterfowl not represented by this classic picture of movement are various seafaring ducks and the Brant. Not surprisingly, the Brant, the scoters, the Long-tailed Duck, and the Red-breasted Merganser are closely associated with the Gulf of California. Movement through the Salton Sink by each of these is by far most conspicuous in spring (mid-March to mid-May), when birds wintering in the gulf move northward through the region. Based on the thousands of scoters detected on inland lakes in San Diego County (Unitt 1984) and the numerous records of migrant flocks of Brant over the Anza-Borrego Desert, it would appear that most traverse a path to the west of the Salton Sink (Fig. 21).

The Salton Sea also serves as an important wintering location for both the Eared Grebe (sometimes in the millions) and the American White Pelican (typically in the tens of thousands).

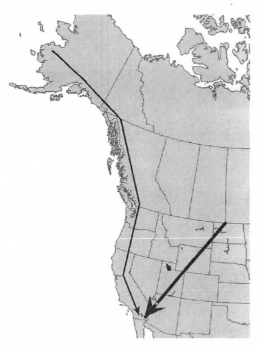

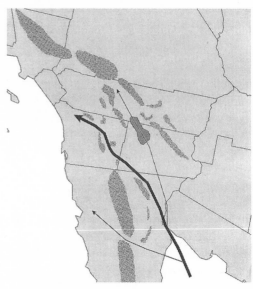

FIGURE 21. Principal spring migration route of the Brant, the scoters, and various other seaducks through southeastern California. Most migrants follow a path that takes them west of the Salton Sink, over the Anza-Borrego Desert, to the coast in northern San Diego County.

FIGURE 20. Main geographic places of origin of waterfowl wintering in the Salton Sink.

Indeed, a substantial percentage of the North American population of the former occurs at the Salton Sea during some part of the year (Jehl 1988). Similarly, the Imperial Valley apparently hosts one-third of the world population of the Mountain Plover each winter (Shuford et al. 2000). This valley is also one of the few regular wintering areas in southern California for the Sandhill Crane, including small numbers of the rare Greater Sandhill Crane (*Grus canadensis tabida*). Like the majority of the waterfowl occurring in the Salton Sink, populations of these species come mainly from the Canadian prairies, the Great Plains, or the Great Basin.

The strong relationship between the Salton Sink and the Gulf of California is revealed by the various species that push northward into the sink each year (Patten and McCaskie 2003). Outside of the regular breeding seabirds noted above, the Yellow-footed and Laughing Gulls appear annually in large numbers as postbreeding visitors from June through November. A similar pattern is shown by rarer species from the gulf,

from the Magnificent Frigatebird and the Least Tern (annual in small numbers) to the Blue-footed and Brown Boobies and the Roseate Spoonbill (rare and irregular in occurrence) to the Neotropic Cormorant and the Elegant Tern (few records for either). Numerous other species associated with the Gulf of California occur rarely to casually at the Salton Sea during late summer and early fall. The only location in the western United States that now hosts regular numbers of the Wood Stork is the south end of the Salton Sea, around the Alamo River delta. Like the population of the Fulvous Whistling-Duck, the Wood Stork population has declined drastically in the past several decades, dropping from thousands in the 1960s to hundreds by the 1980s and a few dozen in the 1990s (see Patten et al. 2003). At this rate of decline, the Wood Stork and the Fulvous Whistling-Duck are vying with the Northern Cardinal and the Elf Owl (both of which occur on the lower Colorado River) for the dubious distinction of being the next species extirpated from California.

MIGRATORY SHOREBIRDS

Aside from the Great Salt Lake, a vastly larger body of water, few places in the interior of western North America and few coastal locales support the diversity or the abundance of migratory shorebirds that the Salton Sea does (Page and Gill 1994; Shuford et al. 2000, 2002). The Salton Sea is a crucial stopover site on the Pacific Flyway (Page et al. 1992; Shuford et al. 1999, 2002, 2003). It is especially important for the Black-necked Stilt, the American Avocet, the Western Sandpiper, and the Long-billed Dowitcher, each occurring in the tens of thousands. In addition, thousands of Whimbrels, Marbled Godwits, Least Sandpipers, and Wilson's and Red-necked Phalaropes migrate through the Salton Sink each spring, and large numbers of the phalaropes are often detected in the region during fall (especially at Campo Geotérmico Cerro Prieto). The only regular wintering population of the Stilt Sandpiper in the West consists of several hundred birds at the south end of the sea. Finally, although few shorebirds breed at the Salton Sea, it is an important nesting site for the Black-necked Stilt, the Snowy Plover, the American Avocet, and the Killdeer (Grant 1982).

As might be expected, the occurrence of shorebirds in the Salton Sink is closely related to their status in the Gulf of California. Particularly noteworthy in this regard are the typically coastal species that move through the Salton Sea each spring but are otherwise largely unknown in the interior Southwest: the Ruddy Turnstone, the Red Knot, and the Sanderling. Movements of a few strongly coastal species, such as the Whimbrel and the Short-billed Dowitcher, through additional locales in the interior of California give a glimpse of the principal path taken by shorebirds migrating northward in spring from wintering grounds along the Gulf of California or elsewhere in western Mexico (Fig. 22). Huey (1927) postulated that species such as the Surfbird and various loons must cross over to the Pacific Ocean through a low-lying pass between the Sierra Juárez and the Sierra San Pedro Mártir.

Although this hypothesis has certain merits, it appears that many coastal shorebirds (and a healthy number of loons) migrate northward across the Río Colorado floodplain into the Salton Sink and thence move through the San Gorgonio Pass. Many migrants probably reach the coast by skirting the southern edge of the Transverse Ranges from there, but many shorebirds must cross over the Transverse Ranges and continue their northward journey through the western Mojave Desert and probably through California's vast Central Valley (see Fig. 22) or perhaps along the eastern edge of the Sierra Nevada (see Warnock and Bishop 1998). This predominant pattern of movement contrasts with the movement of seagoing waterfowl, most of which cut to the coast through the southerly portion of the Salton Sink and thence through the Anza-Borrego Desert and the interior of San Diego County (see Fig. 21).

In contrast to waterfowl occurring in the Salton Sink, most of which originate in central Canada and the Great Plains rather than in western Canada and Alaska, the shorebirds occurring

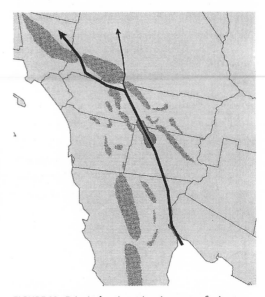

FIGURE 22. Principal spring migration route of migratory shorebirds through the Salton Sink. The majority of migrants move through the region, with many using the Salton Sea as a migratory stopover. Most continue over the Transverse Ranges, through California's Antelope and Central Valleys.

in the sink are largely from the latter region. In these regions the High Arctic supports distinct subspecies of several species of sandpipers, such as the Ruddy Turnstone, the Red Knot, the Dunlin, and the Short-billed Dowitcher. Apart from a single Asian specimen of the Red Knot, all Salton Sink specimens of these four species are of the Alaskan subspecies *Arenaria interpes interpes, Calidris canutus roselaari, C. alpina pacifica,* and *Limnodromus griseus caurinus,* respectively. By contrast, although there are two sight reports of the central Canadian Short-billed Dowitchers, no regional specimens are of the central Canadian subspecies *A. i. morinella, C. c. rufa, C. a. hudsonia,* and *L. g. hendersoni,* respectively. Thus, whereas the Salton Sink might be considered part of the Intermountain or Central Flyway for waterfowl, it indisputably rests in the Pacific Flyway for shorebirds.

BIRDS OF PREY

The Imperial Valley undoubtedly supports the largest extant population of the Burrowing Owl in California, just as the adjacent Mexicali Valley supports the largest population of the species in Baja California (Palacios et al. 2000). The Burrowing Owl joins the Barn and Great Horned Owls as the only regularly encountered owls in the region, although a small breeding population of Western Screech-Owls persists locally in the valleys. A scarce but annual species is the Short-eared Owl, a small population of which winters in the Imperial Valley. All three valleys support substantial populations of the American Kestrel and the Loggerhead Shrike, with both species being fairly common to common breeders (and common winter visitors) throughout the agricultural portions of the Salton Sink.

Agricultural areas throughout these valleys are a haven for wintering raptors, particularly the Northern Harrier, which seems abundant at times, and including such uncommon species as the Merlin and the Ferruginous Hawk, both of which are regular in small numbers. A few Bald Eagles winter around the Salton Sea annually (or nearly so), particularly around the Whitewater River delta at the north end. At times the Salton Sink supports impressive numbers of both the Osprey and the Peregrine Falcon. Although neither species breeds in the Salton Sink, each is present in small numbers through the summer; thus, they might colonize in the future, although the falcon would find a dearth of suitable nest sites. Perhaps the most interesting recent trend in raptor use of the Salton Sink is the marked increase in numbers of migrant Swainson's Hawks. Despite being listed as a threatened species in California, this hawk is seen increasingly as a migrant through much of southern California, is occurring earlier each spring (often by mid-February), and has begun to winter in some parts of the state.

PASSERINES AND OTHER SMALL LANDBIRDS

The breeding passerine fauna of the Salton Sea is undeniably that of the lower Colorado River valley and thus is a bit at odds with that of other locales in the California portion of the Sonoran Desert. Such species as the Common Ground-Dove, the Gila Woodpecker, Abert's Towhee, the Song Sparrow (*Melospiza melodia fallax*), and the Bronzed Cowbird breed regularly in the Salton Sink (with the dove, towhee, and sparrow being common), yet none can be readily found elsewhere in the Sonoran Desert of California, away from the Colorado River. Breeding landbirds more typical of the Sonoran Desert avifauna include the Lesser Nighthawk, the Verdin, the Cactus Wren, the Black-tailed Gnatcatcher, and the Crissal Thrasher. All but the last are common, with the nighthawk being surprisingly abundant at times. Species typical of the Mojave Desert and higher elevations of the Sonoran Desert (e.g., the Black-throated Sparrow) are strangely scarce.

Marsh and riparian habitats, especially along the rivers, at various lakes, large lagoons, and ponds, at Wister, and at the Salton Sea National Wildlife Refuge support breeding Marsh Wrens, Common Yellowthroats, Song Sparrows, Blue Grosbeaks, Red-winged Blackbirds, Yellow-headed Blackbirds, and other mesic-adapted breeders that are generally scarce or local in deserts. Although scarce elsewhere in the United

States, the Large-billed Savannah Sparrow (*Passerculus sandwichensis rostratus*), endemic as a breeder to the northern Gulf of California, is a fairly common postbreeding and winter visitor to the Salton Sink that recently began breeding at Campo Geotérmico Cerro Prieto and may soon colonize the Salton Sea.

When one delves deeper into the taxonomic affinities of the landbirds of the Salton Sink, however, one finds a paradox. At the species level the link with the Colorado Desert is obvious, but at the subspecies level polytypic species show as many, if not more, links with cismontane southern California (Patten et al. 2003). It has been suggested for some time that many coastal and montane breeders in the Pacific Northwest and in California migrate from their wintering grounds in western Mexico through the deserts of southwestern California (Howell 1923; Miller 1957). This idea is borne out by the predominant subspecies of polytypic species moving through the region. Among longer-distance migrants Pacific Coast forms are more numerous than corresponding inland forms (Patten et al. 2002), from only moderately (e.g., *Contopus sordidulus saturatus* Western Wood-Pewee, *Vermivora celata lutescens* Orange-crowned Warbler) to considerably (e.g., *Thryomanes bewickii charienturus* Bewick's Wren, *Wilsonin pusilla chryseola* Wilson's Warbler) to exclusively (e.g., *Oporornis tolmiei tolmiei* MacGillivray's Warbler, *Pheucticus melanocephalus maculatus* Black-headed Grosbeak). Surprisingly, this pattern prevails even among short-distance dispersers and vagrants like the Mountain Chickadee, the Red-winged Blackbird (excluding the resident breeding population), and the American Goldfinch. Similarly, recent colonizations of landbirds have come from the west, as exemplified by the Loggerhead Shrike, the Horned Lark, and Bewick's Wren (Patten et al. 2003). There are a few exceptions—for example, in the Northern Flicker, *Colaptes auratus canescens*, from the Great Basin and Sierra Nevada, outnumbers *C. a. collaris*, from the coast—but the general pattern holds.

On a different but somewhat related level, for species with southwestern and northwestern subspecies (rather than Pacific Coast and interior West subspecies) the southwestern subspecies is not always the one that occurs most often in the Salton Sink. In the case of the Willow Flycatcher, for example, *Empidonax traillai brewsteri*, of the Pacific Northwest and central California, is a common migrant, but *E. t. extimus*, of the Southwest, has been recorded but once. Similarly, there are far more specimens of *Sayornis saya saya*, the Say's Phoebe of the coast and Northwest, than of *S. s. quiescens*, of the Southwest, even though the latter breeds, albeit in small numbers. In other cases the pattern is more in line with geographic expectations. The Northern Rough-winged Swallow is chiefly represented by *Stelgidopteryx serripennis psammochrous*, of the Southwest, with the nominate subspecies, of the Northwest, occurring in lesser numbers and only during migration. Similarly, most wintering Chipping Sparrows are the western *Spizella passerina arizonae*, not the northern *S. p. boreophila*.

Migrant landbirds in general, regardless of species or subspecies, tend to follow similar

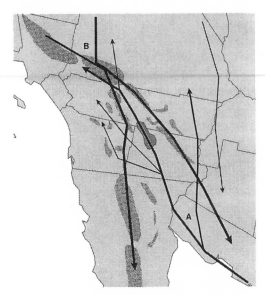

FIGURE 23. Spring (A) and fall (B) migration routes of passerines and other landbirds (e.g., Vaux's Swifts, hummingbirds) through the Salton Trough. Spring migrants travel through the Salton Sink, whereas fall migrants either skirt the Transverse Ranges to the Colorado River or follow the Peninsular Ranges into Baja California.

migratory pathways through the Salton Sink in spring and fall. In spring the principal route appears to be along the eastern edge of the Gulf of California (K. L. Garrett pers. comm.), through the Río Colorado delta, and northward along the Colorado River and through the Salton Sink. By contrast, in fall many passerines avoid the hot, dry Sonoran Desert, instead following the spine of the Peninsular Ranges or skirting the northern flank of the Transverse Ranges and then moving south along the Colorado River (Fig. 23). Many probably fly nonstop over the western Sonoran Desert. This different flight path explains the much higher abundances of migrant landbirds in spring relative to fall in the Salton Sink and the much lower number of fall migrant landbirds in the California portion of the Sonoran Desert relative to the Mojave Desert portion (M. A. Patten unpubl. data).

A Checklist of the Birds of the Salton Sea

THE PRIMARY CHECKLIST

TABLE 3 LISTS all species and subspecies reliably recorded in the Salton Sink. In the few cases in which a subspecific designation could not be made (because a specimen is lacking) or the subspecies is only presumed on the basis of its migratory habits, biogeography, and records from adjacent regions, taxa are given in brackets. Abbreviations for museum holdings and other sources are given in Table 4. We have attempted to verify the validity of all subspecific designations. When a particular specimen is assumed to represent a given taxon but we did not directly check the specimen, the museum is given in brackets. Table 3 is intended to provide only a rough idea of relative abundance and seasonal occurrence; the species accounts should be consulted for details and explanations. Typically, the season during which the species is most common is listed first in the table (and often in the species accounts).

Species accounts should be consulted for more detail about any species or subspecies in this list. We do not provide a site guide, directing the interested observer to locales in the Salton Sink where various species might be found. Information about status, distribution, and habitat gleaned from species accounts will go a long way toward meeting the need for such a guide. Birders wanting more conventional site guides or specific directions to the better birding sites should consult Brandt (1977), Dunn (1977), Mlodinow and O'Brien (1996:413), and Schram (1998).

THE NEXT TWENTY

Biogeography and records from neighboring regions (Grinnell 1928; Unitt 1984; Rosenberg et al. 1991; Russell and Monson 1998; Patten et al. 2001) imply that a few species and subspecies are conspicuously absent from the Salton Sea checklist. In the case of some subspecies, only targeted collecting will confirm the presence of ones already presumed to occur in the Salton Sink (e.g., *Certhia americana zelotes* Brown Creeper). The following 20 taxa might reasonably be expected to occur in the Salton Sink, and we dare say that they will be recorded in the not too distant future.

Continued on page 65

TABLE 3
Checklist of All Bird Taxa Reliably Recorded in the Salton Sea Region

COMMON NAME	SCIENTIFIC NAME	STATUS[a]	VERIFICATION	COLLECTION/REFERENCE[b]
GAVIIFORMES _Gaviidae_				
Red-throated Loon	_Gavia stellata_	xTs, xW	Skeleton	LSUMZ
Pacific Loon	_Gavia pacifica_	xT	Specimen	SDNHM
Common Loon	_Gavia immer_	rT, xW	Specimen	FMNH, SDNHM
PODICIPEDIFORMES _Podicipedidae_				
Least Grebe	_Tachybaptus dominicus bangsi_	xTa	Photograph	CBRC, _AB_ 43:166
Pied-billed Grebe	_Podilymbus podiceps podiceps_	cR*	Specimen	MVZ, SDNHM
Horned Grebe	_Podiceps auritus_	rT, xW, xS	Photograph	_American Birds_ 39:1061
Eared Grebe	_Podiceps nigricollis californicus_	cW	Specimen	DMNH, HSU, LACM, LSUMZ, MVZ, SBCM, SBMNH, SDNHM, SDSU, USNM, YPM
Western Grebe	_Aechmophorus occidentalis occidentalis_	uS*, fW	Specimen	SBCM, UCLA
Clark's Grebe	_Aechmophorus clarkii transitionalis_	fR*	Specimen	MVZ, SDNHM
PROCELLARIIFORMES _Diomedeidae_				
Laysan Albatross	_Phoebastria immutabilis_	xTs	Specimen	LACM, SBCM
Procellariidae				
Cook's Petrel	_Pterodroma cookii_	xS	Photograph	CBRC
Wedge-tailed Shearwater	_Puffinus pacificus_	xS	Photograph	CBRC, _AB_ 42:1225
Buller's Shearwater	_Puffinus bulleri_	xS	Specimen	SBCM
Sooty Shearwater	_Puffinus griseus_	xS	Specimen	SDNHM
Hydrobatidae				
Leach's Storm-Petrel	_Oceanodroma leucorhoa_ subsp.?	xS	Sight record	SDNHM Salton Sea archive
Black Storm-Petrel	_Oceanodroma melania_	xS	Sight record	SDNHM Salton Sea archive

TABLE 3 *(continued)*

COMMON NAME	SCIENTIFIC NAME	STATUS[a]	VERIFICATION	COLLECTION/REFERENCE[b]
Least Storm-Petrel	*Oceanodroma microsoma*	xS	Sight record	SDNHM Salton Sea archive
PELECANIFORMES *Sulidae*				
Blue-footed Booby	*Sula nebouxii nebouxii*	irTp	Specimen	SBCM, SDNHM
Brown Booby	*Sula leucogaster brewsteri*	rTp	Specimen	HSU, LACM, SBCM
Pelecanidae				
American White Pelican	*Pelecanus erythrorhynchos*	fW, uS, euS*	Specimen	LACM, MVZ, SBCM, SDNHM, USNM
Brown Pelican	*Pelecanus occidentalis californicus*	fS, rS*	Specimen	SBCM
Phalacrocoracidae				
Brandt's Cormorant	*Phalacrocorax penicillatus*	xTs	Specimen	SDNHM
Neotropic Cormorant	*Phalacrocorax brasilianus* subsp.?	xT, xS, xS*	Photograph	CBRC
Double-crested Cormorant	*Phalacrocorax auritus albociliatus*	cP*	Specimen	MVZ, SBCM, USNM
Fregatidae				
Magnificent Frigatebird	*Fregata magnificens*	rTp, xW	Photograph	*AB* 26:904
CICONIIFORMES *Ardeidae*				
American Bittern	*Botaurus lentiginosus*	uW, xS+	Specimen	SBCM, SDNHM, SDSU, UMMZ
Least Bittern	*Ixobrychus exilis hesperis*	fS*, rW	Specimen	LACM, MVZ, SBCM, SDNHM, UCLA, USNM
Great Blue Heron	*Ardea herodias wardi*	cP*	Specimen	MVZ, SBCM, SDNHM, USNM
Great Egret	*Ardea alba egretta*	cP*	Specimen	LACM, SBCM, USNM, WFVZ
Snowy Egret	*Egretta thula candidissima*	cP*	Specimen	PSM, SBCM, SBMNH, SDNHM, UCLA, USNM

TABLE 3 *(continued)*

COMMON NAME	SCIENTIFIC NAME	STATUS[a]	VERIFICATION	COLLECTION/REFERENCE[b]
	Egretta thula brewsteri	xS*	Specimen	SDNHM
Little Blue Heron	*Egretta caerulea*	rTsa, xS*	Photograph	SDNHM Salton Sea archive
Tricolored Heron	*Egretta tricolor ruficollis*	xP, xS+	Photograph	CBRC
Reddish Egret	*Egretta rufescens [dickeyi]*	xS, xW	Photograph	CBRC
Cattle Egret	*Bubulcus ibis ibis*	cR*	Specimen	LACM, SBCM, SDNHM, WFVZ
Green Heron	*Butorides virescens anthonyi*	fP*	Specimen	FMNH, MVZ, SBCM, SBMNH, SDNHM, UCLA, USNM
Black-crowned Night-Heron	*Nycticorax nycticorax hoactli*	xcP*	Specimen	LSUMZ, MVZ, SBCM
Yellow-crowned Night-Heron	*Nyctanassa violacea bancrofti*	xTs	Photograph	CBRC
Threskiornithidae				
White Ibis	*Eudocimus ruber albus*	xS	Sight record	CBRC
Glossy Ibis	*Plegadis falcinellus*	xS	Photograph	CBRC, Patten and Lasley 2000
White-faced Ibis	*Plegadis chihi*	fS*, cW	Specimen	LACM, LSUMZ, MVZ, SBCM, SBMNH, SDNHM, UCLA, USNM, WFVZ
Roseate Spoonbill	*Platalea ajaja*	xS	Specimen	MVZ, SBCM, SDNHM, WFVZ
Ciconiidae				
Wood Stork	*Mycteria americana*	ecTp, lrTp	Specimen	HSU, LACM, SBCM, UCLA
Cathartidae				
Turkey Vulture	*Cathartes aura meridionalis*	cT, uW	Specimen	LACM, SBCM, SDNHM
ANSERIFORMES *Anatidae*				
Black-bellied Whistling-Duck	*Dendrocygna autumnalis [fulgens]*	xS, xT	Photograph	CBRC, *NAB* 54:422
Fulvous Whistling-Duck	*Dendrocygna bicolor*	lrS*, xT	Specimen	MVZ, SBCM, SDNHM, USNM

TABLE 3 *(continued)*

COMMON NAME	SCIENTIFIC NAME	STATUS[a]	VERIFICATION	COLLECTION/REFERENCE[b]
Greater White-fronted Goose	*Anser albifrons frontalis*	uT, rW	Specimen	SDNHM
Snow Goose	*Chen caerulescens caerulescens*	cW, xS	Specimen	LACM, LSUMZ, PSM, SBMNH, SDNHM, UCLA, UMMZ, WFVZ
Ross's Goose	*Chen rossii*	cW, xS	Specimen	LSUMZ, SBCM, SDNHM
Canada Goose	*Branta canadensis moffitti*	cW, xS	Specimen	SBCM, SDNHM, SDSU
	Branta canadensis parvipes	uW	Specimen	SDNHM
	Branta canadensis minima	rW	Specimen	HSU
	Branta canadensis leucopareia	xW	Sight record	SDNHM Salton Sea archive
Brant	*Branta bernicla nigricans*	uTs, rS, rTa	Photograph	Fig. 33
Tundra Swan	*Cygnus columbianus columbianus*	xW	Specimen	SBCM, SBMNH, SDNHM
Wood Duck	*Aix sponsa*	xW	Photograph	SDNHM Salton Sea archive
Gadwall	*Anas strepera strepera*	fW, rS	Specimen	MVZ, SBCM, SBMNH, SDNHM, USNM
Eurasian Wigeon	*Anas penelope*	rW	Specimen	FMNH, UCLA
American Wigeon	*Anas americana*	cW, rS	Specimen	LACM, MVZ, SBCM, SDNHM, SDSU
Mallard	*Anas platyrhynchos platyrhynchos*	cW, uS*	Skeleton	SDNHM
Blue-winged Teal	*Anas discors*	rT, xW, xS	Specimen	SBCM, SDNHM, USNM
Cinnamon Teal	*Anas cyanoptera septentrionalium*	cW, fS*	Specimen	FMNH, MVZ, SBCM, SDNHM, USNM, WFVZ
Northern Shoveler	*Anas clypeata*	cW, rS	Specimen	LACM, MCZ, MVZ, SBCM, SBMNH, SDNHM, UCLA, USNM, WFVZ
Northern Pintail	*Anas acuta*	cW, rS	Specimen	LACM, MCZ, MVZ, SBCM, SBMNH, SDNHM, SDSU, USNM, WFVZ

TABLE 3 *(continued)*

COMMON NAME	SCIENTIFIC NAME	STATUS[a]	VERIFICATION	COLLECTION/REFERENCE[b]
Baikal Teal	*Anas formosa*	xW	Specimen	MVZ
Green-winged Teal	*Anas crecca carolinensis*	cW, rS	Specimen	BMNH, FMNH, MVZ, SBCM, SBMNH, SDNHM, SDSU
	Anas crecca crecca	xTs, xS	Sight record	SDNHM Salton Sea archive
Canvasback	*Aythya valisineria*	uW, xS	Specimen	LACM, MVZ, SBCM, SBMNH, WFVZ
Redhead	*Aythya americana*	cP*	Specimen	FMNH, LACM, MVZ, SBCM, SDNHM, USNM
Ring-necked Duck	*Aythya collaris*	uW, xS	Photograph	SDNHM Salton Sea archive
Tufted Duck	*Aythya fuligula*	xW	Photograph	CBRC
Greater Scaup	*Aythya marila nearctica*	rW, xS	Specimen	MVZ, SBCM, SDNHM
Lesser Scaup	*Aythya affinis*	uW, rS	Specimen	AMNH, LACM, MVZ, SBCM, SDNHM, WFVZ
Surf Scoter	*Melanitta perspicillata*	rTsa, rS, xW	Specimen	LACM, SBCM, SDNHM
White-winged Scoter	*Melanitta fusca deglandi*	rTsa, rS, xW	Specimen	SDNHM
Black Scoter	*Melanitta nigra americana*	xP	Specimen	UCLA
Long-tailed Duck	*Clangula hyemalis*	xT, xW	Specimen	MVZ, SBCM
Bufflehead	*Bucephala albeola*	cW, rS, xS*	Specimen	MVZ, SBCM
Common Goldeneye	*Bucephala clangula americana*	uW	Specimen	LACM, MVZ, SBCM
Barrow's Goldeneye	*Bucephala islandica*	xW	Sight record	SDNHM Salton Sea archive
Hooded Merganser	*Lophodytes cucullatus*	rW	Specimen	SBCM
Common Merganser	*Mergus merganser americanus*	rW	Specimen	LACM
Red-breasted Merganser	*Mergus serrator*	uW, xS	Specimen	LACM, MVZ, SBCM
Ruddy Duck	*Oxyura jamaicensis rubida*	cP*	Specimen	LACM, LSUMZ, MVZ, SBCM, SBMNH, SDNHM, SDSU, UCLA, USNM, WFVZ, YPM (skeleton)

TABLE 3 *(continued)*

COMMON NAME	SCIENTIFIC NAME	STATUS[a]	VERIFICATION	COLLECTION/REFERENCE[b]
FALCONIFORMES				
Accipitridae				
Osprey	*Pandion haliaetus carolinensis*	uP	Specimen	SBCM
White-tailed Kite	*Elanus leucurus majusculus*	rP*	Specimen	SDNHM
Bald Eagle	*Haliaeetus leucocephalus alascanus*	rW	Specimen	SBCM
	Haliaeetus leucocephalus leucocephalus	xTp	Sight record	SDNHM Salton Sea archive
Northern Harrier	*Circus cyaneus hudonius*	cW, rS	Specimen	LACM, LSUMZ, SBCM, UCLA, WFVZ
Sharp-shinned Hawk	*Accipiter striatus velox*	uW	Specimen	FMNH, LACM, SBCM, UCLA
Cooper's Hawk	*Accipiter cooperi*	uW	Specimen	LACM, SBCM, WFVZ
Common Black-Hawk	*Buteogallus anthracinus [anthracinus]*	xTs	Sight record	CBRC
Harris's Hawk	*Parabuteo unicinctus superior*	efR*, xW, xS	Specimen	LACM, SBCM
Red-shouldered Hawk	*Buteo lineatus elegans*	rP	Specimen	AMNH
Broad-winged Hawk	*Buteo platypterus platypterus*	xW	Photograph	SDNHM Salton Sea archive
Swainson's Hawk	*Buteo swainsoni*	rTsa, xW	Photograph	Fig. 34
Zone-tailed Hawk	*Buteo albonotatus*	xW	Photograph	CBRC, *NAB* 55:355
Red-tailed Hawk	*Buteo jamaicensis calurus*	cW,rS*	Specimen	LSUMZ, SBCM, USNM, WFVZ
	Buteo jamaicensis harlani	xW	Photograph	SDNHM Salton Sea archive
	Buteo jamaicensis fuertesi	xW	Sight record	SDNHM Salton Sea archive
	Buteo jamaicensis kriderii	xW	Photograph	SDNHM Salton Sea archive
Ferruginous Hawk	*Buteo regalis*	uW	Specimen	LACM
Rough-legged Hawk	*Buteo lagopus sanctijohannis*	rW	Photograph	SDNHM Salton Sea archive
Golden Eagle	*Aquila chrysaetos canadensis*	xT, xW	Specimen	UCLA

TABLE 3 *(continued)*

COMMON NAME	SCIENTIFIC NAME	STATUS[a]	VERIFICATION	COLLECTION/REFERENCE[b]
Falconidae				
American Kestrel	*Falco sparverius sparverius*	cP*	Specimen	LACM, MCZ, SBCM, SDNHM, UCLA, WFVZ
	Falco sparverius peninsularis	rP*	Specimen	SDNHM
Merlin	*Falco columbarius columbarius*	rW	Specimen	MVZ, SBCM, SDNHM
	Falco columbarius suckleyi	xW	Specimen	SBCM
	Falco columbarius richardsonii	xW	Photograph	SDNHM Salton Sea archive
Peregrine Falcon	*Falco peregrinus anatum*	uS, rW	Specimen	LSUMZ, MVZ, PSM, SBCM, SDNHM
Prairie Falcon	*Falco mexicanus*	rW, xS	Specimen	LACM, MVZ, UCLA
GALLIFORMES *Odontophoridae*				
Gambel's Quail	*Callipepla gambelii gambelii*	cR*	Specimen	AMNH, FMNH, HSU, LACM, MVZ, PSM, SBCM, SBMNH, SDNHM, UCLA, USNM, WFVZ
GRUIFORMES *Rallidae*				
Black Rail	*Laterallus jamaicensis coturniculus*	rS*	Photograph	SDNHM Salton Sea archive
Clapper Rail	*Rallus longirostris yumanensis*	uR*	Specimen	MVZ, PSM (egg set), SBCM, SDNHM, USNM (egg set)
Virginia Rail	*Rallus limicola limicola*	fR*	Specimen	LACM, MVZ, SBCM, SBMNH, UMMZ
Sora	*Porzana carolina*	cW, xS	Specimen	MVZ, SBCM, SDNHM, UCLA, UMMZ
Common Moorhen	*Gallinula chloropus cachinnans*	fR*	Specimen	LACM, LSUMZ, MVZ, SBCM, SBMNH, SDNHM, WFVZ
American Coot	*Fulica americana americana*	cR*	Specimen	LSUMZ, MVZ, SBCM, SDNHM, USNM, WFVZ

TABLE 3 *(continued)*

COMMON NAME	SCIENTIFIC NAME	STATUS[a]	VERIFICATION	COLLECTION/REFERENCE[b]
Gruidae				
Sandhill Crane	*Grus canadensis canadensis*	luW	Specimen	LSUMZ, SDNHM, WFVZ
	Grus canadensis tabida	lrW	Specimen	SDNHM
CHARADRIIFORMES *Charadriidae*				
Black-bellied Plover	*Pluvialus squatarola*	cT, fW, uS	Specimen	SBCM, SDSU, UCLA
American Golden-Plover	*Pluvialus dominica*	xT	Specimen	SBCM
Pacific Golden-Plover	*Pluvialus fulva*	xT, xW	Photograph	Fig. 37
Snowy Plover	*Charadrius alexandrinus nivosus*	cS*, uW	Specimen	LACM, MVZ, SBCM, SDNHM
Wilson's Plover	*Charadrius wilsonia beldingi*	xS*	Egg set	SBCM
Semipalmated Plover	*Charadrius semipalmatus*	cT, uW, rS	Specimen	LACM, MVZ, SBCM, SDNHM, USNM
Killdeer	*Charadrius vociferus vociferus*	cR*	Specimen	FMNH, LACM, MVZ, SBCM, SBMNH, UCLA, UMMZ, USNM, WFVZ
Mountain Plover	*Charadrius montanus*	fW	Skeleton	SBCM, SDNHM
Eurasian Dotterel	*Charadrius morinellus*	xW	Photograph	CBRC
Haematopodidae				
American Oystercatcher	*Haematopus palliatus* subsp.?	xTa	Photograph	CBRC, Roberson 1980:114
Recurvirostridae				
Black-necked Stilt	*Himantopus mexicanus mexicanus*	cR*	Specimen	FMNH, HSU, LACM, LSUMZ, MVZ, SBCM, SBMNH, SDNHM, SDSU, UCLA, USNM, WFVZ
American Avocet	*Recurvirostra americana*	cR*	Specimen	FMNH, LACM, LSUMZ, MCZ, MVZ, SBCM, SBMNH, SDNHM, UCLA, USNM, WFVZ

TABLE 3 *(continued)*

COMMON NAME	SCIENTIFIC NAME	STATUS[a]	VERIFICATION	COLLECTION/REFERENCE[b]
Scolopacidae				
Greater Yellowlegs	*Tringa melanoleuca*	fW, rS	Specimen	HSU, LACM, MVZ, SBCM, SBMNH, SDNHM, USNM, WFVZ
Lesser Yellowlegs	*Tringa flavipes*	uW, xS	Specimen	MVZ, SBCM, USNM
Spotted Redshank	*Tringa erythropus*	xTs	Photograph	CBRC, Morlan 1985
Solitary Sandpiper	*Tringa solitaria cinnamomea*	rT, xW	Specimen	MVZ, SBCM, SDNHM, UCLA, USNM
Willet	*Catoptrophorus semipalmatus inornatus*	fP	Specimen	LACM, LSUMZ, SBCM, UCLA
Wandering Tattler	*Heteroscelus incanus*	xT	Specimen	USNM
Spotted Sandpiper	*Actitis macularia*	fW	Specimen	MVZ, SBCM, USNM
Whimbrel	*Numenius phaeopus hudsonicus*	cT, rS, xW	Specimen	MVZ, SBCM, SDNHM, SDSU
Long-billed Curlew	*Numenius americanus*	cW, luS	Specimen	SBCM, SDNHM, USNM
Hudsonian Godwit	*Limosa haemastica*	xT	Skeleton	LSUMZ
Marbled Godwit	*Limosa fedoa fedoa*	fW, uS	Specimen	MVZ, SBCM, UCLA
Ruddy Turnstone	*Arenaria interpes interpes*	uTs, rTa, xW, xS	Specimen	SBCM, SDNHM
Black Turnstone	*Arenaria melanocephala*	rTs, xTa, xW, xS	Specimen	SBCM
Surfbird	*Calidris virgata*	rTs	Specimen	SDNHM, USNM
Red Knot	*Calidris canutus roselaari*	fTs, rTa	Specimen	LACM, SBCM, SDNHM
	Calidris canutus rogersi	xTs	Specimen	SBCM
Sanderling	*Calidris alba*	uT, rW	Specimen	LACM, LSUMZ, SBCM, MVZ
Semipalmated Sandpiper	*Calidris pusilla*	rT	Specimen	SBCM, SDNHM
Western Sandpiper	*Calidris mauri*	cT, fW, rS	Specimen	FMNH, KU, LACM, LSUMZ, MVZ, SBCM, SBMNH, SDNHM, SDSU, UCLA, USNM, UWBM
Little Stint	*Calidris minuta*	xTs	Photograph	CBRC

TABLE 3 *(continued)*

COMMON NAME	SCIENTIFIC NAME	STATUS[a]	VERIFICATION	COLLECTION/REFERENCE[b]
Least Sandpiper	*Calidris minutilla*	cW, rS	Specimen	AMNH, DMNH, LACM, MVZ, SBCM, SBMNH, SDNHM, UMMZ, USNM, UWBM
White-rumped Sandpiper	*Calidris fuscicollis*	xTs	Specimen	SDNHM
Baird's Sandpiper	*Calidris bairdii*	rTa, xTs	Specimen	LACM, SBCM
Pectoral Sandpiper	*Calidris melanotos*	xTas, xW	Specimen	SBCM
Dunlin	*Calidris alpina pacifica*	uW	Specimen	FMNH, MVZ, SBCM, SBMNH, SDNHM
Curlew Sandpiper	*Calidris ferruginea*	xT	Specimen	SBCM
Stilt Sandpiper	*Calidris himantopus*	lcTs, luW	Specimen	SBCM, SDNHM
Ruff	*Philomachus pugnax*	xTa, xW	Photograph	Fig. 44
Short-billed Dowitcher	*Limnodromus griseus caurinus*	cT	Specimen	MVZ, SBCM, SDNHM, WFVZ
	Limnodromus griseus hendersoni	xTs	Sight record	SDNHM Salton Sea archive
Long-billed Dowitcher	*Limnodromus scolopaceus*	cW, rS	Specimen	DMNH, FMNH, MVZ, SBCM, SBMNH, SDNHM, UCLA, UMMZ, USNM, WFVZ
Wilson's Snipe	*Gallinago delicata*	uW	Specimen	FMNH, LACM, MCZ, MVZ, SBCM, SDNHM, UCLA, USNM, WFVZ
Wilson's Phalarope	*Phalaropus tricolor*	cT, xW	Specimen	KU, MVZ, SBCM, USNM
Red-necked Phalarope	*Phalaropus lobatus*	cT	Specimen	LSUMZ, MVZ, SBCM, SBMNH, SDNHM, UCLA, UMMZ
Red Phalarope	*Phalaropus fulicarus*	rTa, xTs, xS, xW	Specimen	SBCM, SDNHM, UMMZ
Laridae				
Pomarine Jaeger	*Stercorarius pomarinus*	xS, xT	Specimen	SBCM
Parasitic Jaeger	*Stercorarius parasiticus*	rTa, xTs	Specimen	SBCM

TABLE 3 *(continued)*

COMMON NAME	SCIENTIFIC NAME	STATUS[a]	VERIFICATION	COLLECTION/REFERENCE[b]
Long-tailed Jaeger	*Stercorarius longicaudus pallescens*	xT	Specimen	LACM, SDNHM
Laughing Gull	*Larus atricilla*	fTp, xW, xS*	Specimen	LACM, MVZ, SBCM, UCLA
Franklin's Gull	*Larus pipixcan*	uTs, rTa, xS, xW	Specimen	SBCM, SDNHM
Little Gull	*Larus minutus*	xW, xT	Specimen	LACM, SDNHM (skeleton)
Bonaparte's Gull	*Larus philadelphia*	uW, rS	Specimen	MVZ, SBCM, SBMNH (skeleton), SDNHM, USNM
Heermann's Gull	*Larus heermanni*	xTp, xW, xS	Specimen	LSUMZ, SBCM
Mew Gull	*Larus canus brachyrhynchus*	rW	Specimen	SBCM, SDNHM
Ring-billed Gull	*Larus delawarensis*	cW, uS	Specimen	DMNH, LSUMZ, MVZ, SBCM, SDNHM, UCLA, WFVZ
California Gull	*Larus californicus californicus*	cW, uS, lrS*	Specimen	DMNH, LACM, LSUMZ, SBCM, SDNHM
	Larus californicus albertaensis	uW	Specimen	DMNH, LSUMZ
Herring Gull	*Larus argentatus smithsonianus*	fW	Specimen	DMNH, SBCM, SDNHM
Thayer's Gull	*Larus glaucoides thayeri*	rW, xS	Specimen	SBCM
Lesser Black-backed Gull	*Larus fuscus graellsii*	xW, xTa	Photograph	CBRC, Dunn 1988
Yellow-footed Gull	*Larus livens*	cS, rW	Specimen	DMNH, HSU, LACM, LSUMZ, PSM, SBCM, SDNHM
Western Gull	*Larus occidentalis wymani*	rP	Specimen	[LSUMZ], UCSB
	Larus occidentalis occidentalis	xW	Sight record	SDNHM Salton Sea archive
Glaucous-winged Gull	*Larus glaucescens*	rW, xS	Specimen	SBCM
Glaucous Gull	*Larus hyperboreus barrovianus*	xP	Specimen	SBCM
Black-legged Kittiwake	*Rissa tridactyla pollicaris*	xW, xS, xTa	Specimen	SDNHM

TABLE 3 *(continued)*

COMMON NAME	SCIENTIFIC NAME	STATUS[a]	VERIFICATION	COLLECTION/REFERENCE[b]
Sabine's Gull	*Xema sabini*	rTa, xS, xTs	Specimen	LACM, SBCM
Gull-billed Tern	*Sterna nilotica vanrossemi*	fS*	Specimen	FMNH, LACM, LSUMZ, MVZ, PSM, SBCM, SDNHM, UCLA, WFVZ
Caspian Tern	*Sterna caspia*	fS*, rW	Specimen	LACM, LSUMZ, MVZ, SBCM, SBMNH, SDNHM, UCLA
Royal Tern	*Sterna maxima maxima*	xS	Photograph	SDNHM Salton Sea archive
Elegant Tern	*Sterna elegans*	xS	Photograph	SDNHM Salton Sea archive
Common Tern	*Sterna hirundo hirundo*	cTa, rTs	Specimen	HSU, LSUMZ, SBCM
Arctic Tern	*Sterna paradisaea*	xTsa, xS	Sight record	SDNHM Salton Sea archive
Forster's Tern	*Sterna forsteri*	fS*, uW	Specimen	LACM, MVZ, SBCM, SBMNH, SDNHM, UMMZ
Least Tern	*Sterna antillarum* subsp.?	rTs, xS	Photograph	SDNHM Salton Sea archive
Black Tern	*Chlidonias niger surinamensis*	fT, uS	Specimen	LACM, MVZ, SBCM, SBMNH, SDNHM, SDSU, UCLA, UMMZ, WFVZ, YPM
Black Skimmer	*Rynchops niger niger*	fS*, xW	Specimen	LACM, SBCM, SBMNH (skeleton), SDNHM
Alcidae				
Ancient Murrelet	*Synthliboramphus antiquus*	xTs	Sight record	SDNHM Salton Sea archive
COLUMBIFORMES *Columbidae*				
Band-tailed Pigeon	*Columba fasciata [monilis]*	xT, xW	Sight record	SDNHM Salton Sea archive
White-winged Dove	*Zenaida asiatica mearnsi*	cS*	Specimen	FMNH, HSU, LACM, MVZ, PSM, SBCM, SBMNH, SDNHM, WFVZ
Mourning Dove	*Zenaida macroura marginella*	cR*	Specimen	FMNH, HSU, LACM, MVZ, PSM, SBCM, SDNHM, WFVZ

TABLE 3 *(continued)*

COMMON NAME	SCIENTIFIC NAME	STATUS[a]	VERIFICATION	COLLECTION/REFERENCE[b]
Inca Dove	*Columbina inca*	uR*	Photograph	Fig. 49
Common Ground-Dove	*Columbina passerina pallescens*	cR*	Specimen	LACM, LSUMZ, SBCM, SBMNH, SDNHM, UCLA, WFVZ
Ruddy Ground-Dove	*Columbina talpacoti eluta*	xW	Photograph	CBRC
CUCULIFORMES *Cuculidae*				
Yellow-billed Cuckoo	*Coccyzus americanus*	xS	Sight record	SDNHM Salton Sea archive
Greater Roadrunner	*Geococcyx californianus*	fR*	Specimen	AMNH, FMNH, LACM, MVZ, PSM, SBCM, SDNHM, UMMZ, USNM
Groove-billed Ani	*Crotophaga sulcirostris*	xTa	Sight record	CBRC
STRIGIFORMES *Tytonidae*				
Barn Owl	*Tyto alba pratincola*	uR*	Specimen	LACM, SBCM, SDNHM
Strigidae				
Flammulated Owl	*Otus flammeolus* subsp.?	xTa	Sight record	SDNHM Salton Sea archive
Western Screech-Owl	*Otus kennicottii yumanensis*	lrR*	Specimen	SBCM, WFVZ
Great Horned Owl	*Bubo virginianus pallescens*	uR*	Specimen	LACM, MVZ, SBCM, SDNHM, UCLA, UMMZ
Elf Owl	*Micrathene whitneyi whitneyi*	xTa	Photograph	SDNHM Salton Sea archive
Burrowing Owl	*Athene cunnicularia hypugaea*	cR*	Specimen	LACM, MVZ, SBCM, SBMNH, SDNHM, UCLA
Long-eared Owl	*Asio otus wilsonianus*	rW	Specimen	SBCM
Short-eared Owl	*Asio flammeus flammeus*	rW	Specimen	LACM, SBCM, SDNHM
Northern Saw-whet Owl	*Aegolius acadicus acadicus*	xW	Specimen	SBCM

TABLE 3 (continued)

COMMON NAME	SCIENTIFIC NAME	STATUS[a]	VERIFICATION	COLLECTION/REFERENCE[b]
CAPRIMULGIFORMES *Caprimulgidae*				
Lesser Nighthawk	*Chordeiles acutipennis texensis*	cS*, xW	Specimen	FMNH, LACM, LSUMZ, MLZ, MVZ, SBCM, SDNHM, SDSU, UCLA, USNM, WFVZ
Common Poorwill	*Phalaenoptilus nuttallii nuttallii*	xT	Specimen	LACM, MVZ, SBCM, SDNHM, USNM
Whip-poor-will	*Caprimulgus vociferus arizonae*	xTa	Specimen	SBCM
APODIFORMES *Apodidae*				
Black Swift	*Cypseloides niger borealis*	xTsa	Specimen	SDNHM
Chimney Swift	*Chaetura pelagica*	xTs	Sight record	SDNHM Salton Sea archive
Vaux's Swift	*Chaetura vauxi vauxi*	cTs, uTa	Specimen	MVZ, SBCM, SDNHM, UCLA, USNM
White-throated Swift	*Aeronautes saxatalis saxatalis*	uW	Specimen	FMNH, MVZ, SDNHM, WFVZ
Trochilidae				
Black-chinned Hummingbird	*Archilochus alexandri*	fS*	Specimen	MVZ, SBCM, USNM
Anna's Hummingbird	*Calypte anna*	fR*	Specimen	SBCM
Costa's Hummingbird	*Calypte costae*	fR*	Specimen	LACM, SBCM, SDNHM, USNM
Calliope Hummingbird	*Selasphorus calliope*	rTs	Specimen	SBCM
Rufous Hummingbird	*Selasphorus rufus*	uT	Specimen	SBCM, SDNHM
Allen's Hummingbird	*Selasphorus sasin sasin*	xTa	Photograph	Fig. 53
CORACIFORMES *Alcedinidae*				
Belted Kingfisher	*Ceryle alcyon*	uW	Specimen	MVZ, SBCM, SDNHM, USNM
PICIFORMES *Picidae*				
Lewis's Woodpecker	*Melanerpes lewis*	rW	Specimen	SBCM

TABLE 3 *(continued)*

COMMON NAME	SCIENTIFIC NAME	STATUS[a]	VERIFICATION	COLLECTION/REFERENCE[b]
Red-headed Woodpecker	*Melanerpes erythrocephalus*	xS	Photograph	CBRC, Cardiff and Driscoll 1972
Acorn Woodpecker	*Melanerpes formicivorus* subsp.?	xTa, xW, xS	Photograph	Fig. 54
Gila Woodpecker	*Melanerpes uropygialis uropygialis*	uR*	Specimen	PSM, SBCM, SDNHM, USNM
Williamson's Sapsucker	*Sphyrapicus thyroideus*	xTa, xW	Specimen	WFVZ
Yellow-bellied Sapsucker	*Sphyrapicus varius*	xW	Specimen	SBCM
Red-naped Sapsucker	*Sphyrapicus nuchalis*	rW	Specimen	LACM, SDNHM
Red-breasted Sapsucker	*Sphyrapicus ruber daggetti*	xW	Photograph	SDNHM Salton Sea archive
Ladder-backed Woodpecker	*Picoides scalaris cactophilus*	uR*	Specimen	AMNH, FMNH, LACM, MCZ, MVZ, SBCM, SDNHM, UCLA, UMMZ, USNM, WFZV
Nuttall's Woodpecker	*Picoides nuttallii*	xTa	Sight record	SDNHM Salton Sea archive
Downy Woodpecker	*Picoides pubescens* subsp.?	xTs	Sight record	SDNHM Salton Sea archive
White-headed Woodpecker	*Picoides albolarvatus* subsp.?	xTa	Sight record	*AFN* 10:58
Northern Flicker	*Colaptes auratus canescens*	fW	Specimen	FMNH, [LSUMZ], MCZ, SBCM, SDNHM, WFVZ
	Colaptes auratus collaris	uW	Specimen	FMNH, SBCM, SDNHM, WFVZ
	Colaptes auratus cafer	xW	Specimen	SDNHM
	Colaptes auratus luteus	xW	Specimen	SBCM
PASSERIFORMES *Tyrannidae*				
Olive-sided Flycatcher	*Contopus cooperi cooperi*	uTs, xTa	Specimen	MVZ, SBCM, SDNHM
	Contopus cooperi majorinus	rTs	Specimen	MVZ, SDNHM

TABLE 3 *(continued)*

COMMON NAME	SCIENTIFIC NAME	STATUS[a]	VERIFICATION	COLLECTION/REFERENCE[b]
Greater Pewee	*Contopus pertinax pallidiventris*	xTa	Specimen	SBCM
Western Wood-Pewee	*Contopus sordidulus veliei*	fT	Specimen	[LSUMZ], MVZ, SBCM, SDNHM, [USNM]
	Contopus sordidulus saturatus	cT	Specimen	[FMNH], [LSUMZ], SBCM, SDNHM
Alder Flycatcher	*Empidonax alnorum*	xTa	Specimen	SDNHM
Willow Flycatcher	*Empidonax traillii brewsteri*	cT, xS	Specimen	LACM, LSUMZ, MVZ, SBCM, SDNHM
	Empidonax traillii extimus	xTs	Specimen	SDNHM
Least Flycatcher	*Empidonax minimus*	xW	Specimen	SDNHM
Hammond's Flycatcher	*Empidonax hammondii*	fTs, rTa	Specimen	FMNH, LSUMZ, MVZ, SBCM, SDNHM, USNM
Gray Flycatcher	*Empidonax wrightii*	uT, rW	Specimen	SBCM, SDNHM
Dusky Flycatcher	*Empidonax oberholseri*	rTs, xTa	Specimen	MVZ, SBCM, SDNHM
Western Flycatcher	*Empidonax difficilis difficilis*	fT, xW	Specimen	FMNH, MVZ, SBCM, SDNHM, USNM
Black Phoebe	*Sayornis nigricans semiatra*	cW, uS*	Specimen	DMNH, LACM, LSUMZ. MVZ, PSM, SBCM, SDNHM, USNM
Eastern Phoebe	*Sayornis phoebe*	xTa, rW	Specimen	SBCM, SDNHM
Say's Phoebe	*Sayornis saya quiescens*	uR*	Specimen	SDNHM
	Sayornis saya saya	cW	Specimen	FMNH, LACM, MVZ, PSM, SBCM, SBMNH, SDNHM, UMMZ, USNM, WFVZ
Vermilion Flycatcher	*Pyrocephalus rubinus flammeus*	leuS*, rW, xS+	Specimen	DMNH, FMNH, MCZ, MVZ, PSM (egg set), SBCM, UCLA, USNM, UMMZ
Dusky-capped Flycatcher	*Myiarchus tuberculifer olivascens*	xW	Photograph	CBRC
Ash-throated Flycatcher	*Myiarchus cinerascens*	uT, rW	Specimen	CHAS, LSUMZ, MVZ, SBCM, SDNHM, UCLA

TABLE 3 *(continued)*

COMMON NAME	SCIENTIFIC NAME	STATUS[a]	VERIFICATION	COLLECTION/REFERENCE[b]
Brown-crested Flycatcher	*Myiarchus tyrannulus magister*	xS*	Sight record	SDNHM Salton Sea archive
Tropical Kingbird	*Tyrannus melanocolichus [satrapa]*	xTa, xW	Sight record	SDNHM Salton Sea archive
Cassin's Kingbird	*Tyrannus vociferans*	xTas, xW	Sight record	van Rossem 1911, SDNHM Salton Sea archive
Western Kingbird	*Tyrannus verticalis*	cS*	Specimen	LACM, MVZ, PSM, SBCM, SDNHM, UCLA, WFVZ
Eastern Kingbird	*Tyrannus tyrannus*	xT, xS	Photograph	SDNHM Salton Sea archive
Scissor-tailed Flycatcher	*Tyrannus forficatus*	xTa	Sight record	CBRC
Laniidae				
Northern Shrike	*Lanius excubitor borealis*	xW	Photograph	*AB* 43:367
Loggerhead Shrike	*Lanius ludovicianus gambelii*	fR*	Specimen	FMNH, [HSU], LACM, MVZ, SBMNH, SDNHM, UCLA, UMMZ
	Lanius ludovicianus excubitorides	fP*	Specimen	AMNH, [CU], [FMNH], LACM, MCZ, MVZ, [PSM], SBCM, SDNHM, UCLA, [UMMZ]
Vireonidae				
Bell's Vireo	*Vireo bellii pusillus*	xW, xT	Specimen	UCLA
Plumbeous Vireo	*Vireo plumbeus plumbeus*	rW, xTa	Specimen	SDNHM
Cassin's Vireo	*Vireo cassinii cassinii*	uTs, rTa	Specimen	FMNH, LSUMZ, MVZ, SBCM, UCLA, USNM
Warbling Vireo	*Vireo gilvus swainsonii*	cT	Specimen	FMNH, LSUMZ, MVZ, SBCM, SDNHM, UCLA, USNM
Red-eyed Vireo	*Vireo olivaceus*	xTa	Sight record	CBRC
Corvidae				
Western Scrub-Jay	*Aphelocoma californica woodhouseii*	xW	Specimen	SDNHM

TABLE 3 *(continued)*

COMMON NAME	SCIENTIFIC NAME	STATUS[a]	VERIFICATION	COLLECTION/REFERENCE[b]
	Aphelocoma californica obscura	xTa	Specimen	SDNHM
Pinyon Jay	*Gymnorhinus cyanocephalus subsp.?*	xW	Photograph	SDNHM Salton Sea archive
Clark's Nutcracker	*Nucifraga columbiana*	xTa	Sight record	Esterly 1920, Clary and Clary 1936b
American Crow	*Corvus brachyrhynchos hesperis*	rW	Sight record	SDNHM Salton Sea archive
Common Raven	*Corvus corax clarionensis*	fR*	Specimen	MCZ, MVZ, SDNHM
Alaudidae				
Horned Lark	*Eremophila alpestris leucansiptila*	fR*	Specimen	[CHAS], FMNH, LACM, MCZ, MVZ, [PSM], SBCM, SDNHM, UCLA
	Eremophila alpestris actia	rR*	Specimen	MVZ, SDNHM, UCLA
	Eremophila alpestris leucolaema	fW	Specimen	FMNH, UCLA
	Eremophila alpestris ammophila	fW	Specimen	FMNH, SDNHM, UCLA, UMMZ, WFVZ
	Eremophila alpestris enthymia	rW	Specimen	UCLA
	Eremophila alpestris lamprochroma	xW	Specimen	FMNH, MVZ
Hirundinidae				
Purple Martin	*Progne subis subis*	xT, xS	Photograph	SDNHM Salton Sea archive
Tree Swallow	*Tachycineta bicolor*	cT, fW	Specimen	MVZ, SBCM, SBMNH, SDNHM
Violet-green Swallow	*Tachycineta thalassina thalassina*	fT, xW	Specimen	MVZ, SBCM
Northern Rough-winged Swallow	*Stelgidopteryx serripennis psammochrous*	cT, uW	Specimen	MVZ, SBCM, SBMNH, SDNHM, WFVZ
	Stelgidopteryx serripennis serripennis	rT	Specimen	SDNHM

TABLE 3 *(continued)*

COMMON NAME	SCIENTIFIC NAME	STATUS[a]	VERIFICATION	COLLECTION/REFERENCE[b]
Bank Swallow	*Riparia riparia riparia*	fT, xW	Specimen	MVZ, SBCM
Cave Swallow	*Petrochelidon fulva pallida*	xT	Photograph	CBRC, Patten and Erickson 1994
Cliff Swallow	*Petrochelidon pyrrhonota tachina*	cT, fS*, xW	Specimen	[LSUMZ], SBCM, SDNHM, WFVZ
	Petrochelidon pyrrhonota pyrrhonota	cT	Specimen	MVZ
Barn Swallow	*Hirundo rustica erythrogaster*	cT, lrS*, xW	Specimen	LACM, MVZ, SBCM, SDNHM
Paridae				
Mountain Chickadee	*Poecile gambelii baileyae*	xW, xTa	Specimen	SDNHM
Oak Titmouse	*Baeolophus inornatus affabilis*	xW	Specimen	CSULB, SBCM
Remizidae				
Verdin	*Auriparus flaviceps acaciarum*	cR*	Specimen	AMNH, ANSP, DMNH, FMNH, HSU, LACM, MVZ, PSM, SBCM, SBMNH, SDNHM, UCLA, UMMZ, WFVZ
Aegithalidae				
Bushtit	*Psaltriparus minimus melanurus*	xW, xT	Specimen	SBCM
Sittidae				
Red-breasted Nuthatch	*Sitta canadensis*	rW	Specimen	SBCM
White-breasted Nuthatch	*Sitta carolinensis aculeata*	xTa	Photograph	Fig. 59
Certhiidae				
Brown Creeper	*[Certhia americana montana]*	rW		Presumed to occur
	[Certhia americana zelotes]	rW		Presumed to occur
	Certhia americana americana	xW	Specimen	SDNHM

TABLE 3 *(continued)*

COMMON NAME	SCIENTIFIC NAME	STATUS[a]	VERIFICATION	COLLECTION/REFERENCE[b]
Troglodytidae				
Cactus Wren	*Campylorhynchus brunneicapillus anthonyi*	fR*	Specimen	AMNH, FMNH, LACM, LSUMZ, MCZ, MVZ, PSM, SBCM, SDNHM, UCLA, UMMZ
Rock Wren	*Salpinctes obsoletus obsoletus*	rW	Specimen	LACM, SBCM
Canyon Wren	*Catherpes mexicanus subsp.?*	xW	Sight record	SDNHM Salton Sea archive
Bewick's Wren	*Thryomanes bewickii charienturus*	uW, rS+	Specimen	LACM, SBCM, SDNHM, [UCLA]
	Thryomanes bewickii eremophilus	rW	Specimen	SDNHM
House Wren	*Troglodytes aedon parkmanii*	fW	Specimen	MVZ, SBCM, SDNHM, UCLA
Winter Wren	*Troglodytes troglodytes [pacificus]*	xW	Sight record	SDNHM Salton Sea archive
	Troglodytes troglodytes hiemalis	xW	Sight record	SDNHM Salton Sea archive
Marsh Wren	*Cistothorus palustris aestuarinus*	cR*	Specimen	[DMNH], [LSUMZ], MVZ, SBCM, SDNHM, UCLA, UMMZ
	Cistothorus palustris pulverius	rW	Specimen	SDNHM
	Cistothorus palustris plesius	fW	Specimen	FMNH, MVZ, SDNHM, UMMZ
Regulidae				
Golden-crowned Kinglet	*Regulus satrapa subsp.?*	rW	Sight record	SDNHM Salton Sea archive
Ruby-crowned Kinglet	*Regulus calendula calendula*	cW	Specimen	FMNH, LACM, MVZ, PSM, SBCM, SDNHM, UMMZ
Sylviidae				
Blue-gray Gnatcatcher	*Polioptila caerulea obscura*	cW	Specimen	FMNH, LACM, LSUMZ, MVZ, PSM, SBCM, SDNHM, UCLA, UMMZ

TABLE 3 *(continued)*

COMMON NAME	SCIENTIFIC NAME	STATUS[a]	VERIFICATION	COLLECTION/REFERENCE[b]
Black-tailed Gnatcatcher	*Polioptila melanura lucida*	cR*	Specimen	AMNH, ANSP, FMNH, LACM, LSUMZ, MCZ, MVZ, PSM, SBCM, SDNHM, UCLA, UMMZ, USNM, WFVZ
Turdidae				
Western Bluebird	*Sialia mexicana occidentalis*	rW	Specimen	LACM, UCLA, UMMZ
	Sialia mexicana bairdi	xW	Specimen	SDNHM
Mountain Bluebird	*Sialia currucoides*	ifW	Specimen	LACM, SBCM, SDNHM
Townsend's Solitaire	*Myadestes townsendi townsendi*	rW, rTa	Specimen	SBCM, SDNHM
Swainson's Thrush	*Catharus ustulatus ustulatus*	fTs, xTa	Specimen	LSUMZ, MVZ, SBCM, SDNHM, UCLA
	[Catharus ustulatus oedicus]	uTs		Presumed to occur
Hermit Thrush	*Catharus guttatus guttatus*	uW	Specimen	SBCM, SDNHM
	Catharus guttatus nanus	rW	Specimen	SDNHM
	Catharus guttatus slevini	xTa	Specimen	SDNHM
American Robin	*Turdus migratorius propinquus*	ifW, xS*	Specimen	LACM, LSUMZ, MCZ, MVZ, SBCM, SDNHM, USNM
Varied Thrush	*Ixoreus naevius [meriloides]*	xTa, xW	Sight record	SDNHM Salton Sea archive
Mimidae				
Gray Catbird	*Dumetella carolinensis* subsp.?	xTa	Sight record	CBRC
Northern Mockingbird	*Mimus polyglottos polyglottos*	cR*	Specimen	FMNH, LSUMZ, MLZ, MVZ, PSM, SBCM, SBMNH, SDNHM, UCLA, WFVZ
Sage Thrasher	*Oreoscoptes montanus*	uTs, rW, xTa	Specimen	AMNH, FMNH, LACM, LSUMZ, MLZ, MVZ, PSM, SBCM, SDNHM, UCLA, UMMZ

TABLE 3 *(continued)*

COMMON NAME	SCIENTIFIC NAME	STATUS[a]	VERIFICATION	COLLECTION/REFERENCE[b]
Brown Thrasher	*Toxostoma rufum longicauda*	xTa, xW	Sight record	SDNHM Salton Sea archive
Bendire's Thrasher	*Toxostoma bendirei*	xW	Photograph	CBRC
Curve-billed Thrasher	*Toxostoma curvirostre palmeri*	xT, xW	Photograph	CBRC
Crissal Thrasher	*Toxostoma crissale coloradense*	uR*	Specimen	FMNH, LACM, LSUMZ, MVZ, SBCM, SDNHM, UCLA, UMMZ, USNM
Le Conte's Thrasher	*Toxostoma lecontei lecontei*	euR*, xTp	Specimen	FMNH, LACM, MVZ, SBCM, SDNHM, UCLA
Motacillidae				
American Pipit	*Anthus rubescens pacificus*	cW	Specimen	DMNH, FMNH, MVZ, SBCM, SBMNH, SDNHM
Sprague's Pipit	*Anthus spragueii*	xW, xTa	Photograph	CBRC
Bombycillidae				
Cedar Waxwing	*Bombycilla cedrorum*	iuW	Photograph	Fig. 61
Ptilogonatidae				
Phainopepla	*Phainopepla nitens lepida*	uP*	Specimen	AMNH, DMNH, FMNH, LACM, MCZ, MVZ, PSM, SBCM, SBMNH, SDNHM, UCLA, UMMZ, USNM, WFVZ, YPM
Parulidae				
Blue-winged Warbler	*Vermivora pinus*	xTa	Sight record	CBRC
Tennessee Warbler	*Vermivora peregrina*	xTa, xW	Photograph	SDNHM Salton Sea archive
Orange-crowned Warbler	*Vermivora celata lutescens*	cW	Specimen	MVZ, SBCM, SDNHM, UMMZ, [USNM]
	Vermivora celata orestera	fW	Specimen	LACM, MVZ, SBCM, SDNHM, [UCLA]
	Vermivora celata celata	xTa	Specimen	SDNHM
Nashville Warbler	*Vermivora ruficapilla ridgwayi*	cT, xW	Specimen	LACM, MVZ, SBCM, SDNHM, USNM

TABLE 3 *(continued)*

COMMON NAME	SCIENTIFIC NAME	STATUS[a]	VERIFICATION	COLLECTION/REFERENCE[b]
Virginia's Warbler	*Vermivora virginiae*	xTa, xW	Sight record	SDNHM Salton Sea archive
Lucy's Warbler	*Vermivora luciae*	leuS*, xS	Specimen	MVZ, UCLA
Northern Parula	*Parula americana*	xW, xT	Specimen	SBCM, SDNHM
Yellow Warbler	*Dendroica petechia morcomi*	cT, rW	Specimen	FMNH, MVZ, SBCM, SDNHM, UCLA, USNM
	Dendroica petechia rubiginosa	rT	Specimen	SDNHM
Chestnut-sided Warbler	*Dendroica pensylvanica*	xW, xTa	Specimen	SBCM
Magnolia Warbler	*Dendroica magnolia*	xTa	Specimen	SDNHM
Cape May Warbler	*Dendroica tigrina*	xW, xTa	Photograph	SDNHM Salton Sea archive
Black-throated Blue Warbler	*Dendroica caerulescens caerulescens*	xT	Sight record	SDNHM Salton Sea archive
Yellow-rumped [Myrtle] Warbler	*Dendroica [coronata] coronata*	rW	Specimen	SBCM
[Audubon's Warbler]	*Dendroica [coronata] auduboni*	cW	Specimen	LACM, MVZ, PSM, SBCM, SBMNH, SDNHM, UMMZ, USNM
Black-throated Gray Warbler	*Dendroica nigrescens*	fT, xW	Specimen	CHAS, FMNH, MVZ, SBCM, SDNHM, UCLA, USNM
Black-throated Green Warbler	*Dendroica virens*	xW	Sight record	SDNHM Salton Sea archive
Townsend's Warbler	*Dendroica townsendi*	fT, xW	Specimen	FMNH, LSUMZ, MVZ, SBCM, SDNHM, USNM
Hermit Warbler	*Dendroica occidentalis*	rTs	Specimen	LSUMZ, MVZ, SBCM, SDNHM, UCLA
Blackburnian Warbler	*Dendroica fusca*	xTa	Specimen	SDNHM
Yellow-throated Warbler	*Dendroica dominica dominica*	xW	Shotograph	CBRC
Pine Warbler	*Dendroica pinus pinus*	xTa	Specimen	SDNHM
Prairie Warbler	*Dendroica discolor discolor*	xTa	Photograph	SDNHM Salton Sea archive

TABLE 3 *(continued)*

COMMON NAME	SCIENTIFIC NAME	STATUS[a]	VERIFICATION	COLLECTION/REFERENCE[b]
Palm Warbler	*Dendroica palmarum palmarum*	xW, xTa	Sight record	SDNHM Salton Sea archive
Bay-breasted Warbler	*Dendroica castanea*	xTs	Sight record	SDNHM Salton Sea archive
Blackpoll Warbler	*Dendroica striata*	xTa	Sight record	SDNHM Salton Sea archive
Cerulean Warbler	*Dendroica cerulea*	xTa	Specimen	SBCM
Black-and-white Warbler	*Mniotilta varia*	xT, xW	Specimen	LACM, SBCM
American Redstart	*Setophaga ruticilla*	rW, rT	Specimen	SBCM, SDNHM
Prothonotary Warbler	*Protonotaria citrea*	xTa	Sight record	CBRC
Ovenbird	*Seiurus aurocapilla aurocapilla*	xT, xW	Specimen	SBCM
Northern Waterthrush	*Seiurus noveboracensis*	xTa, xW	Sight record	SDNHM Salton Sea archive
Louisiana Waterthrush	*Seiurus motacilla*	xTa	Specimen	MVZ
MacGillivray's Warbler	*Oporornis tolmiei tolmiei*	uT	Specimen	LACM, MVZ, SBCM, SDNHM, UCLA, USNM
Common Yellowthroat	*Geothlypis trichas occidentalis*	cP*	Specimen	DMNH, FMNH, LACM, MVZ, SBCM, SDNHM, UCLA, UMMZ, USNM
	Geothlypis trichas yukonicola	xW	Specimen	SDNHM
Wilson's Warbler	*Wilsonia pusilla chryseola*	cT, xW	Specimen	CHAS, FMNH, LACM, MVZ, SBCM, SBMNH, SDNHM, UCLA, [USNM], YPM
	Wilsonia pusilla pileolata	uT	Specimen	MVZ, SBCM, SDNHM, UCLA
Painted Redstart	*Myioborus pictus pictus*	xW, xT	Photograph	Fig. 62
Yellow-breasted Chat	*Icteria virens auricollis*	rS*	Specimen	MVZ, SBCM, SDNHM, USNM
Thraupidae				
Summer Tanager	*Piranga rubra rubra*	xW, xT	Specimen	SBCM, SDNHM
	Piranga rubra cooperi	erS*	Sight record	SDNHM Salton Sea archive

TABLE 3 *(continued)*

COMMON NAME	SCIENTIFIC NAME	STATUS[a]	VERIFICATION	COLLECTION/REFERENCE[b]
Western Tanager	*Piranga ludoviciana*	fT	Specimen	FMNH, LACM, LSUMZ, MVZ, SBCM, SDNHM
Emberizidae				
McCown's Longspur	*Calcarius mccownii*	xTa, rW	Specimen	SBCM
Lapland Longspur	*Calcarius lapponicus alascensis*	rW	Specimen	SBCM
Smith's Longspur	*Calcarius pictus*	xW	Sight record	CBRC
Chestnut-collared Longspur	*Calcarius ornatus*	xW	Sight record	SDNHM Salton Sea archive
Green-tailed Towhee	*Pipilo chlorurus*	rW, rT	Specimen	MVZ, SBCM, SDNHM, UMMZ
Spotted Towhee	*Pipilo maculatus curtatus*	rW	Specimen	SBCM, SDNHM
	Pipilo maculatus megalonyx	xTa	Specimen	SBCM
California Towhee	*Pipilo crissalis senicula*	xW	Specimen	FMNH
Abert's Towhee	*Pipilo aberti aberti*	cR*	Specimen	AMNH, DMNH, FMNH, HSU, LACM, LSUMZ, MCZ, MVZ, PSM, SBCM, SBMNH, SDNHM, UCLA, UMMZ, USNM, WFVZ
Cassin's Sparrow	*Aimophila cassinii*	xTs	Sight record	CBRC
American Tree Sparrow	*Spizella arborea ochracea*	xW	Specimen	SDNHM
Chipping Sparrow	*Spizella passerina arizonae*	uW	Specimen	FMNH, LACM, MCZ, MVZ, SBCM, SDNHM, UCLA, WFVZ
	Spizella passerina boreophila	xTa	Specimen	SDNHM
Clay-colored Sparrow	*Spizella pallida*	xTs, xW	Sight record	SDNHM Salton Sea archive
Brewer's Sparrow	*Spizella breweri breweri*	cW	Specimen	FMNH, MCZ, MVZ, SBCM, SDNHM, UCLA
Black-chinned Sparrow	*Spizella atrogularis cana*	xT	Specimen	SBCM

TABLE 3 *(continued)*

COMMON NAME	SCIENTIFIC NAME	STATUS[a]	VERIFICATION	COLLECTION/REFERENCE[b]
Vesper Sparrow	*Pooecetes gramineus confinis*	uW	Specimen	LACM, MCZ, MVZ, SBCM, SDNHM, UCLA, UMMZ
Lark Sparrow	*Chondestes grammacus strigatus*	uT, luS*	Specimen	LACM, LSUMZ, MVZ, SBCM, SDNHM, UCLA, WFVZ
Black-throated Sparrow	*Amphispiza bilineata deserticola*	rW	Specimen	MVZ, SDNHM
Sage Sparrow	*Amphispiza belli nevadensis*	fW	Specimen	AMNH, ANSP, FMNH, LACM, MCZ, MVZ, PSM, SBCM, SDNHM, UCLA, UMMZ
Lark Bunting	*Calamospiza melanocorys*	rT, xW	Specimen	LACM, SBCM, SDNHM, UCLA
Savannah Sparrow	*Passerculus sandwichensis nevadensis*	cW	Specimen	LACM, LSUMZ, MVZ, PSM, SBCM, SDNHM, UCLA
	Passerculus sandwichensis anthinus	rW	Specimen	SDNHM
	Passerculus sandwichensis brooksi	xTa	Specimen	SDNHM
	Passerculus sandwichensis rostratus	lfW, xS*	Specimen	FMNH, LACM, LSUMZ, MCZ, MVZ, SDNHM, SDSU, UCLA
Grasshopper Sparrow	*Ammodramus savannarum perpallidus*	xT, xW	Specimen	SDNHM
Le Conte's Sparrow	*Ammodramus leconteii*	xW	Photograph	CBRC
Fox Sparrow	*Passerella iliaca schistacea*	xW	Specimen	SBCM, UMMZ
	Passerella iliaca megarhynchus	xTa	Specimen	SBCM
	Passerella iliaca [unalaschcensis]	xW	Sight record	SDNHM Salton Sea archive
	Passerella iliaca zaboria	xW	Sight record	SDNHM Salton Sea archive
Song Sparrow	*Melospiza melodia fallax*	cR*	Specimen	FMNH, LSUMZ, MCZ, MVZ, PSM, SBCM, SBMNH, SDNHM, UCLA, UMMZ, YPM

TABLE 3 *(continued)*

COMMON NAME	SCIENTIFIC NAME	STATUS[a]	VERIFICATION	COLLECTION/REFERENCE[b]
	Melospiza melodia heermanni	rP*	Specimen	SDNHM
	Melospiza melodia montana	rW	Specimen	[LSUMZ], MVZ, SBCM, SDNHM, UMMZ
Lincoln's Sparrow	*Melospiza lincolnii lincolnii*	fW	Specimen	DMNH, LACM, LSUMZ, MCZ, MVZ, PSM, SBCM, SDNHM, UCLA
	Melospiza lincolnii gracilis	xW	Specimen	FMNH, SDNHM
Swamp Sparrow	*Melospiza georgiana ericrypta*	rW	Specimen	SBCM, SBMNH
	Melospiza georgiana georgiana	xW	Specimen	SBCM
White-throated Sparrow	*Zonotrichia albicollis*	xW	Specimen	SBCM
Harris's Sparrow	*Zonotrichia querula*	xW	Specimen	SBCM
White-crowned Sparrow	*Zonotrichia leucophrys gambelii*	cW	Specimen	DMNH, FMNH, LACM, MVZ, PSM, SBCM, SBMNH, SDNHM, UMMZ
	Zonotrichia leucophrys oriantha	rT, xW	Specimen	LSUMZ, MVZ, SDNHM, UCLA
	Zonotrichia leucophrys pugetensis	xW	Specimen	SDNHM
Golden-crowned Sparrow	*Zonotrichia atricapilla*	rW	Specimen	SBCM
Dark-eyed Junco	*Junco hyemalis montanus*	uW	Specimen	SBCM, SDNHM
	Junco hyemalis thurberi	uW	Specimen	MVZ, SDNHM, UCLA
	Junco hyemalis shufeldti	xW	Specimen	SDNHM
	Junco hyemalis hyemalis	xW	Specimen	SBCM, UCLA
	Junco hyemalis cismontanus	rW	Specimen	SBCM, SBMNH, SDNHM, UCLA
	Junco hyemalis caniceps	xTa	Specimen	SBCM, SDNHM
	Junco hyemalis mearnsi	xW	Specimen	SBCM

TABLE 3 *(continued)*

COMMON NAME	SCIENTIFIC NAME	STATUS[a]	VERIFICATION	COLLECTION/REFERENCE[b]
Cardinalidae				
Pyrrhuloxia	*Cardinalis sinuatus [fulvescens]*	xP	Photograph	CBRC
Rose-breasted Grosbeak	*Pheucticus ludovicianus*	xT, xW	Specimen	SDNHM
Black-headed Grosbeak	*Pheucticus melanocephalus maculatus*	cT	Specimen	LSUMZ, MVZ, SBCM, SDNHM, UCLA, USNM
Blue Grosbeak	*Passerina caerulea salicaria*	fS*	Specimen	FMNH, MVZ, SBCM, SDNHM, UCLA, USNM
Lazuli Bunting	*Passerina amoena*	fT	Specimen	LSUMZ, MVZ, SBCM, SDNHM, UCLA, USNM
Indigo Bunting	*Passerina cyanea*	xT, xW	Specimen	MVZ, SBCM
Painted Bunting	*Passerina ciris*	xTa	Sight record	CBRC
Dickcissel	*Spiza americana*	xTa	Sight record	SDNHM Salton Sea archive
Icteridae				
Bobolink	*Dolichonyx oryzivorus*	xTa	Sight record	SDNHM Salton Sea archive
Red-winged Blackbird	*Agelaius phoeniceus sonoriensis*	cR*	Specimen	ANSP, FMNH, LACM, MVZ, PSM, SBCM, SDNHM, SDSU, UCLA, UMMZ, USNM, WFVZ
	Agelaius phoeniceus neutralis	xTs	Specimen	MVZ, SDSU
	Agelaius phoeniceus californicus	xTa	Specimen	USNM
Tricolored Blackbird	*Agelaius tricolor*	xTs	Sight record	SDNHM Salton Sea archive
Western Meadowlark	*Sturnella neglecta neglecta*	cR*	Specimen	FMNH, HSU, LACM, MVZ, PSM, SBCM, SDNHM, SDSU, UCLA, UMMZ, WFVZ
Yellow-headed Blackbird	*Xanthocephalus xanthocephalus*	fS*, uW	Specimen	FMNH, LACM, MCZ, MVZ, PSM, SBCM, SBMNH, SDNHM, SDSU, UCLA, UMMZ

TABLE 3 *(continued)*

COMMON NAME	SCIENTIFIC NAME	STATUS[a]	VERIFICATION	COLLECTION/REFERENCE[b]
Brewer's Blackbird	*Euphagus cyanocephalus minusculus*	cW, luS*	Specimen	MVZ, SBCM, SDNHM, WFVZ
Great-tailed Grackle	*Quiscalus mexicanus nelsoni*	cR*	Specimen	[FMNH], [LSUMZ], MLZ, SBCM, SDNHM, WFVZ
	Quiscalus mexicanus monsoni	uR*	Specimen	SDNHM
Bronzed Cowbird	*Molothrus aeneus loyei*	lrS*	Specimen	SBCM
Brown-headed Cowbird	*Molothrus ater obscurus*	cP*	Specimen	FMNH, LACM, MVZ, SBCM, SDNHM, SDSU, UCLA, WFVZ
	Molothrus ater artemisiae	uW	Specimen	SDNHM
	Molothrus ater ater	xW	Specimen	SDNHM
Orchard Oriole	*Icterus spurius spurius*	xT, xW	Photograph	MVZ photograph archive
Hooded Oriole	*Icterus cucullatus nelsoni*	uS*, xW	Specimen	MVZ, SBCM, SDNHM, WFVZ
Bullock's Oriole	*Icterus bullockii*	cS*, xW	Specimen	FMNH, MVZ, PSM, SBCM, SDNHM, UCLA, WFVZ
Baltimore Oriole	*Icterus galbula*	xT	Specimen	SBCM, SDNHM
Scott's Oriole	*Icterus parisorum*	xTa, xW	Sight record	SDNHM Salton Sea archive
Fringillidae				
Purple Finch	*Carpodacus purpureus californicus*	rW	Specimen	LACM
Cassin's Finch	*Carpodacus cassinii*	xW	Specimen	SBCM
House Finch	*Carpodacus mexicanus frontalis*	cR*	Specimen	LACM, PSM, SBCM, SBMNH, SDNHM, UCLA, UMMZ, WFVZ
Red Crossbill	*Loxia curvirostra* subsp.?	irW	Sight record	SDNHM Salton Sea archive
Pine Siskin	*Carduelis pinus pinus*	iuW	Specimen	SBCM, SDNHM, UCLA
Lesser Goldfinch	*Carduelis psaltria psaltria*	uR*	Specimen	MVZ, SBCM, SDNHM, UCLA

TABLE 3 *(continued)*

COMMON NAME	SCIENTIFIC NAME	STATUS[a]	VERIFICATION	COLLECTION/REFERENCE[b]
Lawrence's Goldfinch	*Carduelis lawrencei*	rW	Specimen	MVZ, SBCM, SDNHM, UCLA, YPM
American Goldfinch	*Carduelis tristis salicamans*	uW	Specimen	SBCM, SDNHM
Evening Grosbeak	*Coccothraustes vespertinus* subsp.?	xT, xW	Sight record	SDNHM Salton Sea archive
Nonnative Species				
Ring-necked Pheasant	*Phasianus colchicus*	luR*	Specimen	SDNHM
Feral Pigeon	*Columba "livia"*	cR*	Sight record	SDNHM Salton Sea archive
Spotted Dove	*Streptopelia chinensis*	lrR+, xS	Sight record	SDNHM Salton Sea archive
European Starling	*Sturnus vulgaris*	cR	Specimen	CSULB, SBCM, SDNHM, SDSU, WFVZ
House Sparrow	*Passer domesticus*	cR	Specimen	LACM, MVZ, PSM, SBCM, SDNHM, UCLA, WFVZ

Note: For taxa supported only by photographs or sight records, documentation is on file in the CBRC archives at WFVZ for those species sufficiently rare in California to be reviewed by the California Bird Records Committee. For other such taxa, we reference published photographs when applicable or reproduce a photograph herein with the species account. We archived documentation at SDNHM for all other taxa supported only by photographs or sight records.

[a]Status and seasonal codes below are based on DeSante and Pyle 1986:

STATUS CODES

c	Common (occurs in large numbers and/or is widespread in the region)
f	Fairly common (occurs in modest numbers)
u	Uncommon (occurs in small numbers)
r	Rare (occurs infrequently but regularly; a few records per year)
x	Extremely rare (few records for the region; unexpected)
l	Local (occurs regularly but in less than 10% of region)
e	Extirpated/former status
i	Irruptive/irregular (peak commonness status is listed)

SEASONAL CODES

S	Summer visitor
T	Transient (if unmodified [see below], then both spring and fall)
W	Winter visitor
R	Resident
P	Perennial visitor (individuals migrate, but species present all year)

MODIFIERS

*	Breeding confirmed
+	Breeding suspected
a	Autumn records only (for transients)
s	Spring records only (for transients)
as	Autumn records predominate (for transients)
sa	Spring records predominate (for transients)
p	Postbreeding visitor (for transients)

[b]See Table 4 for a list of abbreviations.

TABLE 4
Abbreviations for Journals and Museums Cited in the Text

AB	*American Birds* (continued *Audubon Field Notes* in 1971)
AFN	*Audubon Field Notes*
AMNH	American Museum of Natural History, New York
ANSP	Academy of Natural Sciences, Philadelphia
BMNH	Bell Museum of Natural History, University of Minnesota, Minneapolis
CAS	California Academy of Sciences, San Francisco
CBRC	California Bird Records Committee (records archived at WFVZ)
CHAS	Chicago Academy of Sciences
CSULB	California State University, Long Beach
CU	Cornell University, Ithaca, New York
DEL	Delaware Museum of Natural History, Greenville
DMNH	Denver Museum of Natural History
FMNH	Field Museum of Natural History, Chicago
FN	*Field Notes* (continued *American Birds* in 1994)
HSU	Humboldt State University, Arcata, California
KU	University of Kansas, Lawrence
LACM	Natural History Museum of Los Angeles County, Los Angeles
LSUMZ	Museum of Natural Science, Louisiana State University, Baton Rouge
MCZ	Museum of Comparative Zoology, Harvard University, Cambridge, Massachusetts
MLZ	Moore Laboratory of Zoology, Occidental College, Los Angeles
MVZ	Museum of Vertebrate Zoology, University of California, Berkeley
NAB	*North American Birds* (incorporated *Field Notes* in 1999)
PRBO	Point Reyes Bird Observatory, Stinson Beach, California
PSM	Slater Museum of Natural History, University of Puget Sound, Tacoma
SBCM	San Bernardino County Museum, Redlands, California
SBCM/WCH	San Bernardino County Museum, Wilson C. Hanna Collection
SBMNH	Santa Barbara Museum of Natural History, Santa Barbara
SDNHM	San Diego Natural History Museum
SDSU	San Diego State University, San Diego
UA	University of Arizona, Tucson
UAM	University of Alaska Museum, Fairbanks
UCLA	Dickey Collection, University of California, Los Angeles
UCSB	University of California, Santa Barbara
UMMZ	University of Michigan Museum of Zoology, Ann Arbor
USNM	National Museum of Natural History, Washington, D.C.
WBM	Burke Museum, University of Washington, Seattle
WFB	Wildlife and Fisheries Biology Collection, University of California, Davis
WFVZ	Western Foundation of Vertebrate Zoology, Camarillo, California
YPM	Peabody Museum of Natural History, Yale University, New Haven, Connecticut

BLACK-VENTED SHEARWATER
Puffinus opisthomelas Coues, 1864

This species is not very common in the northern Gulf of California, but it is common enough in summer during periods of strong southerly winds that records have long been expected at the Salton Sea. Since the recent summary of Procellariiform records for the Sonoran Desert (Patten and Minnich 1997), this shearwater was recorded at Lake Havasu in the wake of tropical storm Nora (Patten 1998; Jones 1999).

RED-BILLED TROPICBIRD
Phaethon aethereus mesonauta Peters, 1930

A common breeder in the Gulf of California, including north to Rocas Consag and Isla San Jorge, this species has been recorded at the mouth of the Colorado River 25 April 1925 (van Rossem and Hachisuka 1937); twice inland in California, in Morongo Valley and near Palo Verde; and five times in Arizona (Rosenberg and Witzeman 1998; Jones 1999).

GARGANEY
Anas querquedula Linnaeus, 1758

With more than 20 California records and several more for Arizona, it seems only a matter of time before this small duck is found in the Salton Sink. California records are mainly for mid-March through April and from late August to early November, mostly of birds associating with flocks of Cinnamon and Blue-winged Teals. The species was first recorded in North America only in 1957 (see Spear et al. 1988), but more than 100 records have accumulated since.

HARLEQUIN DUCK
Histrionicus histrionicus (Linnaeus, 1758)

Although there are few records for Nevada and but one for Arizona (Rosenberg and Witzeman 1998), this sea duck has occurred twice (four individuals) at Puerto Peñasco, Sonora (Russell and Monson 1998), suggesting that birds move through the Salton Sink.

SHARP-TAILED SANDPIPER
Calidris acuminata (Horsfield, 1821)

Small numbers of this Asian sandpiper are recorded annually in late fall on the northern and central coasts of California. Although it is rare in southern California, there are records for Arizona, Nevada, New Mexico, and the western Mojave Desert of California (Small 1994; AOU 1998; Rosenberg and Witzeman 1998).

SOUTH POLAR SKUA
Stercorarius maccormicki Saunders, 1893

This skua is recorded regularly in the northern Gulf of California from July through November (Tershy et al. 1993), with high counts of up to 35 birds (R. L. Pitman). That none have been recorded at the Salton Sea is surprising.

BROAD-BILLED HUMMINGBIRD
Cynanthus latirostris magicus
  (Mulsant and Verreaux, 1872)

Although a vagrant to California, this hummingbird has accumulated more than 50 fall and winter records in the state, including nearly 10 for the southeastern deserts. There are 6 records for the lower Colorado River (Rosenberg et al. 1991; B. Mulrooney), and the species has reached the Anza-Borrego Desert (Luther 1980).

HAIRY WOODPECKER
Picoides villosus hyloscopus (Cabanis and Heine, 1863)

This woodpecker generally occurs in coniferous forests in southern California. It rarely wanders to the desert lowlands but has been recorded in the northern Coachella Valley (Glenn 1983), at Morongo Valley (Patten 1995a), and in Joshua Tree National Park (Prescott et al. 1997). The southern Coachella Valley, and thus the northern end of the Salton Sink, seems a likely locale for a vagrant to occur.

WHITE-EYED VIREO
Vireo griseus griseus (Boddaert, 1783)

California records of this vireo of the Southeast have gone from none to nearly 40 since 1970 (Patten and Marantz 1996). This species occurs mainly in spring, a season when large numbers of landbirds pass through the Salton Sink. Furthermore, records have increased in adjacent Arizona, and there is a recent spring record from El Golfo de Santa Clara, Sonora, at the head of the Gulf of California (K. L. Garrett).

YELLOW-THROATED VIREO
Vireo flavifrons Vieillot, 1808

Like the White-eyed Vireo and the Hooded Warbler, the Yellow-throated Vireo has enjoyed a sharp increase in records in California since 1970 (Patten and Marantz 1996). Also like those species, the majority (ca. two-thirds) of California's roughly 75 records are from spring, coincident with peak landbird migration through the Salton Sink.

STELLER'S JAY
Cyanocitta stelleri frontalis (Ridgway, 1873) /
   C. s. macrolopha Baird, 1854

Given irregular irruptions of this montane jay into the desert Southwest (see Phillips et al. 1964) and records of other montane species (e.g., Williamson's Sapsucker, Clark's Nutcracker, Cassin's Finch, etc.), this species' absence from the Salton Sink list is surprising. Presumed vagrants of *C. s. frontalis* have reached the Anza-Borrego Desert (Massey 1998), and *C. s. macrolopha* has reached the lower Colorado River and elsewhere in southeastern California (Phillips et al. 1964; Rosenberg et al. 1991).

BROWN CREEPER
Certhia americana zelotes Osgood, 1901 /
   C. a. montana Ridgway, 1882

Creepers are rare in the Salton Sink, with the single specimen being of the nominate subspecies from eastern North America (Unitt and Rea 1997). Neither of the expected subspecies has been recorded definitely in the region, yet both undoubtedly occur. See the Brown Creeper account for details.

SWAINSON'S THRUSH
Catharus ustulatus oedicus (Oberholser, 1899)

This species is uncommon in the Salton Sink, where all specimens are the nominate subspecies

of the Pacific Northwest. However, it seems likely that *C. u. oedicus,* the breeder in cismontane California, moves through in small numbers. See the Swainson's Thrush account for details.

RUFOUS-BACKED ROBIN
Turdus rufopalliatus grisior van Rossem, 1934

There are only nine California records of this subtropical species, but it is increasing in Arizona and has been recorded at several locales surrounding the Salton Sink (see the Hypothetical List for details).

KENTUCKY WARBLER
Oporornis formosus (Wilson, 1811)

California records of this warbler of the southeastern United States, long an exceedingly rare vagrant in the state, have increased nearly tenfold since 1980 (Patten and Marantz 1996). Many records are from southeastern California in spring, making the lack of records for the Salton Sink conspicuous.

HOODED WARBLER
Wilsonia citrina (Boddaert, 1783)

Now one of the more numerous eastern wood-warblers straying to the West, this species has been recorded nearly 350 times (through 1997) in California, mainly in spring in the southern half of the state (Dunn and Garrett 1997). Records have increased steadily since 1980 (Patten and Marantz 1996). It is overdue in the Salton Sink and was only recently recorded in Imperial County (28 May 2000 at In-ko-pah; M. A. Patten).

SONG SPARROW
Melospiza melodia merrilli Brewster, 1896

This subspecies has occurred not far to the east and west of the Salton Sink. There are also records for the Mojave Desert (e.g., Furnace Creek Ranch, Yucca Valley), mainly in fall. See the Song Sparrow account for details.

RED-WINGED BLACKBIRD
Agelaius phoeniceus nevadensis Grinnell, 1914

It has long been supposed that an influx from the Great Basin augments this species' wintering numbers in the Salton Sea region (e.g., Garlough 1922). Birds from the Great Basin subspecies *A. p. nevadensis* definitely reach southern Arizona and the lower Colorado River in winter (Phillips et al. 1964; Rosenberg et al. 1991) but as yet are unknown from the Salton Sink. See the Red-winged Blackbird account for details.

RUSTY BLACKBIRD
Euphagus carolinus (Müller, 1776)
Another eastern stray to the West, this blackbird is regular in California in late fall and winter (mid-October to March). It is recorded more frequently in the desert than along the coast (McCaskie 1971b; Roberson 1980).

COMMON GRACKLE
Quiscalus quiscula versicolor Vieillot, 1819
This grackle is increasing in the West, although not nearly as dramatically as the Great-tailed. There was but one California record of the Common Grackle by the close of the 1960s, but there were nearly ten by the end of the 1970s, about 25 by the end of the 1980s, and nearly 50 by the end of the 1990s. One should look for this species in massive blackbird and grackle flocks around cattle feed lots in winter.

The Species Accounts

I N THIS SECTION we present detailed accounts for all the species for which there are reliable records in the Salton Sink. After describing the broad status of each avian order or family, we detail the occurrence of each component species. We head each account with the species' English and scientific names, in almost all cases following the American Ornithologists' Union's *Check-list of North American Birds* (see below). The scientific name is followed by the name of the author who described the species to science and the year that description was published. Following convention, if the species was originally referred to a different genus from the one in which it is currently classified, the author's name and the year are enclosed in parentheses. We likewise provide authorities for all trinomial (subspecies) names.

A capsule description of the species' seasonal status in the Salton Sink is given next. We then detail the species' temporal and spatial occurrence, typically commencing with the season in which the species is most numerous and placing its status in a broader geographic context. We supply a reference for every specific record; documentation for significant records previously unpublished can be found in the archives at the San Diego Natural History Museum. The summary of the species' status is followed by a summary of its habitat use and autecology in the region, which in turn is followed by a summary of the abundance and seasonal status of various subspecies known to occur in the region. Museum specimens are the basis for nearly all of our statements about subspecies occurrence. We use the patterns of occurrence of various subspecies to draw inferences about the biogeography of various taxa.

THE GEOGRAPHIC REGION

The Salton Sink essentially coincides with the bed of historical Lake Cahuilla, which covered the area to a maximum elevation of approximately 15 m above sea level. The Salton Sink includes all of the Salton Sea, the southern Coachella Valley (north to southern Indio), the Imperial Valley, San Sebastian Marsh and the San Felipe Creek drainage, and the Mexicali Valley south to Campo Geotérmico Cerro Prieto (Fig. 24). The logic behind our choice of this region is twofold. First, Lake Cahuilla, last filled in the mid-1600s, was a

FIGURE 24. The Sonoran Desert of southeastern California, northeastern Baja California, and the Colorado River valley. The heavy solid line represents the shoreline of Lake Cahuilla, the natural boundary of the Salton Sink. The dotted line represents sea level.

precursor of the Salton Sea. Second, the approximate sea level line is a well-defined biogeographic boundary for birds in the region, as it strongly affects habitat and climate.

We have made every effort to use the most exact place names possible for individual records (Fig. 25). In some cases the feature referred to is no longer extant (e.g., Silsbee, Volcano Lake), whereas in others a specific place name could not be determined. Particularly problematic in literature on the region are references to the *north end* and the *south end* of the Salton Sea. These are shorthand terms for vast regions: *north end* for the region bounded by Oasis to the west and North Shore or Corvina Beach to the east, though most references are probably to the Whitewater River delta; *south end* for the region bounded by Poe Road to the west and Bombay Beach to the east, thus including Unit 1 of Salton Sea National Wildlife Refuge, the New River and Alamo River deltas, Obsidian Butte, Sonny Bono Salton Sea

National Wildlife Refuge (abbreviated below as "Salton Sea NWR"), Red Hill, and the Wister Unit of the Imperial Wildlife Area. Given this array of places where birds concentrate, the terms are not helpful in identifying localities. *Mexicali Valley* is also ambiguous, as some authors mean by it the entire Río Colorado delta rather than just the valley. We use *Mexicali Valley* only to refer to the valley itself, from the international boundary south to Cerro Prieto, bounded to the west by the Sierra Cocopah and to the east by Gravel Mesa (see Fig. 25).

DATA COLLECTION

We compiled data for this book from three primary sources: (1) museum records for all species collected in the region (see Table 4); (2) published literature, particularly regional reports in *American Birds, Field Notes,* and *North American Birds,* records in which we often made critical evaluations (see Van Tyne 1956), and also notes in journals such as *Condor* and *Western Birds;* and (3) our own unpublished field notes, as well as those of other birders and biologists working in the Salton Sink. Our own fieldwork amounts to more than sixteen hundred field days since the early 1960s. Data were compiled in a manner allowing for clear, concise descriptions of typical status (e.g., seasonal occurrence, breeding) and abundance in the region as well as for unusual dates, numbers, annual fluctuations, and so on. Ecological data were compiled in a similar manner. Accounts are intended to be synthetic, not comprehensive. We completed the subspecific taxonomy largely through work with the collection at the San Diego Natural History Museum, including 1,015 specimens collected in or near the Salton Sink since 1983, largely as part of our investigation of polytypic landbirds. We are especially indebted to Roger Higson for his help in this endeavor. Our studies also included direct examination of specimens housed at the American Museum of Natural History in New York, the Denver Museum of Natural History, the Dickey Collection at the University of California, Los Angeles, the Field Museum in Chicago, the

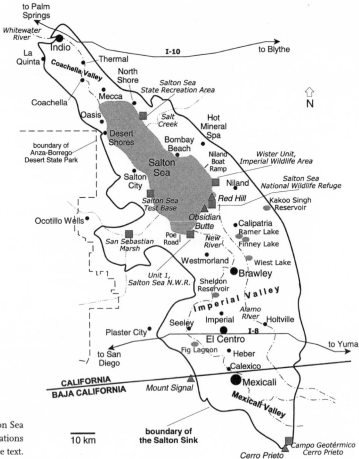

FIGURE 25. The Salton Sea
region, showing locations
mentioned in the text.

10 km

boundary of
the Salton Sink

Moore Laboratory of Zoology at Occidental College, the Museum of Comparative Zoology at Harvard, the Museum of Vertebrate Zoology at the University of California, Berkeley, the Natural History Museum of Los Angeles County, the San Bernardino County Museum, San Diego State University, and the Western Foundation of Vertebrate Zoology in Camarillo, California. All mensural characters were acquired through standard measurement of avian skins (see Baldwin et al. 1931). We include all records through January 2002, significant ones through December 2002.

TAXONOMY AND NOMENCLATURE

The taxonomy, nomenclature, and phylogenetic sequence employed here follow the seventh edition of the American Ornithologists' Union's (1998) *Check-list of North American Birds*, through

its most recent supplement (Banks et al. 2002), with six exceptions:

1. We treat the White and Scarlet Ibises as one species, following Hancock et al. (1992:153).

2. The Surfbird is placed in the genus *Calidris*. The species is known to be closely related to that genus (Jehl 1968; AOU 1998), and based on recent genetic work (*fide* Robert E. Gill) it appears to be a sister species of the Great Knot, *C. tenuirostris* (Horsfield, 1821).

3. Thayer's Gull is not treated as a species distinct from the Iceland Gull since there is no solid evidence to suggest that these forms are reproductively isolated or consistently morphologically distinct. We tentatively treat *Larus glaucoides thayeri* as a subspecies of *L. glaucoides*. Even this level of distinction may not be valid, for much of the east-to-west variation from

L. g. kumlieni to *L. g. thayeri* appears to be smoothly clinal. Thus, perhaps all North American birds should be treated as either *L. g. kumlieni* or *L. kumlieni*, with *L. glaucoides* being an Old World taxon (see Weir et al. 2000).

4. The name Feral Pigeon is used for the widespread domesticated forms of the Rock Dove (*Columba livia*). Referring to them as Rock Doves makes no more sense than does referring to a Cocker Spaniel (*Canis familiaris*) as a Gray Wolf (*Canis lupus*).

5. The Pacific-slope and Cordilleran Flycatchers are not treated as distinct species. The splitting of these cryptic species is contentious (Johnson 1980, 1994b; Phillips 1986:xxviii, 1991:l, 1994c; Johnson and Marten 1988; Howell and Cannings 1992; Gilligan et al. 1994). We do not take sides but merely implore researchers to engage in fieldwork and quantitative analyses to determine whether the various populations actually constitute different species. Since the reviews are presently mixed, the split was premature.

6. The longspurs are basal to the emberizid radiation (Patten and Fugate 1998; Yuri and Mindell 2002) and are thus listed ahead of the towhees and the New World sparrows.

SUBSPECIES

Probably as many as half of the subspecies of birds are not valid. An unfortunate legacy of the ornithologists of the first half of the twentieth century is a litany of names applied to populations of birds that differ from other populations, sometimes trivially so, but whose differences from those populations cannot be diagnosed. For this reason the very existence of the subspecies as a biological entity has been questioned (see, e.g., Wilson and Brown 1953; and Selander 1971), when in fact the concept has merely been poorly applied by many taxonomists. Subspecies described solely on the basis of mean differences could be named indefinitely along a smooth cline, defeating the purpose of a trinomial.

Joseph T. Marshall said it best when he noted that subspecies "constitute whole *populations*

[that] are 'marked' by their peculiarities of color, size, and proportions" (Phillips et al. 1964:x). By paying attention to subspecies we can learn a great deal about migration, movements, and biogeography, but only if individual specimens can be assigned to particular geographic populations. One hundred percent diagnosability will seldom be achieved, because subspecies interbreed freely (by definition under any species concept based on the biology of the organisms), but the goal should be to define subspecies such that correct classification to a particular population (or set of populations) will be highly (e.g., 95%) likely.

We doubt neither the biological reality nor the marvelous utility of subspecific taxonomy, provided it is properly quantified and applied with common sense. To that end we present data on all subspecies known to have occurred in the Salton Sink. When reviewing existing trinomials quantitatively, in principle we followed the so-called 75 percent rule (Amadon 1949; Rand and Traylor 1950; Mayr 1969), the standard for defining subspecies that is most widely accepted in the literature. In practice, however, we tended to employ a stricter, 95 percent rule. By the standard rule, at least 75 percent of the subspecies population must be differentiated in at least one character (but generally several) from 99 percent of other populations. Obviously, this differentiation must hold in both directions. Where we present our own taxonomic treatment, we generally determined diagnosability statistically, following the 75 percent (or higher) rule under standard assumptions of normality and taking sampling effort into account (Patten and Unitt 2002). For quantifiable characters we sometimes present a diagnosability index, with $D_{ij} > 0$ meaning that the subspecies are valid at the 75 percent (or higher) level and $D_{ij} < 0$ meaning that they are not (see Patten and Unitt 2002). In other cases we follow published work that we judged thorough, critical, and conservative in its definition of subspecies. Thus, we follow no single author or treatment. In cases where critical published works differ, we almost always judged the merits of opposing treatments based on our own examination of available museum specimens.

THE MAIN LIST

GAVIIFORMES • *Loons*

Loons are rare on and in the vicinity of the Salton Sea. Even the Common Loon, the only species routinely found inland, is scarce in the region, the majority being flyovers during spring migration. The Red-throated and Pacific Loons are coastal species. Their pattern of occurrence at the Salton Sea mirrors their pattern elsewhere in the interior of western North America (see Garrett and Dunn 1981; Monson and Phillips 1981; Behle et al. 1985; and DeSante and Pyle 1986). There are many more records of the Pacific Loon, especially during migration, whereas the few records for the Red-throated Loon are only from winter and spring. Although the Pacific and Common Loons can occur in large numbers in the northern Gulf of California, particularly in spring (Patten et al. 2001), many may not pass through the Salton Sink, instead reaching the Pacific Ocean via a flyway between the Sierra Juárez and the Sierra San Pedro Mártir in northern Baja California (Huey 1927).

GAVIIDAE • *Loons*

RED-THROATED LOON
Gavia stellata (Pontoppidan, 1763)

Casual visitor during migration (late April–May, mid-November) and in winter. There are nine records of the highly coastal Red-throated Loon for the Salton Sea region. Three records are from winter, when all loons are scarce in the Salton Sink. The first is of a mummified bird in basic plumage at the mouth of the Whitewater River 16 January 1986 (skeleton, LSUMZ 130829), estimated to have been shot 1 January (*AB* 40:1257). Two additional records are supported by photographs, of a juvenile at Wister 25–27 January 1992 (*AB* 46:313) and a bird at Wiest Lake 20 December 2001 (L. W. Cole). Four spring records presumably involved birds moving north after wintering in the Gulf of California. Birds in first-summer plumage were observed near the south end 24 April 1993 (*AB*

47:453) and at Sheldon Reservoir 29 April 1995 (*FN* 49:308), and one was photographed near the mouth of the Whitewater River 22 April 1996 (*FN* 50:332). A late bird in basic plumage was seen at Ramer Lake 28 and 29 May 1994 (*FN* 48:340). In fall, one was seen at Ramer Lake 19 November 1994 (R. Higson; S. von Werlhof), and a juvenile was photographed at Sheldon Reservoir 5–14 November 2000 (*NAB* 55:102).

ECOLOGY. Save for the mummified bird found at the edge of the sea, all regional records of the Red-throated Loon are from freshwater ponds and lakes.

PACIFIC LOON
Gavia pacifica (Lawrence, 1858)

Casual spring (mid-May to mid-June) and fall (mid-October to late November) transient and winter visitor; 3 summer records. The Pacific Loon, a rare migrant inland, has been recorded more than 25 times at the Salton Sea, with records for every season. Most records are for spring (ca. 13), with half as many in the fall; half of all records are from the north end. Spring occurrences extend from 14 May (1999, Fig Lagoon; *NAB* 53:328) to 22 June (1974, 2 in alternate plumage at the north end; *AB* 28:948) and include a specimen of a bird found dead on Interstate 8 about 8 km west-southwest of Seeley 9 June 1984 (A. M. Rea; SDNHM 49451). An exceptionally early adult in alternate plumage was at Fig Lagoon 13 April 2000 (G. McCaskie). Fall records extend from 13 October (1984, north end until 18 October; *AB* 39:101) to 27 November (1977, Salton City; *AB* 32:256), with the others from Obsidian Butte 24 October 1999 (*NAB* 54:103), 26 October 1992, and 2 November 1992 (*AB* 47:148), Fig Lagoon 26 October 2001 (G. McCaskie), the north end 3 November 1985 (*AB* 40:157), and photographed at Sheldon Reservoir 8 November 2000 (*NAB* 55:102). Four winter records involved lone birds at the north end 6–20 March 1965 (*AFN* 19:415) and 9 December 1985 (*AB* 40:333), two at Ramer Lake 22 December 1975–25 January 1976 (*AB* 30:764), and one at an unspecified location on the

Salton Sea 19 February 1955 (*AFN* 9:284). All three summer records are from the Whitewater River delta: 4 July 1968 (*AFN* 22:647), 7 July 2001 (*NAB* 55:482), and 25 August 1977 (*AB* 32:256).

ECOLOGY. Almost all records of this loon are from the Salton Sea itself, but a few are from freshwater lakes and reservoirs in the Imperial Valley.

COMMON LOON

Gavia immer (Brünnich, 1764)

Uncommon to rare spring transient (mid-March through June); rare fall transient (mid-October to mid-December); casual in winter and summer. The Common Loon occurs at the Salton Sea primarily in spring. It has been recorded on occasion during the fall and a few times in summer, but there are no unequivocal records of overwintering birds. This loon is rare to uncommon in spring from 18 March (1995, Poe Road; G. McCaskie) to 30 June (1984, bird in alternate plumage at the north end; *AB* 38:1061). Usually fewer than 5 birds are seen in a day, but small concentrations have been noted at the north end, including 20 on 11 May 1996 (G. McCaskie), 18 on 4 May 1985 (*AB* 39:349), and "fully a dozen" on 19 April 1908 (Grinnell 1908). The maximum single-day count is of 25 at various locales around the Salton Sea 4 May 1996 (G. McCaskie). Also of note were 16 in alternate plumage on the Highline Canal 10 km east of Calipatria 11 April 1994 (*FN* 48:340). Spring migrants originate in the Gulf of California; most are in full alternate plumage. The only four recent summer records are of two at the south end 13 July 1996 (G. McCaskie) and one each near Obsidian Butte 18 August 1990 (*AB* 44:1184), at the north end 3 August 1997 (*FN* 51:1052), and at Cerro Prieto 1 September 2000 (Patten et al. 2001).

Common Loons are decidedly rarer at the Salton Sea in the fall. Records for fall extend from 17 October (1984, 1 at the south end; *AB* 39:101) to 23 December (1978, north end; *AB* 33:666), with seldom more than one bird seen, although five were scattered around the south end 23 October–20 November 1976 (*AB* 31:222). Individuals at Ramer Lake 19 and 20 December 1985 (*AB* 40:333) and Wiest Lake 10 December 1990 (*AB* 45:319) were away from the Salton Sea, as were birds furnishing several mid- to late November records for Sheldon Reservoir. There are only three records after late December, of single birds at the north end 14 January 1978 (*AB* 32:328), Fig Lagoon 23 January 1993 (S. von Werlhof), and the south end 11 March 1978 (*AB* 32:398), the last of which may have been an early spring migrant. The paucity of records from January through March leads us to conclude that December sightings pertain to late fall migrants.

ECOLOGY. Most regional records for the Common Loon are from the Salton Sea, where birds are often seen flying over during the spring migration. A few have been recorded on lakes in the Imperial Valley (e.g., Fig Lagoon and Ramer Lake) and even on small reservoirs.

PODICIPEDIFORMES • Grebes

Grebes are a common, conspicuous component of the avifauna of the Salton Sea. Three species breed regularly: the Pied-billed Grebe and the two species of *Aechmophorus,* with Clark's being more common than the Western. Breeding sites are at the mouths of the three rivers and, particularly for the Pied-billed Grebe, at freshwater ponds and lakes in both the Coachella and the Imperial Valley. Breeding species, however, take a back seat to the Eared Grebe. A principal wintering site and staging area for this species is the Salton Sea, which supports hundreds of thousands. More than 2 million Eared Grebes—the majority of the North American population—pass through the Salton Sea during the spring and fall migrations on their way to and from wintering sites in the Gulf of California. Unfortunately, the Eared Grebe has suffered some of the largest mass mortalities ever recorded among North American birds (Jehl 1996), underscoring the importance of the Salton Sea to the long-term viability of this species.

PODICIPEDIDAE · Grebes

LEAST GREBE
Tachybaptus dominicus bangsi
(van Rossem and Hachisuka, 1837)

One fall record. A Least Grebe photographed (*AB* 43:166) and observed regularly at the Imperial Warm Water Fish Hatchery 19 November–24 December 1988 furnished the only record for the Salton Sea and one of only two valid records for California (Pyle and McCaskie 1992). A report from Salton City likely pertained to another species (Luther et al. 1983). The Least Grebe normally occurs in Mexico, ranging north into southern Texas. It is a casual visitor to southeastern Arizona (Monson and Phillips 1981), where it has appeared as far west as Quitobaquito in Organ Pipe Cactus National Monument (Rosenberg and Witzeman 1998).

TAXONOMY. McMurray and Monson (1947) identified the sole California specimens, an adult male (USNM 393392) and a chick (USNM 393393) taken from a group of five adults and four downy young at Imperial Dam 23 October 1946, as *T. d. bangsi*. This small, pale subspecies (see Storer and Getty 1985) of Baja California and Sonora presumably accounts for the single Salton Sea record.

PIED-BILLED GREBE
Podilymbus podiceps podiceps (Linnaeus, 1758)

Fairly common breeding resident. The marsh-dwelling Pied-billed Grebe is widespread throughout the Americas. It is a fairly common resident at the Salton Sea, where it breeds in marshes and along the rivers. It is particularly numerous at Wister; along the Whitewater River near its mouth, at the north end; at Finney and Ramer Lakes; and at Fig Lagoon. This species apparently expanded into the Salton Sea region sometime after the early 1940s, as Grinnell and Miller (1944) did not note it as a breeder in the region. Numbers are slightly higher at the end of the breeding season (Shuford et al. 2000), presumably because of an infusion of fully grown young, and in winter, presumably because of an influx of birds that bred farther north. Even so, the

20,000 allegedly counted during an aerial survey 9 November 1949 (*AFN* 4:34) were undoubtedly Eared Grebes. Exact numbers through summer can be difficult to determine as this species is secretive when nesting. It nests over much of the year, but breeding activity peaks from February through August. Downy young accompanying adults are a common sight on fresh water in late summer.

ECOLOGY. This species generally occupies fresh water; it seldom occurs on the open salt water of the Salton Sea except fairly close to a river mouth. It reaches maximum abundance in freshwater marshes supporting dense stands of Southern Cattail, but it is also fairly numerous at sloughs and lagoons lined with Saltcedar and along slow-moving rivers and channels lined with Saltcedar and Common Reed.

TAXONOMY. The small, pale nominate subspecies is the only one occurring north of central Mexico.

HORNED GREBE
Podiceps auritus (Linnaeus, 1758)

Rare spring (April to mid-June) and fall (late October through November) transient and winter visitor; 3–6 summer records. As elsewhere in the interior West, the Horned Grebe is a rare migrant and winter visitor at the Salton Sea, with a few summer records. Most records involve single birds. Spring records are typically of alternate-plumaged adults moving north through the Salton Sea from wintering grounds in the Gulf of California. Records extend from 2 April (1988, north end; G. McCaskie) to 10 June (1984, alternate-plumaged adult photographed at the north end from 5 May; *AB* 38:958, 1061), with a peak in late April. Groups of six at the south end 23 April 1989 (*AB* 43:536) and four at the north end 25 April 1987 (*AB* 41:487) are the largest spring concentrations and suggest a regular, albeit small, passage through the area. There are three summer records for the vicinity of the Whitewater River mouth, with single birds 11–21 August 1976 (*AB* 30:1002), 10 July 1977 (*AB* 31:1188), and 30 June–7 July 1990 (*AB* 44:1184). Alternate-plumaged birds near Mecca 4–18 June

1999 (*NAB* 53:432), near Obsidian Butte 27 June 1999 (*NAB* 53:432), and at the south end 28 June 1991 (*AB* 45:1160) perhaps summered on the sea, although they were more likely extremely late spring migrants.

Like those in spring, fall Horned Grebes are apparently birds in transit to the Gulf of California. There are fewer fall records, extending from 25 October (1977, 2 at the north end; *AB* 31:256) to 25 November (1988, north end; *AB* 43: 167). Some fall migrants linger into winter, when individuals are rarely (and less than annually) found among the massive number of Eared Grebes. Winter records fall between 14 December (1980, south end; *AB* 35:335) and 18 February (1979, south end from 28 December 1978; *AB* 33:666). One examined in hand at the south end 23 March 1942 (Grinnell and Miller 1944) was probably a late wintering bird but may have been an extremely early spring migrant. Generally only one or two are found in any given winter, but four were found at the south end 22 December 1975 (*AB* 30:608).

ECOLOGY. Virtually all records of the Horned Grebe are from the Salton Sea proper, typically at the river mouths, bays, or inlets. A few have been found on freshwater lakes in the Imperial Valley.

TAXONOMY. This species is monotypic, *P. a. cornutus* (Gmelin, 1789) being insufficiently distinct to warrant recognition (Mayr and Short 1970:29; Cramp and Simmons 1977:105; cf. Parkes 1952).

EARED GREBE

Podiceps nigricollis californicus Heermann, 1854

Abundant winter visitor and transient (mainly November through March); fairly common summer visitor; has bred. There can be no denying that the Salton Sea has become important to the Eared Grebe in North America (Fig. 26). Although other large saline lakes in the West also serve as major staging grounds, especially Mono Lake and the Great Salt Lake (Jehl 1988; Boyd et al. 2000), the sheer number of Eared Grebes that use the Salton Sea annually, for example, the estimate of 2.5–3 million Eared Grebes on the sea 23 January 1988 (*AB* 42:320), is mind-boggling.

Numbers peak from January through March, when "most [probably 75%] of the New World population" passes through the sea en route to breeding areas farther north from wintering areas around the Gulf of California (Jehl 1994, 1996). Some of these same birds also stage at Mono Lake (Boyd et al. 2000). Even during winter the population is about 1.5 million birds (Jehl 1988:32), although from late November to early January it probably numbers no more than several hundred thousand (J. R. Jehl Jr. *in litt.*). Concentrations tend to be lower in fall, when many overfly the region. Since the 1960s the Eared Grebe has apparently increased dramatically at the Salton Sea (Jehl and McKernan 2002) and elsewhere (Jehl 2001). Given the importance of the region to the Eared Grebe, several massive die-offs (Jehl 1996), the largest of about 150,000 birds in January–March 1992, have caused great concern. In spring numbers drop substantially by April, but dozens to hundreds spend the summer every year. This species is a rare, sporadic breeder in southern California. It has bred at least three times at the Salton Sea, beginning with more than 40 nests at the Whitewater River delta 27 July 1978 (*AB* 32:1207) and a single pair at the south end during summer 1979 (*AB* 33: 896). The last known nesting was of three pairs at the Whitewater River delta during summer 1990 (*AB* 44:1184).

ECOLOGY. Wintering Eared Grebes dot the entirety of the Salton Sea, sometimes as far as the eye can see but more typically within 500 m of the shoreline. Infrequently, some birds, possibly unhealthy, enter agricultural drains or river courses (Jehl 1996). During peak abundance small numbers can be found on just about every substantial body of water in the region. The steep increase in numbers since the 1960s coincides with the spread of the pile worm *Neanthes succinea,* which constitutes more than 95 percent of the grebe's diet from January through April (Jehl and McKernan 2002). These worms occur mainly in shallow water within 1 km of the shore, an area where most grebes concentrate.

TAXONOMY. North American birds are of the black-necked subspecies *P. n. californicus.*

FIGURE 26. The Eared Grebe is the most abundant bird on the Salton Sea, with millions wintering or stopping during migration. Photograph by Kenneth Z. Kurland, November 2000.

WESTERN GREBE

Aechmophorus occidentalis occidentalis

(Lawrence, 1858)

Common breeding resident, less numerous in summer (April–September). The long-term status and distribution of the Western Grebe is clouded. It was considered conspecific with Clark's Grebe until the early 1980s. Before the split, field ornithologists rarely attempted to distinguish between the taxa, which were considered mere color morphs. In general, the Western Grebe outnumbers Clark's Grebe throughout their ranges in western North America (Ratti 1981), but such is not the case at the Salton Sea. During the winter these two species occur in roughly equal proportions, but in most years Clark's Grebe is more common during the summer months at the north end. This change in abundance likely reflects an influx of wintering Western Grebes concomitant with little seasonal movement of Clark's Grebes. Together the *Aechmophorus* grebes totaled up to about 8,600 individuals in late winter of 1999 (Shuford et al. 2000), presumably the result of migrants that wintered on the Gulf of California stopping at the sea; these grebes number in the several thousand at other times of the year.

Aechmophorus grebes of unknown species were found nesting at the Whitewater River delta in 1976 (*AB* 30:1002; Garrett and Dunn 1981). During 1980 the two species nested sympatrically at the Whitewater River (*AB* 34:929). Claims of roughly equal numbers of nests in summer 1990 (*AB* 44:1184) notwithstanding, Clark's Grebe is the more numerous species in summer at the Whitewater River delta. Ratios of Clark's to Western range from 3:2 (8–9 July 2000) to 3:1 (30 June 1990) to an amazing 8:1 (29 June 1991; all M. A. Patten), with the first two ratios being more typical. The Western Grebe tends to be equally numerous elsewhere in the Salton Sink, often predominating along the southern shoreline (e.g., 3:1 in favor of Western in 1999; Shuford et al. 2000). Curiously, at the larger lakes in the Imperial Valley the Western Grebe outnumbers Clark's Grebe as a breeder in a ratio of 4:1 or greater, a pattern also typical in northeastern Baja California (Patten et al. 2001). Breeding begins in early spring, with young appearing by midspring (e.g., an adult with a chick on its back at Ramer Lake 6 May 1995; M. A. Patten). Some breed well into fall (e.g., an adult with a downy chick at Ramer Lake 26 November 1999; B. Miller).

ECOLOGY. Both species of *Aechmophorus* are most frequently encountered near the mouths of the rivers on the open water of the Salton Sea, particularly at the Whitewater River delta. About

80 percent occur within 1 km of the shoreline (Shuford et al. 2000). Breeders frequent river courses and freshwater lakes (especially Fig Lagoon and Finney and Ramer Lakes) with thick Southern Cattail and Saltcedar along their edges. Some birds enter river courses even when they are not breeding. As yet no one has discerned consistent differences in habitat use at the Salton Sea, although Clark's may feed closer to shore (and thus in slightly shallower water), a pattern opposite of what has been reported elsewhere (Nuechterlein and Buitron 1989; J. N. Davis pers. comm.).

TAXONOMY. With the naming of *A. o. ephemeralis* Dickerman, 1986, a small subspecies of central western Mexico, *A. o. occidentalis* now refers only to breeders in Baja California, the United States, and Canada.

CLARK'S GREBE

Aechmophorus clarkii transitionalis Dickerman, 1986
Uncommon breeding resident; generally outnumbers the Western Grebe in summer, at least at the north end. Clark's Grebe has been considered a species distinct from the Western Grebe only since the early 1980s, so the status of Clark's Grebe has become clear only since the taxonomic change. In general, this species is an uncommon breeding resident at the Salton Sea, with numbers seemingly changing little over the course of the year. During the winter it is about equal in abundance to the Western Grebe, but it distinctly outnumbers that species during the summer months (mid-April through July) and thus presumably as a breeder. Clark's Grebe initially was noted breeding at the Whitewater River delta in summer 1980 but probably began breeding there at least as early as 1976. See the Western Grebe account for more detailed information about Clark's Grebe and the relative abundance and status of the two *Aechmophorus* grebes. The nesting schedule of the two species is the same.

ECOLOGY. See the Western Grebe account.

TAXONOMY. Breeders in northern Baja California (*contra* Dickerman 1986) and the United States are the larger subspecies (especially in wing length) *A. c. transitionalis.*

PROCELLARIIFORMES •
Tube-nosed Swimmers

In defiance of common sense and expectation, eight species of the highly pelagic order Procellariiformes have been recorded at the Salton Sea. Records show a tight pattern, with all arriving from very late April through late September, coinciding with strong northward monsoonal winds from the Gulf of California (Patten and Minnich 1997). Aside from two spectacular tropical storms, Kathleen in 1976 (Kaufman 1976) and Nora in 1997 (Patten 1998), the appearance of these birds has been independent of Pacific cyclones. Instead, tubenoses that regularly occur off western mainland Mexico in summer move northward into the Gulf of California, aided by monsoonal winds and increased sea-surface temperatures, which reduce the difference between the waters of the gulf and the normally cooler waters of western Mexico (Patten and Minnich 1997). Continued movement northward across the low elevations of the Mexicali and Imperial Valleys brings these birds to the Salton Sea, an oasis of suitable habitat in the Sonoran Desert.

DIOMEDEIDAE • Albatrosses

LAYSAN ALBATROSS

Phoebastria immutabilis (Rothschild, 1893)
Four records (May–June). Although the Laysan Albatross is rare off the Pacific coast of southern California, this species has been recorded eight times in the Sonoran Desert, four of these at the Salton Sea. One was seen regularly 21 May–20 June 1984 (AB 38:857, 1061) from Desert Shores north to the mouth of the Whitewater River, and two were photographed together 9 June. One hit a power line near the headquarters of Salton Sea NWR 9 May 1991 (SBCM 54388; AB 45:495). Finally, one was seen soaring over the sea at North Shore 2 May 1993 (AB 47:453). There are two additional records of birds in the vicinity of San Gorgonio Pass not far north of the Salton Sea. One flying west near Desert Hot Springs 5 May 1976 (Dunn and Unitt 1977; Binford 1985) was the first recorded inland. One

was photographed before it was seen to "collide with power lines" in the San Gorgonio Pass 6 May 1985 (*AB* 39:349; LACM uncatalogued). There are also two records (14 May 1981 and 18 July 1988) for the Arizona side of the Colorado River (Rosenberg et al. 1991). Nearly all records for the Salton Sea and vicinity are from the first three weeks of May. This species has been observed in the northern Gulf of California (Newcomer and Silber 1989), coinciding with northward movement in the eastern Pacific during late April and early May (Sanger 1974).

PROCELLARIIDAE • Shearwaters and Petrels

COOK'S PETREL
Pterodroma cookii (Gray, 1843)
Three records (July). Cook's Petrel occurs regularly off western Mexico (Pitman 1986) and has proven to be regular in deep water off California (Roberson and Bailey 1991), mostly from mid-May to early December, with a peak in August. The California Bird Records Committee treated a *Pterodroma* seen daily at the Whitewater River mouth from 24 to 29 July 1984 (*AB* 38:1061) as *P. cookii* on the basis of that species' known range, although the similar De Filippi's (not "DeFillipe's"), *P. defilippiana* (Giglioli and Salvadori, 1869), and Pycroft's, *P. pycrofti* Falla, 1933, Petrels could not be eliminated by the descriptions, sketches, and photographs (Dunn 1988). A Cook's Petrel was carefully identified off the mouth of the Whitewater River 10 July–6 August 1993, and what was likely the same bird was seen off Poe Road, at the south end, 17 July 1993 (Patten and Minnich 1997). Two Cook's Petrels were seen off the mouth of the Whitewater River 15 July 1995, and what was probably one of these birds was seen off Corvina Beach later that day (Patten and Minnich 1997); one remained through 25 July 1995 (*FN* 49:979). A distant, unidentified *Pterodroma* near the mouth of the Whitewater River 19 July 1998 (G. McCaskie, P. A. Ginsburg) may have belonged to this species.
TAXONOMY. The species is monotypic, *P. c. orientalis* Murphy, 1929, being a synonym (Mayr and Cottrell 1979).

WEDGE-TAILED SHEARWATER
Puffinus pacificus (Gmelin, 1789)
One summer record. A dark-morph Wedge-tailed Shearwater photographed at the mouth of the Whitewater River 31 July 1988 (McCaskie and Webster 1990; *AB* 42:1225) was only the second recorded in California and the continental United States (Pyle and McCaskie 1992). At sea this species ranges north to near the southern tip of Baja California (Pitman 1986), and large numbers of dark-morph birds have been recorded off western Mexico in July (King 1974). An all-dark, slow-flapping procellariid flying south well off Salton City 10 July 1993 (*AB* 47:1149) possibly belonged to this species (M. A. Patten).
TAXONOMY. The species is monotypic, *P. p. chlororhynchos* Lesson, 1831, being a synonym (Mayr and Cottrell 1979).

BULLER'S SHEARWATER
Puffinus bulleri Salvin, 1888
One summer record. The only inland record for North America of Buller's Shearwater, a species of the south-central Pacific Ocean, is of a sick bird captured at the mouth of the Whitewater River 6 August 1966 (SBCM 31447; *AFN* 20: 599).

SOOTY SHEARWATER
Puffinus griseus (Gmelin, 1789)
Seven summer records (mid-June to mid-August); 1 late spring record. The Sooty Shearwater is one of the most abundant birds in the world. It has been recorded eight times in the Salton Sea region between late April and mid-August. Individuals were observed at Desert Shores 14 August 1971 (*AB* 25:905), near the mouth of the White-water River 16 June 1984 (*AB* 38:1061), and off Salton City 13 July 1991 (*AB* 45:1160). One was seen at the mouth of the New River 14 July 1990 (*AB* 44:1184), and what was probably the same individual was recorded at Desert Shores 20 July 1990, at the mouth of the Whitewater River 21 July 1990, and near Red Hill 25 July 1990 (*AB* 44:1184). The bird recorded at Salton City

in 1991 was not seen well enough for the Short-tailed Shearwater, *P. tenuirostris* (Temminck, 1835), to be excluded; however, a Short-tailed Shearwater anywhere in California in July would be remarkable. A Sooty Shearwater found 4 km south-southwest of Seeley 6 July 1991 (R. Higson; SDNHM 47577) was fat, with moderate contour feather molt; it was struck by Higson's vehicle as it attempted to take flight from a road adjacent to Fig Lagoon. Two found dead on the beach near Oasis 24 August 1996 (Patten and Minnich 1997) were, unfortunately, not preserved, although the carcasses were photographed. Last but certainly not least, the earliest seasonal record is perhaps the most bizarre: one was seen flying across Interstate 10 near Indio 28 April 1989 (*AB* 43:536). Also of note are one found dead about 58 km east of Yuma, Arizona, 6 June 1971 (Quigley 1973; UA 10316) and one near Blythe, California, 19 May 2001 (*NAB* 55:354).

HYDROBATIDAE • Storm-Petrels

LEACH'S STORM-PETREL
Oceanodroma leucorhoa (Vieillot, 1818) subsp.?
Two records, 1 each in summer and fall. Two color morphs of Leach's Storm-Petrel have reached the Salton Sea. A dark-rumped individual was carefully observed near Red Hill 15 September 1976 in the aftermath of tropical storm Kathleen (*AB* 31:222), and a white-rumped individual was seen irregularly at the mouth of the Whitewater River 30 June–21 July 1984 (*AB* 38:1061).

TAXONOMY. Geographic variation in Leach's Storm-Petrel is remarkably complex and subspecific taxonomy remains unresolved (Bourne and Jehl 1982). Along the Pacific Coast, the transition from the white-rumped population breeding north of Point Conception (nominate *O. l. leucorhoa*) to the dark-rumped population of Islas San Benito (*O. l. chapmani* Berlepsch, 1906) is gradual, yielding great variability at intermediate colonies. Birds breeding at Isla Guadalupe (*O. l. socorroensis* Townsend, 1890) are smaller, and populations breeding there in summer and winter may be specifically distinct (Ainley 1980). Because of these complexities, and because rump

color "tells us little about the origins of birds" reaching California (Unitt 1984:28), subspecies names applied to sight records are inappropriate. Even so, at the Salton Sea, the dark-rumped individual was considered *O. l. chapmani* (*AB* 31:222), the white-rumped one "probably *O. l. beali*, but perhaps *O. l. willetti*" (Patten and Minnich 1997).

BLACK STORM-PETREL
Oceanodroma melania (Bonaparte, 1852)
Three fall records, 1 involving multiple individuals. A Black Storm-Petrel seen very near the mouth of the Whitewater River 28 September 1986 (*AB* 41:142) was the first recorded inland in North America. The only other occurrences at the Salton Sea were as many as 17 birds recorded between the mouth of the New River and Red Hill 27 September–9 November 1997, in the wake of tropical storm Nora (Patten 1998; Jones 1999), and 1 on Morton Bay 14 August 1999 (*NAB* 54:104). This species is the most common storm-petrel off southern California, and it commonly nests in the Gulf of California (Anderson et al. 1976; Wilbur 1987).

LEAST STORM-PETREL
Oceanodroma microsoma (Coues, 1864)
Three records, 2 in fall involving multiple birds and 1 in summer. The Least Storm-Petrel, a common nesting species in the Gulf of California (Anderson et al. 1976), has been recorded three times at the Salton Sea, twice in the wake of major tropical storms. Following the passage of tropical story Kathleen 500–1,000 were deposited on the Salton Sea (Kaufman 1976). Storm-petrels were first detected 12 September 1976 at many locations around the sea. Most departed within the next week, but eight were still present 21 October 1976 (*AB* 31:222). By comparison, no more than three individuals were found between the mouth of the New River and Wister 27 September–20 October 1997, following tropical storm Nora, although substantial numbers were seen on Lake Havasu, along the Colorado River, during this same period (Patten 1998; Jones 1999). The third record was of one at the mouth

of the Whitewater River 10 July 1993 (*AB* 47:1149); although not associated with a storm, it fits the pattern of occurrence shown by the various Procellariiformes recorded at the Salton Sea (Patten and Minnich 1997).

PELECANIFORMES • *Totipalmate Birds*

Images of the Salton Sea are strongly associated with various Pelecaniformes, whether massive nesting colonies of Double-crested Cormorants, unique inland occurrences of Brown Pelicans, or such rarities as boobies and Magnificent Frigatebirds. The Salton Sea is the only locale where both the American White Pelican and the Brown Pelican have bred, the former before the late 1950s and the latter in the late 1990s. Furthermore, it is by far the most likely locale in the United States for the Blue-footed Booby, however rare and highly irregular in occurrence, and it is the only inland locale in the western United States where the Magnificent Frigatebird is regular in small numbers.

SULIDAE • *Boobies and Gannets*

BLUE-FOOTED BOOBY
Sula nebouxii nebouxii Milne-Edwards, 1882
Casual, highly irregular vagrant in summer and fall (mid-July through late November), sometimes in modest numbers; 2 winter records. The Blue-footed Booby is an enigmatic symbol of the Salton Sea because this locality is virtually the only place where the species may be seen in the United States. Nevertheless, its true status at the sea has been greatly misunderstood and misrepresented. Rather than being an annual visitor, it is actually a highly irregular postbreeding visitor from the Gulf of California (McCaskie 1970c), with incursions averaging about once every five years. With rare exceptions, records (involving several hundred individuals) fall between 12 July (1990, 2 photographed at Salton City; Patten and Erickson 1994) and 29 November (1997, photographed at Salton City; *FN* 52:125), the majority being of immatures in August and September. The Whitewater River delta and Salton City are favored locations. Except for a male collected

7 km southwest of Seeley 2 September 1990 (Patten and Erickson 1994; SDNHM 46903), all area records are from the Salton Sea. Similar records just outside the Salton Sea region include two at Thousand Palms 3 September 1965 (*AFN* 20:91), one near Whitewater 20 September 1965 (*AFN* 20:91), and a female at Ocotillo Wells 4 August 1968 (SDNHM 36707).

After the first California record, of an immature photographed at the north end of the Salton Sea 1–11 November 1929 (Clary 1930), there were no regional records for more than 20 years. The next records, of one found dead in Thermal in late August 1953 and up to two at the sea 18–31 October 1953 (*AFN* 8:41), were again followed by a long gap. Misunderstandings regarding the species' status at the Salton Sea probably stemmed from the large incursions in the 1960s and 1970s. Records in the mid-1960s (McCaskie 1970c) were mostly from the north end, with one 24 July–21 August 1965; 2–5 photographed 4 August–16 October 1966, with one taken 13 August (SBCM M3844); and one 10 August–1 September 1968. There was a major incursion in 1969, when some 30 were recorded around the Salton Sea 31 August–23 November, with many photographed (see *AFN* 24:98 and McCaskie 1970c), a female collected 14 September (SDNHM 37266), and two individuals taken 28 September (SBCM 31467, 31469).

The next decade began modestly, with but one at the north end 15 August 1970 (McCaskie 1970c). The following two years, however, saw the largest invasions on record, with as many as 48 per day beginning 8 August 1971 (Garrett and Dunn 1981), up to 40 seen and photographed at the north end 22 July–24 September 1972 (including SBCM M5071, taken 5 August), and 5 more seen at the south end 12 August 1972 (Winter and McCaskie 1975). Smaller incursions took place in 1977, with one photographed at Salton Sea Beach 23 August and up to 11 observed at the north end 24 August–9 October (Roberson 1993), and in 1980, with at most 4 at the mouth of the Whitewater River 12 September–23 October 1980 (Roberson 1993). Another report in the 1970s lacks documentation (Patten and Erickson

1994). Following another decade-long gap in records, the Blue-footed Booby appeared in four different years in the 1990s, beginning with up to 4 photographed (see *AB* 44:1185) at various locations around the Salton Sea 12 July–3 October 1990 (Patten and Erickson 1994). Since that time one was observed near Oasis 25 July 1993 (Erickson and Terrill 1996) and up to 3 were photographed at various locales around the Salton Sea 1 September–6 October 1996 (McCaskie and San Miguel 1999). An immature photographed at Mullet Island 14 February and found dead on the nearby shoreline 1 March 1997 was considered one of the three that arrived in fall 1996 (Rottenborn and Morlan 2000). A year later, an immature was observed at the Whitewater River delta 28 September–6 October 1997 and then photographed at Salton City 29 November 1997 (Rottenborn and Morlan 2000). An immature observed at Obsidian Butte 19–28 February 1998 (Erickson and Hamilton 2001) was likely this same individual, providing the second recent winter record.

ECOLOGY. Blue-footed Boobies roost on nearshore rocks, breakwaters, or beaches of the Salton Sea but do not perch on snags (unlike Brown Boobies). They forage by plunge diving but do so slightly closer to shore than do Brown Boobies.

TAXONOMY. California specimens are of the small nominate subspecies, distributed off western Mexico and breeding commonly in the Gulf of California.

BROWN BOOBY

Sula leucogaster brewsteri Goss, 1888

Casual, highly irregular visitor in summer and fall (mid-July to mid-November), sometimes in small numbers; 1 winter record. The Brown Booby is a casual postbreeding visitor to the Salton Sea from the Gulf of California and now has a similar status along the coast of California. Most records are from years of Blue-footed Booby invasions (McCaskie 1970c). Despite the species' postbreeding dispersal pattern, a concentration of more than 2,000 Brown Boobies (and only 3 Blue-footed Boobies) in the Gulf of California off Puerto Peñasco, Sonora, 10 January 1987 (Dun-

ning 1988) shows how common it can be in the northern gulf even during midwinter. Regional records of the Brown Booby extend from 12 July (1990, adult at the New River and an immature at the Alamo River, eventually building to eight birds by 29 September; Heindel and Garrett 1995) to 11 November (1969, up to 8 photographed around the sea from 6 September; McCaskie 1970c, Dunn 1988). The single spring bird, at the north end 25 April 1970 (Dunn 1988), presumably had lingered from the fall 1969 invasion. Other occurrences involved lone individuals, except for up to two at the mouth of the Whitewater River 5–26 August 1972 (Winter and McCaskie 1975). Additional records for the Whitewater River delta are of a first-year male 28 July–13 August 1966 (SBCM 31470), an individual 28 August–7 September 1971 (Bevier 1990), and an adult 24 August–2 September 1974 (Binford 1985). Other records for the south end are of birds collected on unknown dates in October 1967 (HSU 1388) and in 1966 (skeleton, LACM 52054) and birds at Rock Hill 15–24 August 1970 (Bevier 1990; SBCM M4621) and 29 August–18 September 1971 (Roberson 1993). The remaining two records, of one in Calexico 15 July 1972 (Roberson 1993) and a first-year bird photographed perched on a power line 6 km southwest of Niland 28 August 1996 (*FN* 51:120; CBRC 1996-111), involved the only two individuals found away from the sea. Additional reports from the south end (Pyle and McCaskie 1992; CBRC 2000-059) and Salton City (*AB* 27:120) were not sufficiently documented.

ECOLOGY. Although the two booby species have been seen together at the Salton Sea, they typically choose different habitats, with Brown Boobies perching on submerged trees and utility poles protruding from the Salton Sea and Blue-footed Boobies roosting on rocks and embankments along the edge of the sea. The Brown Booby forages in deep water a few hundred meters offshore.

TAXONOMY. The only specimen of an adult male Brown Booby from the coast of California (Imperial Beach, San Diego County, 2 April 1990; SDNHM 46566) belongs to the subspecies *S. l.*

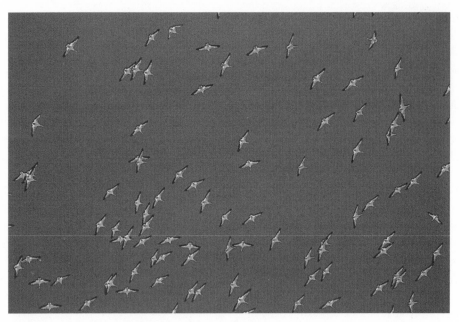

FIGURE 27. One-third to one-half of the entire population of the American White Pelican winters on the Salton Sea. Photograph by Kenneth Z. Kurland, January 2001.

brewsteri, which breeds throughout the Gulf of California and presumably disperses northward. None of the Salton Sea specimens is an adult male, however, and the Brown Boobies of the eastern Pacific vary geographically only in the head and breast of adult males (darkest in *S. l. etesiaca* Thayer and Bangs, 1905, of Central and South America; paler in *S. l. brewsteri* of Mexico; very pale in *S. l. nesiotes* Heller and Snodgrass, 1905, of Clipperton Atoll).

PELECANIDAE • *Pelicans*

AMERICAN WHITE PELICAN
Pelecanus erythrorhynchos Gmelin, 1789
Common winter visitor and transient (mid-October to mid-April); fairly common in summer; formerly bred. Although its numbers have declined throughout its range (Sloan 1982; Sidle et al. 1985), the American White Pelican (Fig. 27) remains a common winter visitor and spring and fall migrant in the Salton Sea region. It formerly bred at the Salton Sea (Fig. 28; Bartholomew et al. 1953) but now is only a fairly common nonbreeding summer visitor. The Salton Sea is an important wintering site for this species, harboring a substantial percentage of the world population. Aerial survey counts of 32,000 on 23 January 1988 (*AB* 42:320) and 40,000 on 7 March 1987 (*AB* 41:487) show how numerous it can be in that season and are particularly impressive considering that a 1964 census (Lies and Behle 1966) reported only 40,067 breeding birds in the whole of the United States and Canada. More recent estimates of the breeding population are nearer 100,000 birds (Sidle et al. 1985). Counts at the Salton Sea were somewhat lower in the 1990s, with numbers "greatly reduced on the Salton Sea from those reported in recent years" during the winter of 1989–90 (*AB* 44:327), but still exceeded 10,000 individuals (Shuford et al. 2000). Wintering birds, which breed mainly in the northern Great Basin (Keith and O'Neill 2000), arrive in mid-October, with concentrations of 1,450 at the south end 20 October (1972; *AB* 27:120) and 4,000 there 21 October (1976; *AB* 31:222) being the earliest large influxes. Most birds depart in March and April, when huge kettles, sometimes of thousands of birds, can be seen circling over the north end before heading north through the San Gorgonio Pass or over the Transverse Ranges.

FIGURE 28. A breeding colony of American White Pelicans at the south end of the Salton Sea, 9 April 1908. Photograph by Joseph Grinnell, courtesy of the Archives of the Museum of Vertebrate Zoology, University of California, Berkeley.

The American White Pelican colonized the Salton Sea as a breeder shortly after the sea's creation, when a set of eggs "far advanced in incubation" was collected from a nest on "Echo Island" (at the south end) 19 April 1908 (Grinnell 1908). At the time there were "980 occupied nests, besides many others in process of construction. At the very minimum there were 2000 pelicans here assembled." Pairing took place as early as 8 January (van Rossem 1911), with egg dates ranging from 12 April (1946, south end; WFVZ 45290) to 9 June (1928, south end; MVZ 3589–91). At least 450 pairs still nested on this island and two nearby islands 20 May 1927 (Pemberton 1927). The number had decreased to 50 breeding pairs by 1932 as "summer recreation near the island colonies had resulted in most of the pelicans leaving" (Thompson 1933). Numbers built up again in the 1940s, with 300 nests by April 1949 (*AFN* 3:224), but decreased again rather sharply. The last breeding at the Salton

Sea was during 1956 and 1957 (Lies and Behle 1966); increased human activity, erosion of nesting islands, and fluctuating water levels forced nesters to abandon the area. Even since local breeding ceased, thousands of nonbreeders (mostly immatures) have persisted at the Salton Sea through the summer months.

ECOLOGY. Nesting American White Pelicans occupied sandy islands surrounded by shallow water. Wintering birds congregate at the river mouths, where they loaf on sandbars and mudflats and forage in shallow, brackish water. Some forage in shallow impoundments, frequently swimming while dipping their heads in grand, synchronous movements.

BROWN PELICAN

Pelecanus occidentalis californicus Ridgway, 1884
Common summer and fall visitor (mainly May through mid-November); rare (but increasing) through winter; recently began breeding at the south

end. A coastal species, the Brown Pelican is a common postbreeding visitor to the Salton Sea, with numbers steadily increasing over the past two decades and the first records only in the early 1950s (at the south end 27 August 1951 and 10 October 1952; *AFN* 7:36). Nowhere else does the species occur inland in such numbers or with such regularity. There are few early spring records, and although it has become regular in large numbers into January, it rarely overwinters. Because few are thus present on the sea from mid-January through much of April, the occurrence of 30 off Salton City 7 March 1987 (*AB* 41: 487) was exceptional. Prior to 1980 there were no records before 15 May (McCaskie 1970c; Garrett and Dunn 1981). Now the first spring arrivals generally appear by early May, with six at the south end 24 April (1992; *AB* 46:479) being the earliest; numbers thereafter build up quickly, as evidenced by at most 400 at the sea by 8 May 1992 (*AB* 46:479). In addition to earlier arrival, numbers build to an order of magnitude greater than they did 30 years ago. Numbers never rose above about 100 during the 1970s (McCaskie 1970c; Anderson et al. 1977), but single flocks of this size or greater are commonplace today. In 1990 the maximum was more than 2,000 at various locales around the Salton Sea 4 August, a number that prompted McCaskie (*AB* 44:1185) to write that "thirty years ago Brown Pelicans were classified as accidental inland; twenty years ago counts of 25 on the Salton Sea were considered remarkable, and ten years ago the largest numbers recorded were still under 100." Since the mid-1990s, single-day counts have reached 2,000 individuals (Shuford et al. 2000) and probably exceed 3,000. Numbers used to decrease significantly by the end of September, with 134 noted at the north end 12 October 1986 (M. A. Patten) being a large count for so late in the season at the time. Birds now linger later in the year, frequently with large numbers persisting through December; for example, 988 were recorded on the 26 December 1996 Salton Sea (south) Christmas Bird Count (*FN* 51:643). The majority of the birds appearing in the region are immatures, but large numbers of adults are sometimes seen.

Brown Pelicans in the Salton Sink undoubtedly originate in the Gulf of California, "a 'stronghold' area of [the] species' distribution" (Anderson et al. 1976). Indeed, birds banded in the gulf have wandered north to the Salton Sea during summer (Anderson et al. 1977). Some pelicans may move through the San Gorgonio Pass to the Pacific coast (e.g., a flock of 17 photographed as they headed northwest through Thousand Palms 30 September 2000; M. A. Patten). The prediction that the species would begin to breed at the Salton Sea was fulfilled in July 1996, when nine large young were discovered in nests at the mouth of the Alamo River (*FN* 50:996). The species made unsuccessful attempts to nest at Obsidian Butte in 1997 and 1998.

ECOLOGY. Brown Pelicans occur almost anywhere along the shoreline of the Salton Sea, most often around rock outcrops and embankments. They forage by plunge diving and thus occur across the Salton Sea. Curiously, birds with flocks of American White Pelicans forage in the manner of that species yet still pull back their wings (a fixed action pattern?) before submerging the head (M. A. Patten). The Brown Pelican has nested on small islands of volcanic rock with a sandy base and at the mouth of the Alamo River on beds of matted reeds. From June through September it can be found at least occasionally on virtually every substantial body of water in the Imperial Valley. There are also winter records for this valley, from near Brawley 11 February 1989 (*AB* 43:365) and 26 January 1991 (G. McCaskie) and at Finney Lake 5 January 1998 (K. M. Burton).

TAXONOMY. Birds from central Mexico northward along the Pacific coast, including throughout the Gulf of California (and thus the Salton Sea), are of the large, dark-necked subspecies *P. o. californicus.*

PHALACROCORACIDAE • Cormorants

BRANDT'S CORMORANT
Phalacrocorax penicillatus (Brandt, 1837)
One spring record. One of only two truly inland records for North America of Brandt's Cormorant is of an injured adult or subadult male that K. Z. Kurland found in his yard 3 km southeast

of El Centro 19 April 1996 (*FN* 50:332; SDNHM 49488). The other inland record is of a bird captured and photographed in hand near Fowler, in Fresno County, California, 13 March 1977 (McCaskie et al. 1988). This species is common in the Gulf of California (Anderson et al. 1976; Wilbur 1987), at least north to Isla San Jorge (Cervantes-Sanchez and Mellink 2001), so future area records should be expected. A photograph in the August 2000 issue of *Birder's World* of nearly 40 "Brandt's Cormorants" at the Salton Sea is clearly in error; all of the cormorants in the photograph are actually Double-cresteds.

NEOTROPIC CORMORANT

Phalacrocorax brasilianus (Gmelin, 1789) subsp.?
Casual summer vagrant (late April through early September); 1 long-staying but unpaired individual nested. The eleven California records of the Neotropic Cormorant are from the Salton Sea and the lower Colorado River Valley, with all but one of the Salton Sea records in the 1990s. This species likewise has been recorded with increasing frequency in southeastern Arizona and central New Mexico (Hundertmark 1978; Rosenberg and Witzeman 1998), and Nevada recorded its first in the late 1990s (*FN* 52:230). Except for an influx of six immatures during spring 1996 (see below), every record has been of an adult, perhaps reflecting the greater difficulty of identifying immatures (Patten 1993). All area records but one are from 24 April (1996, 3 immatures photographed at Fig Lagoon through 13 July; McCaskie and San Miguel 1999) to 1 September (1996, adult at Obsidian Butte; McCaskie and San Miguel 1999). Other records during this period are of two immatures at Salton Sea NWR 27 April 1996 (McCaskie and San Miguel 1999), an immature near Oasis and the Whitewater River delta 4 May–1 June 1996 (McCaskie and San Miguel 1999), and adults photographed near Obsidian Butte 28 April–12 July 1998 and near Mecca 14 June–16 July 1998 (both Erickson and Hamilton 2001). The last bird returned 4 July 1999 (Rogers and Jaramillo 2002).

The exception was a long-staying or frequently returning bird at the Salton Sea during a six-year period in the mid-1980s. This bird, only the third recorded in California, was first detected (and photographed) at the mouth of the Whitewater River 1 August–10 September 1982 (Morlan 1985). It was next observed as it attempted to nest among a Double-crested Cormorant colony near the mouth of the New River 27 February–5 March 1983 (Morlan 1985); the nest failed because the bird was apparently unpaired. It subsequently returned to the Whitewater River delta 30 July 1983 (Roberson 1986), 27 July–31 August 1985 (photographed; Dunn 1988), 23 March–20 April 1986 (Bevier 1990), 19 July–23 August 1986 (Bevier 1990), and 20 June 1987 (Pyle and McCaskie 1992). It was last seen near Red Hill 15–29 August 1987 (Pyle and McCaskie 1992).

ECOLOGY. This species routinely roosts on drowned snags, particularly of Athel and mesquite. Records come from the immediate shoreline of the Salton Sea, ponds adjacent to the sea, and freshwater lagoons in the Imperial Valley.

TAXONOMY. There are no specimens for California, and specimens from Arizona apparently have not been identified to subspecies. Geographic variation of this species in Mexico awaits critical revision. *P. b. chancho* van Rossem and Hachisuka, 1939, was described from coastal Sonora south through Guanajuato and is thus the geographically closest to California. However, *P. b. mexicanus* (Brandt, 1837), from southern and eastern Mexico, has rapidly increased its range northward and cannot be excluded without a specimen. These two subspecies differ from the nominate, of South America, in being smaller and paler but are extremely similar to each other. The validity of *P. b. chancho* has been questioned, with some (e.g., Harrison 1985) treating it as a synonym of *P. b. mexicanus*. If these subspecies are merged, then on the basis of geography California records can be safely attributed to *P. b. mexicanus*.

DOUBLE-CRESTED CORMORANT

Phalacrocorax auritus albociliatus Ridgway, 1884
Common permanent resident and breeder, with higher numbers in winter (September to early April).

The Double-crested is the only cormorant found throughout most of the interior West, although the Neotropic is making headway in southeastern Arizona and central New Mexico. The Double-crested Cormorant breeds widely in California, with the Salton Sea region being one of the most important sites in the West (Carter et al. 1995). Mullet Island, the Red Hill vicinity, and the Whitewater River delta and vicinity are the most heavily used breeding locations. Scattered pairs can be found breeding at many locations around the Salton Sea and at large bodies of water in the Imperial Valley (e.g., Finney and Ramer Lakes). No thorough censuses have been conducted of cormorants breeding in the region (Carter et al. 1995), but the number is probably 4,000–6,000 pairs annually (Shuford et al. 1999, 2000) and fluctuates down to lows near zero (W. D. Shuford in litt.). Indeed, the species may well have been extirpated as a breeder by the mid-1980s, only to recolonize in force in 1995 (K. C. Molina pers. comm.; W. D. Shuford in litt.). Rangewide numbers of this species have declined over the past several decades as a result of loss of habitat and birds' being shot at aquaculture farms (Carter et al. 1995). Breeding begins early, with nest building commencing in late January, the earliest eggs laid by 6 February (1913, "some 500 nests" near Mecca; WFVZ 95475), and young fledging as early as late April (K. C. Molina), although many are still on eggs in mid-April (MVZ 54–57, 97–98). In addition, a few thousand nonbreeders, mostly subadults, summer in the region each year. Numbers build to the tens of thousands in winter, as many breeders from farther north move into the region. Concentrations at this time of the year can be staggering; for example, 11,400 were counted around the Whitewater River delta 21 February 1999 (M. A. Patten), and counts across the sea approach 20,000 birds (Shuford et al. 2000). Wintering birds begin arriving in September, with most departing by the beginning of April.

ECOLOGY. Pairs of cormorants nest in submerged snags near the edge of the Salton Sea and on rocky islands at the south end (Grinnell 1908). A few nest on crossbeams of submerged utility poles at the north end. The species forages virtually everywhere on the sea and on large to modest bodies of water in the adjacent valleys, including at aquaculture farms. Snags, rock outcrops, embankments, and submerged utility poles and pilings are preferred roosting sites.

TAXONOMY. California specimens are of the widespread subspecies *P. a. albociliatus,* of the Pacific coast (including the Gulf of California), which is slightly larger than eastern birds and in which the crest is extensively white in definitive alternate plumage.

FREGATIDAE · Frigatebirds

MAGNIFICENT FRIGATEBIRD
Fregata magnificens Mathews, 1914

Rare summer visitor (mainly mid-June through August); 1 winter record. Magnificent Frigatebirds are rare postbreeding visitors from Mexico (where they are common in the southern Gulf of California) to the Southwest (Mlodinow 1998a). Most are encountered along the southern coast of California and at the Salton Sea. At the latter, two to four birds are recorded in an average summer, with virtually all records involving first-summer birds with clean white heads and pale bluish bills (McCaskie 1970c; Winter 1973; Winter and McCaskie 1975; Luther et al. 1979). Records extend from 4 June (1984, immature at the north end; *AB* 38:1061) to 23 September (1979, 3 around the Salton Sea; *AB* 34:200), with the majority from the north end from late June to late August and a peak in mid-July. The largest flock (for the Salton Sea and California) was of 22 at the north end 29 July 1979, a year that saw an incursion into California and Arizona of more than 100 individuals (*AB* 33:896; Mlodinow 1998a). Other noteworthy concentrations were 16 at the north end 9 July 1998 (*FN* 52:502), 13 around the Salton Sea 16 September 1984 (*AB* 39:102), 9 there 14 July 1979 (*AB* 33:896), and 8 at the north end 10 August 1985 (*AB* 40:157).

The single frigatebird record outside this window of occurrence, of an adult male at the north end 20 February 1972 (Winter 1973), is best regarded as a probable *F. magnificens* given that the second California record of the Great Frigatebird,

TABLE 5
Examples of Relative Nesting Densities of Ciconiiformes
in the Salton Sea Region, 1992, 1998, and 1999

	NUMBER OF PAIRS		
SPECIES	1992	1998	1999
Great Blue Heron	12	364	888
Great Egret	140	227	165
Snowy Egret	150	337	170
Cattle Egret	>25,000	11,138	6,600
Black-crowned Night-Heron	1,300	282	102
White-faced Ibis	370	0	—

Sources: Data for summer 1992 are from a rookery at Finney Lake (*AB* 46:1178), and those for summer 1998 and summer 1999 are estimated from the whole of the Salton Sea (Shuford et al. 1999, 2000).

F. minor Gmelin, 1789, was photographed at Southeast Farallon Island on the unseasonable date (for a Magnificent) of 14 March 1992 (Heindel and Patten 1996). All unseasonal records for the western United States should be critically reevaluated with the Great Frigatebird in mind, as these two species are difficult to distinguish in the field (Howell 1995).

ECOLOGY. Frigatebirds are remarkably powerful fliers, so it comes as no surprise that most seen at the Salton Sea are single birds flying over. On occasion individuals roost on snags protruding from the sea (see, e.g., the photograph in *AB* 26:904).

TAXONOMY. The Magnificent Frigatebird is monotypic, with *F. m. rothschildi* Mathews, 1915, being a synonym (Palmer 1962).

CICONIIFORMES • *Herons, Ibises, Storks, American Vultures, and Allies*

The Salton Sea region is one of the most important areas for Ciconiiformes, both wintering birds and breeders, in western North America. The sea has been identified as a key wintering area for the Snowy Egret (Mikuska et al. 1998) and the White-faced Ibis (Shuford et al. 1996), and some of the largest heron, egret, and ibis

rookeries in California are in the Salton Sea region. Numbers vary from year to year, but densities in the Imperial Valley (Table 5) and along the immediate edge of the Salton Sea are similar. Relative compositions in the two habitats differ, however, with many more Great Blue Herons, Great Egrets, and Snowy Egrets breeding along the edge of the sea but Cattle Egrets predominating (by far) at Imperial Valley rookeries. More nightherons and ibises occur in the latter colonies too.

The Alamo River delta is now the only locale in the southwestern United States where the Wood Stork regularly occurs, and such vagrants as the Roseate Spoonbill have occurred more frequently in this region than in any other else in the West. Finally, the Salton Sink's agricultural lands are important both as a migratory corridor and as foraging habitat for the ecologically distinct Turkey Vulture.

ARDEIDAE • *Herons, Bitterns, and Allies*

AMERICAN BITTERN
Botaurus lentiginosus (Rackett, 1813)
Uncommon winter visitor (mid-August to early May); casual in summer. Of the herons regular in the Salton Sea region, the American Bittern is the least often recorded. This species is secretive and, unlike the diminutive Least Bittern, not

especially vocal. It winters throughout southern California but may reach its peak abundance
at the Salton Sea. Even so, it is an uncommon
bird that is missed more often than not. Winter
records extend from 14 August (1976, Wister;
G. McCaskie) to 8 May (1978, mouth of the
Whitewater River; K. L. Garrett), with most present from late September to mid-April. A bird at
Unit 1 from 4 to 11 May 1996 (G. McCaskie) was
later still but may have been attempting to summer given its prolonged stay so late in spring.
Lone adults at Wister 13 June 1992 (*AB* 46:1178),
at Finney Lake 16 July 1997 (*FN* 51:1052), and at
the Whitewater River delta 17 July 1976 (*AB* 30:
1002), as well as two at the last locale 12 July
1978 (*AB* 32:1027), were clearly summering; at
least one also summered in 1999 (Shuford et al.
2000). Garrett and Dunn (1981) suggested that
this species nests around the Whitewater River
delta, but breeding evidence is lacking; nonetheless, an apparent pair was at Wister 6 May 1995
(M. A. Patten).

ECOLOGY. This species frequents dense freshwater marshes, generally thick with cattail and
bulrush. Favored locales are marshes at and
around Wister and at Unit 1.

LEAST BITTERN
Ixobrychus exilis hesperis Dickey and van Rossem, 1924
*Fairly common breeder (late March–early October);
uncommon in winter.* The skulking, inconspicuous Least Bittern is heard more often than it is
seen. It is local in the Southwest, its populations
being frequently underestimated or even undetected. However, it is a fairly common breeder
around the Salton Sea and in parts of the Imperial Valley. Migrants return by 24 March (1967,
mouth of the Whitewater River; *AFN* 21:457), and
most depart by September, with three calling at
the mouth of the Whitewater River 2 October
1999 (M. A. Patten) representing a late date for
presumed breeders, although exact arrival and
departure dates are clouded by the presence of
some birds throughout the year. Because Least
Bitterns call infrequently in winter, the species'
status in that season is difficult to gauge, but it

appears to be uncommon (Salton Sea [north] and
Salton Sea [south] Christmas Bird Count data
from the 1990s). Breeding is concentrated in
May and June, with birds on eggs as late as 24
June (1970, near Westmorland; WFVZ 148866).

ECOLOGY. Least Bitterns reach peak abundance along the rivers and wide irrigation ditches,
particularly in dense stands of Southern and
Broad-leaved Cattails. Most nest in cattail, but
some use Common Reed and even dense Saltcedar if cattail is nearby. They also occupy edges
of lakes in the Imperial Valley supporting similar habitat (e.g., Finney Lake, Ramer Lake, Fig
Lagoon).

TAXONOMY. Least Bitterns west of the Rocky
Mountains are *I. e. hesperis* (cf. Dickerman 1973),
larger than the nominate subspecies, of eastern
North America, and larger and paler than *I. e.
pullus* van Rossem, 1930, of southern Sonora.
The validity of *I. e. hesperis* has been questioned
(Blake 1977; Mayr and Cottrell 1979; Hancock
and Kushlan 1984), although there is almost
no overlap in wing chord, sex for sex (Palmer
1962:493).

GREAT BLUE HERON
Ardea herodias wardi Ridgway, 1882
Common breeding resident. North America's
largest heron, the Great Blue is a common sight
across the continent. It is a common breeding
resident around the Salton Sea, where a minimum of about 888 pairs recently nested (Shuford et al. 2000) and perhaps as many as 1,000
pairs nest in some years. Breeding sites dot the
shoreline of much of the Salton Sea but are
concentrated across the north end, centered on
the Whitewater River delta, and along the southern shoreline around Wister and the New River
delta. Some breed in sporadic rookeries at lakes
in the Imperial Valley (see Table 5) and at Cerro
Prieto (Molina and Garrett 2001). Most breeders
are on eggs by mid-April (Grinnell 1908; MVZ
49–53, 100–103), although nesting frequently
starts by early January (M. A. Patten; Shuford et
al. 2000). Thousands of nonbreeders remain
through the year, and there is likely an influx of

northerly breeders each winter, as individuals can move vast distances between breeding and wintering sites (Palmer 1962:396).

ECOLOGY. This heron's main breeding sites are snags of partly submerged dead Saltcedar, although many pairs around the north end nest on partly submerged utility poles. Unlike other herons and egrets in the region, Great Blue Herons do not form dense nesting colonies in the Salton Sink, although they occasionally form loose colonies. They forage along ditches, around shallow pools, ponds, and lake shores, along the edge of the Salton Sea, and in well-flooded fields. Apart from the Cattle Egret, the Great Blue is more apt to forage in dry agricultural fields than are other species of Ciconiiformes in the region.

TAXONOMY. Geographic variation in this species is complicated, with general west-east clines from large to small and pale to dark. Birds in California have been variously attributed to *A. h. treganzai* Court, 1908, and *A. h. hyperonca* Oberholser, 1912, but Payne (1979) synonymized both of these subspecies. The nominate subspecies, of the Northeast, the northern Great Plains, and much of Canada, averages smaller than birds of the West and the Southeast; if the last two populations are merged, then the name *A. h. wardi* has precedence (R. W. Dickerman pers. comm.).

GREAT EGRET
Ardea alba egretta Gmelin, 1789
Common breeding resident; more numerous in winter (mid-September to late April). Like the Great Blue Heron, the Great Egret is common across North America, although it has a more southerly distribution; it too is a common breeding resident around the Salton Sea, where about 230 pairs nest (see Table 5). A few also breed in the Mexicali Valley (Mora 1989) and at Cerro Prieto (Molina and Garrett 2001). Breeding begins in mid-March and continues through August. Unlike the Great Blue Heron's, the Great Egret's nesting tends to be more colonial, with sites concentrated along the shorelines at Wister and Morton Bay, around the deltas of the White-

water and New Rivers, and at various lakes in the Imperial Valley. Numbers are augmented in winter (mid-September to late April) by an influx of birds that breed to the north. This species is often the most numerous ardeid along the shoreline of the Salton Sea during winter (e.g., ca. 750 along the northern shore 13 November 1999; M. A. Patten), generally outnumbering the Great Blue Heron and often outnumbering or roughly equaling the Snowy Egret.

ECOLOGY. This egret nests in partially submerged snags, but unlike the Great Blue Heron, it avoids utility poles. Nests are generally over water, often with banks of cattail near the nest tree. These birds forage in shallow water, salt or fresh, throughout the region, including in partially flooded fields.

TAXONOMY. Birds of the Western Hemisphere are of the yellow-billed, black-legged *A. a. egretta*.

SNOWY EGRET
Egretta thula (Molina, 1782) subspp.
Common breeding resident; more numerous in winter (mid-September to late April). Like its larger cousins, the graceful Snowy Egret is widespread across North America, although it is even more southerly in its distribution. It is likewise a common breeding resident around the Salton Sea, where about 335 pairs nest (see Table 5); a few also breed in the Mexicali Valley (Mora 1989, 1991) and at Cerro Prieto (Molina and Garrett 2001). Its breeding begins later than that of other herons in the region, with many building nests only by late May and the first young hatching by late June (Shuford et al. 2000). Nesting sites are much like those of the Great Egret. The Snowy Egret is more numerous in winter, when breeders from the north come into the region, mainly from mid-September through April. Numbers actually peak in late summer and early fall (Shuford et al. 2000), when young out of the nest and early migrants supplement the population. During winter this species' numbers are generally similar to those of the Great Egret, although that species frequently outnumbers the Snowy along the shoreline of the sea.

ECOLOGY. The Snowy Egret frequently nests near the Great Egret but more readily occupies Cattle Egret rookeries (Platter 1976). Nesting almost always takes place in partially submerged Saltcedar, although some birds nesting with Cattle Egrets occupy sites with no standing water (e.g., Brunt's Corner in 1991). The Snowy Egret forages in ditches, along riverbanks and lakeshores, and in shallow freshwater pools and ponds, seldom along the shoreline of the Salton Sea or in flooded fields.

TAXONOMY. The above account refers to *E. t. candidissima* (Gmelin, 1789) of most of North America, Central America, and northern South America. The nominate subspecies, of southern South America, is the smallest, whereas *E. t. candidissima* is medium-sized. There is one California record of the largest subspecies, *E. t. brewsteri* Thayer and Bangs, 1909, of Baja California and western Sonora, of a breeding male near Niland 18 June 1977 (Rea 1983:143; SDNHM 41527, bill depth 11.6 mm).

LITTLE BLUE HERON
Egretta caerulea (Linnaeus, 1758)

Casual spring and summer vagrant (early May to mid-September); 2 nesting records; 1 winter record. Although the Little Blue Heron is a resident breeder in small numbers at San Diego and winters regularly on the southern coast, it is predominantly a rare spring and summer vagrant to California (Unitt 1977); its status at the Salton Sea fits this pattern well. Since one was seen at the mouth of the New River 22 July 1972, the first inland record for California (Winter and McCaskie 1975), the Little Blue has occurred in the Salton Sink roughly every other year in spring or summer. Dates of occurrence extend from 7 May (1977, adult at the north end; *AB* 31:1046) to 15 September (1989, immature at Ramer Lake; *AB* 44:161), with one summering bird lingering at the north end 4 June–10 September 1973 as it molted from immature to adult plumage (*AB* 27: 917, 28:107). Almost all spring and summer records are of adults and involve single birds, although in addition to the breeding records, at most two adults were together at Wister 6 June–

18 July 1992 (*AB* 46:1178). Up to three adults in a heron rookery near Seeley after 10 June 1979 (*AB* 33:896), with two building a nest, laying four eggs, and raising two young, were the first to nest in California. Two adults were observed there again 20 June 1981 (*AB* 35:978). An adult at a nest in a rookery at Brunt's Corner 7 June (not July) 1991 (*AB* 45:1160) was tending two large, healthy nestlings by 5 July (*AB* 45:1160). An adult at the south end 7 December 1974–8 February 1975 (*AB* 29:741) provided the only winter record for the Salton Sea. A report of two near Westmorland (*AFN* 24:98) lacks documentation.

ECOLOGY. Little Blue Heron nests have been found in Saltcedar snags in rookeries dominated by Cattle Egrets. Elsewhere the birds have occurred in freshwater marshes with emergent Southern Cattail and along rivers and canals.

TAXONOMY. This species is monotypic, *E. c. caerulescens* (Latham, 1790) of South America being a synonym (Dickerman and Parkes 1968).

TRICOLORED HERON
Egretta tricolor ruficollis Gosse, 1847

Rare spring and summer vagrant (mid-April through September); casual winter vagrant (October through March); 1 potential nesting record. Like the Little Blue Heron, the Tricolored Heron wanders to the Salton Sea in spring and summer from its normal range in Baja California, only a few hundred kilometers south of the border (Wilbur 1987). In the 1970s and 1980s the Little Blue Heron was much more frequent, but the Tricolored occurred more frequently in the 1990s (with 1–2 annually), including a number of winter records and one circumstantial breeding record. Since the first regional record, at the north end 21 May 1967 (*AFN* 21:540), there have been about 35 spring and summer records from 17 April (1989, adult photographed at Wister; *AB* 43:536) to 1 October (1993, adult at the Whitewater River delta; Erickson and Terrill 1996), the majority from early May through mid-August. Lone individuals account for all records, except for 2 at the Whitewater River 10 May–13 July 1981 (*AB* 35:978), 3 together around Obsidian Butte 16 July–4 August 1994 (Howell and Pyle 1997), and 3–4 at Ramer Lake 6–14 August 1994, the last group comprising 1–2 adults and 2 fresh juveniles at the large heron rookery at that lake (Howell and Pyle 1997), implying local breeding. Remarkably, two nesting pairs were located in a rookery in the southern Mexicali Valley in July 2002 (*NAB* 56:489).

There are seven winter records from 30 September (2001, first-winter bird at Obsidian Butte to 3 December; G. McCaskie, A. Eisner) and 1 April (1996, first-winter bird near Red Hill from 2 March; McCaskie and San Miguel 1999). Other records are of birds at the south end 22 December 1969–20 March 1970 (*AFN* 24:538), 22 November–14 December 1972 (*AB* 27:120, 27:662), and 19–27 February 1979 (*AB* 33:312) and of a first-winter bird near Oasis 2–22 January 1994 (Howell and Pyle 1997). An adult near Mecca 31 January–7 March 1998 (Erickson and Hamilton 2001) returned 14–27 February 1999 and 30 December 1999–12 January 2000 (Rogers and Jaramillo 2002). Additional reports from Red Hill and Finney Lake lack convincing documentation (Heindel and Garrett 1995; Patten, Finnegan, et al. 1995).

ECOLOGY. Virtually all Tricolored Herons are found in freshwater habitats, typically rivers, channels, and ponds immediately adjacent to the Salton Sea, but a few have been seen at lakes in the Imperial Valley. Only once has one occurred along the immediate shoreline of the sea. They usually roost at the edges of ponds against cattail backdrops or in low, partially submerged snags.

TAXONOMY. California specimens are of the large *E. t. ruficollis*, which has the foreneck white and includes *E. t. occidentalis* (Huey, 1927) as a synonym. *E. t. ruficollis* is the only subspecies on the North American continent (Palmer 1962).

REDDISH EGRET

Egretta rufescens [dickeyi] (van Rossem, 1926)
Casual summer and fall vagrant (mid-July through mid-October); 2 winter records. Native to salt marshes of western Mexico and Baja California, the Reddish Egret has wandered to the Salton Sea on 12 occasions in summer and once in win-

ter. Along the southern coast of California, where this species is a rare, annual visitor, most records are from fall and winter. In both areas this species is strongly tied to salt water. Virtually all regional records are of immatures, generally juveniles. Summer records extend from 7 July (2001, photographed at the mouth of the Whitewater River; CBRC 2001-115) to 12 October (2001, photographed at Obsidian Butte from 17 August; CBRC 2001-146). Other records are of individuals at the mouth of the Whitewater River 28 July 1990 (Patten and Erickson 1994), near Salton City 1 August 2001 (CBRC 2001-126), at the south end 8 August 1964 (Roberson 1993), photographed at Salton City 15 August 1981 (McCaskie and San Miguel 1999) and 16–31 August 1997 (Rottenborn and Morlan 2000), and observed at Obsidian Butte 4 September 1994 (Howell and Pyle 1997) and 7–18 September 2002 (L. Wolfe). Two were together at the mouth of the Whitewater River 31 August–28 September 1969 (Roberson 1993). A first-summer bird with a distinctive pinkish base to its bill seen 11 July–8 August 1998 first appeared near Mecca but later moved to Obsidian Butte (Erickson and Hamilton 2001). A second-year Reddish Egret photographed at Obsidian Butte 28 July–19 August 2001 (CBRC 2001-125) furnished the only summer record of a nonjuvenile. Birds in first-winter plumage photographed near Oasis 19 December 1993–20 March 1994 (Erickson and Terrill 1996) and at Obsidian Butte 14 November–17 December 2002 (K. Z. Kurland et al.) provided the only winter records for the Salton Sea and one of only four winter records for the interior Southwest, following immatures along the Colorado River at Imperial Dam 30 September 1954–3 March 1955 (Rosenberg et al. 1991) and photographed there 11 February–3 March 1979 (Binford 1983, Roberson 1993). Reports from the mouth of the Whitewater River (Heindel and Garrett 1995) and along the "northeastern edge" of the Salton Sea (*AB* 26:904) lack sufficient documentation.

ECOLOGY. This egret forages along barnacle beaches and in shallow water off the immediate shoreline of the Salton Sea.

TAXONOMY. California specimens, including one from the Colorado River at Lake Havasu 4–9 September 1954 (Monson 1958; Roberson 1993; MVZ 135902), have been attributed to the geographically expected *E. r. dickeyi,* the pale-headed (in adults) subspecies of Baja California and the Gulf of California. However, an adult specimen from Camp Verde, Arizona, 27 August 1886 is the nominate subspecies, of the Gulf of Mexico (Phillips et al. 1964; Monson and Phillips 1981), so this possibility cannot be excluded without a local adult specimen (immatures cannot be diagnosed).

CATTLE EGRET

Bubulcus ibis ibis (Linnaeus, 1758)

Abundant breeding resident; a colonist since the mid-1960s. With millions of Cattle Egrets now ranging across North America, few people ignorant of the history of bird distribution would guess that this species is a recent colonist. The Cattle Egret was first detected in the United States (Florida) in the early 1940s (Palmer 1962). In California it was first reported in December 1962 (*AFN* 18:386) and first collected at Imperial Beach, San Diego County, 7 March 1964 (McCaskie 1965). Three at the south end in early November 1963 (*AFN* 18:386) were the first reported in the Salton Sea region; one northwest of Westmorland 22 February 1965 was the first collected (SBCM 31477; *AFN* 19:416), and the species was regular in the Imperial Valley by 1966 (*AFN* 21:77). Breeding was first noted 17 May 1970, when 120 nests were found near the mouth of the New River (*AFN* 24:716), and was well established within a few years (Platter 1976). The Cattle Egret has nested annually since, with numbers topping 10,000 pairs around the Salton Sea (see Table 5) and another 1,000 pairs in the Mexicali Valley (Mora 1991), making it the most numerous ciconiiform in the region. Nesting takes place from late April through September (Platter 1976; Mora 1991; Shuford et al. 2000). In addition to the sizable breeding population, tens of thousands winter around the Salton Sea; an influx of northerly breeders may result in higher numbers in winter.

ECOLOGY. Breeding Cattle Egrets establish massive rookeries, generally numbering thousands of pairs and often including many fewer pairs of the Great Blue Herons, Great Egrets, Snowy Egrets, Black-crowned Night-Herons, and sometimes White-faced Ibises (Platter 1976; Mora 1990; see Table 5). Apparently unlike other herons in the region, the Cattle Egret is frequently double brooded (*AB* 46:1178), perhaps in part explaining its population explosion. A large component of its success is the abundance of its foraging habitat in agricultural and flooded fields (Mora 1990).

TAXONOMY. Birds in the Americas are of the short-legged, thin-billed nominate subspecies, of the Western Palearctic and Africa.

GREEN HERON

Butorides virescens anthonyi (Mearns, 1895)

Fairly common breeding resident. The small Green Heron is fairly common around the Salton Sea. Contrary to the bar graph in Garrett and Dunn (1981), but correctly reported in their text, there is no evidence that the Green Heron is more numerous at the Salton Sea in winter than in summer. Its small size and retiring habits hide its true status. Without a concerted effort it is unusual to record double digits, yet about 80 were noted around the south end 19–27 June 1999 (M. A. Patten). Breeding takes place mainly in spring, with nesting activity beginning in February and juveniles appearing by June.

ECOLOGY. Unlike other herons in the region except the bitterns, the Green Heron is infrequently detected away from cover. It reaches peak abundance along rivers and freshwater ditches, especially where there is open mud draped with Common Reed, cattail, or Saltcedar. It also occurs at the edges of marshes and lakes, provided there is suitable cover. Breeders occupy patches of riparian habitat, whether along rivers or in marshes. They do not nest colonially and thus are seldom encountered around large heronries.

TAXONOMY. *B. v. anthonyi,* larger, paler, and with the neck rustier than in the nominate subspecies, occurs west of the Rocky Mountains. It accounts for all California records, though eastern birds could occur as migrants (Phillips 1975b). The type locality is along the Alamo River near Seven Wells, Baja California, barely east of the Salton Sink (Mearns 1907; Deignan 1961). There are no U.S. records of the distinctive *B. v. frazari* (Brewster, 1888) of Baja California Sur, adults of which have the neck purplish, not rufous, but it may occur on occasion.

BLACK-CROWNED NIGHT-HERON
Nycticorax nycticorax hoactli (Gmelin, 1789)
Common breeding resident. The Black-crowned Night-Heron, a cosmopolitan species, is fairly common through much of its range. Belying its common name, it is frequently encountered around the Salton Sea during the day. It is a common breeding resident in the region, although breeding is concentrated in heronries in the Imperial Valley and around the Whitewater River delta. Breeding takes place in spring and summer, beginning in April and with most pairs still tending nests into August. Fledged young appear from mid-June (Shuford et al. 2000) to late July (*AB* 46:1178) and are a common sight by the end of August.

ECOLOGY. Night-Herons roost in dense tangles of cattail and Saltcedar during the day, although dozens sometimes actively feed well into midday. They frequently forage in the company of egrets on mudflats and at the mouths of rivers. This species nests colonially, often with other herons and egrets (see Table 5). Breeders occupy sites with water and snags sufficient to support their nests. Many nest in proximity to Cattle Egret heronries, where they depredate egret nestlings (Mora 1991).

TAXONOMY. Much of South America and all of North America hosts only the small, pale, white-breasted subspecies *N. n. hoactli*.

YELLOW-CROWNED NIGHT-HERON
Nyctanassa violacea bancrofti Huey, 1927
One spring record. A Yellow-crowned Night-Heron in first-summer plumage photographed at Fig Lagoon 27 April–18 June 1996 furnished the only record for the interior of California (McCaskie and San Miguel 1999). Because of pre-

vious records from Arizona (Rosenberg and Witzeman 1998), this species was expected to occur at the Salton Sea. Prior claims for the region (California Department of Fish and Game 1979) are baseless.

TAXONOMY. The only specimen for California, from the Tijuana River estuary 22–25 October 1963 (SDNHM 30758), is the paler, larger-billed western Mexican subspecies *N. v. bancrofti* (McCaskie and Banks 1966; Unitt 1984); all California records are presumed to be to the same.

THRESKIORNITHIDAE • Ibises and Spoonbills

WHITE IBIS
Eudocimus ruber albus (Linnaeus, 1758)
One summer record. One of California's two records of the White Ibis is from the Salton Sea. An adult was present at the Whitewater River mouth 10–24 July 1976 and at the south end 5 August 1976 (Luther et al. 1979). Presumably this same adult returned to Unit 1 25 June–14 July 1977 (Luther et al. 1983).

TAXONOMY. Although the AOU (1998) treated the White Ibis as a species distinct from the Scarlet Ibis, *E. ruber* (Linnaeus, 1758), we find the reasons for merging these species more compelling. We thus follow Hancock et al. (1992: 153) in lumping the species together and in recognizing a larger, monomorphic (white only) North American subspecies, *E. r. albus*, and a smaller, dimorphic South American one.

GLOSSY IBIS
Plegadis falcinellus (Linnaeus, 1766)
Two or 3 summer records (late May through early August). Two of three California records of the Glossy Ibis are from the Imperial Valley (Patten and Lasley 2000): An adult in alternate plumage and full breeding condition was photographed in a flooded field about 4 km northeast of Calipatria 27 May 2000. Apparently this same individual was relocated and photographed in flooded fields 10–12 km west of Calipatria 1–2 July (not 8 August) 2000. Another alternate-plumaged adult was photographed in that same area 1–15 July 2000 (see Patten and Lasley 2000), as was a first-summer bird 1–8 July 2000. *Plegadis* ibises

not in full breeding condition can be difficult to identify to species, so the last record, though apparently of a Glossy, might best be treated as tentative. The Glossy Ibis has expanded its range steadily since the 1970s, reaching Texas and southern Mexico in the 1980s, New Mexico, Colorado, and west Mexico in the 1990s, and California and Arizona by the millennium's end (Patten and Lasley 2000; G. H. Rosenberg pers. comm.).

WHITE-FACED IBIS

Plegadis chihi (Vieillot, 1817)

Common perennial visitor (mainly in winter, September through April); uncommon (and irregular), local breeder. After serious declines throughout the U.S. portion of its range in the 1970s, reaching levels low enough to cause widespread concern, populations of the White-faced Ibis have recovered dramatically. A westward shift in its breeding range, coupled with improved breeding habitat, has led to a significant increase in the number occurring in California (Shuford et al. 1996). The White-faced Ibis is a common winter visitor, migrant, and nonbreeding summer visitor to the Salton Sink; it is an uncommon breeder. Numbers now generally exceed 15,000 birds year-round (Shuford et al. 2000). This species first attempted nesting in the region in 1954, with 5 pairs tending nests near the south end; 28 pairs in 1956 were the first to rear young (Ryder 1967). Nesting was also successful in 1960 and probably in 1961, but "since 1961 no nesting has occurred there, apparently because suitable habitat is lacking" (Ryder 1967). Through the 1970s the White-faced Ibis bred only occasionally in the Salton Sink, for example, 3–4 pairs at the south end in 1977 (*AB* 31:1189) and 10 pairs at the north end in 1978 (*AB* 32: 1207). By the early 1990s breeding colonies had been established at Finney and Ramer Lakes, where 100 pairs nested in 1991 (*AB* 45:1160) and 370 pairs nested in 1992 (see Table 5). Several hundred pairs nested at these lakes, beginning in April and continuing into August, through the mid-1990s. The species was not known to nest anywhere in the region in 1998

and 1999 (Shuford et al. 2000; see Table 5), but more than 100 pairs nested at Finney Lake in 2000 and 2001 (G. McCaskie).

It is in winter that the prevalence of the White-faced Ibis is most impressive, particularly in the Imperial Valley. Winter numbers have risen sharply, with several thousand wintering in the Imperial Valley in the 1950s (Ryder 1967), increasing to about 16,000 by the mid-1990s (Shuford et al. 1996) and about 37,500 in 1999 (Shuford et al. 2000). Winter flocks in the Imperial Valley often number in the thousands. This species is less numerous in the Coachella Valley than in the Imperial and Mexicali Valleys (Patten et al. 1993, 2001).

ECOLOGY. This ibis nests mainly in tall stands of Southern Cattail and emergent snags of drowned Saltcedar. A few dozen pairs also nest in Saltcedar snags around the Whitewater River delta. Foraging birds frequent flooded fields (especially), marshes with shallow water, river edges, and irrigation canals with soil floors.

ROSEATE SPOONBILL

Platalea ajaja Linnaeus, 1758

Casual, highly irregular summer and fall vagrant (mid-May through October), sometimes in flocks; only 3 records since the late 1970s; 2 winter records. The Roseate Spoonbill is a highly irregular postbreeding visitor to the Salton Sink (Table 6). The region has hosted about three-fourths of the records for California (Garrett and Dunn 1981), though the species also occurred in the lower Colorado River valley during five of the larger flight years (Rosenberg et al. 1991). The first regional record is of a bird collected at Volcano Lake 8 June 1915 (Grinnell 1928). Although published as the first California specimen record, an immature male collected at the south end 22 May 1927 (Pemberton 1927; MVZ 51702) was actually predated by a juvenile female taken near Calipatria 2 October 1924 (H. E. Siraub; WFVZ 1451). After an interval of nearly 25 years this species occurred in the Salton Sink in 12 years in the period 1951–83 (see Table 6).

All regional records have been of immatures, all but one being birds in their first summer. The

TABLE 6

Occurrences of the Roseate Spoonbill (Platalea ajaja) *in the Salton Sink, 1915–1994*

YEAR	DATE(S)	LOCATION	NO.	SOURCE(S)
1915	8 June	Volcano Lake	1	Grinnell 1928; USNM 259911
1924	2 Oct.	South end	1	WFVZ 1451
1927	22 May	South end	1	Pemberton 1927; MVZ 51702
1951	23 June–8 Oct.	South end	8	Wooten 1952
1956	28 Oct.	South end	4	*AFN* 6:37
1966	23 Sept.–18 Nov.	South end	2	*AFN* 21:77
1969	20 July–1 Sept.	South end	2	*AFN* 23:694, 24:98
1970	30 Aug.–19 Sept.	South end	2	*AB* 25:107 (photographed)
1972	8 July–8 Oct.	New River mouth	7	Winter 1973; Winter and McCaskie 1975
	20 Aug.–8 Sept.	North end	1	*AB* 27:120
1973	14 June–26 Oct.	South end	35	*AB* 27:918, 28:107; SDNHM 38573
	14 June–late Oct.	North end	14	*AB* 27:918, 28:107; SBCM M5186
	1–16 July	Fig Lagoon	16	*AB* 26:918; G. McCaskie
	21 July	Finney Lake	1	*AB* 26:918
1976	5 May	North end	1	*AB* 30:889
	13 June–19 Sept.	South end	1	*AB* 30:1002, 31:222
1977	3 June–19 Oct.	South end	17	*AB* 31:1189, 32:256
	25 Aug.–23 Sept.	North end	1	*AB* 32:256
1978	19 May–19 Aug.	North end	1	*AB* 33:213
1980	27 July–7 Sept.	Alamo River mouth	1	Roberson 1993
1983	4–10 Sept.	Wister	1	Roberson 1993
1994	27–30 Dec.	Fig Lagoon	1	Howell and Pyle 1997

exception was a brightly colored second-summer bird (mistakenly called an adult by Garrett and Dunn 1981) at Red Hill 27 July–7 September 1980 (Roberson 1993). All but four records, involving more than 100 individuals, occurred during the period from 19 May (1978, 1 at the north end that stayed until 19 August; *AB* 33:213) to 28 October (1956, 4 at the south end; *AFN* 6:37). An exceptionally early individual was at the north end 5 May 1976 (*AB* 30:889). Likewise, one of two birds that appeared at the south end 23–25 September 1966 stayed to the exceptionally late date of 18 November (*AFN* 21:77). Also of note was one that lingered from the 1977 influx at the south end 29 January–31 July 1978 (Roberson 1993), providing the first winter record for the region. The 1970s saw the largest invasions on record; the species occurred at the Salton Sea six years during that decade (see Table 6). The exceptional invasion of 1973, bringing nearly 70 spoonbills to the Salton Sink, took birds as far northeast as Phoenix, Arizona (*AB* 28:87), and as far northwest as Marina del Rey, near Los Angeles (Winter and McCaskie 1975). The invasion

FIGURE 29. Numbers in the Salton Sink of the Wood Stork, a postbreeding visitor from western mainland Mexico, continue to shrink. Photograph by Kenneth Z. Kurland, July 2001.

of 1977, bringing nearly 20 spoonbills to the region, scattered birds as far north as southern Nevada (*AB* 31:1167) and the central coast of California (Roberson 1993). There has been but one record (in winter) since 1983 (see Table 6), coinciding with possible declines in the breeding population in western Mexico (Russell and Monson 1998).

A report of three adults at Wister (Rogers and Jaramillo 2002) likely referred to three Chilean Flamingoes known to be frequenting the area (see the Hypothetical List).

ECOLOGY. The spoonbill occurs in fresh water, feeding along rivers, lakeshores, and canals or in flooded fields, often associating with flocks of Wood Storks. It generally requires somewhat shallow water, where it stalks slowly while sweeping its head from side to side.

CICONIIDAE • *Storks*

WOOD STORK

Mycteria americana Linnaeus, 1758

Formerly common but now rare summer and fall visitor to the south end (primarily May through October); casual summer vagrant to the north end; 1 winter record. Wood Stork populations in north-

western Mexico have declined precipitously over the past half-century (see, e.g., Russell and Monson 1998), a decline reflected at the Salton Sea (Fig. 29). This species was formerly a "fairly common to abundant" postbreeding visitor to the south end and the adjacent Imperial Valley (Abbott 1935; McCaskie 1970b). Impressive high counts for that era were 2,000 on 18 July 1974 (G. McCaskie) and 1,000 on 18 September 1964 (Garrett and Dunn 1981). The vast majority of individuals were first-year birds (i.e., hatched that spring or summer). Even when the Wood Stork was common, numbers at the sea varied from year to year, but hundreds still occurred in the early 1980s, such as 300 on 30 July 1983 (G. McCaskie) and 250 on 30 August 1980 (M. A. Patten, P. Clark). By the late 1980s the story was different, as numbers dropped off to double digits (Fig. 30). Numbers continue to fluctuate, but now most birds are adults, and usually no more than a few dozen occur, with a low count of "no more than 20 during July" 1989 (*AB* 43:1367) and similar low maxima of 21 in 1993 (8 August; *AB* 47:1149) and 23 in 1992 (23 July; *AB* 46:1178). Sadly, about 100 storks, 30 percent of which were immatures, near Red Hill 14 August 1994

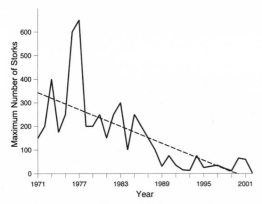

FIGURE 30. Decline of the Wood Stork at the Salton Sea, 1971–2002 (data from McCaskie's field notes). The species occurred in the thousands around the south end in the mid-1960s. The downward trend is significant ($r_s = -0.68$, $P < .001$).

was the highest count in well over ten years (*FN* 48:988), but numbers have been even lower since, with perhaps no more than 14 individuals in 1998 (near Red Hill 4 July; G. McCaskie) and 15 in 1999 (*NAB* 53:432), although 23 were located during summer (5 July–1 August) 2002 (*NAB* 56:486).

The Wood Stork occurs as a postbreeding visitor from 29 April (1995, Ramer Lake; *FN* 49:309) to 29 October (1953, 100 at the south end; *AFN* 8:41), with July and August being typical months of occurrence. Exceptionally early was one at the north end beginning 10 April 1980 (*AB* 34:815). Eleven at Calexico 29 and 30 November 1919 (Howell 1920) were exceptionally late, and one wintered at the south end 18 February–23 April 1978 (*AB* 32:398, 1054). In contrast to the past, storks now seldom linger into September, let alone October; individuals at Wister 19 September 1999 (B. Miller) and Red Hill 30 September 1995 (S. von Werlhof) provided the latest recent records.

The Wood Stork has always been surprisingly rare at the north end of the Salton Sea, with ten records during the period from 2 May (1986, adult; *AB* 40:523) to 19 September (1974, 1 from 8 July; *AB* 28:948, 29:120) and the early 10 April 1980 bird cited above.

ECOLOGY. Storks forage in bays and freshwater channels with submerged trees, especially

around the deltas of the New and Alamo Rivers, although they have rarely been found away from the latter river (including adjacent Morton Bay) since the mid-1980s. They occur occasionally in flooded fields with flocks of White-faced Ibises, Cattle Egrets, and Ring-billed Gulls; for example, 35 were recorded between El Centro and Calexico 2 October 1946 (Hill and Wiggins 1948), and an adult was photographed near Brawley 11 August 1990 (M. A. Patten).

CATHARTIDAE • *New World Vultures*

TURKEY VULTURE
Cathartes aura meridionalis Swann, 1921

Fairly common spring transient (mid-January to mid-May), fall transient (mid-July to early November), and winter visitor; uncommon summer visitor. Even though individuals may be encountered throughout the year, the Turkey Vulture is principally a transient in the Salton Sea region. The presence of birds year-round confounds arrival and departure dates. The Turkey Vulture is most numerous during spring, with dates from 18 January (1998, 25 in the Imperial Valley; M. A. Patten) to 23 May (1999, 15 in the Imperial Valley; M. A. Patten) and peak movement from mid-February through March. It is less numerous in fall because the main migration route lies well to the east of the region (Watkins 1976). In fall, as in spring, it is an early migrant, with records from 9 July (1988, 4 at the south end; M. A. Patten) to 10 November (1996, 3 in the Imperial Valley; M. A. Patten) and a peak from mid-August through September (Shuford et al. 2000). Small numbers (generally fewer than 20) remain through the summer; a recent maximum was 25 in the Imperial Valley 19–27 June 1999 (M. A. Patten). Numbers in winter have increased in recent decades throughout the species' range (Brown 1976); it was considered generally uncommon in the Salton Sink in the late 1970s (Garrett and Dunn 1981), but concentrations now exceed 200 birds at single roosting sites, such as 3 km southeast of El Centro during the winter of 1999–2000 (K. Z. Kurland).

ECOLOGY. Most vultures scavenge over agricultural fields, particularly where free-roaming livestock (cattle or sheep) graze. A few occur over the open desert. Roosting sites are generally in large eucalyptus trees in ranch yards and towns; some birds roost in large Fremont Cottonwoods, and others alight in Date Palms.

TAXONOMY. The large subspecies *C. a. meridionalis*, of which *C. a. teter* Friedmann, 1933, is a synonym (Wetmore 1964; Rea 1983), occurs over most if not all of California. Salton Sea specimens from the south end 18 May 1934 (LACM 18385) and 24 February 1940 (SDNHM 18102) and from near Niland 16 June 1999 (LACM uncataloged) are large (wing chord ca. 510 mm). However, the small nominate subspecies, of Mexico and the desert Southwest east of California, has been collected west to the lower Colorado River valley during migration (Rea 1983:129). It probably extends west into Imperial County and perhaps occasionally forages in the Salton Sea region, which lacks rocky hills suitable for nesting. Rea (1983) assigned the breeding population of southern California to *C. a. aura* (Linnaeus, 1758), but all specimens from San Diego County since 1983 are large.

ANSERIFORMES • *Screamers, Swans, Geese, and Ducks*

The Salton Sea is one of the most important locales on the Pacific Flyway for wintering and migrating waterfowl (Heitmeyer et al. 1989) and probably has been for a century, since heavy irrigation began in the Imperial Valley (Phillips and Lincoln 1930). Setmire et al. (1993) estimated waterfowl at the Salton Sea to number 125,000 birds annually, but the real number, for the whole of the Salton Sink, may be closer to double that estimate. Winter brings large numbers of Snow and Ross's Geese to the Imperial Valley, freshwater marshes host hundreds of thousands of dabbling ducks (Table 7), and the Salton Sea itself supports large numbers of diving ducks, including half of the Pacific Flyway population of the Ruddy Duck (Jehl 1994). Apart from the white geese, most waterfowl wintering in the region breed in the southern Prairie Provinces

rather than in Alaska and western Canada (Rienecker 1976).

Among the dabbling ducks, the Northern Pintail was formerly the most numerous, but range-wide declines have been reflected in the Salton Sink as well (Banks and Springer 1994). Now the Northern Shoveler accounts for half the total dabblers wintering in the region, with the Northern Pintail a distant second, accounting for about a fourth of the total (Shuford et al. 2000). Sixteen percent of dabblers are Green-winged Teals, and 8 percent of dabblers are American Wigeons (Shuford et al. 2000). Among the diving ducks, the Ruddy Duck accounts for more than 80 percent of the wintering population. Snow and Ross's Geese have both increased over the past several decades, especially the latter, but the Greater White-fronted and Canada Geese have declined, the former precipitously so.

Most species of geese and ducks that use the Salton Sea each winter show slightly different peaks of maximum occurrence (Table 8), although virtually all peak sometime during the midwinter period of mid-December to early February. Two notable exceptions among the regular species are the Greater White-fronted Goose and the Cinnamon Teal. Occurrence of both peaks closer to March because both are more strongly represented by spring migrants than by wintering birds. The Blue-winged Teal shows a similar pattern but occurs in much smaller numbers annually. The Brant and the various species of sea ducks also peak in spring, in April and early May, when many move northward out of the Gulf of California as they head overland to reach the Pacific Ocean (see Fig. 21).

ANATIDAE • *Ducks, Geese, and Swans*

BLACK-BELLIED WHISTLING-DUCK
Dendrocygna autumnalis fulgens Friedmann, 1947
Casual summer and fall visitor (primarily late May through August); records increasing since the late 1980s. The Black-bellied Whistling-Duck is a fairly common breeder in southeastern Arizona (Monson and Phillips 1981) and has been increasing in northwestern Mexico with the expan-

TABLE 7
Approximate Mean Numbers of Waterfowl
Wintering Annually in the Salton Sink

SPECIES	MEAN NUMBER	SPECIES	MEAN NUMBER
Fulvous Whistling-Duck	<5	Northern Pintail	14,000
Greater White-fronted Goose	<5	Baikal Teal	Vagrant
Snow Goose		Green-winged Teal	
White morph	20,000	*A. c. carolinensis*	5,000
Blue morph	4–5	*A. c. crecca*	Vagrant
Ross's Goose		Canvasback	500
White morph	7,500	Redhead	350
Blue morph	Vagrant	Ring-necked Duck	100
Canada Goose		Tufted Duck	Vagrant
B. c. moffittii	1,000	Greater Scaup	50
B. c. parvipes	<25	Lesser Scaup	1,500
B. c. minima	<5	Surf Scoter	5
B. c. leucopariea	Vagrant	White-winged Scoter	<1
Tundra Swan, *C. c. columbianus*	1	Black Scoter	Vagrant
Wood Duck	<2	Long-tailed Duck	<1
Gadwall	750	Bufflehead	250
Eurasian Wigeon	2–3	Common Goldeneye	60
American Wigeon	6,500	Barrow's Goldeneye	Vagrant
Mallard, *A. p. platyrhynchos*	500	Hooded Merganser	1
Blue-winged Teal	5	Common Merganser	<1
Cinnamon Teal	325	Red-breasted Merganser	20
Northern Shoveler	25,000	Ruddy Duck	75,000

Sources: Censuses in 1978–87 (Heitmeyer et al. 1989) and 1999 (Shuford et al. 2000), modified using Christmas Bird Counts, data published in *American Birds* or *Field Notes*, and notes from Michael A. Patten and Guy McCaskie.

sion of agriculture (Russell and Monson 1998: 43). California records seemingly reflect this increase, with 1 in the 1930s, 1 in the 1950s, 4 in the 1970s, 3 in the 1980s, 9 in the 1990s, and 3 thus far in the first decade of the new century. Nineteen of 20 valid California records of this species are from the Salton Sink, principally from the south end and Imperial Valley. Salton Sea records are of postbreeding visitants from Mexico

or Arizona, largely from 27 May (2000, 2 photographed at Elmore Ranch north of Poe Road until 22 June; *NAB* 54:422, CBRC 2000-090) to 28 August (1994, Ramer Lake from 27 August; *FN* 48:988, CBRC 1994-157). Exceptions are a flock of nine at Finney Lake 20 April 1990 (Patten and Erickson 1994), one photographed near Obsidian Butte 29 April 2000 (CBRC 2000-076), and a flock of three photographed at Wister 15 October–

TABLE 8

Peak Waterfowl Numbers at Salton Sea National Wildlife Refuge
during the Winter of 1969–70

SPECIES	PEAK NUMBER	DATE
Snow Goose	19,000	7 December
Ross's Goose	50	11 January
Canada Goose	3,290	11 January
American Wigeon	45,500	25 January
Mallard	325	4 January
Cinnamon Teal	1,465	15 March
Northern Shoveler	11,990	11 January
Northern Pintail	84,000	25 January
Green-winged Teal	23,300	25 January
Canvasback	260	25 January
Ruddy Duck	29,000	19 February

Source: AFN 24:538.

4 November 1973 (Luther et al. 1979). The April birds were further exceptional in being probable spring overshoots, as the species returns to its breeding areas in Sonora in early April (Russell and Monson 1998). Two records from the mouth of the Whitewater River, 10 July 1992 (Heindel and Patten 1996) and 12 July–20 August 1993 (Erickson and Terrill 1996), are the only ones for the region away from the Imperial Valley. The lack of records for the Mexicali Valley (see Patten et al. 2001) is likely an artifact of the much lighter observer coverage in that area.

The first report of the Black-bellied Whistling-Duck for California was of one collected in the "fall of 1912" at an unknown location in the Imperial Valley (Bryant 1914); since we cannot locate the specimen, we cannot verify the record. Also, "up to 4" were reported at the south end 2 June–5 August 1972 (*AB* 26:904), but the two observed 29 July 1972 were the only ones adequately documented.

ECOLOGY. Most Black-bellied Whistling-Ducks have been located in freshwater ponds with a thick cattail and Saltcedar edge or in flooded fields. About half have been associated with flocks of Fulvous Whistling-Ducks.

TAXONOMY. The sole California specimen, a life mount from Buena Vista Lake, Kern County, 19 June 1938 (SBMNH 7202), is of the northern *D. a. fulgens*. Salton Sink records are undoubtedly the same. It is this subspecies, not the nominate as reported by Palmer (1976a), Blake (1977), and Madge and Burn (1988), that breeds in Mexico and north to Arizona and Texas (Mayr and Cottrell 1979). *D. a. fulgens* differs from the nominate subspecies, from Panama southward, and *D. a. discolor* Sclater and Salvin, 1873, of South America, in that its breast, foreneck, and back are a deeper reddish brown.

FULVOUS WHISTLING-DUCK
Dendrocygna bicolor (Vieillot, 1816)
Rare and declining breeder (arriving early April); rare summer and fall visitor (May through early November); casual in winter and away from the south end; formerly an uncommon to fairly common resident, peaking in summer and fall. The Fulvous Whistling-Duck (Fig. 31) is nearly cosmopolitan,

FIGURE 31. The Fulvous Whistling-Duck is on the verge of extirpation at the Salton Sea and thus from all of California. Photograph by Kenneth Z. Kurland, June 1998.

with populations in East Africa, India, Myanmar, Hawaii (where it may be nonnative), and tropical America. In the Americas it breeds north as far as the southwestern United States. Although it is abundant in parts of its range, it has declined considerably in the western United States, especially in California and Arizona. In California it formerly bred commonly in the Central Valley and along the coast of southern California, as well as in marshes in the southeastern deserts (Willett 1912; Grinnell and Miller 1944). Even in the mid-1960s at least 20 pairs nested in the Imperial Valley (*AFN* 20:546). Subsequently the Fulvous Whistling-Duck has become alarmingly scarce. In California this elegant species is currently restricted as a breeder to the Imperial Valley, where it is on the verge of extirpation. Most breeding records are from Finney and Ramer Lakes, although the species has bred around the Alamo River delta near Red Hill (*AB* 42:1339; see below). During most of the 1990s there were fewer than five breeding pairs in the region. By the end of that decade the story was even more grim: a female with three ducklings near Red Hill 27 June 1998 (G. McCaskie) and a female with ten ducklings at Finney Lake

27 July 1999 (*NAB* 53:432) provided the only nesting evidence during those summers. Only four adults were detected in summer 2000, with no evidence of nesting (*NAB* 54:422), and only one bird was found in summer 2001 (*NAB* 56:356), although as many as nine were in the Imperial Valley from 10 to 28 June 2002 (*NAB* 56:486).

Breeders used to arrive in mid-March, exceptionally as early as 1 March (1942, Wister; R. Reedy). Following the species' precipitous decline, breeders arrive in mid-April, with the earliest typical recent arrival being up to seven at Finney and Ramer Lakes 3 and 4 April 1985 (*AB* 39:349) and the only recent March arrival being two at Finney Lake 12 March 1995 (M. A. Patten). Nest building and egg laying take place in April and May. Young are frequently out of the nest by late June or early July, with four adults and seven ducklings at Finney Lake 16 June 1990 being the earliest (M. A. Patten). Some nest later, such that young are not out of the nest until fall, such as a female with three young at Salton Sea NWR 10 September 1949 (*AFN* 4:35). Departure dates for breeders are clouded by the appearance of nonbreeders in late summer and early fall. Formerly numbers built to hundreds in early fall, such as 240–300 at Salton Sea NWR 15 August–9 September 1949 (*AFN* 4:35) and about 460 there in early September 1951 (*AFN* 6:38). Many of these birds lingered into late fall, such as 90 still present 15 October 1949 and 5 remaining until 9 November (*AFN* 4:35). Post-breeding influxes still occurred in the 1980s (e.g., more than 100 individuals around the south end during July 1988 and 70 still present 21 August 1988; *AB* 42:1339, 43:167) and into the early 1990s (e.g., more than 50 at Ramer Lake 27 August 1994; S. von Werlhof) but were nonexistent by the close of the 1990s, when there were no fall records after late August.

Before 1980 the Fulvous Whistling-Duck wintered annually in the Imperial Valley, from 28 November (1964, 3 near Niland; *AFN* 19:78) to 28 February (1959, 80 at Finney Lake from at least 18 February; *AFN* 13:322). It is now casual in winter, averaging one record (often of small flocks) every three to four years. Surprisingly, this species has been recorded at the north end on only seven occasions, all at the mouth of the Whitewater River. Dates of occurrences at the north end mirror those in the Imperial Valley, with single individuals 5 April 1908 (MVZ 761), 13 April 1908 (MVZ 787), 2 May 1971 (Garrett and Dunn 1981), and 28 June 1986 (*AB* 40: 1255); 2 on 12 May 1984 (*AB* 38:958); 19 on 15 July 1977 (Garrett and Dunn 1981); and 3 on 22 August 1992 (*AB* 47:148). A winter report from the north end (*AFN* 20:459) is questionable.

ECOLOGY. This species breeds in freshwater marshes with a patchwork of open water and dense patches of Southern Cattail, although it also takes advantage of Common Reed. It seldom forages far from shore or protective vegetation, although it feeds in stubble fields near water. Mated birds congregate in small groups, whereas unmated birds and nonbreeders gather in larger groups. Postbreeding birds often occupy flooded fields.

TAXONOMY. Despite its wide range, this species is monotypic, with *D. b. helva* Wetmore and Peters, 1922, being a synonym (Palmer 1976a). Of note, the type locality for *D. b. helva*

is along the New River about 4 km northwest of Calexico (Mearns 1907:131).

GREATER WHITE-FRONTED GOOSE
Anser albifrons frontalis Baird, 1858
Rare (formerly fairly common) transient and winter visitor (mid-October to early March); 3 summer records; 2 spring records. Scarcest of the geese reaching the Salton Sea, the Greater White-fronted Goose is a rare winter visitor to the region, where it occurs chiefly in the Imperial Valley. This species is a rare to casual transient through the deserts of California, so its status at the sea is not surprising. Nevertheless, it was formerly a common winter visitor, with hundreds annually in the Imperial Valley, including up to 1,800 at Salton Sea NWR 3 December 1954 (*AFN* 9:285). More recent maxima are much lower, with 350 during the winter of 1965–66 (*AFN* 20:458) and fewer than 50 through most of the 1980s and 1990s. Records extend from 7 October (1971, 50 at Salton Sea NWR; *AB* 26:120) to 5 March (1949, 40 at Salton Sea NWR; *AFN* 3:224), although since the late 1970s most winter records are from December to mid-February. Three flying south 5 km southwest of Brawley 11 September 2000 (G. McCaskie), an adult near Red Hill 25 September 1994 (M. A. Patten), and at most 33 at Obsidian Butte 26 September–2 October 1993 (M. A. Patten, S. von Werlhof) were fall migrants. Before 1970, maxima occurred in fall, with 2,000 at the south end in mid-October 1962 (*AFN* 17:67) and 1,800 there 3 November 1951 (*AFN* 6:37). Today, usually lone individuals or small groups are seen, with more than 100 over Salton City 31 December 1998 (M. B. Stowe) and 41 at Unit 1 on 26 January 1992 (G. McCaskie) being the largest flocks since the late 1960s. Because this species is an extremely early spring migrant, it is likely that this latter flock was northbound from wintering grounds in western Mexico; even pre-1970 abundances peaked in late fall and late winter rather than in midwinter. The two spring records are of an immature near Mecca 15 April 1989 (*AB* 43:536) and at most 2 at the mouth of the White-

water River 12 April–10 May 1981 (*AB* 35:863). In summer, a crippled bird was found near the mouth of the New River 16 May 1992 (*AB* 46: 479); two birds, one crippled, remained at the mouth of the Whitewater River 19 May–24 August 1974 (*AB* 28:948); and an apparently healthy adult was about 6 km northwest of Calipatria 3–30 July 2000 (G. McCaskie et al.).

ECOLOGY. In the Imperial Valley, Greater White-fronted Geese occur in Alfalfa fields hosting thousands of Snow and Ross's Geese. At the north end they have occurred on mudflats around the mouth of the Whitewater River.

TAXONOMY. Greater White-fronted Geese wintering in southern California (and in most of the West) are the small, moderately marked *A. a. frontalis*, the breeder in northern North America. Geese passing through the Salton Sink apparently originate in southwestern Alaska (Ely and Takekawa 1996). The large, coarsely marked Tule Goose, *A. a. elgasi* Delacour and Ripley, 1975, breeds in south-central Alaska and winters in the Sacramento Valley of California. (The name *A. a. gambelii* Hartlaub, 1852, was misapplied to this large subspecies by Swarth and Bryant 1917; see Phillips et al. 1964 for details.) An individual suggesting *A. a. elgasi* was reported at the south end 13–30 November 1975 (*AB* 30:125). Given identification difficulties, we consider this sight record inconclusive. Extralimital records of *A. a. elgasi* are essentially unknown; for example, a bird from Tucson, Arizona, collected 15 February 1935 was identified by Swarth as a Tule Goose (Vorhies et al. 1935; AOU 1957) but proved to be *A. a. frontalis* upon reexamination (Phillips et al. 1964).

SNOW GOOSE

Chen caerulescens caerulescens (Linnaeus, 1758)
Common winter visitor to the south end (early October through April); casual in summer and at the north end. Large flocks of white geese, Snow and Ross's (Fig. 32), draw birders and hunters to the south end of the Salton Sea in winter. The Snow Goose is the more numerous of the two species and has long been common, but its numbers have

FIGURE 32. Both Snow and Ross's Geese winter by the thousands at Salton Sea National Wildlife Refuge. Usually only four or five of the Snow Geese are of the blue morph. Photograph by Kenneth Z. Kurland, November 1998.

increased steadily over the past half-century. The increase was first noted in the late 1940s, when numbers totaled 8,000 individuals (*AFN* 3:185). Numbers are now close to triple that amount (see Table 7), perhaps reflecting the species' mid-continent increase (Bellrose 1976), a hypothesis supported by band recoveries. Of 2,459 Snow Geese banded in northern California in the early 1960s, only 2 were recovered in the Imperial Valley, despite a wintering population 10,000 strong (Kozlik et al. 1959). Rienecker (1965) thus concluded that "this group probably migrates through Montana, Utah, Nevada, and Arizona into the Imperial Valley, missing the northern part of California."

Wintering birds arrive in October, as early as 2 October (1949, 35 at Salton Sea NWR; *AFN* 4:34). Most depart by March, with a few detected as late as 7 May (1994, healthy immature at Salton Sea NWR; M. A. Patten). A flock of 75 at Salton Sea NWR 10 April 1994 (G. McCaskie) was a large concentration for so late a date. In contrast to its prevalence east of the continental divide (Bellrose 1976; Root 1988), the blue morph (the Blue Goose) constitutes only a tiny percentage of the wintering population in the Salton Sink. The first regional record was of an adult male near the mouth of the Alamo River 5–11 February 1945 (Jewett 1945), but the species was not recorded again until Sams (1958) observed one at Salton Sea NWR 14 December 1957. In a typical winter there are usually 4–5 Blue Geese in the Imperial Valley, with maxima of 8 at Unit 1 and Salton Sea NWR 10 and 11 January 2000 (P. E. Lehman) and 6 in the winter of 1991–92 (AB 46:314). Away from the south end the Snow Goose is rare in the region, although small flocks are occasionally detected during winter (e.g., 10 at the north end 30 December 1965; AFN 20:376). Likewise, this species rarely summers, from 21 May (1980, healthy adult at the north end; AB 34:815) to 15 September (2000, 3 present all summer near Salton Sea NWR; G. McCaskie). One summer record is of a Blue Goose photographed at Ramer Lake in 1994 (FN 48:988). Most summering birds are crippled or carry shot.

ECOLOGY. Wintering Snow Geese feed largely in Alfalfa and grain fields managed by Salton Sea NWR (both around the headquarters and at Unit 1) and Wister. They occasionally forage in adjacent agricultural fields. Most roost on large freshwater impoundments at the same locales.

TAXONOMY. All records for the West are of the small nominate subspecies, commonly called the Lesser Snow Goose, of which C. c. hyperborea (Pallas, 1769) became a synonym when it was determined that the Blue Goose was merely a color morph (Cooke and Cooch 1968; AOU 1973).

ROSS'S GOOSE

Chen rossii (Cassin, 1861)

Common winter visitor (mid-October through April); casual at the north end; 2 summer records. Like the Snow Goose, Ross's Goose increased greatly

in number at the Salton Sea over the second half of the twentieth century. Its increase has been even more dramatic than that of its larger cousin, as Ross's Goose was unrecorded in the region before the 1940s (Grinnell and Miller 1944) and there were but a few records by the early 1950s (O'Neill 1954; AFN 6:37). Through most of the 1960s wintering flocks of white geese contained only a few dozen Ross's; for example, of more than 16,000 white geese at the south end 28 December 1968 only 11 were Ross's (AFN 23:423). Even so, about 200 were reported at the south end during the winter of 1965–66 (AFN 20:459). By the winter of 1975–76 some 600–700 Ross's Geese were wintering in the Imperial Valley (AB 30:765). The percentage is now vastly different, with 20–33 percent of flocks of Chen geese being Ross's (e.g., 1,500 Ross's of 7,500 white geese at the south end 26 December 1996; FN 51:643). This species may arrive as early as 20 October (1997, 100 at the south end; G. McCaskie), but most arrive during November. The vast majority depart by April. Some individuals straggle into May, with one at the mouth of the Whitewater River 25 April–7 May 1978 (AB 32:1054) being the latest. A healthy adult at the mouth of the same river 25 May 1998 (M. A. Patten) was likely an extremely late spring transient, as it could not be found on subsequent dates. Summer records are of two near Mecca 7 May 1988–1 January 1989 (G. McCaskie, M. A. Patten et al.) and one at the south end 20 June–19 July 1998 (G. McCaskie). Like the Snow Goose, Ross's Goose is seldom recorded in the region away from the south end (centered around Salton Sea NWR and Wister). It is casual at the mouth of the Whitewater River and vicinity, mainly in winter and spring.

A blue-morph Ross's Goose with a massive flock of white Ross's Geese and Snow Geese at Unit 1 24 February–3 March 1990 (AB 434:328; S. von Werlhof) was the first to be recorded in southern California. What was presumably this same individual returned 19 October 1991 (S. von Werlhof), 6 November 1993 (S. von Werlhof), 26 December 1996–10 March 1997 (FN 51: 801; G. Hightower), 17 November 1997–21 February 1998 (FN 52:125, 256), and 9 January

2000 (*NAB* 54:220); it was photographed in January 1998 (Berlijn 1999). This morph was only recently described (McLandress and McLandress 1979), but records of it have been increasing throughout North America (Patten 1998).

ECOLOGY. Wintering Ross's Geese are virtually always found with flocks of the more numerous Snow Goose in Alfalfa and grain fields and in adjacent freshwater ponds and impoundments. Interestingly, Ross's Geese tend to form small pure flocks within the large mixed flocks, and they tend to be proportionately more numerous at Unit 1 than elsewhere in the region.

CANADA GOOSE

Branta canadensis (Linnaeus, 1758) subspp.
Uncommon to fairly common (formerly common) winter visitor (October through April); rare in summer. Few birds on the North American continent are more familiar than the Canada Goose; it is generally a common bird from coast to coast, with hundreds of thousands wintering in California annually. Around the Salton Sea, however, this species is greatly outnumbered by both species of white geese, particularly in the Imperial Valley, and is declining. In past years several thousand Canada Geese wintered at the Salton Sea (e.g., 5,000 during the winter of 1965–66; *AFN* 20:458). Since 1990, by contrast, totals have seldom exceeded a few hundred birds. For example, the figure of 918 on the Salton Sea (south) Christmas Bird Count 18 December 1990 (*AB* 45:998) has not been equaled or surpassed, and the species was missed in 1995 (*FN* 50:852). We lack definitive data, but it appears that the species' winter range has shifted northward, in parallel to the pattern on the Atlantic Coast (Terborgh 1989:22). Wintering Canadas are present from mid-November through early April, with extreme dates from 23 October (1949, 6 at Salton Sea NWR; *AFN* 4:34) to 2 May (1992, 2 at mouth of the New River; M. A. Patten). Two at the mouth of the Whitewater River 20 September 1980 (G. McCaskie) and one at Red Hill 2 October 1993 (S. von Werlhof) were extremely early, and one at the mouth of the Whitewater River 10 May 1986 (G. McCaskie) was extremely late (see below). Small numbers (at most 5), mostly crippled or injured birds, remain through the summer nearly annually, especially around Rock Hill, Wister, and the Whitewater River delta.

ECOLOGY. For foraging the Canada Goose favors fields of sprouting Alfalfa and similar fields with stubble. Some forage on lawns in ranch yards and around small ponds and lakes. All geese in the region roost in fields and on freshwater ponds, but the Canada roosts on the sea far more often than do the white geese.

TAXONOMY. The large forms (*B. c. canadensis*, *B. c. occidentalis*, *B. c. moffitti*, *B. c. parvipes*, etc.) and small forms (*B. c. minima*, *B. c. leucopariea*, *B. c. hutchinsii*) of the Canada Goose may represent separate species (AOU 1998). By far the predominant subspecies in the Salton Sea region is the large, pale-breasted Western Canada Goose, *B. c. moffitti* Aldrich, 1946. As elsewhere in the Southwest, this subspecies accounts for more than 97 percent of the Canada Geese occurring in the region, including all summer records. The next most numerous subspecies in the Southwest is the Lesser Canada Goose, *B. c. parvipes* (Cassin, 1852), of which *B. c. taverneri* Delacour, 1951, is a synonym (AOU 1957; Palmer 1976a; Gibson and Kessel 1997; cf. Johnson et al. 1979). *B. c. parvipes* is smaller and shorter-necked than *B. c. moffitti,* tends to have a slightly darker breast, and sometimes sports a thin white neck collar. Most of the remaining (probably fewer than 3%) of Canada Geese wintering in the region are Lessers, including a specimen from 6 January 1968 (SDNHM 36993) and a notable group of 5 among 25 *B. c. moffittii* at Unit 1 on 2 February 1992 (M. A. Patten). A late *B. c. parvipes* was at the mouth of the New River with a *B. c. moffitti* 2 May 1992 (M. A. Patten).

At most 1 percent of Canada Geese wintering in the region are the tiny Cackling Canada Goose, *B. c. minima* Ridgway, 1885, with records extending from 4 November (1949, 2 at Salton Sea NWR; *AFN* 4:34) to 28 February (1970, Rock Hill; Garrett and Dunn 1981) and including an undated specimen from Salton Sea NWR (HSU 6202). The maximum count for a winter is 3, at

Unit 1 on 11 November 1988 (M. A. Patten); a report of 55 at the south end 28 December 1968 (*AFN* 23:423) is questionable. Individuals near Westmorland 16–30 April 1967 (G. McCaskie) and at the mouth of the Whitewater River 10 May 1986 (G. McCaskie) were exceptionally late, presumed spring transients. This subspecies is little larger than a Mallard and has a dark brown breast (usually with a bronze or dark burgundy cast) and often a thin white collar. The endangered Aleutian Canada Goose, *B. c. leucopareia* (Brandt, 1836), winters almost exclusively in California to the north of the Salton Sink. It is slightly larger than *B. c. minima* and has a paler, medium gray-brown breast (lacking a burgundy cast), a wide white neck collar, and usually black on the throat separating the white cheek patches. Birds showing characters of *B. c. leucopareia* have been noted on five occasions at Salton Sea NWR. One "shot from a family group of six at the south end" 3 December 1975 was said to belong to this subspecies upon in-hand examination (*AB* 30:765), but unfortunately the specimen was not saved. An apparent family group (an adult and 3 immatures) banded on Buldir Island, Alaska, frequented the refuge 12 November 1977–6 January 1978 (Garrett and Dunn 1981; P. F. Springer). Two, an adult female and an immature male, were shot by hunters 24 November 1987 (P. F. Springer *in litt.*). Similarly, hunters shot three out of a family group of eight 18 November 1988 (P. F. Springer); two others later succumbed, but the survivors were seen as late as 18 March 1989 (G. McCaskie). Finally, one was observed at the refuge 25 February 1996 (G. McCaskie).

BRANT

Branta bernicla nigricans (Lawrence, 1846)
Rare to uncommon spring transient (primarily mid-March through May), with a few summering each year; 2 fall records. The highly coastal Brant is generally extremely rare inland in the West. However, in spring many head north overland via the Salton Sea (Fig. 33) to return to their breeding grounds (*AB* 37:911) after wintering in the Gulf of California (Wilbur 1987). The passage of Brant through the Salton Sink is a comparatively

recent phenomenon, beginning in the mid-1960s. Wintering individuals were not noted in the Gulf of California until 1964 (Nowak and Monson 1965), but numbers there have increased substantially since (Smith and Jensen 1970; Russell and Monson 1998). The species was not recorded at the Salton Sea until two were noted at the mouth of the Whitewater River 16 September 1962 (*AFN* 17:67). The Brant is now a rare to uncommon spring transient through the Salton Sink, reaching peak numbers at the north end, where birds gather before the next leg of their journey. Flocks of up to 500 have been recorded at the Whitewater River delta (e.g., 24 April 1965; *AFN* 19:511), with flocks of 100–200 not unusual. Migrants are noted flying northward only on occasion but include flocks of 200 birds 11 km southeast of Ocotillo (just west of the region) 26 April 1987 (P. Unitt) and 64 at Glamis (just east of the region) 24 April 1994 (*FN* 48:340).

Spring migrants appear in mid-March, with records as early as 4 March (1989 and 1992; S. von Werlhof, *AB* 46:479), exceptionally 16 February (1978, south end; *AB* 32:398) and 24 February (1974, 2 at the south end; *AB* 28:691). Numbers build into April and peak in the second half of that month, usually with many fewer by mid-May, although up to 100 are occasionally present into June (e.g., at the north end 5 June 1988; M. A. Patten). A few attempt to summer at the north end each year (see Fig. 33), although most succumb to the extreme heat before season's end; for example, 60 on 1 June 1990 were reduced to a mere 15 by 31 July (*AB* 44: 1185). Summering flocks usually do not number more than 10–20 birds, but 80–100 spent the summer at the north end in 1988 (*AB* 42:1340), and an unparalleled 130 birds were on the Salton Sea 2 July 1992 (*AB* 46:1178). Some survive into early September, with known summering birds remaining into October (e.g., 3 at Salton Sea NWR 12 October 1963 were first noted 22 April; Nowak and Monson 1965). We therefore presume that two at Salton City 26 September–16 October 1993 (G. McCaskie, M. A. Patten) and two at the south end 21 October 1976 (*AB*

FIGURE 33. Flocks of Brant visit the Salton Sea annually on the northward migration out of the Gulf of California (A). Some are stranded on the sea, especially at its north end, through the summer (B). Photographs by Richard E. Webster, 5 May 1984 and 3 June 1984, respectively.

31:222) summered. The species is virtually unknown as a fall migrant in the Salton Sink. Individuals at Salton City 11 November 1969 (*AFN* 24:99) and 11 November 1988 (M. A. Patten) represent the only two records of apparent fall migrants.

ECOLOGY. Virtually all Brant in the region occur along the immediate shoreline of the Salton Sea, reflecting the species' saltwater affinities. Birds roost on open water, mudflats, or barnacle beaches. A few have been recorded on freshwater reservoirs and lakes in the Imperial Valley;

for example, two were seen at Sheldon Reservoir 10 May 1996 (G. McCaskie), and there are multiple records for Ramer and Finney Lakes and for Cerro Prieto (Patten et al. 2001). In addition, one was seen at San Sebastian Marsh 24 March 1989 (P. Unitt).

TAXONOMY. The subspecies migrating through and wintering in California and western Mexico is the Black Brant, *B. b. nigricans* (cf. Browning 2002), and all birds recorded at the Salton Sea have belonged to this black-bellied taxon. The pale-bellied Atlantic Brant, *B. b. hrota* (Müller, 1776), has been recorded casually in winter in coastal California, but there are no records of this form for the Salton Sea. Occasional claims of Atlantic Brant in the region have invariably proved to be based on Black Brant that were heavily worn or bleached.

TUNDRA SWAN
Cygnus columbianus columbianus (Ord, 1815)
Casual (but almost annual) winter visitor (mid-November to early March). The Tundra Swan is scarce anywhere south of the Great Basin, with records extending as far south as northern Baja California and Sonora (Howell and Webb 1995; Russell and Monson 1998), barely south of the Salton Sink. The species was reported in the sink more frequently from 1910 to 1930, but it now occurs about once every other year on average. Records extend from 10 November (1956, at most 4 at Salton Sea NWR to 25 November; *AFN* 11:60) to 2 March (1996, immature at Unit 1 from 20 February; *FN* 50:222), with a distinct peak in December. Generally only 1 or 2 individuals are encountered, but a remarkable 13 were seen in the Imperial Valley 3–7 December 1991 (*AB* 46:314). Other notable concentrations were groups of 7 at the north end 7 December 1976 (*AB* 30:765), 6 at Salton Sea NWR 8 December 1971 (*AB* 26:654), 5 at the south end 26 November 1975 (*AB* 30:125), 5 at the north end 19 November 1994 (*FN* 49:100), and 3 "at the Salton Sea" in early December 1930 (Hanna 1931).

ECOLOGY. A few swans have been noted on the Salton Sea itself, but most records come from freshwater ponds, flooded fields, or the larger bodies of water in the Imperial Valley.

TAXONOMY. All birds carefully studied in the region have shown the characters of the nominate subspecies, the Whistling Swan, which is widespread in temperate North America. One or two Bewick's Swans, *C. c. bewickii* Yarrell, 1830, of the Old World and of which *C. c. jankowskii* Alphéraky, 1904, is a synonym (Vaurie 1965; Cramp and Simmons 1977), are recorded annually in California's Central Valley, but this subspecies has not been reported in the Salton Sink. The extent of yellow on the base of the bill of this subspecies does not overlap the maximum amount shown by the Whistling Swan (Evans and Sladen 1980), rendering confusion in the field unlikely.

WOOD DUCK
Aix sponsa (Linnaeus, 1758)
Casual winter visitor (mid-December through March); 3 late spring or summer records. That the stunning Wood Duck continues to be a rarity in the Salton Sink is surprising given its marked increases and range expansion in recent decades, including into the Colorado River valley (Banks and Springer 1994). Fall records extend from 30 October (1976, south end; *AB* 31:222) to 13 November (1932, male at the south end; van Rossem 1933b). Winter records are few, the species being recorded only once every two to three years on average, from 22 December (1985, north end; *AB* 40:1007) to 29 March (1997, female near Mecca; *FN* 51:927). Most records involve single birds, but four were present at Wister 26 December 1996 (*FN* 51:643), and six were there 6 February 1999 (B. Miller). There are two late spring or summer records from the north end, of a female 31 May 1985 (*AB* 39:349) and one 18 June 1973 (*AB* 27:918) and of a pair along the New River 7 km southwest of Brawley 11 May–9 June 2000 (G. McCaskie, G. Hazard), hinting at the possibility of local nesting.

ECOLOGY. Wood Ducks have occurred on freshwater ponds, including those at commercial catfish farms, particularly those with dense vegetation along the banks.

GADWALL

Anas strepera strepera Linnaeus, 1758

Fairly common winter visitor (October through April); rare in summer. Wintering flocks of the Gadwall arrive in the Southwest later than do other regularly occurring *Anas* ducks. Winter records for the Salton Sink extend from the end of September (e.g., 2 at mouth of the Whitewater River 30 September 1990; M. A. Patten) to early May (e.g., male at the eastern edge of the Salton Sea 8 May 1954; SBCM 31377), with most from mid-October to mid-April. Because this species is not particularly common at the Salton Sea, at least relative to other dabbling ducks, arrival and departure dates are clouded by the presence of small numbers of summering birds each year. This species has bred on multiple occasions along the lower Colorado River (Rosenberg et al. 1991), but breeding is unknown for the Salton Sea. See the American Wigeon account for information on a hybrid recorded in the region.

ECOLOGY. Dabbling ducks at the Salton Sea rely on extensive freshwater marshlands along the edge of the sea and in the Coachella, Imperial, and Mexicali Valleys, particularly those abounding in underwater vegetation. These ducks are generally most numerous in such settings (especially around Wister, at Unit 1, and at the Hazard Units of Salton Sea NWR), yet the species use subtly different habitats (see the accounts of food habits in Bellrose 1976). The Gadwall frequents marshlands, fish-hatchery ponds, and freshwater impoundments. It is more numerous on the Salton Sea itself, particularly around the river mouths, than many of the other dabbling ducks. It frequently associates with wigeons.

TAXONOMY. Birds throughout the Holarctic region belong to the nominate subspecies. The small, drab *A. s. couesi* (Streets, 1876), of the central Pacific, has been extinct for more than 100 years but is generally treated as a subspecies of the Gadwall (Palmer 1976a; Madge and Burn 1988; Sibley and Monroe 1990).

EURASIAN WIGEON

Anas penelope Linnaeus, 1758

Rare winter visitor (early November to mid-April); 1 spring record. The Eurasian Wigeon is a rare winter visitor to the West Coast but is annual (2–3 per year) among large concentrations of American Wigeons at the Salton Sea, particularly around the south end. A pair collected at Brawley 8 December 1917 was purchased at a Los Angeles bird market several days later (Wyman 1918). Closer examination showed the female to be an American Wigeon (Wyman 1920), but the male stands as the first record for the Salton Sea (FMNH 129328); it was followed by another male in the Imperial Valley 9 January 1924 (L. Miller; UCLA 38052). Records extend from 5 November (1976, first-winter male near Wister; *AB* 31:222) to 3 April (1985, adult male at the south end; *AB* 39:349), with one wintering male lingering to 6 May (2001, Rock Hill; L. Lina et al.). Exceptionally late was a male at Wister 13 April– 11 May 1991 (*AB* 45:495); this likely spring migrant provided one of the latest records for California.

ECOLOGY. As noted above, this wigeon invariably is found in large concentrations of the American Wigeon. Almost all records are from freshwater ponds and impoundments near agricultural fields.

AMERICAN WIGEON

Anas americana Gmelin, 1789

Common winter visitor (mid-August to early May); rare in summer. Many dabbling ducks have declined, at least to some extent, as winter visitors in southern California, but few as much as the American Wigeon. It remains common, but wintering numbers at the Salton Sea have decreased dramatically since the early 1970s (Rienecker 1976), with peak counts formerly near 50,000 birds (*AFN* 24:538) but now more typically at fewer than 10,000 birds (Heitmeyer et al. 1989; Shuford et al. 2000). Records of wintering birds extend from mid-August (e.g., south end 11 August 1979; G. McCaskie) to early May (e.g.,

12 km northwest of Calipatria 5 May 1937; MVZ 71333), with most from mid-October to early April. Most wigeons wintering in the region breed in the northern Great Plains, many moving through the Great Salt Lake region of northern Utah en route to the Salton Sea (Rienecker 1976). A few nonbreeders summer at the Salton Sea, mainly around the Whitewater River delta. Numbers are generally quite small (fewer than 5 birds), but up to 35 were seen at the north end during the summer of 1987 (*AB* 41:1487). A female taken along the New River 10 October 1899 (F. Stephens; SDNHM 83) shows that this species occurred in the region even prior to formation of the Salton Sea.

An adult male hybrid American × Eurasian Wigeon, regular on both sides of the Pacific Ocean in winter (Carey 1993; Merrifield 1993), was at the south end 23 January 1983 (T. Clotfelter; SDNHM 42139). Several more unusual hybrids have been detected in the Salton Sink. A male American Wigeon × Northern Pintail was seen 6 km south of Calipatria 9 (not 6) January 1954 (*AFN* 8:269; MVZ 130985). A male American Wigeon × Gadwall at Ramer Lake 31 January 1956 (*AFN* 10:281) returned 7 December 1956 (G. S. Suffel; LACM 29042). And a putative American Wigeon × Mallard was observed at the south end 28 January 1973 (*AB* 27:662).

ECOLOGY. This wigeon often forms large, pure flocks on freshwater ponds and impoundments near the fringe of the Salton Sea, particularly around the south end. It feeds in partly flooded fields and shallow marshes; the wigeon, the Northern Pintail, and the Green-winged Teal are the dabblers most likely encountered in the former habitat. The wigeon, however, is the only dabbler that feeds readily on open lawns, such as in parks with ponds or small lakes. Some wigeons occur on the open water of the Salton Sea, especially near river mouths, and on lakes and reservoirs.

MALLARD

Anas platyrhynchos platyrhynchos Linnaeus, 1758
Fairly common winter visitor (September to mid-April); rare in summer (has bred). The Mallard is common and familiar throughout its Holarctic range, particularly at more temperate latitudes. Numbers drop substantially toward the south, so much so that the familiar green-headed ducks are uncommon in Mexico (Howell and Webb 1995). Notwithstanding a claim of 3,000 (*AFN* 4:34), a count we consider to be erroneously high and to possibly even represent a misidentification, the Mallard is seldom tallied above the low hundreds in the Salton Sink (e.g., 250 at the north end 21 December 1980; *AB* 35:727) and is greatly outnumbered by several other *Anas* species (see Tables 7 and 8). Wintering birds occur from early September (e.g., at Salton Sea NWR 1 September 1949; *AFN* 4:34) to mid-April (e.g., 4 at Salton Sea NWR 23 April 1949; *AFN* 3:224), with the majority present from late September through March. Small numbers (at most 10 birds) remain through the summer each year, with some occasionally breeding, for example, an adult with ducklings at Unit 1 on 5 June 1983 (G. McCaskie) and a female with eight ducklings at Obsidian Butte 16 May 1999 (M. A. Patten).

Of note is an apparent hybrid Mallard × American Black Duck at the south end 15–20 May 1977 (*AB* 31:1047). See the American Wigeon account for additional information on hybrid Mallards in the Salton Sink.

ECOLOGY. Like all dabbling ducks in the region, this species is most numerous at freshwater habitats, especially marshes, impoundments, and small lakes, reservoirs, and lagoons. A few congregate on the Salton Sea at river mouths.

TAXONOMY. The nominate subspecies of the Mallard, of which *A. p. neoboria* Oberholser, 1974, is a synonym (Browning 1974), is the only one definitely recorded in California. It is strongly dimorphic sexually and has the familiar green-headed male. However, there are three California records possibly of the Mexican Duck, *A. [p.] diazi* Ridgway, 1886, a monotypic (Aldrich and Baer 1970), hen-plumaged semispecies from Mexico and the interior Southwest (southeastern Arizona to west Texas). The first two records are of specimens, a female at Grafton, Yolo County, in July 1900 (Phillips 1923:56) that Grinnell and Miller (1944:560) considered to be of questionable natural occurrence and a male at Alviso, Santa Clara County, 27 July 1927 (CAS 14153) that Hubbard (1977) described as being "very dark and heavily marked" with "rectrices similar to those of *diazi*" but that he nonetheless treated as a hybrid *A. p. diazi* × *A. p. platyrhynchos,* as he did the vast majority of birds in the United States exhibiting the Mexican Duck phenotype. The third record is of a male carefully studied at Lake Tamarisk at Desert Center, Riverside County, 17 October 1998 (M. A. Patten). These records suggest the possibility of the Mexican Duck's occurring in the Salton Sea region.

BLUE-WINGED TEAL
Anas discors Linnaeus, 1766

Uncommon spring (mid-February to early May) and fall (August to mid-October) transient; rare winter visitor (late November to early March); casual in summer. In contrast to its status elsewhere in the desert Southwest, the Blue-winged Teal is more numerous at the Salton Sea in spring than in fall. Furthermore, the Salton Sea is the only desert locale where the species winters regularly, although a few are sometimes found in that season in the lower Colorado River valley (Rosenberg et al. 1991). Wintering and occasional summering birds cloud determination of migration dates. Apparent spring migrants have been noted from 14 February (1988, male at the north end; *AB* 42:320) to 11 May (1985, male at the mouth of the Whitewater River; G. McCaskie), with a peak from mid-March through April. Numbers are generally small (2–4 birds), but flocks of more than 5 are occasionally noted; for example, 10 were seen at the mouth of the Whitewater River 15 April 1984 (G. McCaskie), 10 were seen at the south end 21 April 1991 (G. McCaskie), and 14 were seen at Wister 24 February 1951 (*AFN* 5:226). The Blue-winged Teal summers nearly annually, from 28 May (1990, male at the mouth of the Whitewater River; G. McCaskie) to 30 July (2000, 2 at Wister; G. McCaskie). Generally only one bird is found, but six males were seen at Wister 20 June 1981 (G. McCaskie).

Fall migrants are seldom reported, their detection hindered by males' being in cryptic eclipse plumage; females are always difficult to identify. Still, the Blue-winged Teal is apparently uncommon from 1 August (2001, south end; G. McCaskie) into November, with a peak from late September through October and a high count of 20 at the south end from 14 to 16 October 2002 (J. L. Dunn). A male taken along the New River 16 October 1899 (F. Stephens; SDNHM 88) predated formation of the Salton Sea. The species is regular in winter, mainly in the vicinity of the south end (particularly Red Hill and Salton Sea NWR). Records extend from 26 November (1978, male at the Imperial Warm Water Fish Hatchery; G. McCaskie) to 11 March (1978, male at the New River from 18 February; G. McCaskie). Only 2–3 birds are usual at that season, but 15 were seen "around the Salton Sea" 10 December 1990 (*AB* 45:320). A slight increase beginning in late January (e.g., 28 January 1989, male at El Centro; *AB* 43:365) suggests that some "winter" birds are early spring transients. Such records coincide with the onset of spring migration of the Cinnamon Teal, although, as in fall, the Blue-winged appears to migrate slightly later than the Cinnamon.

ECOLOGY. Almost all Blue-wingeds are found among large flocks of the Cinnamon Teal; indeed, they are rarely found in the absence of their sister species, although some pure flocks have been noted. These teals typically occupy shallow freshwater habitats, mainly freshwater impoundments, ponds, and marshes fringing the Salton Sea.

TAXONOMY. This teal is monotypic, as *A. d. orphna* Stewart and Aldrich, 1956, cannot be diagnosed (Palmer 1976a).

CINNAMON TEAL
Anas cyanoptera septentrionalium
Snyder and Lumsden, 1951

Common transient (late January through April, mid-July through October); uncommon breeder and winter visitor. A characteristic duck of western wetlands, the Cinnamon Teal is common in the Southwest. It is an early migrant. Spring movement begins in mid-January (e.g., south end 18 January 1998; M. A. Patten), reaches a peak from late February through March (e.g., 1,000 at Salton Sea NWR 28 February 1949; *AFN* 3:185), and tails off by early May. Exact arrival and departure dates cannot be determined because of the presence of wintering birds and a large number of breeders (see below). Fall movements are also early, the first birds appearing in early July (e.g., 100 at mouth of the Whitewater River 9 July 1989; M. A. Patten). Fall migration peaks from late July through September (e.g., 1,000 around the sea 23 July 1988 and 21 August 1988; M. A. Patten and G. McCaskie, respectively). Movement tails off in October, with the species mostly gone by 15 November (*AFN* 4:34). Winter numbers are generally low (fewer than 100 birds), but 290 were recorded at Salton Sea NWR 1 December 1948 (*AFN* 3:185). Paralleling its migratory habits, this species also breeds early; for example, a nest in a canal near Imperial Valley College already contained one egg by 28 March (1998; E. R. Caldwell). Young are out of the nest by early summer; for example, females with ducklings were seen at the south end 2 June 1990 (G. McCaskie) and at Finney Lake 16 June 1991 (M. A. Patten).

Hybrid male Cinnamon × Blue-winged Teals were at the south end 12 March 1954 (*AFN* 8: 269), 12 June 1976 (*AB* 30:1003), and 28 April 1984 (G. McCaskie). Even though this combination is fairly common for a hybrid (Bolen 1978; Cooper and Graham 1985), these three records are the only ones for the Salton Sink, although it is probably overlooked or unreported.

ECOLOGY. Breeding Cinnamon Teals require freshwater marshes, although they occupy a wide variety of locales fitting this broad description, ranging from bulrush marshes at Wister to reed-lined ditches in the Imperial Valley to cattail-strewn fringes of Finney and Ramer Lakes. Wintering birds have a penchant for the river channels, irrigation ditches, small ponds, and marshes with dense cover at the edge. Migrants are more catholic but frequently reach peak numbers on large freshwater impound-

ments at the edge of the Salton Sea. This species is almost never seen on open waters of the sea.

TAXONOMY. North American Cinnamon Teals are *A. c. septentrionalium*, smaller, less rufescent, and with the flanks less spotted than in the four South American subspecies.

NORTHERN SHOVELER
Anas clypeata Linnaeus, 1758
Common winter visitor (mid-July to mid-May); rare to uncommon in summer. Among the dabbling ducks in the Salton Sink only the Northern Pintail is nearly as abundant as the Northern Shoveler, which generally holds true over much of the Southwest. Winter records extend from mid-July (e.g., 50 around the sea 16 July 1977; G. McCaskie) to mid-May (e.g., 10 at mouth of the Whitewater River 23 May 1999; M. A. Patten), with the largest numbers of birds present from mid-August through April. This species has increased tremendously in the region over the past half-century (see Tables 7 and 8), now making it, with sharp declines of the pintail, the most numerous dabbling duck wintering in the Salton Sink. The maximum count was 4,600 individuals 2 October 1949 (*AFN* 4:35), but counts in the tens of thousands have become routine, such as 12,000 near Wister 21 November 1992 (M. A. Patten) and 13,264 around the sea 18 November 1999 (Shuford et al. 2000). The Northern Shoveler is now the most numerous summering *Anas* duck at the Salton Sea, with typical counts of a few dozen birds in summer, particularly around Wister. Despite the prevalence of summering birds, there is no evidence of breeding.

ECOLOGY. The Northern Shoveler is the most numerous dabbling duck along and close to the shore of the Salton Sea. Most shovelers use shallow fresh or brackish water, particularly freshwater impoundments, ponds, and marshes bordering the sea. They forage by straining food through their long bill, which is armed with comblike filters. Unlike the Northern Pintail, the American Wigeon, and the Green-winged Teal, the shoveler is scarce in flooded fields with grain stubble.

NORTHERN PINTAIL

Anas acuta Linnaeus, 1758

Common winter visitor (mid-July to May); rare in summer (has bred). The Northern Pintail remains common across North America, despite suffering rangewide declines (Banks and Springer 1994). Formerly, this species was easily the most numerous dabbling duck wintering in the Salton Sink, occurring in numbers that were sometimes astounding (e.g., 78,000 at Salton Sea NWR 22 December 1948; *AFN* 3:185). That situation has changed in the past several decades, however, as the number of pintails has decreased and the number of shovelers has increased (see Table 7). Even so, the pintail remains a common winter visitor from early July (e.g., 150 around the sea 8 July 1972; G. McCaskie) to early May (e.g., female at the Whitewater River delta 6 May 1984; SBCM 39387), with most present from late August to mid-April. Small numbers (at most 15) of nonbreeders regularly summer at the Salton Sea, particularly around the mouths of the Whitewater and Alamo Rivers and along the shoreline south of Wister. During that season its numbers are roughly equivalent to those of the Green-winged Teal, but the pintail is less numerous than the Northern Shoveler. Two pairs reportedly bred at the south end in summer 1962 (*AFN* 16:447), and it was reported as a breeder at the south end by Garrett and Dunn in 1981, but we have been unable to verify nesting records, and there have been no documented breeding records since about 1980.

In addition to the hybrid American Wigeon × Northern Pintail cited above (see the American Wigeon account), a male Northern Pintail × Green-winged Teal was collected at Wister in December 1974 (J. Dawson; HSU 3345).

ECOLOGY. The Northern Pintail is generally outnumbered by the Northern Shoveler along the edge of the Salton Sea, but it is the most numerous duck wintering in freshwater marshes (e.g., Wister, Salton Sea NWR, duck clubs) scattered across the Imperial Valley. Some occur around the river mouths and on lakes and reservoirs in the region and forage in partly flooded stubble fields (Kalmbach 1934). This species is one of the *Anas* ducks most likely encountered on the Salton Sea, along with the Gadwall, the American Wigeon, and Northern Shoveler.

TAXONOMY. Although various subspecies have been described, and although North American birds have been claimed to be larger than those in the Old World, this species is monotypic (Palmer 1976a:445).

BAIKAL TEAL

Anas formosa Georgi, 1775

One winter record. A male Baikal Teal, a rare Siberian species casual in North America, was collected 8 km west of Niland 29 December 1946 (Laughlin 1947a; MVZ 97120). It represents one of only five valid records for California (Morlan 1985; Pyle and McCaskie 1992) and one of only a few for the continent outside of Alaska.

GREEN-WINGED TEAL

Anas crecca Linnaeus, 1758 subspp.

Common winter visitor (August to mid-May); rare in summer. North America's smallest dabbling duck, the Green-winged Teal is a common winter visitor to the Southwest, including the Salton Sea. Within the genus *Anas* it is outnumbered in the region only by the Northern Pintail, the Northern Shoveler, and the American Wigeon (see Table 7). Wintering birds occur from early August (e.g., 100 around the sea 6 August 1977; G. McCaskie) to mid-May (e.g., male near Niland 20 May 1978; SDNHM 40743), but mostly from late August through April, when thousands are present (Shuford et al. 2000). Small numbers of nonbreeders (at most 20 birds; e.g., *AB* 41: 1487) regularly summer at the Salton Sea, mainly at the Whitewater River delta and at Wister and vicinity. See the Northern Pintail account and the Taxonomy section below for information on hybrids recorded in the region.

ECOLOGY. This species is found at all locales that harbor dabbling ducks, especially freshwater marshes with shallow water. A few occur on the Salton Sea at the mouths of rivers, particularly the Whitewater and Alamo Rivers. It often forms large, mixed flocks with other species of *Anas*.

TAXONOMY. The account above refers to the North American subspecies *A. c. carolinensis* Gmelin, 1789, the males of which are easily identified by the vertical white bar on the sides of the breast. There are two regional records of male Common Teals, *A. c. crecca* Linnaeus, 1758, the Old World form with a horizontal white stripe along the sides and bolder tertial fringes, which was recently accorded species status by the British Ornithologists' Union's Records Committee (BOU 2001). A bird seen at the south end 17 April 1989 (*AB* 43:536) probably wintered locally, although records are most numerous in spring in northwestern California (Harris 1991), suggesting limited spring passage through the state. One at the north end 3–27 June 1987 (*AB* 41:1487) molted into eclipse plumage and thus could not be identified beyond 27 June, but it presumably remained for the entire summer. Even though the subspecies is recorded annually in winter, there is but one specimen of a male *A. c. crecca* for California (Harris and Gerstenberg 1970; HSU 6109). The larger but otherwise similar *A. c. nimia* Friedmann, 1948, from the Aleutian Islands, reportedly has been collected in Oregon (Gilligan et al. 1994), but this subspecies is probably best treated as a synonym of *A. c. crecca* (Gibson and Kessel 1997). A male Green-winged Teal at the south end 14 January 1987 was identified as a hybrid *A. c. crecca* × *A. c. carolinensis* (*AB* 41:328).

CANVASBACK
Aythya valisineria (Wilson, 1814)
Formerly common, now uncommon winter visitor (mid-October through April); rare in summer. The Canvasback has declined at the Salton Sea, as it has at many locales across North America (Miles and Ohlendorf 1993). It was formerly a common winter visitor; for example, 2,500 were counted at the south end 10 February 1977 (*AB* 31:373). Counts since the mid-1980s have seldom reached triple digits, although 500 were recorded at the south end 29 November 1986

(G. McCaskie) and 379 were censused around the sea 18 November 1999 (Shuford et al. 2000). Records extend from 7 October (1955, male 6.5 km south of Mecca; E. N. Harrison, WFVZ 1495) to 28 April (1990, south end; M. A. Patten). Four at the mouth of the Whitewater River mouth 7 May 1977 (G. McCaskie) were exceptionally late. The Salton Sea is the only locale in southern California where this species is regular during summer. Still, it is rare in that season from 12 May (1984, 2 at mouth of the Whitewater River until at least 10 June; G. McCaskie) to 2 September (1979, mouth of the Whitewater River from 19 August; *AB* 33:896). Generally only one or two birds summer, but seven were at the north end in June–July 1981 (*AB* 35:978), and ten were there 20 June 1987 (M. A. Patten).

ECOLOGY. The vast majority of Canvasbacks are encountered on the Salton Sea, with congregations forming around the mouths of the three rivers. A few occur on small lakes and reservoirs in the region. Small flocks on the sea regularly mix with other *Aythya* ducks, particularly scaup and Redheads.

REDHEAD
Aythya americana (Eyton, 1838)
Fairly common breeding resident. The Redhead is the most numerous breeding diving duck in California, barely edging out the Ruddy Duck for that claim (Saunders and Saunders 1981). This scenario is reversed at the Salton Sea, with the Redhead slightly less numerous as a breeder, yet it remains one of the most common breeding ducks in the region. Seasonal movements are apparently limited, with the majority of Salton Sea birds hatched locally or in northern California (Rienecker 1968). Numbers are slightly higher in summer (mid-April to early October) at the Salton Sea, but not significantly so (*contra* Garrett and Dunn 1981:45). For example, 350 tallied around the Salton Sea 29 and 30 June 1985 (*AB* 39:962) provided a high count for summer, yet the mean winter count from 1978 to 1987 for the Coachella and Imperial Valleys was 336 (Heitmeyer et al. 1989). Breeding takes

place in late spring, with ducklings a common sight by midsummer, such as eight at the south end 25 June 1988 and five at Ramer Lake 16 July 1994 (both M. A. Patten).

ECOLOGY. Breeding Redheads occupy freshwater habitats with dense cover of Southern Cattail or Common Reed at the margins, as around Wister, Salton Sea NWR, Finney and Ramer Lakes, and Fig Lagoon. Some even nest in drainage canals with slow-moving water and adequate cover. Wintering birds often move onto the Salton Sea, but many remain on larger bodies of fresh water in the valleys.

RING-NECKED DUCK
Aythya collaris (Donovan, 1809)

Uncommon winter visitor (late October through April); 5 summer records. Elsewhere in the California desert the Ring-necked Duck is one of the species of *Aythya* encountered more frequently, but at the Salton Sea it is comparatively less frequent. Following the first area records near Mecca from 11 November to 20 December in the early 1930s (van Rossem 1933b), this species became an uncommon winter visitor, with small numbers present annually from 6 October (2000, 7 at Sheldon Reservoir; G. McCaskie) to 27 April (1996, near Oasis; G. McCaskie). Double-digit counts are unusual. There are five records after mid-May, of an adult male near Oasis 26 May 1996 (G. McCaskie) and individuals at the north end about 30 May 1980 (McKernan et al. 1984), at the mouth of the Whitewater River 3 June 1987 (G. McCaskie), at an unspecified Salton Sea location 7 June 1969 (Garrett and Dunn 1981), and at Ramer Lake 15 July 1995 (G. McCaskie).

ECOLOGY. Most records are from fresh water, typically catfish ponds and other fish-hatchery ponds skirting the edge of the Salton Sea. The Ring-necked Duck also frequents reservoirs and lakes in the Imperial Valley. It is seldom found on the Salton Sea itself, which perhaps explains its rarity in the region in comparison with other species of *Aythya* and its status elsewhere in the desert. It often forms pure flocks, even when other diving ducks are nearby.

TUFTED DUCK
Aythya fuligula (Linnaeus, 1758)

One winter record. The Tufted Duck, an Old World species, is a rare, regular winter visitor to the Pacific states, with most records from the coastal slope north of Los Angeles. A female photographed at the Imperial Warm Water Fish Hatchery 1–22 February 1986 (Roberson 1993) furnished the only record for southeastern California, although the species has reached Arizona (Rosenberg and Witzeman 1998). A male reported at Red Hill (*FN* 50:222) lacks documentation.

GREATER SCAUP
Aythya marila nearctica (Stejneger, 1885)

Rare, occasionally uncommon winter visitor (late October to mid-April); casual in summer. Normally a coastal species, the Greater Scaup is generally greatly outnumbered inland by its sister species, the Lesser Scaup. This observation tends to hold at the Salton Sea as well, but the situation there is sometimes reversed, presumably as a result of the extensive saline habitat. Winter records extend from 24 October (1977, 26 at the south end through 25 October; *AB* 32:257) to 27 April (1963, mouth of the Whitewater River; G. McCaskie), with most departing by late March. High counts, from the south end, include more than 100 on 19 January 1983 (*AB* 37:338) and 18 December 1986 (G. McCaskie) and more than 150 on 1 February 1986 (*AB* 40:333), but usually fewer than 10 are found. This species is casual in summer from 8 May (1999, female at mouth of the Whitewater River; M. A. Patten) to 31 August (1989, south end; *AB* 43:1367), including a male from the mouth of the Whitewater River 16 July 1977 (skeleton, SDNHM 40451). Summer records have involved only one to two individuals, typically of apparent immatures.

ECOLOGY. This scaup is found principally on the open Salton Sea, generally concentrating in protected bays. It is only rarely found on fresh water, but the first regional record, of a female collected 1.5 km northeast of Cerro Prieto 7 February

1928 (Grinnell 1928; MVZ 52068), is from such a setting.

TAXONOMY. North American birds are *A. m. nearctica,* which includes *A. m. mariloides* (Vigors, 1839) as a synonym (Banks 1986b) and differs from the nominate subspecies, of the Old World, in the coarser barring on the upperparts of the male and less extensive white on the primaries.

LESSER SCAUP
Aythya affinis (Eyton, 1838)
Fairly common winter visitor (mid-October through April); rare in summer. At the Salton Sea, as in most of the Southwest, the Lesser Scaup is the most numerous species of *Aythya.* It is outnumbered only by the Ruddy Duck among the various diving ducks (see Table 7). The scaup is a fairly common to common winter visitor from 13 October (2000, male at Sheldon Reservoir; G. McCaskie) to 16 May (1999, 25 near Mecca; M. A. Patten). Most are present from early November to mid-April, with up to 100 in the southwest portion of the Salton Sea 30 April 1949 (*AFN* 3:224) providing an impressive late concentration. Each year a few (at most 10) nonbreeders summer on the Salton Sea, although in the genus *Aythya* the Lesser Scaup is greatly outnumbered by the Redhead, a local breeder, in that season. The Lesser Scaup typically outnumbers the Greater Scaup, even though the Greater is more numerous in the Salton Sink than anywhere else in the interior Southwest (see the Greater Scaup account).

ECOLOGY. Being a diving duck, this species is found only on the deeper water of the Salton Sea and on large freshwater bodies in the valleys, both lakes (e.g., Finney Lake, Ramer Lake, Fig Lagoon) and reservoirs (e.g., Sheldon Reservoir). At the sea it is most numerous around river mouths and in protected bays.

SURF SCOTER
Melanitta perspicillata (Linnaeus, 1758)
Rare (to uncommon) spring transient (mid-March through May); rare fall transient (October–November) and winter visitor; rare but annual in summer. A sea duck that winters commonly in the Gulf of California, the Surf Scoter is rare in the interior of North America, including at the Salton Sea. The substantial numbers of this species in the gulf led Grinnell (1928) to ponder its migratory route in spring. We now know that it moves overland across southeastern California. Although there are records for every month of the year, the vast majority of Surf Scoter records for the Salton Sea are during spring, from 6 March (1982, Salton Sea NWR; *AB* 36:330) to early May, with stragglers to 5 June (1983, 2 at the north end until 25 June; *AB* 37:1027). Small congregations are frequently found on the sea during spring, such as up to 25 at the north end 14 April 1990 (*AB* 44:496), 17 at various locations around the Salton Sea 15 April 1989 (*AB* 43:536), and 15 at the sea 30 May 1998 (*FN* 52:390). Most spring records come from the north end, especially the vicinity of the Whitewater River delta. Many migrants that arrive after late April attempt to summer, such as up to 3 at the north end 14 May–1 August 1988 (*AB* 42:1340) and up to 3 at Salton City 29 May–4 July 1998 (*FN* 52: 503), although summering birds have arrived as early as 27 April (1999, at most 4 near Mecca into September; B. Mulrooney, M. A. Patten). In a typical year only 1–4 birds attempt to summer, but 25 at the north end in early June 1990 had declined to 10 by July (*AB* 44:1186), 16 were at the north end 25 June 1989 (*AB* 43:1266) with 5 surviving well into July, up to 12 were on the Salton Sea May–July 1977 (*AB* 31:1189), and 12 were at the north end through June and July 1998 (*FN* 52:503). Known summering birds have been found as late as 30 September (1990, extremely worn male at mouth of the Whitewater River; M. A. Patten), but by July most succumb owing to the lack of proper food and high temperatures.

This scoter is annual in fall and winter, but in much smaller numbers. Fall records extend from 7 October (1989, 7 at the north end; *AB* 44:161) to 26 November (1988, North Shore from 20 November; *AB* 43:167) and include a specimen from the north end 8 November 1990 (SBCM 52491). Generally only 1–3 birds are noted, but a remarkable 50 were at North Shore 16 November 2002 (T. Benson). The next largest group recorded in that season was 8–10 at the north end 27 October 1934 (Clary and Clary, *Surf Scoters*, 1935). Winter records are scant, although the first Surf Scoter recorded at the Salton Sea occurred in that season, when a freshly dead male was found near Mecca 25 December 1934 (Clary and Clary, *Surf Scoters*, 1935; LACM 86749). Definite winter records extend from 25 November (2000, female near Obsidian Butte through 19 December; *NAB* 55:102) to 21 February (1998, male and female together at Obsidian Butte; *FN* 52:257). Some birds recorded in mid- to late November could have wintered. One or two birds are the rule in winter, but there are records of 10 on the Salton Sea 4 January 1954 (*AFN* 8:269), 9–10 at Salton City during the period from late December to early January 1971 (*AB* 26:655), 8 at the south end during the winter of 1989–90 (*AB* 44:328), up to 6 at the north end in the winter of 1996–97 (*FN* 51:801), and 5 at the north end 19 December 1982 (*AB* 37:771).

ECOLOGY. Not surprisingly, given its saltwater affinities, when in the region the Surf Scoter usually occurs on the Salton Sea. Most are found near shore in bays and backwaters with submerged snags, whose bases are typically covered with barnacles, on which scoters often feed. Records from freshwater ponds and lakes are of an immature male at Ramer Lake 27 October 1949 (SDNHM 29852), one at El Centro 15 November 1986 (*AB* 41:143), 5 males together at Wiest Lake 17 November 1986 (*AB* 41:143), a female at Ramer Lake 29 November 1966 (*AFN* 21:77), a male on a catfish pond near Mecca 15 December 1991 (*AB* 46:997), 14 at Fig Lagoon 28 March 1998 (E. R. Caldwell), and an immature male at Sheldon Reservoir 23 April 1988 (R. Higson; SDNHM 45026).

WHITE-WINGED SCOTER
Melanitta fusca deglandi (Bonaparte, 1850)
Rare spring transient (mid-March to June); rarer still in fall (late October through November), winter (early December to early March), and summer.
Like the Surf Scoter, the White-winged Scoter is rare in the interior of North America, including at the Salton Sea, where spring records predom-

inate, though there are records for every month. The Surf Scoter is distinctly the more frequent of the two species inland in California and the Southwest; their respective status at the Salton Sea is no different. Spring records of the White-winged extend from 3 March (1957, 4 at the north end; *AFN* 11:290) to early May, with stragglers to 6 June (1991, south end; *AB* 45:1161). Generally one or two birds are found, a group of nine at the mouth of the Whitewater River beginning 14 April 1990 (*AB* 44:496) being the largest spring concentration. Like the Surf Scoter, some White-wingeds stay into summer after arriving in spring (e.g., 1–2 females at the north end 15 April–19 August 1989; *AB* 43:536, G. McCaskie). Similarly, most succumb to the harsh conditions by July, although a male near Mecca 2 October 1999 (*NAB* 54:104) was likely one of at most 5 present from 25 April (*NAB* 53:328). Numbers steadily decrease through the summer; for example, 9 at the mouth of the Whitewater River in early July 1990 declined to 3 by late July (*AB* 44:1186) and but one by 3 September (G. McCaskie). An immature male found dead at the north end 6 July (not 29 June) 1996 (*FN* 50:996; SDNHM 49555) lends support to this idea.

The species is rare to casual in fall and winter, but it is recorded in these seasons nearly annually. Fall records, extending from 22 October (1970, unspecified locale on the Salton Sea; *AB* 25:108) to 25 November (1989, 2 at Red Hill; *AB* 44:161), likely represent migrants heading south to wintering grounds in the Gulf of California. True wintering birds have occurred from 9 December (1994, north end; *FN* 49:196) to 1 March (1992, south end; *AB* 46:480), with four at the south end 24 January 1970 (*AFN* 24:539) being the largest flock. A flock of 19 reportedly seen at the center of the Salton Sea during an aerial survey 25 February 1989 (*AB* 43:366) is questionable considering the distance of observation and the date.

ECOLOGY. Virtually all regional records of this scoter are from the Salton Sea, with the Whitewater River delta and environs hosting the majority. Like other scoters, it most often frequents bays and backwaters with numerous submerged snags.

TAXONOMY. The North American subspecies *M. f. deglandi*, with *M. f. dixoni* (Brooks, 1915) a synonym (Palmer 1976b), is the only one recorded in California. Together with *M. f. stejnegeri* (Ridgway, 1887), of Siberia, this taxon is sometimes treated as a species (the White-winged Scoter) distinct from the nominate, *M. f. fusca* (Linnaeus, 1758), of the Western Palearctic (the Velvet Scoter). Nominate males differ from both *M. f. deglandi* and *M. f. stejnegeri* in nearly lacking a black knob at the base of the yellower bill. The male *M. f. deglandi* differs from the male *M. f. stejnegeri* in having a smaller knob on the bill and browner flanks.

BLACK SCOTER

Melanitta nigra americana (Swainson, 1832)

Casual vagrant in winter (late November through February) and summer (mid-April through July); 1 fall record. The Black Scoter is by far the rarest of the three scoters in the interior of western North America, including at the Salton Sea, where there are 16 records involving 21 individuals. Curiously, this species shows little similarity to the Surf and White-winged Scoters in its pattern of occurrence. For instance, there is only one spring record per se, from near Calipatria 25 April 2001 (*NAB* 55:355), but there are seven summer records, five of which involved birds that arrived in spring and stayed to at least mid-July: an immature male at Desert Shores 13 April–12 July 1990 (*AB* 44:496, 1186); a female at the Whitewater River delta 14 April–13 July 1990 (*AB* 44:496, 1186); immature males near Mecca 28 April–24 July 1999 (*NAB* 53:328, 432) and 4 June–24 July 1999 (*NAB* 53:432); up to five at the north end 1 May–19 September 1977 (*AB* 31:1189, 32:257), an adult female of which was found dead 3 September (UCLA 40669); an immature male at the Whitewater River delta 4 May–13 July 1996 (*FN* 50:332, 996); and an adult male photographed near the mouth of the Alamo River 14 August 1976 (*AB* 30:1003).

Oddly, there is only one record from fall, of two at North Shore 16 November 2002 (T. Ben-

son). The remaining eight records are of wintering birds between 25 November (2000, up to 2 photographed together near Obsidian Butte through 19 December; *NAB* 55:227) and 27 February (1977, adult male at the south end from 9 February; *AB* 31:373). Additional winter records are of one at the south end 9 December 1987 (*AB* 42:321), two females near Oasis 16–18 December 1993 (*FN* 48:247), a female at Rock Hill 16 December 1993–22 January 1994 (*FN* 48:247), one at the north end 31 December 1994 (*FN* 49: 196), and immature males at Oasis 10 January 2001 (S. Glover), at the south end 29 January–11 February 1977 (*AB* 31:373), and at Bombay Beach 13 February 1994 (*FN* 48:247).

ECOLOGY. All regional records of the Black Scoter are from the Salton Sea, principally in areas with numerous exposed Saltcedar snags.

TAXONOMY. Like *M. fusca deglandi*, the North American subspecies of the Black Scoter, *M. n. americana*, is sometimes treated as a species distinct from the nominate, *M. n. nigra* (Linnaeus, 1758), of the Old World (the Common Scoter). The bill pattern of adult males differs considerably, *M. n. americana* having a large, extensively orange-yellow knob at the base of the bill, whereas *M. n. nigra* has a smaller, mostly black knob with the little yellow on the bill confined to the saddle of the maxilla.

LONG-TAILED DUCK

Clangula hyemalis (Linnaeus, 1758)

Casual spring transient (mid-March to early June), winter visitor (mid-December through February), and fall vagrant (November); 1 summer record. Like the three species of scoters, the Long-tailed Duck is a sea duck that is rare in the interior of North America. It too moves through the Salton Sea region en route to and from the Gulf of California. It also winters rarely on the sea; indeed, it is nearly annual in that season, and most regional records are from winter. In contrast to the scoters, there is but one summer record. Across all seasons the Long-tailed Duck is nearly annual at the Salton Sea (recorded in 9 of 10 years on average), with one to three individuals in years of occurrence. It is recorded less frequently than the Surf or White-winged Scoter (much less than the former) but more frequently than the Black Scoter. Lone individuals account for nearly all records, but as many as three have been seen together.

Spring records extend from 16 March (1967, south end; *AFN* 21:457) to 9 June (1995, female at the south end from 20 May; *FN* 49:309, 980), with an early female photographed at Rock Hill 6 March 1982 (R. E. Webster). One at the south end 9 July 1967 (*AFN* 21:604) provided the only summer record. There are only eight fall records, from Red Hill 31 October 1971 (not 1972, *contra* Garrett and Dunn 1981; *AB* 26:120), 16 November 1950 (Small 1959, *AFN* 5:227), 23 November 1995 (*FN* 50:114), 25 November 1989 (*AB* 44: 161), and 1 December 2002 (G. McCaskie); at the south end 25 November 2000 (*NAB* 55:102); near Niland in November 1957 (E. A. Cardiff; SBCM 31425); and at Salton City 13 (not 15) November 2000 (*NAB* 55:102). Winter records extend from 16 December (1950, immature male 11 km northwest of Calipatria; W. C. Russell, MVZ 122600) to 24 February (1990, 2 at Red Hill from 11 February; *AB* 44:328). The first Salton Sea record is from winter, of a female at Desert Beach (north end) 22 February 1948 (*AFN* 2:149). Two at the south end 13 February–17 April 1982 and three 3 April (*AB* 36:893) were likely lingering winter birds.

ECOLOGY. This duck occurs in the same settings as the scoters, although it has perhaps a slightly greater tendency to occur on freshwater reservoirs.

BUFFLEHEAD
Bucephala albeola (Linnaeus, 1758)
Uncommon winter visitor (November to mid-April); casual in summer (has bred). Although the Bufflehead is a widespread winter visitor to the Southwest, it is one of the latest ducks to arrive. Winter records extend from 2 November (1993, 12 at the south end; M. A. Patten) to mid-April, with stragglers to 5 May (1937, female 12 km northwest of Calipatria; MVZ 71343). It is never numerous. Counts are typically in the single digits, but 50 were seen at the south end 5 March 1994 (G. McCaskie). Some late spring birds attempt

to summer, when the species is casual at the Salton Sea. Summering birds occur once every three to four years from 7 June (1987, south end; *AB* 41:1487) to 12 July (1980, south end; *AB* 34: 930). Amazingly, the Bufflehead has bred once at the Salton Sea: a female and two half-grown ducklings were photographed at the mouth of the Whitewater River 1 May 1999 (*NAB* 53:432). This species normally does not breed in southern California; the breeding record from the Salton Sea is the southernmost for the state (San Miguel 1998).

ECOLOGY. Most wintering Buffleheads occur on the Salton Sea, especially around the river mouths and in protected bays. A few appear on freshwater lakes and reservoirs. We do not know what sort of nest cavity the breeders commandeered, for it appears that few suitable sites are available in the region.

COMMON GOLDENEYE
Bucephala clangula americana (Bonaparte, 1850)
Uncommon winter visitor (mid-November to mid-April); casual in summer. The Common Goldeneye was once common at the Salton Sea. For example, 1,200 were reported in late December 1954 (*AFN* 9:286) and in December 1956 (*AFN* 11:29), and an impressive 4,630 were recorded in January 1958 (*AFN* 12:306). The species remained fairly common into the mid-1980s (e.g., 75 at the south end 1 February 1986; G. McCaskie), but it is now uncommon, with fewer than 20 being a typical high count. Records extend from 11 November (1934, female near Mecca; Clary and Clary, *American Golden-eye*, 1935; LACM 86776) to 18 April (1999, female near Mecca; M. A. Patten). This species is now casual in late spring and summer, but it was formerly more numerous in those seasons; for example, eight were seen at the north end 11 May 1963 and five were seen at the south end 12 May 1963 (G. McCaskie; Garrett and Dunn 1981). Summer records extend from 8 May (1999, immature male at Oasis; M. A. Patten) to 1 September (2000, female at Cerro Prieto; Patten et al. 2001). As with the Canada Goose and perhaps the Canvasback, the wintering population of the

Common Goldeneye appears to have shifted north, away from the Salton Sink. In the case of the goldeneye, the wintering population may have shifted east as well, for the species has increased substantially as a wintering bird on the lower Colorado River since 1960 (Rosenberg et al. 1991).

ECOLOGY. Goldeneyes tend to congregate with other diving ducks at river mouths or in bays or inlets near them. Small flocks are sometimes found on freshwater reservoirs or lakes in the Imperial Valley; for example, 11 were found at Brawley 6 December 1958 (Stott and Sams 1959).

TAXONOMY. North American Common Goldeneyes are the small subspecies *B. c. americana.*

BARROW'S GOLDENEYE
Buchephala islandica (Gmelin, 1789)
Casual winter vagrant (December–January). Barrow's Goldeneye is an uncommon and local winter visitor to California, with a small wintering population at Parker Dam on the Colorado River being the only regular one in southern California. It has been recorded only six times at the Salton Sea, from 1 December (1979, female at Salt Creek; *AB* 34:306) to 26 January (1985, south end; *AB* 39:210). The other four records are of one at the mouth of the Whitewater River 10 December 1978 (*AB* 33:312), a female shot by a hunter (specimen not preserved) at Red Hill 20 December 1992 (*AB* 47:300), one at Salton City 21 January 1979 (*AB* 33:312), and a female at the mouth of the Alamo River 24 January 1987 (*AB* 41:328). One reported along the "west side" of the sea (*AFN* 9:56) lacks documentation.

ECOLOGY. Most Barrow's Goldeneyes have been seen at river mouths or in bays near them, and all have been with small flocks of Common Goldeneyes.

HOODED MERGANSER
Lophodytes cucullatus (Linnaeus, 1758)
Rare winter visitor (mid-November to mid-February); 1 record each for spring and summer. The Hooded Merganser is uncommon in winter in California and rare in that season in the desert

Southwest. It is a rare but nearly annual winter visitor to the Salton Sea region, first noted when a female was found dead near Mecca 27 November 1928 (Stevenson 1929). Birds in all plumages have been noted since then, and the species has even reached the Río Colorado delta (Patten et al. 1993). Records extend from 13 November (1978, Niland; *AB* 33:213) to 19 February (1966, near Calipatria; *AFN* 20:459) and include a specimen of a female from the southeastern part of the Salton Sea 20 January 1973 (SBCM 30395). Two unseasonal records are of birds at Salt Creek 6 May 1995 (*FN* 49:309) and at the Whitewater River delta 25 August 1990 (*AB* 45:151), the latter representing perhaps the first true summer record for southern California (Garrett and Dunn 1981).

ECOLOGY. Hooded Merganser records come from freshwater ponds and impoundments, but a couple have been seen on the Salton Sea itself (at river or creek mouths).

COMMON MERGANSER
Mergus merganser americanus Cassin, 1852
Rare winter visitor (mid-November to mid-March); casual in late spring and summer. The Common Merganser does not live up to its moniker at the Salton Sea. Given that this species is a locally common winter visitor to the interior of southern California and elsewhere in the Southwest, it defies expectation that it is so rare at the Salton Sea. Indeed, the Red-breasted is the expected merganser at the Salton Sea, regardless of season. The Common Merganser, by contrast, is rare, especially on the sea itself. Records extend from 26 November (1933, south end; LACM 86781) to 13 March (1993, near Seeley; S. von Werlhof), with most in December, including two additional specimens from that month (Clary and Clary, *American Golden-eye*, 1935). Inexplicable was a concentration of 200 birds at Cerro Prieto 12 November 1995 (Patten et al. 2001); large numbers winter on the lower Colorado River, although relatively few occur south of Lake Havasu (Rosenberg et al. 1991). The largest flock on record for the Salton Sea is of up to 20 at Salt Creek 12 January–9 February 1980 (G. McCaskie). A female at Ramer Lake 12 April

1987 (G. McCaskie) was either a late wintering bird or, more likely, a spring migrant. The Common Merganser is casual in late spring and summer in the Salton Sink, with only seven records from 22 April (1995, Ramer Lake; S. von Werlhof) to 3 August (1968, adult male near Red Hill (*AFN* 22:648). A female at the mouth of the Whitewater River 3 May 1980 (G. McCaskie), one at Red Hill 5 June 1964 (G. McCaskie), a female near Mecca 8 May 1999 (M. A. Patten), a first-spring male photographed at Oasis 20 June 1999 (M. A. Patten), and a male and female together at Oasis 13 June 2001 (B. Miller) provided the only other late records. We have been unable to verify a report from the north end for "mid-August to late September" 1977 (Garrett and Dunn 1981).

ECOLOGY. Almost all records of this species are from freshwater reservoirs, aquafarms, and lakes in the Coachella, Imperial, and Mexicali Valleys. Virtually all mergansers noted on the Salton Sea prove to be the Red-breasted, although that species is found in freshwater settings in the region as frequently (or more so) as the Common. Conversely, the Common occurs on the sea occasionally, especially near river mouths and other inlets.

TAXONOMY. Common Mergansers of the Western Hemisphere are the smaller subspecies *M. m. americanus*.

RED-BREASTED MERGANSER
Mergus serrator Linnaeus, 1758
Uncommon spring transient (early March to early May) and summer visitor; rare fall transient (mid-October to mid-November) and winter visitor. Given that the Red-breasted Merganser is the most coastal of the three North American mergansers, it is not surprising that this species is generally the rarest one inland. As is often the case with coastal species, however, such expectations are not met at the Salton Sea; instead, this species is by far the most numerous and the most frequently encountered of the three mergansers. It is an uncommon spring transient from 4 March (1961, 6 at Ramer Lake; G. McCaskie) to 9 May (1965, 15 at Ramer Lake; G. McCaskie), with a peak in the first half of

April; for example, 60 were seen at the mouth of the Whitewater River 1 April 1978 (G. McCaskie), and 50 were seen there 14 April 1990 (*AB* 44:496). Sixteen at Cerro Prieto 23 February 1998 (Patten et al. 2001) were probably early spring migrants. Some remain to summer annually, sometimes in large numbers, such as 150 at the north end 6 August 1977 (G. McCaskie) and in summer 1978 (Garrett and Dunn 1981); it is not unusual to achieve double-digit counts in that season, particularly in June. The latest known summering birds were recorded 21 September (1975, 2 at mouth of the Whitewater River; G. McCaskie). This merganser is rare in fall from 13 October (1984, mouth of the Whitewater River; G. McCaskie) to 26 November (1933, south end; G. W. Willett, LACM 86781), with a slight peak from late October to mid-November (e.g., 20 at the south end 31 October 1971; G. McCaskie). A few remain to winter each year, when counts are generally low (fewer than 5 birds), although 40 were seen at Salton City 14 December 1978 (*AB* 33:312). Presumed wintering birds have been recorded as late as 22 February (1963, Ramer Lake; G. McCaskie), although some birds in late February may have been early spring transients (Garrett and Dunn 1981).

ECOLOGY. Most records of this species from the region meet the general dogma of habitat segregation for the Red-breasted and Common Mergansers, the former species on salt water, the latter on fresh. But while it is true that the majority of Red-breasteds are found on the Salton Sea, the species is frequently encountered on lakes, reservoirs, and freshwater ponds in the region, and it is still more expected at these locales than is the Common Merganser.

TAXONOMY. *M. s. schioleri* Salomonsen, 1949, is not valid (Palmer 1976b; Cramp and Simmons 1977), so the species is monotypic.

RUDDY DUCK
Oxyura jamaicensis rubida (Wilson, 1814)
Abundant winter visitor (October to mid-April); fairly common breeder. Each winter the Salton Sea hosts half of the Pacific Flyway's population of the Ruddy Duck (Jehl 1994). In that season

only the Eared Grebe outnumbers this species on the open parts of the Salton Sea. The estimated total number is around 75,000 individuals (see Table 7) and may approach 100,000 in some years. The large number of breeding birds (more than 1,000) prohibits an accurate determination of arrival and departure dates of wintering birds, but most are present from early October to mid-April (Garrett and Dunn 1981). As for breeders, males begin displaying in late February and March. Nesting begins in late March but takes place mainly in April and May. Newly hatched ducklings are prevalent by early summer, from 2 June (1993, 2 at mouth of the Whitewater River; G. McCaskie) to 5 July (1991, 3 at Finney Lake; M. A. Patten).

ECOLOGY. Breeding Ruddy Ducks inhabit dense stands of bulrush and Common Reed and freshwater marshes and river mouths with dense Southern Cattail along their edges. Wintering birds are most numerous on the open water of the Salton Sea, the largest concentrations being within a few hundred meters of shore (particularly near river mouths). They also occur at freshwater ponds and lakes throughout the region.

TAXONOMY. The subspecies of continental North America, *O. j. rubida*, is larger and paler than the nominate subspecies, of the West Indies.

FALCONIFORMES • *Diurnal Birds of Prey*

The Salton Sea provides vast foraging opportunities for fish-eating species such as the Osprey and the Bald Eagle, and its immediate shoreline frequently hosts the Peregrine Falcon. The great valleys of the Salton Sink—the Coachella, the Imperial, and the Mexicali—are important wintering areas for various diurnal raptors (Shuford et al. 2000:tables 5–7) and apparently form a migratory corridor for Swainson's Hawk. An observer searching for diurnal raptors on a routine winter day would record several Ospreys, many Northern Harriers, several *Accipiter* hawks, numerous Red-tailed Hawks, perhaps several other *Buteo* species (usually Red-shouldered and Ferruginous), and even four species of falcons.

With luck, in a good year careful searching could further yield a Bald Eagle, a few White-tailed Kites, and a Rough-legged Hawk. Interestingly, however, with the exception of the American Kestrel, no diurnal raptors breed commonly in the region. Both the White-tailed Kite and the Red-tailed Hawk nest locally in the Salton Sink, the former not annually, the latter in small numbers only in the southern Coachella Valley and perhaps on the western fringe of the Mexicali Valley. As a result, save for the kestrel, nonbreeding visitors such as the Osprey and the Peregrine Falcon are more likely to be seen through summer than are other raptor species.

ACCIPITRIDAE • *Kites, Eagles, Hawks, and Allies*

OSPREY

Pandion haliaetus carolinensis (Gmelin, 1789)

Uncommon perennial visitor; more numerous in winter (late October to mid-March). Given the Osprey's impressive comeback across the continent and its abundance as a breeder in the Gulf of California (Henny and Anderson 1979), the low numbers at the Salton Sea are perhaps surprising. Moreover, there are no nesting records (cf. *AB* 37:1027), although small numbers (4–8 individuals) routinely stay through the summer. With an influx of birds that have bred to the north, numbers are slightly higher during winter, primarily from late October through mid-March. Still, even in that season the overall population in the region is probably never more than about 25 individuals. Apparent migrants have been noted as early as mid-September and as late as mid-April, but the presence of summering birds confounds this determination.

ECOLOGY. The Osprey is typically encountered along the immediate shoreline of the Salton Sea, frequently perched on dead cottonwoods or similar large snags protruding from the water or atop utility poles. It also occurs at freshwater lakes and reservoirs in the Imperial Valley and along rivers and large irrigation canals.

TAXONOMY. This cosmopolitan species varies little geographically. Save for the smaller, whiter *P. h. ridgwayi* Maynard, 1887, of the West Indies, birds in North America are the subspecies *P. h. carolinensis,* differing from the nominate subspecies, of the Old World, in having reduced dark brown markings on the breast.

WHITE-TAILED KITE

Elanus leucurus majusculus Bangs and Penard, 1920

Rare transient and winter visitor (mid-August to early March); has bred multiple times, in increasing frequency, suggesting colonization of the region. The White-tailed Kite is primarily a rare migrant and winter visitor to the Salton Sink, but records are scattered throughout the year, and it has nested on 4–5 occasions in the Imperial Valley. As elsewhere in the West, this species has dramatically increased its numbers and expanded its range in the past three decades (Eisenmann 1971; Gilligan et al. 1994; Rosenberg and Witzeman 1998). It may have colonized the Imperial Valley in the 1990s (see below), as it did the Mexicali Valley sometime during the 1980s (Patten et al. 1993). The vast majority of records are from fall and winter, presumably the result of postbreeding dispersal scattering birds through the region, although records are concentrated in the agricultural areas of the Coachella, Imperial, and Mexicali Valleys. In the Coachella Valley, records extend from 11 August (1976, near Mecca; *AB* 31:222) to 6 March (1965, near Mecca from 13 February; *AFN* 19:417); the 1965 record was the first for the region. Numbers throughout the region decrease after mid-February but build again in April. It is difficult to explain the recent increase in spring sightings, when reports extend from 3 April (1985, Finney Lake; *AB* 39:350) to 3 June (1984, north end; *AB* 38:1061). One near Mecca 3 June–30 July 1984 (*AB* 38:1061) was outside the spring and fall windows and provided the only summer record for the Coachella Valley.

By contrast, recently there has been a spate of nesting records in the Imperial Valley, and the kite may now be resident there. It is certainly more numerous than in the Coachella Valley; for example, there were at least 12–15 individuals in the Imperial Valley during the winter of 1993–94 (M. A. Patten). Furthermore, most recent records for the valley are of pairs rather than lone birds. A pair in Brawley after 28 November 1974 (*AB* 29:120) represented the first record for the Imperial Valley. They stayed through spring 1975 and presumably nested in May, when a juvenile accompanied an adult but was found dead 23 May (*AB* 29:1030). An adult remained in the area until 7 December 1975 (*AB* 30:765). No subsequent breeding was recorded until 1993, when a pair nested near the mouth of the New River during June and a fresh juvenile was seen at Poe Road 2 June (*AB* 47:1149). Another juvenile near the south end 14 August 1994 (*FN* 48:988) suggested local breeding that year, although this bird may have dispersed from a more coastal locale. The species definitely nested near the south end in 1995, when an adult and two recently fledged young were observed 5 July (*FN* 49:980), and again in 1996, when a pair tended a nest in May (*FN* 50:996). Two pairs were suspected of nesting along the New River in the Imperial Valley during late spring and summer 2000, one east of Fig Lagoon, the other southwest of Brawley (G. McCaskie et al.).

ECOLOGY. Kites have nested in large Fremont Cottonwoods. They roost in cottonwoods, eucalyptus trees, and on utility wires and poles, and they forage (often by hovering) over agricultural fields, even ones being cultivated.

TAXONOMY. *E. l. majusculus,* a somewhat larger, longer-tailed form, occurs throughout Central America and north through the Southwest. Given the significant overlap in size, however, this subspecies is perhaps doubtfully distinguishable from the nominate subspecies, of South America (Palmer 1988a:134; Brown and Amadon 1989); a quantitative assessment is needed.

BALD EAGLE

Haliaeetus leucocephalus (Linnaeus, 1766) subspp.

Rare winter visitor (late October to mid-March); 2 fall records; 1 summer record. Populations of the Bald Eagle have recovered throughout North America, moving this majestic species away from the brink of extirpation in the continental

United States (Gerrard 1983; White 1994). Still, numbers in California are lower than they were historically, although they too have shown an increase in recent decades. This eagle is a rare annual winter visitor to the Salton Sea, with records extending from 11 October (1977, south end to 12 October; *AB* 32:257) to 22 March (1983, a first-winter bird at Unit 1; *AB* 37:912) but mostly between mid-November and mid-February. Almost all regional records are of birds in first-winter plumage. An exceptionally early individual near Seeley 13 August–29 October 1977 (*AB* 32:257) was banded and fitted with a radio transmitter as a nestling in central Arizona. It was originally recovered in Joshua Tree National Park and was returned to Arizona before it wandered southwest to Seeley (Garrett and Dunn 1981). An early juvenile was at Wister 6 September 1999 (J. L. Dunn). Anomalous was a juvenile flying low over Wister 14 July 2001 (*NAB* 55: 482).

ECOLOGY. Most records of this eagle are from the immediate vicinity of the Salton Sea, with birds usually feeding along and roosting on the shoreline. The bird at Seeley, a first-winter bird 7 km south-southeast of Brawley 27–29 January 1990 (*AB* 44:328), another photographed there 12–21 February 2000 (K. . Kurland, B. Miller), one over Niland 22 December 1998 (G. McCaskie), and one 15 km east-southeast of Calipatria 20 December 1999 (G. McCaskie) were exceptions.

TAXONOMY. Although Palmer (1988a) considered the Bald Eagle monotypic, the two subspecies appear to be valid (Brown and Amadon 1989). The smaller, nominate subspecies is a breeding resident in the Southwest; therefore, the bird at Seeley must have belonged to this subspecies, and the unseasonable juveniles at Wister probably did as well. We presume all wintering birds to be the larger *H. l. alascanus* Townsend, 1897, of Canada and the northern United States, with an unsexed first-winter bird from the southern Coachella Valley near the Salton Sea 21 December 1940 (SBCM 12602) definitely belonging to this subspecies (wing chord 590 mm, tail 340 mm), as it is too large to be a nominate bird of either sex (see Behle 1985:25).

NORTHERN HARRIER
Circus cyaneus hudsonius (Linnaeus, 1766)
Common winter visitor (July to mid-May); casual in summer. The Northern Harrier is a conspicuous and common winter visitor in suitable open habitats of the Southwest. The Coachella, Imperial, and Mexicali Valleys support hundreds each winter, a sizable number for a raptor. For example, 115 individuals were tallied between the Salton Sea (north) and Salton Sea (south) Christmas Bird Counts on 29 December 1990 and 18 December 1990, respectively (*AB* 45:994, 999), yet these counts cover but a small portion of available habitat in the region. The majority of wintering birds arrive from late August to mid-September, although the first individuals usually appear by late July or early August, the earliest being an immature male at the mouth of the Whitewater River 3 July 1994 (M. A. Patten). Most have departed for their northerly breeding grounds by early May, the latest being at Wister 14 May 2000 (G. McCaskie). Nonbreeders have remained through the summer, with records from 1 June (2000, vicinity of Poe Road; G. McCaskie) to 30 June (1984, south end; G. McCaskie) and averaging one every two to three years.

ECOLOGY. The expansive agricultural fields in the region supply ample foraging and roosting habitat for this terrestrial raptor. Favored roosting sites are provided by fallow fields with dense cover, overgrown areas with tall nonnative grasses and weeds, and extensive marshes with dense cattail and bulrush (e.g., Wister, Unit 1). Harriers forage over marshes and fields throughout the region. Between hunting forays individuals frequently perch atop low Saltcedars jutting from marshes or on hay bales, low poles, or other suitably sturdy objects bordering fields.

TAXONOMY. This circumboreal species, called the Hen Harrier in Britain, is represented in North America by the subspecies *C. c. hudsonius*. Compared with Old World subspecies, *C. c. hudsonius* is larger (sex for sex), and females and immatures are more richly colored (rustier) and have bolder tail bands.

SHARP-SHINNED HAWK
Accipiter striatus velox (Wilson, 1812)

Uncommon winter visitor (early August to early May). Neither the Sharp-shinned nor Cooper's Hawk is numerous in the Salton Sink, and the former is the less numerous of the two. The Sharp-shinned is an uncommon winter visitor to the region, with records from 6 August (1988, near Mecca; M. A. Patten) to 3 May (1981, Finney Lake; G. McCaskie). As is typical for much of southern California, wintering Sharp-shinned Hawks return a bit later (2–3 weeks) and leave a bit earlier (1–2 weeks) than do Cooper's Hawks.

ECOLOGY. The *Accipiter* hawks are woodland species seldom found away from large trees. Thus, the Sharp-shinned Hawk is most numerous in the region in older portions of towns with large Fremont Cottonwoods, eucalyptus, and other substantial trees. It is less frequent in riparian areas dominated by Athel and seems to avoid Date Palm groves and orchards more often than does Cooper's Hawk.

TAXONOMY. The widespread North American subspecies *A. s. velox* is the only one recorded in California. This subspecies differs from the various subspecies in the West Indies and tropical America in being generally larger, darker, and less rufescent on the underparts. The extremely dark *A. s. perobscurus* Snyder, 1938, has been mistakenly attributed to California (Clark and Wheeler 1998). The balance of evidence for the single putative record of *A. s. perobscurus* points to the bird's actually having been a melanistic *A. s. velox* (Patten and Wilson 1996).

COOPER'S HAWK
Accipiter cooperi (Bonaparte, 1828)

Uncommon winter visitor (mid-July to mid-May); 3 summer records. Of the two *Accipiter* hawks that occur in the region, Cooper's is the more numerous, albeit still uncommon. It occurs at the Salton Sea in winter, with records extending from 18 July (not June) (1999, adult at Brawley; M. J. San Miguel, *NAB* 53:432) to 15 May (1977, near Westmorland; G. McCaskie). An immature along the Highline Canal east-southeast of Calipatria 27 May 2000 (M. A. Patten) was exceptionally late, whereas adults at Brawley 1 July 2001 (*NAB* 44:482) and 6 July 2002 (G. McCaskie) were exceptionally early or summering locally. Small numbers breed in the northern Coachella Valley outside of the Salton Sea region (e.g., Palm Springs, Palm Desert), and the species has been found three times in summer in the southern Coachella Valley, with individuals at Coachella 31 May 1999 (M. A. Patten) and at the north end 20 June 1981 (G. McCaskie) and two (a pair?) at the north end 8 July 1995 (G. McCaskie). There is no evidence of nesting, however, though an adult male at Coachella 17 March 1991 was engaged in courtship display flights (M. A. Patten).

ECOLOGY. This woodland species is seldom found away from settled areas and ranch yards with mature Fremont Cottonwoods and other large trees, although it is frequent in Date Palm groves and even citrus orchards in the Coachella Valley. It occurs in riparian woodland dominated by mature Goodding's Black Willow and Fremont Cottonwood, but like the Sharp-shinned Hawk, it avoids riparian scrub dominated by Saltcedar.

TAXONOMY. Although Cooper's Hawk is generally considered monotypic (Friedmann 1950; Palmer 1988a), there is some evidence that the small subspecies *A. c. mexicanus* Swainson, 1831, is valid (Whaley and White 1994). The breeding range of small birds is the Pacific Northwest, not the whole of the West (Whaley and White 1994). If recognized, this highly migratory subspecies probably accounts for a substantial portion of migrant and wintering birds. However, potential local breeders and many other migrants and wintering birds would be *A. c. cooperi*. For example, an immature female from 6.5 km south of Mecca 25 October 1952 (WFVZ 1417) has a wing chord measuring 262 mm, outside the range of small birds from the Pacific Northwest but at the mean for large (nominate) birds of the Southwest (Whaley and White 1994:173).

COMMON BLACK-HAWK

Buteogallus anthracinus [anthracinus] (Deppe, 1830)

One spring record. An adult Common Black-Hawk at Oasis 28 March–2 May 1997 was only the second to be recorded in California (Rottenborn and Morlan 2000). This species has wandered in spring north and west of its usual range on a dozen occasions from 23 March into July (Daniels et al. 1989; Rosenberg et al. 1991).

TAXONOMY. The nominate subspecies, which breeds locally in the Southwest, presumably accounts for the two California records. It is markedly larger than all other subspecies.

HARRIS'S HAWK

Parabuteo unicinctus superior van Rossem, 1942

Formerly a common breeding resident in the Imperial Valley; now a casual winter visitor (mid-November through April); 1 recent summer record. Harris's Hawk ranges from the southwestern United States to central South America. In California this species formerly was found in the lower Colorado River valley and the Imperial Valley (Grinnell and Miller 1944); it also occurred in the Mexicali Valley and the Río Colorado delta (Grinnell 1928; WFVZ 127684). By the mid-1960s it had been extirpated from California as a breeder (Remsen 1978; Rosenberg et al. 1991), the last definite wild bird being seen north of Blythe 28 November 1964 (Garrett and Dunn 1981; Rosenberg et al. 1991). Despite a reintroduction effort in the first half of the 1980s (Walton et al. 1988), this species appeared to be lost from the state avifauna. However, California has apparently always been on the fringe of the natural range of Harris's Hawk, for this species has undergone cyclic expansion into and retraction out of the state over the past century (Patten and Erickson 2000). Although there were but few California records by the early 1900s, an influx late in the second decade of the century and early in the third decade brought it to the Imperial Valley, where it was first found nesting in 1920, with a pair seen mating near Brawley 30 March

and three eggs taken from a nest 5 km west of Calipatria 31 March (Bancroft 1920). By the early 1920s it was common, although claims of 400–500 between Calexico and Heber 22 October 1920 (Chambers 1921) and 250 near Calexico 28 August 1923 (Chambers 1924) are perhaps questionable (J. C. Bednarz pers. comm.). Afterward it slowly declined, and it was last recorded in the Imperial Valley 9 March 1954 (at Salton Sea NWR; *AFN* 8:270).

Since the early 1950s it has been difficult to determine to what extent Harris's Hawk still occurs in the region as a natural component of the avifauna. For example, released birds attempted to nest near Niland in 1976 (*AB* 30:1003). Virtually all recent records of the species in California are of birds considered to have escaped from falconers (Luther et al. 1983; Garrett and Dunn 1981; Unitt 1984). Interestingly, however, all recent Salton Sea records but two are during winter from 20 November (1988, banded adult at Finney Lake; K. A. Radamaker, M. A. Patten) to 22 March (1986, adult photographed at Wister; D. A. Leal). Other recent records are of adults at Mecca 17 December 1977–5 March 1978 (Luther et al. 1983; *AB* 32:884), at the north end 17 December 1989 (*AB* 44:989), and at Finney Lake 2 January–24 February 1990 (*AB* 44:328). Harris's Hawks staged an incursion into southern California and northern Baja California in 1994 (Bednarz 1995; Massey 1998; Patten and Erickson 2000; Erickson and Hamilton 2001). This incursion added two more recent winter records, with an adult near Mecca 31 December 1994–29 January 1995 (Erickson and Hamilton 2001) and two adults at Westmorland 7–18 December 1994 (*FN* 49:196). Later, up to three adults were at Indio 11–27 November 1999 (Rogers and Jaramillo 2002). The only recent nonwinter records are of birds found dead near Ramer Lake 29 April 1972 (*AB* 26:808) and a juvenile at the mouth of the New River 25 June 1989 (*AB* 43:1367). These recent records suggest that the species continues to occur naturally in the Salton Sea region, at least as a casual winter visitor, although the June juvenile hints that the species still breeds nearby.

ECOLOGY. This species formerly occurred in woodlands and ranches with extensive Honey Mesquite and Fremont Cottonwood, a habitat extensive in the valley at the time. Recent records have been of birds in more urban settings, often using utility poles as roosting sites. The June immature was foraging along thickets of Saltcedar and Common Reed, but other birds appeared to be foraging around agricultural fields.

TAXONOMY. Harris's Hawks that occur in the southwestern United States and northwestern Mexico are *P. u. superior*, the largest, darkest subspecies. The subspecies is apparently valid (Brown and Amadon 1989), although color differences are not as significant as originally described (Palmer 1988a) and the subspecies has been tentatively merged with *P. u. harrisi* (Audubon, 1837), of Texas, eastern Mexico, and Central America, by some authorities (e.g., Streseman and Amadon 1979).

RED-SHOULDERED HAWK
Buteo lineatus elegans Cassin, 1856
Rare fall and winter visitor (late July through April); multiple summer records in the Coachella Valley, where potentially breeding. The Red-shouldered Hawk is a rare but regular postbreeding and winter visitor to the California deserts, including the Salton Sink. This hawk approaches uncommon status at times in the Coachella Valley (e.g., 5 there during January 1978; *AB* 32:399), where it may breed on occasion (see below). Immatures are most frequent in the region, but a few adults occur, particularly in the Coachella Valley. In recent years records have been increasing outside the species' normal range, not only around the Salton Sea but elsewhere in the arid West. There was only one published Salton Sea record (for the north end) by the early 1970s (Wilbur 1973), but records steadily increased, with an additional seven sightings for the north end by the early 1980s (Garrett and Dunn 1981). The first published record for the Imperial Valley, from Finney Lake 24 September 1977 (*AB* 32:257), is predated by a specimen from Calexico 8 November 1909 (P. I. Osburn; AMNH 750308). The species was regular in the Imperial Valley by the early

1980s. For most of the region records extend from 22 July (2000, immature at Wister until 4 August; G. McCaskie) to 18 April (1999, first-spring bird at Oasis; M. A. Patten), with an early fall immature at Salton City 9 July 1989 (*AB* 43:1367). Adults and immatures have been noted in the vicinity of Coachella, Oasis, and Mecca through the summer (e.g., an adult at Oasis 1 June 1996 and a molting adult there 4 July 1999; M. A. Patten), suggesting that the species may nest locally. Positive evidence of breeding is lacking, but pairs were observed displaying at Oasis in March 1996 and March 1997 (G. McCaskie).

ECOLOGY. The Red-shouldered Hawk is most often found in clumps of trees (mostly nonnative) in settled areas, especially around Date Palm groves. Even though it frequents riparian woodland in its normal California range (Bloom et al. 1993), it is seldom found in such habitats around the Salton Sea.

TAXONOMY. Save for a single specimen, Red-shouldered Hawks in California and Oregon and out-of-range birds in most of the Southwest are the small, dark, heavily marked subspecies *B. l. elegans*, including the Calexico specimen and a specimen from just outside the Salton Sink of a female at Cathedral City 3 September 1987 (WFVZ 43361). The lone exception (P. Pyle *in litt.*), an adult female from Sacramento, California, 21 September 1986 (WFB 4819), is the larger (wing chord 351 mm in this case), paler, more lightly marked *B. l. lineatus* (Gmelin, 1788), a migratory subspecies of northeastern North America that reaches the Gulf of Mexico in winter and has strayed to Utah and Bermuda (Friedmann 1950:282).

BROAD-WINGED HAWK
Buteo platypterus platypterus (Vieillot, 1823)
Three winter records (1 potentially a spring vagrant). In California the Broad-winged Hawk is a casual spring and winter visitor and locally rare but regular as a fall transient (Binford 1979). It has been recorded on three occasions in the Salton Sink in winter, with an immature near Brawley 28 January–18 February 1978 (*AB* 32:

399), an immature near Calipatria 18 December 1979 (*AB* 34:662), and an adult photographed 3 km southeast of El Centro 25 and 26 February 1999 (*NAB* 53:208). The adult was associated with the first returning Swainson's Hawks of the spring and was thus perhaps an early spring vagrant; if so, it would be the earliest for California by two months (Garrett and Dunn 1981; McCaskie et al. 1988).

TAXONOMY. Birds on the North American continent are the widespread nominate subspecies, which in general differs from various island forms in the Caribbean in being larger and darker (Palmer 1988b).

SWAINSON'S HAWK

Buteo swainsoni Bonaparte, 1838

Rare transient in spring (mid-February to mid-June) and in fall (late August through October); casual in winter. Although the species is listed as threatened by the state of California, and despite apparent declines in many populations (White 1994), since the mid-1980s records of Swainson's Hawk have been increasing in many parts of California, including the Salton Sink (Fig. 34). In spring in the Imperial Valley it is regular. Except for one at Salton City 5 May 1984 (*AB* 38:960), all regional records are from the Imperial, Coachella, and Mexicali Valleys, with about 90 percent from the Imperial. Dates for spring migrants range from 14 February (1994, near El Centro; *FN* 48:247) to 11 June (1983, Plaster City; *AB* 37:1027), with most from April (but see below). The largest recent concentrations are flocks of 21 (not 19) near Imperial 7 and 8 April 1990 (*AB* 44:496), 17 in the Imperial Valley 10 March 1995 (*FN* 49:309), and 9 near El Centro 4 March 1989 (*AB* 43:536). This species used to be much more common in California and no doubt moved through in larger numbers in earlier decades.

Swainson's Hawk is rarely detected in fall in the Salton Sink, with all records of definite fall transients during the period from 22 August (2000, 2 dark-morph birds 5 km southwest of Brawley; G. McCaskie) to 2 November (1991, 7 km southeast of Brawley; S. von Werlhof), and the vast majority for October. Remarkable were

FIGURE 34. The Salton Sink is on the migration route of Swainson's Hawk, now rare in California. Photograph by Kenneth Z. Kurland, November 1998.

120–30 roosting 3 km southeast of El Centro 24 and 25 October 1999 (*NAB* 54:104; K. Z. Kurland), by far the largest concentration recorded in the region. Birds at Brawley 26 November 1999 (*NAB* 54:104), near Niland 8 December 2000 (*NAB* 55:227), and near El Centro 11 December 1994 (2; *FN* 49:196), 14 December 1996 (K. Z. Kurland), and 14 December 2000 (*NAB* 55:227) were probably late fall migrants, as were at most two adults about 25 km southeast of the region at Ciudad Victoria 11–15 December 1994 (Patten et al. 2001). Although Swainson's Hawk has begun wintering in small numbers in the Central Valley of California (Herzog 1996), individuals at El Centro 1 January 1999 (K. Z. Kurland) and photographed near Calipatria 11 January 2001 (A. Brees) represent the only records for the Salton Sink of apparent wintering birds. A dark-morph immature was seen just east of the Salton Sea region near Ejido Chiapas, about 45 km east-southeast of Mexicali, 10 January 1994 (Patten et al. 2001). Spring migrants in California seemingly occur earlier every year, often by mid-February. Thus, we cannot ascertain the status of individuals near Seeley 28 January 1995 (*FN* 49:196), near Brawley 2 February 1997 (*FN* 51:801), and in the Imperial Valley 7 February 1998 (*FN* 52:257). Either they were wintering locally or they were exceptionally early spring migrants. In 1999 up to 20 southeast of El Centro during the last half of February (G. H. Rosenberg et al.) and 13 around El Centro and Westmorland 15 February (*NAB* 53:208) were undoubtedly spring migrants.

ECOLOGY. Virtually all Swainson's Hawks noted in the Salton Sink are seen flying over, standing in agricultural fields, or roosting in tall trees, especially Red River Gums.

ZONE-TAILED HAWK
Buteo albonotatus Kaup, 1847
Casual fall and winter vagrant (September to mid-March); 1 summer record. The Zone-tailed Hawk is a rare winter visitor to cismontane southern California from Arizona or Mexico; there are also two breeding records for the state. The species occurs in California's deserts chiefly in spring, but all six unequivocal records for the Salton Sink are from fall and winter. A first-winter bird photographed in a yard 3 km southeast of El Centro 4 December 1996–22 February 1997 (McCaskie and San Miguel 1999) established the first well-documented record for the region. This same location hosted an adult 5 January–28 February 2001 and two immatures beginning 13 December 2000, one departing 18 March 2001, the other 3 May 2001, and one returning 12 July 2001–19 February 2002, when it molted into adult plumage (all 3 photographed; *NAB* 55: 227, 355, 482). Two Zone-tailed Hawks at Brawley 12 October 2001, one an adult clearly videotaped (M. Pollock), were apparently fall migrants, although an adult videotaped in Brawley 25 January–9 March 2002 (T. Harrison et al.) may have been one of these birds. Adults returned to a location 3 km southeast of El Centro 26 and 27 September 2002 (K. Z. Kurland) and to Brawley

29 October–8 December 2002+ (A. Kalin). Three other reports—from Finney Lake, Salton Sea NWR, and near Brawley—lack sufficient documentation (Heindel and Garrett 1995; Patten, Finnegan, et al. 1995), although the first and last were likely correct.

RED-TAILED HAWK
Buteo jamaicensis (Gmelin, 1788) subspp.
Common winter visitor (mid-August to mid-May); rare breeder in the Coachella Valley; casual in summer in the Imperial Valley. The predominant *Buteo* throughout much of the United States and Canada, the Red-tailed Hawk is a common winter visitor in the Salton Sea region, reaching its peak abundance in the Imperial and Mexicali Valleys, although it is also numerous (for a raptor) in the Coachella Valley and in much of the undeveloped desert where roosting sites are available. Many northerly breeders winter in the Salton Sink each year, occurring from 9 August (1986, south end; G. McCaskie) to 23 May (1999, 3 in the Imperial Valley; M. A. Patten), with the majority present from late September through early April. During winter the species is a common sight in the valleys, daily counts easily exceeding several dozen without any special effort to find the species. It remains casually as a nonbreeder in the Imperial Valley, with records from 12 June (1999, Wister; M. A. Patten) to 23 July (1988, south end; M. A. Patten). It formerly bred in the southern Mexicali Valley in the vicinity of Cerro Prieto (e.g., a nest with 1 egg 18 April 1905; E. A. Goldman, USNM B30925), and may still breed at the western fringe of that valley (Patten et al. 2001). Definite breeding is now limited to the Coachella Valley, where only a few pairs nest from late March through midsummer, with young fledging as early as late July but more typically in August.

ECOLOGY. Expansive farmlands in the Coachella, Imperial, and Mexicali Valleys, coupled with an abundance of utility poles and tall eucalyptus, provide seemingly endless habitat for this human-tolerant species. It forages in these open areas and is frequently encountered over open Creosote scrub. Birds nest in tall Red River Gum,

Blue Gum, and Fremont Cottonwood and on utility poles.

TAXONOMY. Virtually all of the Red-tailed Hawks recorded in the Salton Sink and elsewhere in California belong to *B. j. calurus* Cassin, 1856, a subspecies with the light morph showing heavily barred thighs, a dark head and mantle, and a distinct breast band; perhaps 1 in 20 is a rufous-morph bird, and even fewer are true dark-morph (deep chocolate brown) individuals. The account above refers to the regional status of this subspecies. A few individuals showing the characters of *B. j. fuertesi* Sutton and van Tyne, 1935, of the southern Great Plains, winter in the Imperial Valley each year, although there are no specimens to support this identification. Nevertheless, there are two winter specimens for San Diego County to the west of the region (Unitt 1984). The light morph of this subspecies differs from *B. j. calurus* in lacking a distinct breast band and in having only faint markings on the thighs.

A white-headed adult with a mostly white tail, a pale reddish subterminal band to the tail, no dark patagia, broadly white scapulars, and a clean white breast was photographed 10 km southeast of Calipatria 31 December 1987 (M. A. Patten, C. A. Marantz). It thus appeared to be a typical *B. j. kriderii* Hoopes, 1873. This subspecies has yet to be collected in California, but there is a specimen taken from central Arizona 12 October 1931 (Phillips et al. 1964), and the subspecies has reached both northwestern and south-central Utah (Behle 1985). Its status as a valid subspecies is disputable. Red-tailed Hawks exhibiting this phenotype apparently do not occupy a distinct breeding range but rather occur "only in association . . . with *borealis* or *calurus*" (Taverner 1936). Because a valid subspecies must have a distinct, exclusive breeding range, Taverner (1936) treated *B. j. kriderii* as a white morph of *B. j. borealis* (Gmelin, 1788), Phillips et al. (1964) tentatively recognized it but acknowledged that birds dubbed *B. j. kriderii* are "more likely . . . just variant individuals within the northern populations of *calurus* and *borealis*," and Palmer (1988a) synonymized it with *B. j. calurus*. Still, birds showing characters of *B. j. kriderii* breed only in the northeastern Great Plains, so we follow Friedmann (1950), the AOU (1957), Monson and Phillips (1981), Behle (1985), Godfrey (1986), and Brown and Amadon (1989) in recognizing it, while acknowledging that extensive intergradation with *B. j. borealis* clouds diagnosis.

B. j. harlani (Audubon, 1830), a blackish subspecies with a mottled whitish tail that breeds in central Alaska (Mindell 1983), is a casual (but annual) winter visitor to California. Birds showing the characters of this subspecies have been recorded at the Salton Sea on six occasions in winter from 7 November (1997, southeast of Brawley; *FN* 52:125) and 9 March (1996, near Mecca; *FN* 50:332). The other four records are of birds near Brawley 29 January–11 February 1989 (*AB* 43:366) and 10 December 1994–11 February 1995 (*FN* 49:196), at Mexicali 1–15 December 1994 (Patten et al. 2001), and at Salton Sea NWR 15 November 1997 (G. McCaskie). A bird photographed at Thermal 7–14 December 1982 (*AB* 37:338) suggested the near mythical light-morph Harlan's Hawk (see Mindell 1985; and Clark and Wheeler 1987). This individual returned annually (K. A. Radamaker pers. comm.) and was last seen 12 March 1988 (M. A. Patten, K. A. Radamaker). Its identity is debatable, but descriptions and photographs seem to fit what we would expect for a light-morph adult *B. j. harlani*, although the possibility of an intergrade cannot be excluded without a specimen.

FERRUGINOUS HAWK
Buteo regalis (Gray, 1844)
Rare winter visitor (late September through March). The stately Ferruginous Hawk has declined throughout much of its U.S. range, which centers around the Great Basin (Ryser 1985; White 1994). This species is less common in the Salton Sink than it is elsewhere in interior California (Garrison 1990). It is a rare winter visitor to the Salton Sea region, mainly to the Imperial and Mexicali Valleys (Patten et al. 1993). Records extend from 23 September (1973, near Westmorland; G. McCaskie) to 30 March (1975 and 1986, south end; G. McCaskie, P. Unitt), with the

majority from December to mid-February. In most winters fewer than ten individuals occur, so that among the large *Buteo* hawks the Ferruginous is more frequently encountered than the Rough-legged but is greatly outnumbered by the Red-tailed.

ECOLOGY. This species frequents agricultural lands throughout the valleys, often perching on utility poles, although it frequently roosts on the ground.

ROUGH-LEGGED HAWK
Buteo lagopus sanctijohannis (Gmelin, 1788)
Rare winter visitor (late October through March). The Rough-legged Hawk is a Holarctic species with a northerly distribution. In the West in winter this species is scarce south of the Great Basin (see Garrison 1993), but it is a rare though nearly annual winter visitor in small numbers to the Imperial Valley. Records extend from 30 October (1976, 2 at the south end until 30 November; *AB* 31:223) to 2 April (1975; 1–2 at the south end from 15 December 1974; *AB* 29:742), with the majority from late November to mid-February. As many as three individuals have been found in a single winter (i.e., winter of 1988–89; *AB* 43: 366). A few winter in the Coachella Valley on an irregular basis, the maximum there also being a count of three individuals on the Salton Sea (north) Christmas Bird Count 22 December 1985 (*AB* 40:1008).

ECOLOGY. This hawk occurs principally in the vast agricultural lands, using tall trees or utility poles as perches and roosting sites.

TAXONOMY. *B. l. sanctijohannis* is the only one of the three subspecies that occurs in North America. It is the smallest and darkest subspecies but is extremely variable throughout its range (Cade 1955).

GOLDEN EAGLE
Aquila chrysaetos canadensis (Linnaeus, 1758)
Casual vagrant (fall, winter, and spring). Despite being uncommon in rocky, hilly regions to the west and north of the Salton Sea and in the Chocolate Mountains to the east (R. L. McKernan), the Golden Eagle has wandered to the

Salton Sea on only five occasions in fall, winter, and spring. Individuals were seen at the south end 23 January 1963 (Garrett and Dunn 1981), near Brawley 30 September–4 November 1986 (*AB* 41:143), and 3 km north-northwest of Seeley 26 October 1996 (P. Unitt), and adults were seen near Brawley 19 May 1921 (A. J. van Rossem; UCLA 12250) and 3 km northeast of Niland 22 May 1991 (*AB* 45:496).

As noted by Garrett and Dunn (1981), many reports of the Golden Eagle at the Salton Sea likely pertain to first-winter Bald Eagles. For example, reports from 8 km north of Brawley 18 December 1910 (van Rossem 1911), near the south end 10 March 1956 (*AFN* 10:364), and at Salton Sea NWR 22 November 1956 (*AFN* 11:60) are not accompanied by details sufficient to eliminate an immature Bald Eagle. The locality for a report from the south end 29 April–6 May 1962 (Garrett and Dunn 1981) is unclear; it may have been outside the Salton Sink, so we exclude it.

TAXONOMY. There are five subspecies of the Golden Eagle (Vaurie 1965), but only the dark, moderate-sized *A. c. canadensis* occurs in North America.

FALCONIDAE • *Caracaras and Falcons*

AMERICAN KESTREL
Falco sparverius Linnaeus, 1758 subspp.
Common breeding resident; numbers augmented in winter (mid-August to May). The American Kestrel is widespread and common throughout the Americas. It is especially common around the Salton Sea. To be sure, given that predators at the highest trophic level are generally uncommon, the sheer abundance of the kestrel in the open country of the Coachella, Imperial, and Mexicali Valleys is astounding. Even during midsummer, when they are least abundant, it is easy to tally double digits (e.g., 20 between Wister and the New River 11 and 12 July 1999; M. A. Patten). Although the kestrel is common in midsummer, counts are much lower than those during winter. Hundreds of kestrels breeding to the north move into the region, mainly from mid-August to May (exact dates cannot be determined because of the large number of breeders).

This annual influx often yields impressive counts; for example, 339 were censused between the Salton Sea (south) and Salton Sea (north) Christmas Bird Counts on 28 December 1993 and 2 January 1994, respectively (*FN* 48:854, 858).

This species nests throughout the region wherever suitably large nesting trees are found (see below). Breeding numbers are at their highest in the southern Coachella Valley and in portions of the Imperial and Mexicali Valleys near small towns and ranches. Nesting commences in February and March. Activity peaks in April, the first young fledging by early July. Some pairs are still feeding young in midsummer, with fledging taking place in August.

ECOLOGY. Breeding kestrels are confined to areas with suitable nest sites. The species is a cavity nester, so it frequently uses eucalyptus, Fremont Cottonwood, and other trees large enough to support large cavities. It will also occupy dense clumps of dead palm fronds (i.e., not a true cavity) and old holes drilled into palms by Gila Woodpeckers. The extensive agricultural lands in the irrigated valleys provide abundant foraging opportunities. Females prefer open fields that support low vegetation or are nearly bare. This preference may force males into slightly more wooded areas such as orchards or towns (Koplin 1973; Gawlik and Bildstein 1995). For example, during a survey near El Centro 3 January 1974 more than 80 percent of kestrels in agricultural fields were females, but all in the city were males (Mills 1976).

TAXONOMY. Except in Florida, in Baja California, and from west Mexico southward, kestrels throughout North America are of the large, heavily marked nominate subspecies. The small *F. s. peninsularis* Mearns, 1892, of Baja California and western Sonora, has been taken in California in winter, including males at San Diego 6 November 1921 (Unitt 1984; SDNHM 2288, wing 170 mm) and on the lower Colorado River 25 January 1931 (S. G. Harter; SDNHM 14207, wing 169 mm). This subspecies also has been taken in the Río Colorado delta in northeastern Baja California 14 October 1927–4 February 1928 (Grinnell 1928; MVZ 52101–3); specimens of

nonbreeders were mapped in the Mexicali Valley by Bond (1943:178); and a number of winter (September–May) specimens from south-central Arizona are of this form (Rea 1983). Thus, it is likely that *F. s. peninsularis* occurs regularly in the Salton Sink in winter.

Breeders are mostly the nominate subspecies, presumably exclusively so in the Coachella Valley. However, a breeding female at Seeley 25 August 1984 (A. M. Rea; SDNHM 43294) is *F. s. peninsularis* (wing 175 mm, humerus 40.7 mm, ulna 46.0 mm, coracoid 21.8 mm), being substantially smaller than specimens from coastal San Diego County (SDNHM 46530; humerus 43.3 mm, ulna 48.2 mm, coracoid 23.1 mm) and British Columbia (SDNHM 41018; humerus 44.9 mm, ulna 49.7 mm, coracoid 24.7 mm). By contrast, a male found 14 km east of El Centro 21 May 1987 (SDNHM 44741) is *F. s. sparverius* (humerus 41.0 mm, sternum 28.4 mm, coracoid 22.4 mm), being indistinguishable from specimens from coastal San Diego County (SDNHM 46386; humerus 41.6 mm, sternum 29.5 mm, coracoid 22.4 mm) and Siskiyou County, in northern California (SDNHM 42215; humerus 41.4 mm, sternum 28.6 mm, coracoid 22.3 mm). Thus, the Imperial Valley and probably the Mexicali Valley add to areas in the Southwest with apparent population intergradation (see Bond 1943).

MERLIN

Falco columbarius Linnaeus, 1758 subspp.

Rare winter visitor (mid-September to mid-April). Principally a bird of northern latitudes, the Merlin is rare in the Salton Sink. Nevertheless, it is annual in small numbers from 16 September (1995, near Oasis; G. McCaskie) to 17 April (1940, south end; B. Bailey, SDNHM 18163). Relative to other falcons in the region, it is far less common than the American Kestrel, about as numerous as the Peregrine Falcon, and generally more frequently encountered than the Prairie Falcon (excepting during narrow windows in spring and fall when the Prairie Falcon is present but the Merlin is on its breeding grounds to the north).

ECOLOGY. The Merlin is seemingly less tolerant of the expanses of open agricultural fields than is the American Kestrel. It tends to occur along plowed, fallow, or weedy fields, often ones that have not been farmed in over a year. It most often perches on tall snags but will use utility poles.

TAXONOMY. Virtually all Merlins occurring in the region (and in the Southwest) are the widespread nominate subspecies, of which *F. c. bendirei* Swann, 1922, is generally treated as a synonym (Palmer 1988b; Hamilton and Schmitt 2000). A female collected northwest of Westmorland 31 October 1954 is the extremely dark northern subspecies *F. c. suckleyi* Ridgway, 1873 (Cardiff 1956; SBCM 31673). Merlins 3 km north of Niland 31 December 1999 (*NAB* 54:221) and near Calipatria 5–21 November 2000 (G. McCaskie) and 1 December 2001 (G. McCaskie) also showed characters of *F. c. suckleyi*. There are eight records—spanning the period 12 October–15 April—of birds showing characters of the pale Great Plains subspecies *F. c. richardsonii* Ridgway, 1871: near the mouth of the New River 12 October 1975 (G. McCaskie); at the south end 28 December 1978 (*AB* 33:666), 29–30 December 2001 (photographed; C. A. Marantz et al.), and 27 January 1996 (G. McCaskie); near Westmorland 1 January 1993 (P. Unitt); near Brawley 5 March 2000 (*NAB* 54:221); and near Mecca 3 April 1994 (G. McCaskie) and 15 April 1989 (*AB* 43:563).

PEREGRINE FALCON

Falco peregrinus anatum Bonaparte, 1838

Rare perennial visitor, with most records for summer (July to mid-September) and midwinter. After years of precipitous declines across North America (Hickey 1969; Herman 1971), the Peregrine Falcon staged an impressive comeback, especially after 1980 (White 1994). It now breeds in much of coastal California and the Sierra Nevada and is routinely noted at many locales on the coastal slope. In California, except in the Salton Sea region, this falcon is generally extremely rare east of the axis of the Sierra Nevada and the Peninsular Ranges. At the Salton Sea, individuals are

encountered year-round, most often as non-breeding summer visitors. The species is annual in small numbers in that season, with most records probably involving individuals from the Gulf of California, where it still nests in fair numbers (Porter et al. 1988). Apparent spring transients have been recorded as early as mid-April (e.g., 10–17 April 1982, north end; *AB* 36:894). Birds appearing in spring sometimes stay through the summer (e.g., subadult at the north end 5 May–6 July 1984; *AB* 38:1061), but summering birds typically arrive in late June (e.g., 30 June 1990, immature at Salton City; *AB* 44:1186) and depart by mid-September (e.g., 15 September 1949, near Westmorland; SBCM M1135). The majority of summer reports involve immatures or subadults, but a few adults have been noted. After the fall exodus few Peregrines are recorded, although the species does winter rarely (and apparently increasingly), especially around the south end, with records extending from 18 December (1979, south end, and 1988, near Red Hill; *AB* 34:662, 43:1176) to 15 March (1942, Imperial Valley; C. A. Harwell).

ECOLOGY. Interestingly, during migration and summer virtually all Peregrines are along the immediate edge of the Salton Sea, but they are more dispersed in winter, with nearly as many records for agricultural lands of the Imperial Valley as for the sea itself. For example, two of the earliest records for the area are of birds collected 16 km south of Westmorland 23 December 1922 (Grinnell and Miller 1944; MVZ 144768) and at Calipatria 13 February 1938 (S. G. Jewett; SDNHM 20062).

TAXONOMY. The account above refers to the widespread North American subspecies *F. p. anatum*. This form has the mantle medium bluish, contrasted with the pale blue *F. c. tundrius* White, 1968, of the far north, and the large, midnight blue *F. p. pealei* Ridgway, 1873, of the Pacific Northwest (White and Boyce 1988). The latter is uncommon in winter along the northern coast of California (Earnheart-Gold and Pyle 2001). Two have occurred as far south as San Diego (Swarth 1933; CAS 11694; Anderson et al. 1988:515), but the subspecies is unreported from inland southern California. *F. c. tundrius* has reached southern California more frequently, but a pale individual with a bold supercilium and narrow facial bar at the Whitewater River delta 1 May 1993 (K. F. Campbell, M. A. Patten) is the only Peregrine Falcon recorded in the Salton Sink that showed the characters of this subspecies. Even so, we consider the record inconclusive without a specimen. This individual may have been a pale, well-marked *F. p. anatum* from Baja California, as *F. p. anatum* breeding in the Gulf of California more closely resemble *F. p. tundrius* (N. J. Schmitt *in litt.;* Earnheart-Gold and Pyle 2001; cf. White 1968).

PRAIRIE FALCON

Falco mexicanus Schlegel, 1851

Rare winter visitor (late August to early June); 2 summer records. Despite never gracing the endangered species list, as its more famous cousin has, the Prairie Falcon is actually the rarer of the two large falcons occurring in the Salton Sea region, particularly in summer. In surrounding desert regions, by contrast, the Prairie Falcon is substantially more common than the Peregrine. The Prairie Falcon is a rare winter visitor to the Salton Sink from 25 August (1990, north end; M. A. Patten) to 9 June (1990, mouth of the Alamo River; G. McCaskie), with one at Unit 1 on 16 June 1979 (*AB* 33:897) being exceptionally late. Of note was one along Interstate 8 just west of sea level 20 August 1999 (G. McCaskie). Most records are from mid-September to mid-May, so any large falcon encountered in summer at the Salton Sea will invariably be a Peregrine, as that species reaches its maximum abundance during summer. Nonetheless, there are two summer records of the Prairie, of individuals at the north end 5 July 1984 (*AB* 38:1061) and at Salton Sea NWR 9 July 1994 (S. von Werlhof).

ECOLOGY. The Prairie Falcon is usually seen perched atop utility poles surrounded by extensive agricultural fields. Indeed, the extensive hunting habitat in the Coachella, Imperial, and Mexicali Valleys seems to provide ideal foraging habitat for this species, so its rarity in the region is perplexing.

GALLIFORMES • *Gallinaceous Birds*

The grouse, quail, and other chickenlike birds are poorly represented in the Salton Sink. Were it not for a persistent population of the non-native Ring-necked Pheasant in the Imperial and Mexicali Valleys (Patten et al. 2001; see the Nonnative Species), the sole representative of this order would be Gambel's Quail. The quail is a common breeding resident, making the lack of diversity a little less apparent.

ODONTOPHORIDAE • *New World Quail*

GAMBEL'S QUAIL

Callipepla gambelii gambelii (Gambel, 1843)

Common breeding resident. Gambel's Quail (Fig. 35) is emblematic of the avifauna of the Mojave and Sonoran Deserts. It is a common and conspicuous breeding resident in the Salton Sea region, often encountered in large flocks with one to several males standing sentinel on an open perch. Breeding commences in late winter and early spring, with chicks fledging as early as mid-March (e.g., at Frink Spring 9 March 1988 and at Mecca 12 March 1981; R. L. McKernan) but more typically by April or May (e.g., a pair with ca. 10 chicks at Finney Lake 6 May 1995; M. A. Patten). Some are still on eggs in summer, such as a pair near Mecca 24 June 1916 (MVZ 2734), but by midsummer broods from different parents combine to form large congregations (e.g., more than 20 small chicks near the mouth of the New River 18 July 1999; M. A. Patten). Just to the east of the region, at Glamis in the Algodones Dunes, Gambel's Quail has nested as late as mid-September (R. L. McKernan).

ECOLOGY. This quail occurs in virtually any scrubby habitat not heavily developed for agriculture or houses. Even then, this species occurs in yards and fallow fields. It is particularly common along rivers, around lakes and lagoons in the Imperial Valley, or at other locales with water for drinking and bathing. It reaches its peak abundance, fittingly, in Quail Brush scrub and mesquite thickets (van Rossem 1911; Gullion 1960) but generally avoids pure stands of Salt-cedar (Garrett and Dunn 1981).

FIGURE 35. Gambel's Quail remains common in the Salton Sink wherever thickets of native shrubs persist. Photograph by Kenneth Z. Kurland, January 1999.

TAXONOMY. The pale nominate subspecies is the only one recorded in California or anywhere west of extreme eastern Arizona (Phillips et al. 1964).

GRUIFORMES • *Rails, Cranes, and Allies*

Extensive marshes skirting the Salton Sea and scattered throughout the Coachella, Imperial, and Mexicali Valleys host a variety of rails. Two taxa, the California Black Rail and the Yuma Clapper Rail, are threatened subspecies. Although the former species has declined locally, counts of it have exceeded 60 near Holtville; by contrast, the Salton Sink is home to about one-third of the world population of the latter, one of only two populations away from the Colorado River (the other is Ciénaga de Santa Clara, Sonora). The other four species are relatively common, all but the Sora breeding (at least locally) in the

region. The Common Moorhen and the American Coot are particularly well represented, the latter abundant as a winter visitor. Coots have a long history in the Salton Sink, being the most common species captured by Native Americans at Lake Cahuilla (see Wilke 1978; Beezley 1995; Patten and Smith-Patten 2003).

RALLIDAE • *Rails, Gallinules, and Coots*

BLACK RAIL
Laterallus jamaicensis coturniculus (Ridgway, 1874)
Rare and local breeding resident. The diminutive, seldom seen Black Rail is an uncommon and local resident at scattered locales in the desert Southwest (Evens et al. 1991). It was first recorded at the Salton Sea when a specimen was taken at Calipatria 5 January 1947 (Laughlin 1947b). Records for the region were few for the next two decades but included one in winter at Heise Springs 24 February 1971 (Suffel 1971). Indeed, this species was presumed to be rare and infrequent in the region until the late 1970s, when small populations were discovered in the Imperial Valley and elsewhere around the sea. In 1977, for example, up to seven were seen at Finney Lake 19 April–13 June, and another was seen at the south end 24 May (*AB* 31:1189). In the same year, two were seen at the mouth of the Whitewater River 14 June, and another was seen at Salt Creek 8 March (*AB* 31:1189). Whether 1977 was anomalous is open to question, however, as there have been no subsequent records for Salt Creek and one at the mouth of the Whitewater River 20 June 1981 (*AB* 35:978) is the only other recorded at that locale. Even so, the species persisted at Finney Lake through the 1980s, disappearing when the California Department of Fish and Game drained the lakes for renovation, with the last recorded in April 1989 (Evens et al. 1991). Other regional records from the late 1970s through the 1980s are from the vicinity of the mouth of the New River and Fig Lagoon.

Sadly, there are virtually no records of the Black Rail for any of these areas since 1989, when eight were located around the mouth of the New River 10–19 April (Evens et al. 1991), although one was heard along this river northwest of

Seeley 3 April 1994 (P. Unitt). In 1999 the Point Reyes Bird Observatory failed to find the species during focused surveys for it around the south end (Shuford et al. 2000). Yet this rail persists in the region along the All American Canal near Calexico (e.g., 11 birds 6 April–13 May 1979; *AB* 33:897) and Holtville (e.g., more than 60 birds during April and May 1984; *AB* 38:1061), at extensive marshes near Seeley (G. McCaskie), and at Hot Mineral Spa (Conway et al. 2002), albeit at extremely low densities and with a total population of under 50 individuals (Evens et al. 1991). Most records are from spring, but this apparent seasonality is presumably merely an artifact of detectability—rails are vocal chiefly in that season. Six individuals near Niland 18–25 October 1974 (*AB* 29:129) and records from January and February (cited above) suggest that the Black Rail is resident in the Salton Sink, just as it is in similar habitat along the lower Colorado River (Rosenberg et al. 1991); nonetheless, it may be only sporadic in winter (Garrett and Dunn 1981; Evens et al. 1991).

ECOLOGY. The Black Rail has fairly specific habitat requirements, thus occurring in fewer marshy areas than do other rails. Cattail and bulrush with a thick understory and moist mud or a thin veil of water constitute the best habitat (Flores and Eddleman 1995), although in some occupied areas Saltcedar and Common Reed predominate (Evens et al. 1991).

TAXONOMY. Birds in California and the West are small subspecies *L. j. coturniculus,* further differing from other subspecies in the East, the West Indies, and Mexico in its thinner bill, its more extensive patch of rust on the nape, and its deeper mouse-gray to slate-gray coloration.

CLAPPER RAIL

Rallus longirostris yumanensis Dickey, 1923

Uncommon breeder and rare winter visitor, principally at the south end. The Clapper Rail is primarily a species of tidal marine estuaries, with one subspecies, the endangered *R. l. yumanensis,* occupying freshwater marshes in the interior of the southwestern United States and north-eastern Baja California. This subspecies breeds only in the lower Colorado River valley and in the Salton Sink, the latter area holding about one-third of the world population (Setmire et al. 1990). It was thought to be confined to the Colorado River valley until small numbers were detected 10 and 11 June 1931 in marshes near Niland and adjacent to the Alamo and New Rivers (Moffitt 1932). Within a short time the species was documented as a breeder in the region (Abbott 1940). Despite representing a sizable proportion of the subspecies' population, numbers at the Salton Sea are modest. For example, a mere 96 individuals were censused around the south end during the summer of 1993 (*AB* 47:1149), and only 279 were located during extensive surveys in 1999 (Shuford et al. 2000), although the actual population may exceed 400 birds (Shuford et al. 1999). *R. l. yumanensis* is partially migratory, many wintering in brackish marshes along the Gulf of California (Banks and Tomlinson 1974; Anderson and Ohmart 1985). Thus, although the Clapper Rail is perennial at the Salton Sea, numbers are augmented in spring. Principal regional sites are Wister, Unit 1, and adjacent marshes around the New River. The species breeds from March through July, building its nest on a raised platform of vegetation concealed in dense marshes.

Formerly, the Clapper Rail was fairly common at the mouth of the Whitewater River, where extensive cattail marshes persisted into the mid-1980s. A maximum of 13 individuals was observed 29 March 1982 (R. L. McKernan). Since clearing of cattail and changes in water level, this species is only occasional at this location, primarily in spring (e.g., 16 May 1992; M. A. Patten). It persists in small numbers in some agricultural drains and small marshes near Oasis, at North Shore, and near Bombay Beach (R. L. McKernan).

ECOLOGY. At the Salton Sea the Clapper Rail occurs in heavily vegetated freshwater marshes of near monotypic stands of Southern Cattail (Bennett and Ohmart 1978). Vegetation density is more important than species composition, as some rails occur in dense stands of Common Reed (Anderson and Ohmart 1985). In marshes along the Colorado River it feeds primarily on crayfish, *Procambarus* spp. and *Oropectes* spp. (Ohmart and Tomlinson 1977); similar crustaceans are taken at the Salton Sea, and the abundance of these animals may be a better predictor of rail population densities than is vegetation (Anderson and Ohmart 1985).

TAXONOMY. *R. l. yumanensis* is the only subspecies occurring away from the coast. Compared with Clapper Rails of the Pacific Coast, it has plumage less richly colored (paler, with more olive and gray tones) and the bill slenderer (Dickey 1923). Its plumage and habitat spawned the hypotheses that *R. l. yumanensis* derived from a hybrid between the Clapper Rail and the closely related King Rail, *R. elegans* Audubon, 1834, and that large rails on the Pacific Coast are actually subspecies of *R. elegans,* not *R. longirostris* (Olson 1997).

VIRGINIA RAIL

Rallus limicola limicola Vieillot, 1819

Uncommon breeding resident; probably more common in winter. The Virginia Rail is both small and secretive, making determining its occurrence and abundance difficult. Furthermore, not only does it occupy dense marsh habitat, but it is typically vocal only during spring and summer. This rail occurs at scattered desert oases throughout the Mojave and Sonoran Deserts, Garrett and Dunn (1981) calling it a common breeding resident along the Colorado River and around the Salton Sea, although breeding commenced at the former locale only 30 years ago (Rosenberg et al. 1991). The Virginia Rail is doubtfully common at the Salton Sea; rather, it is an uncommon breeding resident whose numbers are probably augmented substantially in winter. Seasonal population augmentation was suggested by Garrett and Dunn (1981) and confirmed along the lower Colorado River (Rosenberg et al. 1991), but this species is notoriously difficult to census, so direct evidence is lacking. Nevertheless, it is partly migratory, with occa-

sional records of transients and wintering birds away from breeding areas in southern California from September to March (Garrett and Dunn 1981), so it seems likely that winter numbers are higher. Breeding in the Salton Sink takes place in spring in marshes with suitable cover, such as the Whitewater River delta, Wister, the New River delta, Fig Lagoon, and San Sebastian Marsh. Chicks have been noted as early as 14 June (1999, photographed at Wister; B. Mulrooney), with most records from late June and July and the latest being 3 August (1981, mouth of the Whitewater River; G. McCaskie).

ECOLOGY. Breeding Virginia Rails are largely restricted to areas supporting dense cattail and standing fresh water, but they will use dense bulrush and perhaps even Common Reed. Nonbreeders are more widespread, having been noted in flooded areas with mesquite or Saltcedar overstory (Clary and Clary 1936c; M. A. Patten).

TAXONOMY. Virginia Rails north of central Mexico are of the nominate subspecies, more richly colored and darker than any of three forms from Central America and South America.

SORA

Porzana carolina (Linnaeus, 1758)

Fairly common winter visitor (late July through early May); 2 summer records. Like the Virginia Rail, the Sora is small and secretive, masking its true abundance. Fortunately, this species is much more vocal, often calling at all seasons even into midmorning, and more readily seen than many other marsh rails. It is a common winter visitor to the Salton Sink from 21 July (1990, near Red Hill; G. McCaskie) to 8 May (1937 and 1999, 12 km northwest of Calipatria and at the mouth of the Whitewater River from 4 May; MVZ 71369, M. A. Patten). Most occur from mid-August to the end of April. It is rarely recorded elsewhere in the California desert, except along the lower Colorado River, where it is likewise a common winter visitor, with virtually identical dates of occurrence (Rosenberg et al. 1991). There is no evidence that this species breeds at the Salton Sea, but it has been recorded twice in summer at the Whitewater River delta, 7 June 1981 and 4 July 1998 (G. McCaskie).

ECOLOGY. The Sora occupies virtually any freshwater marsh with dense cattail, bulrush, Common Reed, or scrubby Saltcedar.

COMMON MOORHEN
Gallinula chloropus cachinnans Bangs, 1915
Fairly common breeding resident. The Common Moorhen, widespread across the southern Holarctic, is fairly common in the Salton Sea region and is more easily seen there than over much of the West. Indeed, it is relatively easy to accumulate double-digit counts by searching through suitable habitat. Breeding takes place in early spring, with young appearing as early as midspring (e.g., an adult with 2 chicks at Wister 2 May 1992; M. A. Patten) and continuing to fledge through the summer (e.g., 2 juveniles in the Mexicali Valley 2 August 1997; R. A. Erickson). Because populations in the West are partly migratory (Garrett and Dunn 1981), it is possible that numbers at the Salton Sea are higher during winter.

ECOLOGY. Breeding moorhens occupy clumps of Southern Cattail along rivers and wide irrigation ditches and around freshwater ponds and lakes (e.g., Wister, Unit 1). Some even nest in old agricultural drains in the city of Mexicali (E. Mellink). Foraging occurs principally in fresh water and generally takes place near cover at the edge of ponds or rivers.

TAXONOMY. The subspecies *G. c. cachinnans* occurs throughout the species' North American range south to northwestern South America. In general it is distinguished from other subspecies (11 others worldwide) by its large size and extensively brown mantle and wing coverts.

AMERICAN COOT
Fulica americana americana Gmelin, 1789
Common breeding resident; abundant in winter (September through March). Few bird species are more conspicuous around the Salton Sea than the American Coot, particularly during the winter (September through March), when tens of thousands are present (Shuford et al. 2000).

It is thus the most common rallid in the region, as over much of North America, and it is certainly the most easily seen. It is also a common breeding resident, with nesting in spring and adults accompanied by downy chicks being a common sight by early and midsummer (mid-June through July).

ECOLOGY. Breeding coots occupy dense cover, typically Southern Cattail and bulrush, at the edge of ponds, lakes, rivers, and major irrigation canals. Foraging and wintering birds dot most freshwater ponds, lakes, and reservoirs in the region, particularly in winter. Some forage on the Salton Sea around freshwater inflows, particularly at the mouths of the Alamo and Whitewater Rivers.

TAXONOMY. The coot exhibits north-to-south clinal variation toward long legs, toes, and bill, a more extensive frontal shield, and blacker plumage. Birds across continental North America are of the nominate subspecies, the smallest, grayest form, generally having the most restricted frontal shield (see Roberson and Baptista 1988).

GRUIDAE • *Cranes*

SANDHILL CRANE
Grus canadensis (Linnaeus, 1758) subspp.
Uncommon winter visitor (mid-September to early April) to the Imperial Valley; casual in the Coachella Valley. The lower Colorado River valley and the Salton Sea region host the only regularly wintering Sandhill Cranes (Fig. 36) in California south of the Central Valley. Several flocks winter in agricultural fields of the Imperial Valley, mainly to the south and east of Brawley and to the west of Imperial. The number of birds wintering in the Imperial Valley was stable at about 100 individuals for decades, but by the winter of 1991–92 the number had tripled, and it continues to increase; for example, more than 300 were recorded near Brawley 26 January 1992 (P. E. Lehman), about 350 were there during the winter of 1999–2000 (B. Miller), and 396 were there 17 November 2000 (B. Miller). Sandhill Cranes arrive as early as 8 September (2001, 22 birds 7 km south of Brawley; P. A. Ginsburg)

FIGURE 36. The Imperial Valley is one of three regions in southern California where the Sandhill Crane winters, the others being the Carrizo Plain and the Palo Verde Valley. Photograph by Kenneth Z. Kurland, February 2000.

and 15 September (2000, 4 birds 7 km south of Brawley; *NAB* 55:103), with the majority returning by early October (e.g., 82 birds 3 October 1974; *AB* 29:121). They generally depart by early April (Abbott 1940), with a flock of 15 flying north over the south end 10 April 1984 (*AB* 38:960) providing the latest record for southern California.

There are few records for the Coachella Valley. A flock of more than 100 flying over the north end 5 April 1981 (*AB* 35:863) had probably wintered in the Imperial Valley. A group of five there 25 October 1977 (*AB* 32:257) was likewise probably part of the flock wintering in the Imperial Valley. There are three winter records for the Coachella Valley, with an immature photographed along the Whitewater River near Mecca 29 December 1990 (*AB* 45:320) and single birds at the north end 27 December 1969 (*AFN* 24:453) and 4 January 1997 (*FN* 51:638). An August report from the north end (*AB* 25:906) is undoubtedly mistaken.

ECOLOGY. Wintering crane flocks are a bit nomadic, as they move to different fields of stubble and dry grass to forage, although they favor roosting sites southeast of Brawley, to which they return nightly. The Sandhill Crane frequently uses recently burned fields, particularly those with residual grain stubble.

TAXONOMY. For decades the vast majority (more than 95%) of the cranes wintering in the Imperial Valley were the Lesser Sandhill Crane, *G. c. canadensis* (Linnaeus, 1758), the small form that breeds in northeastern Siberia, Alaska, and northern Canada and accounts for the majority of wintering cranes in southern California (Garrett and Dunn 1981). However, the Greater Sandhill Crane, *G. c. tabida* (Peters, 1925), the largest form and a breeder south to northern California, is the predominant subspecies wintering in the lower Colorado River valley (Rosenberg et al. 1991). It winters in small numbers in the Imperial Valley, the first being a female taken 2 March 1940 (Abbott 1940; SDNHM 18110). The status of this subspecies may be changing, however, as a flock of 47 cranes east of Imperial 25 October 1998 contained 46 *G. c. tabida* and only a single *G. c. canadensis* (M. A. Patten). A few individuals in the Imperial Valley (e.g., 2 in a flock of 41 southeast of Brawley 10 December 1989; M. A. Patten) have appeared to be intermediate in size between *G. c. tabida* and *G. c. canadensis* and so might be *G. c. rowani* Walkinshaw, 1965, an apparently valid subspecies of intermediate size (Browning 1990). Pogson and Lindstedt (1991) claimed that *G. c. rowani* winters in small numbers in California's Central Valley; however, there are no supporting specimens for the state.

CHARADRIIFORMES • *Shorebirds, Gulls, Auks, and Allies*

The Salton Sea is one of the most important stopover points on the Pacific Flyway for migrating shorebirds, and together with the Great Salt Lake it hosts on average more migratory shorebirds than any other area in the intermountain and desert regions of the West (Page and Gill 1994; Shuford et al. 2000). The Salton Sea is particularly important for the Black-necked Stilt, the American Avocet, the Western Sandpiper, and the Long-billed Dowitcher (Table 9). Spring numbers are particularly high, with some species uncommon to rare inland in the Southwest, such as the Willet, the Whimbrel, and the Marbled Godwit, regular at the Salton Sea and often occurring in large numbers. Furthermore, the sea is the only place in the interior West that hosts regular numbers of generally coastal species, such as the Red Knot, the Sanderling, and the Ruddy Turnstone (see Table 9), and it is the only location in California hosting regular flocks of the Stilt Sandpiper. By contrast, fall migration is more protracted, with far lower numbers of the coastal species, but species such as the Western Sandpiper and the Red-necked Phalarope pass through then in equally large or larger numbers. In winter the sea supports the largest population of the Snowy Plover in the interior West and more than 30 percent of the world population of the Mountain Plover (Shuford et al. 2000). Curiously, the sea is not a major staging area for Wilson's Phalarope, in stark contrast to some other large saline lakes in the interior West (Jehl 1988).

CHARADRIIDAE • *Lapwings and Plovers*

BLACK-BELLIED PLOVER
Pluvialis squatarola (Linnaeus, 1758)

Common transient and winter visitor (late June to early June); uncommon summer visitor. The Salton Sea is a magnet for migratory shorebirds, so much so that a species like the Black-bellied Plover, normally rare to uncommon in the desert Southwest, is common at the sea. The plover joins the Marbled Godwit, the Western Sandpiper, and the Long-billed Dowitcher in an unusual distinction for migratory shorebirds through the region: the latest spring migrants nearly meet the earliest fall migrants. Returning fall transients arrive as early as 25 June (1989, south end; *AB* 43:1368), but birds in full alternate plumage, and thus presumably spring transients, have been noted as late as 6 June (1999, mouth of the Whitewater River; M. A. Patten). As is typical for this set of species, mid-June is the nadir for the plover's abundance at the sea. Even so, it is an uncommon nonbreeding visitor through summer, generally numbering a few dozen, with a maximum of 90 at the north end 7 June 1987 (*AB* 41:1487). It is common as a migrant and winter visitor, with numbers differing little in these seasons. Winter tallies can be impressive, such as 1,310 and 1,381 counted in surveys conducted 22 January 1999 and 11 November 1998, respectively (Shuford et al. 2000), and 800 at Salton Sea NWR alone in January 1951 (*AFN* 5:227).

ECOLOGY. Black-bellied Plovers forage along the immediate shoreline and in large impoundments of fresh or brackish water, backwaters, and fields, particularly if recently irrigated. They have broad preferences, using beaches of sand, barnacles, or mud; mudflats at river mouths; or alkaline flats near water.

TAXONOMY. The Black-bellied Plover is generally regarded as monotypic. Engelmoer and Roselaar (1998) recognized the smaller *P. s. cyanosurae* (Thayer and Bangs, 1914), of North America, and named the larger *P. s. tomkovichi*, of Wrangel Island. Their table 24, however, shows considerable overlap among all measurements,

TABLE 9
Approximate Mean Numbers of Shorebirds Migrating
Annually through or to the Salton Sea Region in Spring

SPECIES	MEAN NUMBER	SPECIES	MEAN NUMBER
Black-bellied Plover	2,000	Black Turnstone	1–2
American Golden-Plover	Vagrant	Surfbird	1
Pacific Golden-Plover	Vagrant	Red Knot	1,500
Snowy Plover	250	Sanderling	400
Semipalmated Plover	1,000	Semipalmated Sandpiper	5
Killdeer	5,000	Western Sandpiper	75,000
Black-necked Stilt	35,000	Little Stint	Vagrant
American Avocet	50,000	Least Sandpiper	5,000
Greater Yellowlegs	600	White-rumped Sandpiper	Vagrant
Lesser Yellowlegs	200	Baird's Sandpiper	Vagrant
Spotted Redshank	Vagrant	Pectoral Sandpiper	Vagrant
Solitary Sandpiper	1–2	Dunlin	2,000
Willet	6,500	Curlew Sandpiper	Vagrant
Wandering Tattler	<1	Stilt Sandpiper	500
Spotted Sandpiper	300	Short-billed Dowitcher	5,000
Whimbrel	7,500	Long-billed Dowitcher	60,000
Long-billed Curlew	5,000	Wilson's Snipe	20
Hudsonian Godwit	Vagrant	Wilson's Phalarope	5,000
Marbled Godwit	10,000	Red-necked Phalarope	15,000
Ruddy Turnstone	500	Red Phalarope	Vagrant

Sources: Information published in *American Birds* or *Field Notes* and field observations from Michael A. Patten and Guy McCaskie.

with no population being diagnosable at even the 75 percent level ($D_{ij} < -1.5$ for all pairwise comparisons).

AMERICAN GOLDEN-PLOVER
Pluvialis dominica (Müller, 1776)
Casual spring (early April to mid-June) and fall (July–November) vagrant; 1 summer record. Because the American and Pacific Golden-Plovers were regarded as conspecific until recently (Connors 1983; Connors et al. 1993; AOU 1993), and because of difficulty in field identification of these taxa (Dunn et al. 1987), the status of

each species at the Salton Sea is clouded and uncertain. Golden-plovers are casual transients at the Salton Sea, with records of both species in spring and fall. The majority are apparently of the American Golden-Plover, although most individuals involved were not critically identified to subspecies prior to the taxonomic split. Even so, this species is decidedly the more frequent of the two golden-plovers elsewhere in the interior Southwest, including in the Mojave Desert of California (M. T. Heindel pers. comm.) and in Arizona (Rosenberg and Witzeman 1998). We thus infer, with the proper caveats, that most records of migrants pertain to the American Golden-Plover. In spring such records extend from 6 April (1985, north end; *AB* 39:350) to 13 June (1970, mouth of the Whitewater River; *AFN* 24:716), with a peak from late April to mid-May, coinciding with the temporal window in which golden-plovers have since been identified specifically as Americans, including a specimen of a male in basic or first-alternate plumage from the Niland Boat Ramp 1 May 1994 (R. L. McKernan; SBCM 54199). Lone individuals account for almost all spring records, exceptions being two photographed at the mouth of the Whitewater River 15 May 1982 (*AB* 36:894) and two south of Wister 4 May 1996 (*FN* 50:332). A female in alternate plumage at Wister 11 June–18 July 1999 (*NAB* 53:432) furnished the only summer record for California. The American Golden-Plover is equally scarce in fall, with records from 11 July (1970, near the mouth of the Alamo River; *AFN* 24:716) to 12 November (1983 and 1999, 4 at Brawley and 1 photographed near the mouth of the New River from 11 November; *AB* 38:246, *NAB* 54:104). A single Pacific Golden-Plover joined the flock of four. Small flocks of American Golden-Plovers are occasionally detected in fall; in addition to the flock of four, six were present at the south end 26 October 1986 (*AB* 41:143).

ECOLOGY. Golden-plovers of either species are generally found with flocks of Black-bellied Plovers on mudflats at the edge of the Salton Sea or in large freshwater impoundments (as at Wister).

PACIFIC GOLDEN-PLOVER

Pluvialis fulva (Gmelin, 1789)

Casual winter vagrant (late November to early April); 4 spring records (late April to mid-May); 2 fall records. The Pacific Golden-Plover is a rare migrant through coastal California, where small numbers also winter annually. By contrast, this species is virtually unrecorded in the interior of western North America, with only two valid records for the Mojave Desert of California (*NAB* 55:103, *fide* G. McCaskie) and only one for Ari-

zona (*NAB* 53:84). For this reason we assume that nearly all Salton Sea records of unidentified golden-plovers pertain to the American (see the American Golden-Plover account). However, we infer that all winter records of golden-plovers to pertain to the Pacific because it is the only golden-plover known to winter in the West (cf. Paulson and Lee 1992). There are nine records from the Salton Sink in that season (specifically mid-November to mid-April), with lone birds southeast of Brawley 12 November 1983 (with 4 American Golden-Plovers; *AB* 38:246), in February 1998 (Fig. 37; K. Z. Kurland), 16 February–13 April 2000 (*NAB* 54:221), and 2 February 2002 (M. San Miguel), up to three at Red Hill 29 November–26 December 1986 (*AB* 41:328), lone birds at Red Hill 12 January–6 March 1982 (*AB* 36:331) and 30 January 1983 (*AB* 37:338), one at the south end 24 January 1970 (*AFN* 24:539), and three near Westmorland 13–19 February 1966 (*AFN* 20:459). Other records attributed to the south end during the winter of 1969–70 (*AFN* 24:539) were the result of a typesetting error (they actually pertain to the Stilt Sandpiper).

There are four spring records of single birds critically identified as Pacific Golden-Plovers: at the mouth of the Whitewater River 23 April 1988 (*AB* 42:481), at the south end 23 April 1989 (*AB* 43:536), near the mouth of the New River 2 May 1992 (*AB* 46:480), and at the mouth of the Whitewater River 19 May 1999 (*NAB* 53:329). The last two were males in full alternate plumage. Another was reported without details 17 or 18 April 1999 (Shuford et al. 2000:table 5-1). The only fall records are of an adult in alternate plumage at Bruchard Bay 2 July 1988 (*AB* 42:1340) and a juvenile 4 km east of Imperial 17 October 1999 (*NAB* 54:105). One in alternate plumage near Oasis 21 June 1997 (*FN* 51:1053) was either an exceptionally late spring migrant or, more likely, an exceptionally early fall migrant.

ECOLOGY. See the American Golden-Plover account. Wintering Pacifics have occurred in agricultural fields and on the shore of the sea (e.g., at Red Hill).

FIGURE 37. This Pacific Golden-Plover is one of only a dozen or so recorded in the Salton Sink. Photograph by Kenneth Z. Kurland, February 1998.

FIGURE 38. The Salton Sea supports about 200 pairs of nesting Snowy Plovers, a substantial number for this threatened species. Photograph by Kenneth Z. Kurland, December 1999.

SNOWY PLOVER

Charadrius alexandrinus nivosus (Cassin, 1858)

Fairly common breeder (mid-March through October); less common in winter. The Salton Sea is the most important wintering site for the Snowy Plover (Fig. 38) in the interior of western North America. This species breeds widely in the interior West but is more common year-round at the Salton Sea than anywhere else in this vast area save at the Great Salt Lake in northern Utah (Page et al. 1986, 1991). That the population in coastal California was recently listed as threatened under the federal Endangered Species Act underscores the importance of the sea, as banded individuals have moved between the coast and the sea (K. C. Molina pers. comm.). The Snowy Plover is a fairly common breeder at the Salton Sea (Grant 1982:45), where 200–225 pairs breed annually (Page et al. 1991; Shuford et al. 2000), mostly on the west side from Desert Shores to the mouth of San Felipe Creek and on the east side from Bombay Beach to Wister (Page and Stenzel 1981). Breeding was first recorded in the

Salton Sink in 1929 (Page and Stenzel 1981), so it is possible that this species colonized the area at roughly the same time as the Laughing Gull and the Gull-billed Tern. However, the plover has been known from the sea since shortly after its formation, for example, from near Mecca 29 March 1908 (MVZ 736–738) and 26 March 1911 (van Rossem 1911), so it is possible that it bred locally well before the late 1920s.

As with most of the more numerous waterbirds occurring in the region, determination of arrival and departure dates is hindered by the presence of nonmigrant individuals. The Snowy Plover is generally uncommon through the winter (Page et al. 1986), when it is more numerous than the Semipalmated Plover but greatly outnumbered by the Killdeer, the Black-bellied Plover, and even the Mountain Plover. The wintering population is usually between half and three-fourths the size of the breeding population (Page et al. 1986; Shuford et al. 2000), yet it is still the largest in the interior West. Counts are generally under two dozen, but 71 were seen at the south end 19 December 2000 (G. Mc-Caskie et al.) and 56 were there 28 December 1995 (*FN* 50:852). Apparent returning migrants have been noted as early as 12 March (1988, Whitewater River delta; M. A. Patten), although a flock of 25 at the southeastern edge of the sea 1 March 2000 may have been exceptionally early (G. McCaskie). Summer numbers remain high through October, such as 50 at Salton Sea NWR 14 October 1955 (*AFN* 10:56) and 40 along Morton Bay 25 October 1998 (M. A. Patten).

ECOLOGY. Most Snowy Plovers nest close to standing water and generally far from conspecifics (Grant 1982). All Charadriiformes nesting at the Salton Sea avoid shade, presumably to avoid predation (Grant 1982:57). The plover forages along the shoreline of the sea, mostly on sand and barnacle beaches, although it also concentrates in shallow impoundments with exposed mud and on alkaline flats and mudflats around the river mouths.

TAXONOMY. The two subspecies in the Americas, *C. a. nivosus*, of North America and the West Indies, and *C. a. occidentalis* Cabanis, 1872, of South America, differ from the four Old World subspecies in their paler upperparts, shorter legs, and, in alternate plumage, whiter lores and limited rufous in the crown. *C. a. occidentalis* is larger than *C. a. nivosus* and has a somewhat wider black breast band (Blake 1977).

WILSON'S PLOVER

Charadrius wilsonia beldingi Ridgway, 1919
One breeding record. Lloyd F. Kiff, then director of the Western Foundation of Vertebrate Zoology, holder of the largest collection of egg sets in the Western Hemisphere, carefully examined and compared a nest and eggs collected on Mullet Island 20 May 1948 (SBCM 19108) and determined that they belonged to Wilson's Plover, providing California with its only nesting record (Roberson 1993). No birds were reported tending the nest. A report of five Wilson's Plovers at the south end doubtless was mistaken (Langham 1991).

TAXONOMY. Without a specimen of an adult accompanying the nest, the subspecies cannot be known, but it was almost certainly *C. w. beldingi*, which breeds as close as the northern Gulf of California (Patten et al. 2001). It is the darkest subspecies, with a wide breast band and short supercilia, and males have the most extensively black head, especially through the lores (Blake 1977:548). Females show extensive rufous in the breast band, more so than the nominate subspecies, of eastern North America, but less so than *C. w. cinnamominus* Ridgway, 1919, of South America (in which males also have a rufous breast band). The only Wilson's Plover specimen for California (MVZ 31920), taken from Pacific Beach, San Diego County, 24–29 June 1894, is *C. w. beldingi* (Ingersoll 1895).

SEMIPALMATED PLOVER

Charadrius semipalmatus Bonaparte, 1825
Common spring transient (mid-March through May); fairly common fall transient (late June through October); uncommon to rare winter visitor; rare in summer. The Semipalmated Plover is generally an uncommon migrant through the interior Southwest, but it is a common spring and fairly common fall migrant at the Salton Sea, frequently numbering in the hundreds per day. Spring abundance can be particularly impressive, evidenced by the roughly 2,000 around the sea 29 April 1962 (Garrett and Dunn 1981). Spring transients pass through from 17 March (1991, Salton City; M. A. Patten) to 7 June (1969, 6 at the mouth of the Whitewater River; G. McCaskie), with a peak from mid-April to mid-May. Like that of virtually all shorebirds at the Salton Sea, fall migration is more protracted, with lower concentrations of birds. Fall transients pass through from 22 June (1991, 3 at the mouth of the New River; G. McCaskie) to 25 October (1998, 2 at the south end; M. A. Patten), with peak movement in August. A few winter at the sea annually, clouding fall departure dates and spring arrival dates. The Semipalmated Plover is somewhere between uncommon and rare in that season; it is virtually always outnumbered by the Snowy Plover and seldom recorded in double digits. Nonbreeders remaining through the summer are rare, fewer than birds remaining

through the winter. The summer maximum is up to 11, occurring at the Whitewater River delta 6 June–18 July 1999 (M. A. Patten et al.).

ECOLOGY. This plover is seldom encountered away from the immediate shoreline of the sea and its adjacent impoundments, backwaters, and ponds. Wintering birds are especially tied to shallow impoundments. The species favors mud-flats and alkaline flats over beaches. Occasional birds venture into the Imperial Valley, where they occupy the edges of lakes and reservoirs and sometimes even flooded fields.

KILLDEER
Charadrius vociferus vociferus Linnaeus, 1758
Common breeding resident. Few birds are as noisy and conspicuous as the Killdeer. This large plover is common throughout much of North America, although it is absent from most of the noncultivated desert. Irrigation of the Coachella, Imperial, and Mexicali Valleys and flooding of the Salton Sea created an abundance of suitable habitat for this species. Except in open desert scrub the Killdeer is a common breeding resident throughout the Salton Sink. Numbers seem especially high in winter; for example, the total for both the north and south ends both 22 December 1992 and 2 January 1993 was about 500 (*AB* 47:968, 973). Breeding begins in early spring, with egg dates from 17 March to 1 July and peak clutch initiation about 1 May (Grant 1982). Peak hatching is about 1 June, but chicks are noted from 15 April through August (Grant 1982).

ECOLOGY. The Killdeer nests farther from water than most other breeding shorebirds in the region (Grant 1982). Conspecific nests are generally well spaced. Incubating adults soak their belly feathers and wet their feet to cool eggs and limit water loss; this behavior is known from all Charadriiformes that bred at the Salton Sea in the late 1970s (Grant 1982:66). The Killdeer forages along the shoreline of the sea (especially around river mouths), at the edges of lakes and reservoirs, on extensive lawns in towns (e.g., ball fields, parks), and in agricultural fields, be they partly flooded, mostly dry with short vegetation, or recently plowed.

TAXONOMY. The nominate subspecies, widespread in North America, is larger and more uniformly brown on the wings (i.e., there is reduced refuscent edgings to the coverts) than subspecies in the West Indies and South America.

MOUNTAIN PLOVER
Charadrius montanus Townsend, 1837
Fairly common winter visitor (late September through March). The Mountain Plover (Fig. 39) has suffered critical declines across much of its range and is therefore being considered for listing under the federal Endangered Species Act. The Imperial Valley hosts what is undoubtedly the largest wintering population anywhere, with as many as about 3,700 individuals annually (Shuford et al. 2000:25), more than 60 percent of all wintering Mountain Plovers in California (Shuford et al. 1999) and 30–38 percent of the world population of the species (Shuford et al. 2000). Flocks are nomadic, but counts of 200–300 individuals are frequent. Totals from the Salton Sea (south) Christmas Bird Count, whose circle does not include the majority of prime habitat, of 531 on 21 December 1982 (*AB* 37:772), 701 on 22 December 1987 (*AB* 42:1137), and an impressive 1,003 on 18 December 1990 (*AB* 45:999) indicate just how common the species can be. A few wintering birds return by late August, with three at the south end 24 August (1991; *AB* 46:149) being the earliest (but see below). The majority return in late September (e.g., 30, including several juveniles, near the mouth of the New River 27 September 1992; M. A. Patten). Most depart in March, the latest record being 31 March (1979, 10 at Unit 1; G. McCaskie). Three records of exceptionally early fall migrants at the south end are of one 15 July 1979 (*AB* 33:897), two birds 28 July 1969 (*AFN* 23:695), and one at Red Hill 5 August 1967 (*AFN* 21:604; SDNHM 36117). The Mountain Plover is scarce in the Coachella Valley despite what appears to be a good deal of suitable habitat. The species was first detected 17 January–10 February 1936, with two birds 16 km northwest of Indio (Clary and Clary 1936c). It is absent most years, but a high count of 19 was achieved on the Salton Sea

FIGURE 39. The Imperial Valley's agricultural fields have become a major winter habitat for the Mountain Plover, supporting one-third of the species' entire population. Photograph by Kenneth Z. Kurland, March 1999.

(north) Christmas Bird Count 19 December 1982 (*AB* 37:771).

ECOLOGY. This oddly named species breeds in dry upland grasslands (Knopf and Miller 1994) and winters in barren dirt fields, freshly plowed agricultural fields (Knopf and Rupert 1995), and, especially in the Salton Sea region, recently burned agricultural fields (M. A. Patten; Shuford et al. 2000:25). Indeed, the practice of burning fields may be needed to provide suitable habitat for the species and may explain its typical absence from the Coachella Valley, where fields are seldom burned. The Mountain Plover is an opportunistic feeder, though it dines principally on Coleoptera and Orthoptera (Knopf 1998).

EURASIAN DOTTEREL
Charadrius morinellus Linnaeus, 1758
One winter record. In North America the Eurasian Dotterel breeds in western and northern Alaska. This population migrates through eastern Asia, so the species is scarce elsewhere in North America, where there are a dozen records for the Pacific Coast of juveniles in fall, chiefly September. The species wintered once in north-

western Baja California (*FN* 52:257, 269), however, setting a precedent for the lone record for the Salton Sink, of a bird photographed northeast of Calipatria 22 and 23 January 2001 (*NAB* 55:227; CBRC 2001-071).

HAEMATOPODIDAE • *Oystercatchers*

AMERICAN OYSTERCATCHER
Haematopus palliatus Temminck, 1820 subsp.?
One summer record. Three juvenile American Oystercatchers photographed at Salton City 14–19 August 1977 (see Roberson 1980:114) gave California its only inland record. These birds later moved east across the sea to Salt Creek 20–30 August 1977 (Luther 1980). Even though the species breeds as far north as northern Baja California, it remains extremely rare anywhere in California, with fewer than 20 records to date. A sighting from Idaho (Stephens and Stephens 1987) and a tentative one from New Mexico (Hubbard 1978) are the only others reported inland in the West.

TAXONOMY. The only specimens for California are of the subspecies *H. p. frazari* Brewster, 1888, of western Mexico and Baja California. This

FIGURE 40. With numbers in the tens of thousands, the Black-necked Stilt is one of the Salton Sea's most common and conspicuous birds. Photograph by Kenneth Z. Kurland, July 2001.

subspecies differs from the nominate subspecies in having smudgy black below the breast, less extensively white wing stripes, and blackish smudging on the rump. Photographs of the Salton Sea birds, however, show them to be clean white below a sharp breast division with no dusky flank markings, suggesting that they may have been nominate *H. p. palliatus*.

RECURVIROSTRIDAE • *Stilts and Avocets*

BLACK-NECKED STILT

Himantopus mexicanus mexicanus (Müller, 1776)
Abundant breeding resident. Vying with the Killdeer for the title of noisiest and most aggressive shorebird in the West, the Black-necked Stilt (Fig. 40) has been nicknamed the "Marsh Poodle" by some birders for its incessant yipping at all passersby. The stilt has long been common in the region (see Abbott 1931). It is an abundant breeder at the Salton Sea, with winter numbers increasing enough over the past three decades to equal those during the summer. Year-round totals are now fairly stable, reaching into the tens of thousands (Shuford et al. 1999), and show a numerical increase during migration (April, July–August; Shuford et al. 2000). Nesting activity

is concentrated from April through June, with a median laying date of 15 May and a median hatching date of 14 June (Grant 1982). Egg dates range from 2 April to 16 August (Grant 1982), and three downy young at Mexicali 14 April (Murphy 1917) must have come from eggs laid even earlier.

ECOLOGY. The stilt nests in loose colonies close to water, even on a spit or isthmus of barnacles or sand closely bounded by water (Fig. 41). This species risks egg loss through muddying, as do other waterbirds breeding in the region that belly-soak to aid in thermoregulation of eggs (Grant 1982:31). Stilts feed in shallow pools and impoundments at the edge of the sea, concentrated at the river mouths. Impoundments along Morton Bay sometimes hold staggering numbers (e.g., ca. 5,000 birds 16 July 1998; M. A. Patten).

TAXONOMY. The nominate subspecies is widespread in North America. It differs from *H. m. melanurus* Vieillot, 1817, of southern South America, in having a black crown and lacking a white band separating the black nape from the black mantle. It differs from *H. m. knudseni* Stejneger, 1887, of the Hawaiian Islands, in its

FIGURE 41. A Black-necked Stilt nest near Mecca in a typical open setting. Photograph by Michael A. Patten, June 2000.

narrow black nape stripe and reduced black in the auriculars.

AMERICAN AVOCET
Recurvirostra americana Gmelin, 1789
Abundant perennial visitor; rare, irregular breeder (May through July). Aside from the Western Sandpiper and perhaps the Black-necked Stilt and the Long-billed Dowitcher, the American Avocet is the most numerous shorebird at the Salton Sea (see Table 9; Shuford et al. 2000). The sea's lengthy shoreline attracts many thousands of avocets, probably the greatest concentrations in the interior West. The sizes of flocks can be astounding; for example, there were about 10,000 along Morton Bay 16 July 1998 (M. A. Patten). Numbers fluctuate through the year with movements of migrants, although the avocet is generally abundant year-round (though sometimes merely common in May and June). In contrast to the Black-necked Stilt, which is nearly as numerous a perennial visitor, the avocet rarely breeds at the Salton Sea and apparently does not do so annually. The first regional nest was discovered at the south end 27 June 1964 (*AFN*

18:535). Avocets bred again in the late 1970s, when two pairs nested at the south end in 1977 (*AB* 31:1189) and five pairs nested there in 1978 (*AB* 32:1208). Nesting has been sporadic since, both at the south end (most frequently Bruchard Bay and Wister) and at Salton City (e.g., a nest with 2 eggs 31 May 1992; M. A. Patten). Egg dates range from 22 May to 4 July (Grant 1982).

ECOLOGY. Nest sites of the avocet tend to be in or near backwaters or shallow pools isolated from, but adjacent to, the sea. The largest concentrations of migrants, nonbreeders, and wintering birds are attracted to shallow water, especially if it is somewhat brackish; impoundments and backwaters adjacent to the sea are especially attractive.

SCOLOPACIDAE • *Sandpipers, Phalaropes, and Allies*

GREATER YELLOWLEGS
Tringa melanoleuca (Gmelin, 1789)
Fairly common transient and winter visitor (late June to mid-May); uncommon summer visitor. Unlike many other species of shorebirds occurring at the Salton Sea, the Greater Yellowlegs is also

fairly common elsewhere in the interior Southwest. It occurs throughout the year in the Salton Sink. Numbers peak during migration, mainly from mid-March to mid-May and from late June to early November, but it is fairly common through the winter as well. In contrast to most shorebirds moving through the region, this species may be more numerous in fall, when hundreds are sometimes tallied (e.g., 250 at the south end 25 October 1998; M. A. Patten). It is an uncommon nonbreeding summer visitor (e.g., 25 around the sea during June and July 1987; AB 41:1487). The annual presence of summering birds hinders the determination of arrival and departure dates, particularly because some summering birds are first noted in spring (e.g., 1 at the mouth of the Whitewater River 1 May–4 July 1999; M. A. Patten). Nevertheless, spring migrants decrease by late April, with birds in full alternate plumage noted as late as 16 May (1998, mouth of the Whitewater River; M. A. Patten). Fall migrants return as early as 25 June (1989, south end; G. McCaskie), the majority of adults returning in early to mid-July.

ECOLOGY. As noted by van Rossem (1911), this species is scarce at the edge of the Salton Sea, though it does occur there. It prefers flooded fields and shallow ponds and impoundments. It is frequently noted in shallow irrigation canals carrying water to the sea, and some even occur at edges of concrete-lined reservoirs.

LESSER YELLOWLEGS
Tringa flavipes (Gmelin, 1789)
Uncommon transient and winter visitor (late June to mid-May); casual in summer. The Lesser Yellowlegs is generally outnumbered by the Greater Yellowlegs over much of the West, including at the Salton Sea. Peak numbers at the sea seldom reach triple digits even during prime migration periods. Spring migrants pass through roughly from 28 March (1997, north end; M. A. Patten) to 16 May (1998, Oasis; M. A. Patten), with a peak in the latter half of April (e.g., 100 near Calipatria 22 April 1972; AB 26:809). Fall migrants generally return at the end of June, with one in full alternate plumage at the south end

24 June 1995 (G. McCaskie) being the earliest. Fall migration is no more perceptible than spring (contra Garrett and Dunn 1981:48), but there is a slight peak from mid-August through mid-September (e.g., ca. 100 near Calipatria 31 August 2000; G. McCaskie), coincident with the arrival of juveniles. Even during peak movement counts seldom exceed a few dozen birds. Fall migration blends into the winter period, so departure dates cannot be determined. This species was rare around the Salton Sea in winter only two decades ago (Garrett and Dunn 1981), but it is now nearly as frequent in that season as during migration (e.g., 50 at Obsidian Butte 5 February 2000; G. McCaskie). Thus, numbers in the Salton Sink, especially at the south end of the sea, are fairly constant from the arrival of the first fall transients to the departure of the last in spring. Most wintering birds occur around the south end, especially from Wister to Obsidian Butte, but the Lesser Yellowlegs may be found just about anywhere in the region in winter, from small ponds at the north end to flooded fields in the Imperial and Mexicali Valleys. It is casual through the summer, with records once every two to three years from 5 June (1983, Unit 1; G. McCaskie) to 4 July (1987, south end; G. McCaskie).

ECOLOGY. The Lesser Yellowlegs is encountered less frequently than the Greater in part because of its more specialized habitat preference (Garrett and Dunn 1981). Flooded fields and exposed mudflats in shallow ponds and impoundments attract most birds encountered in the region. The latter habitat is especially favored in winter, although some use flooded fields in that season (e.g., at least 10 birds 25 km southeast of Mexicali 29 December 1991; Patten et al. 1993). The Lesser is far less frequent than the Greater in irrigation canals or along the rivers and is virtually never found along the shoreline of the sea or at reservoirs.

SPOTTED REDSHANK
Tringa erythropus (Pallas, 1764)
One spring record. The first Spotted Redshank recorded in California was one in alternate

plumage photographed at the mouth of the Whitewater River 30 April–6 (not 4) May 1983 (Morlan 1985; *AB* 37:912). Outside Alaska there are few records of this Asiatic shorebird for North America, five of them from California (Mlodinow 1999).

SOLITARY SANDPIPER
Tringa solitaria cinnamomea (Brewster, 1890)
Rare fall transient (mid-July through September); casual in spring (mid-April to early May); 2 winter records. The Solitary Sandpiper is an uncommon fall and rare spring migrant through California. Although widespread in southern California, it is less frequent in the Salton Sink than in similar habitat elsewhere, including in the Mojave Desert. As they do elsewhere in the West, fall migrants predominate, adults from July into early August and juveniles thereafter. The species occurs annually in small numbers in that season, from 12 July (1990, south end; *AB* 44:1186) to 30 September (2000, Finney Lake; G. McCaskie), including a flock of eight flying south past Salton City 24 August 1991 (C. McGaugh). There are 12 spring records in the narrow window of 12 April (1949, Westmorland; SBCM 31894) to 9 May (1987, Finney Lake; *AB* 41:488), half of them supported by specimens (MVZ; SDNHM; UCLA; USNM). This species is rare in the United States in winter. There are fewer than ten winter records for California, two of them from the Salton Sea, from the south end 26 December 1957 (not 1958; *AFN* 12:306) and from the north end 12 January 1977 (Garrett and Dunn 1981).

ECOLOGY. This species occurs at small freshwater ponds with weedy edges, at marshes with open water, and along the rivers and irrigation canals.

TAXONOMY. The western North American *T. s. cinnamomea* is the only subspecies recorded in California, including all Salton Sink specimens from spring and a juvenile 6 km northwest of Westmorland 12 September 1954 (SBCM 31895). *T. s. cinnamomea* is slightly larger than the nominate subspecies, of eastern North America, and the spotting on the upperparts is cinnamon-buff

rather than buff-white. Although it is a short-distance migrant, the nominate subspecies is a likely vagrant to California and should be considered. Because *T. s. solitaria* winters farther north, Phillips et al. (1964:34) suggested that wintering birds in Arizona would be of the nominate subspecies, although *T. s. cinnamomea* is the only subspecies definitely recorded in that state.

WILLET
Catoptrophorus semipalmatus
   inornatus (Brewster, 1887)
Common transient and winter visitor (mid-June to mid-May); uncommon nonbreeding summer visitor. Away from its breeding areas in the Great Basin and on the Great Plains, the Willet is largely coastal, yet at the Salton Sea it may be encountered commonly. Substantial numbers occur year-round, so migration dates cannot be determined; nonetheless, numbers reach a low in the narrow window of mid-May to mid-June as the last of the spring migrants trickle through and before the harbingers of fall return. As an example, 125 around the Whitewater River delta 1 May 1999 had declined to 60 by 8 May, 25 by 16 May, and a mere 4 by 31 May. A maximum of a dozen were in the area through 15 June 1999, but numbers had built to 20 by 19–27 June and 75 by 13 July (M. A. Patten). Counts are otherwise fairly constant throughout the year, with modest peaks during spring from April to mid-May (e.g., 300 at the north end 19 April–3 May 1998; M. A. Patten) and during fall from late June to early October (e.g., 200 at the south end 22–25 July 1999; M. A. Patten). Hundreds winter in the region (Shuford et al. 2000). Generally only a few nonbreeders remain through the summer, but 40 were recorded at the south end 17 June 1967 (*AFN* 21:604).

ECOLOGY. The Willet is most frequent on the sea's shoreline, particularly on mudflats, sand spits, and barnacle beaches. It also occurs on rock jetties. It rarely occurs at freshwater lakes and reservoirs in the Imperial Valley.

TAXONOMY. Birds in western North America are the smaller, less heavily barred (in alternate plumage) subspecies *C. s. inornatus,* the Western Willet. Interestingly, although *C. s. inornatus* is a regular migrant in many locales in eastern North America, the nominate subspecies is unrecorded west of the Great Plains. Phillips (1962) argued that the name *C. s. speculiferus* (Cuvier, 1829) had priority over *C. s. inornatus,* but no one has followed this suggestion, and Phillips himself (Phillips et al. 1964; Monson and Phillips 1981) later ignored the subspecies issue all together.

WANDERING TATTLER
Heteroscelus incanus (Gmelin, 1789)

Casual vagrant in spring (late April to mid-June) and fall (August to early October); 1 winter record. During migration the Wandering Tattler is casual anywhere inland, where there are more records for the Salton Sea than for anywhere else. It occurs at the Salton Sea almost every year, particularly in spring during the peak of northward shorebird migration out of the Gulf of California. The first area record is of a bird collected at a playa of the Alamo River 12 km east of Mexicali 27 or 28 April 1894 (Mearns 1907: 130; USNM 133767). Grinnell (1928) mistakenly indicated that "specimens" were collected. Furthermore, he seemingly cast doubt on its whereabouts, but USNM 133767 is from "Baja California; Laguna of Salton River [= Alamo River]. Collected 27–28 April 1894 by E. A. Mearns" (James P. Dean *in litt.*). Spring records since have been exclusively of birds in alternate plumage from 27 April (1974, 1 at the south end; *AB* 28: 852) to 11 June (1978, 1 at the south end; *AB* 32:1208), including one in alternate plumage photographed at the mouth of the Whitewater River 16 (not 17) May 1992 (see *AB* 46:481). All records have been of lone individuals, except for up to four together at Salton City 8 and 9 May 1971 (*AB* 25:800), up to three at Salton City 5 and 6 May 1984 (*AB* 38:960), and two near Salton City 25 May 1996 (*FN* 50:332). There are only eleven fall records, from 3 August (1991, basic-plumaged adult near Obsidian Butte; *AB* 46:149) to 4 October (1999, Obsidian Butte from 3 October; *NAB* 54:105), with one at Wister 20 July

2002 (*NAB* 56:486) being exceptionally early. The other eight records are from the south end 22 August 1977 (*AB* 32:257), 5 September 1998 (*NAB* 53:104), and 7–8 September 1996 (*FN* 51:120); Salton City 25 August 1984 (*AB* 39:102; photograph); near Red Hill 31 August 1961 (McCaskie 1970d); the Niland Boat Ramp 15 September 1990 (*AB* 45:151); and Obsidian Butte 11–15 September 2000 (*NAB* 55:103) and 19 September 2002 (K. Z. Kurland). One in basic plumage at Obsidian Butte 5 March–24 May 2000 (*NAB* 54:221) and what was apparently the same bird there 29 December 2000–2 February 2001 (*NAB* 55:227) provided the only winter records inland in North America.

ECOLOGY. Virtually all area records of the tattler are from the shoreline of the sea, especially on barnacle beaches and rock outcrops.

SPOTTED SANDPIPER
Actitis macularia (Linnaeus, 1766)

Fairly common transient and winter visitor (mid-July through May); casual in summer. The teetering, lively Spotted Sandpiper is widespread throughout North America. Because it is not a flocking species it is generally thought of as uncommon, but it is actually a fairly common migrant throughout the interior West. Its status at the Salton Sea is similar, although in contrast to its habit in much of the interior West, this species also winters fairly commonly in the region. Birds are present from early fall to late spring, with records from 9 July (1988, south end; M. A. Patten) to 25 May (1998, south end; M. A. Patten). Individuals near Mecca 3 June 2001 and at Salton City 6 June 1982 (both G. McCaskie) were either late spring transients or birds attempting to summer locally. There is a slight increase in numbers during migration, particularly in spring, when double-digit counts can be recorded in the last week of April and the first two weeks of May. A smaller peak occurs during fall, from early August into mid-September (see Shuford et al. 2000), partly coincident with the arrival of juveniles in mid-August. One or two individuals occasionally summer, particularly around the south end (e.g., birds seen 19 June

1976 and 11 June 1978; G. McCaskie), but the Spotted Sandpiper is not annual in that season.

ECOLOGY. In general this species avoids open mudflats and beaches, adding to the impression of its rarity. A search of freshwater ditches and the edges of freshwater ponds, impoundments, and lakes, however, often yields a fair number. The species occurs rarely along the shoreline of the Salton Sea, more often around backwaters or beaches with some vegetative cover (e.g., Iodine Bush), and occasionally in flooded fields.

WHIMBREL

Numenius phaeopus hudsonicus Latham, 1790

Common to abundant spring (mid-March through early June) and common fall (late July through September) transient; rare summer visitor; casual in winter. Generally a coastal species, the Whimbrel is a common transient through the Salton Sea yet relatively uncommon elsewhere in the interior Southwest. It is particularly common (sometimes abundant) in spring, when large numbers move northward through the region from the species' wintering grounds around the Gulf of California. Spring records extend from 4 March (1989, near Imperial; G. McCaskie) to 7 June (1987, 15 at the south end; *AB* 41:1487), with a peak from mid-April through early May (e.g., 10,000 near Westmorland 19–25 April 1970 [*AFN* 24:644] and 3,000 flying north at Brawley 3 May 1990 [*AB* 44:496]). Studies of bird migration in the Coachella Valley recorded annual (1979–85) spring flocks of Whimbrels ranging in number from 1,000 to 12,000, with peak movement between 10 and 25 April and a mean bearing of almost due northwest ($\tau = 320°$; Rayleigh's test: $r = 0.79$, $P < .001$) (R. L. McKernan *in litt.*). Small numbers (fewer than 20) linger through the summer each year, particularly around the south end (especially Bruchard Bay, Wister, and parts of the Imperial Valley). A flock of about 100 summered near Obsidian Butte 27 June–9 July 1999 (M. A. Patten), and 50 were at the south end 17 June 1967 (*AFN* 21:604). Fall migrants arrive as early as 15 July (2000, 50 at the south end; G. McCaskie). Although smaller numbers move through in

that season, it is still a common fall migrant; for example, 350 were at the south end 4–11 August 2000 (G. McCaskie), and 400 were there 19 August 1973 (Garrett and Dunn 1981). Most pass through by the end of August, but records of apparent migrants extend to 2 October (1993, at Red Hill, and 1999, near Obsidian Butte; S. von Werlhof, M. A. Patten).

The Whimbrel is casual in the Salton Sea region during winter (*contra* McCaskie 1970d). There are only eight records in that season, with single birds at the south end 29 January 1989 (*AB* 43:366), near Oasis 12 January 1991 (Patten et al. 1993), 25 km southeast of Mexicali 29 December 1991 (Patten et al. 1993), and near Calipatria 23 February 1992 (G. McCaskie) and 8 February 1997 (G. McCaskie); one to two at Bruchard Bay 17 December 1991–23 February 1992 (*AB* 46:315); and three there 18 December 1990 (*AB* 45:999). According to McCaskie, a Whimbrel accompanying a large flock of Long-billed Curlews 5 km north of Calipatria 2 December 2000 (*NAB* 55:228) might have been in the area since September.

ECOLOGY. The Whimbrel occurs on barnacle beaches, on mudflats, at the edge of freshwater impoundments, in flooded fields, and in agricultural fields with low grass or Alfalfa cover. It is most frequent on mudflats and in agricultural fields, however, often in the company of Long-billed Curlews in the latter. In spring by far the largest numbers occur in agricultural fields in the Imperial Valley.

TAXONOMY. Aside from occasional strays of the white-rumped *N. p. variegatus* (Scopoli, 1786) in the West (away from westernmost Alaska) and nominate *N. p. phaeopus* (Linnaeus, 1758) in the East (Heindel 1999), Whimbrels in North America are the distinctive brown-rumped subspecies *N. p. hudsonicus* of the northern Nearctic. This taxon, often called the Hudsonian Curlew, is distinctive enough morphologically and genetically that some have suggested it be accorded full species status (e.g., Zink et al. 1995). Engelmoer and Roselaar (1998) recognized *N. p. rufiventris* Vigors, 1829, as an Alaskan subspecies larger than *N. p. hudsonicus*, which they restricted

FIGURE 42. The thousands of Long-billed Curlews wintering in the Imperial Valley constitute the greatest winter concentration known for this species. Photograph by Kenneth Z. Kurland, February 2000.

to the Canadian Arctic. However, their table 86 demonstrates that mensural overlap is too substantial for recognition of *N. p. rufiventris* at even the 75 percent level.

LONG-BILLED CURLEW

Numenius americanus Bechstein, 1812

Common transient and winter visitor (late June to early June); uncommon nonbreeding summer visitor to the Imperial Valley. Like the Willet, the Long-billed Curlew (Fig. 42) is present throughout the year at the Salton Sea, the area of the species' greater concentration inland. Its numbers too reach their nadir during June, as the last of the spring migrants depart in the first few days of June and the first fall arrivals do not return until that month's end. Migration peaks in July and August (e.g., 7,890 near Red Hill 28 July 1987; R. L. McKernan), with thousands remaining in the Imperial Valley in winter (e.g., 4,490 at the south end 22 December 1987; *AB* 42:1187). The curlew is less common in the Coachella and Mexicali Valleys, but tens to hundreds occur in those areas as well. A few dozen nonbreeders (probably one-year-old birds) remain through the summer each year, mainly around the south end and especially around Red Hill, Obsidian Butte, and Bruchard Bay.

ECOLOGY. Although some curlews occur along the shoreline of the sea and flocks roost in shallow impoundments, the species reaches its peak abundance in agricultural fields in the Imperial Valley. It favors fields with short vegetation, such as Perennial Rye or Alfalfa, particularly if the fields are fallow and somewhat dry. Interestingly, summering birds are most frequent in shallow ponds bordering the sea and are seldom encountered in fields.

TAXONOMY. Despite the naming of *N. a. parvus* Bishop, 1910, breeding in the western Great Basin, the distinguishing character, its shorter bill, is so variable a trait in this species that the "subspecies" frequently cannot be diagnosed (Grinnell and Miller 1944; Phillips et al. 1964; M. A. Patten). We therefore treat the species as monotypic.

HUDSONIAN GODWIT

Limosa haemastica (Linnaeus, 1758)

Two records. There are two Salton Sea records of the Hudsonian Godwit, a casual visitor to California with only about 20 records statewide.

Remains of an "adult primarily in fresh basic plumage with a few alternate feathers remaining in the scapulars" (S. W. Cardiff *in litt.*) found near the mouth of the Whitewater River 11 November (not October) 1980 (*AB* 35:226) were preserved as a skeleton (LSUMZ 126414). The carcass was "virtually mummified," so the bird had died several weeks earlier. A decade later a male in alternate plumage was photographed at Red Hill 21 May 1990 (Patten and Erickson 1994).

MARBLED GODWIT
Limosa fedoa fedoa (Linnaeus, 1758)
Common transient and winter visitor (late June through May); uncommon to fairly common non-breeding summer visitor. Like many other shorebirds, the Marbled Godwit is uncommon as a migrant through most of the interior Southwest but common at the Salton Sea. It also winters commonly at the Salton Sea (more than 1,200 birds; Shuford et al. 2000). Smaller numbers are present each summer, particularly around the south end, especially around Wister and Bruchard Bay. Summer totals tend to reach at most 100, but 300 birds were counted at the south end 17 June 1967 (*AFN* 21:604). Substantial numbers of nonbreeding birds in summer hinder detection of early-returning and late-departing migrants. Even so, numbers generally begin to increase in fall beginning in late June, with three in full alternate plumage near Mecca 27 June 1998 (C. A. Marantz) being the earliest definite migrants. Peak return is in July and August. Similarly, numbers dwindle by late April, generally reaching a low point by the end of May, with two in full alternate plumage at the mouth of the Whitewater River 31 May 1999 (M. A. Patten) being the latest definite migrants. A few late migrants undoubtedly move through as late as the first week of June, as numbers generally drop sharply after 7 June (e.g., 60 at Morton Bay 7 June 1998 had decreased to 25 birds by 25 June; M. A. Patten).

ECOLOGY. The majority of godwits in the region congregate at river mouths, shallow impoundments, ponds, and backwaters. Some feed along the shoreline of the sea, along sandy beaches or on mudflats but avoiding barnacle beaches. The Marbled Godwit is scarce in the region away from the immediate edge of the Salton Sea, although it does use flooded fields in the Imperial and Mexicali Valleys.

TAXONOMY. This species was regarded as monotypic until the description of the short-winged, short-legged, more massive *L. f. beringiae* Gibson and Kessel, 1989, which breeds in Alaska and winters coastally south to San Francisco Bay (Gibson and Kessel 1989). Birds occurring inland and farther south are the nominate subspecies.

RUDDY TURNSTONE
Arenaria interpes interpes (Linnaeus, 1758)
Uncommon to fairly common spring transient (April to early June); rare fall transient (late June to early October) and winter visitor; 2 summer records. The Ruddy Turnstone is an uncommon to fairly common spring transient and rare annual fall transient at the Salton Sea, though it is extremely rare elsewhere in the interior West. This species was first detected at the Salton Sea 17 May 1930, when four were seen at the south end (Willett 1932). It has since proven to be regular in spring, with records extending from 9 April (1978, Salton City; *AB* 32:1055) to 3 June (1980, north end; *AB* 34:815) and peaking from mid-April to mid-May. Generally fewer than 25 individuals are encountered in a single day, but about 200 were seen around the Salton Sea 10 and 11 May 1991 (*AB* 45:496), and a flock of 75 was seen at the mouth of the Whitewater River 8 May 1982 (*AB* 36:394). An adult at the mouth of the New River 8–22 June 1991 (*AB* 45:1161) and at most two in first-alternate plumage at the mouth of the Whitewater River 4 June–25 July 1999 (*NAB* 53:432, M. A. Patten) apparently summered.

The Ruddy Turnstone is scarcer in fall, from 25 June (1989, 4 adults at the mouth of the New River; *AB* 43:1368) to early or mid-October, with departure dates clouded by birds attempting to winter. Juveniles have been recorded from 6 August (1988, Morton Bay; G. McCaskie) to 30 September (1990, mouth of the Whitewater River; *AB* 45:151). Small numbers winter at the Salton Sea, particularly at Salton City and at the south end around Obsidian Butte and near the mouth of the Alamo River. The first to be recorded in winter was one at Salton City 7 February 1982 (*AB* 36:331). The species now appears almost annually in that season between 15 October (1992, basic-plumaged adult female north of Salton City; R. L. McKernan, SBCM 53892) and 8 April (1999, 3 at Obsidian Butte from 27 February; *NAB* 53:209). Maximum winter counts are of 25 near Red Hill 17 December 2001 (G. McCaskie), 17 around the sea 22 January–5 February 1999 (Shuford et al. 2000), and up to 12 near Red Hill 24 October–22 December 1999 (H. Detwiler et al.). A flock of 14 near Red Hill 30 September 2000 probably comprised both migrants and birds attempting to winter, as 10 were still present 8 November and 1 remained until 19 December (G. McCaskie).

ECOLOGY. This species occurs anywhere along the immediate shoreline of the Salton Sea, favoring barnacle beaches and rock outcrops. One near the mouth of the New River 6 August 1994 (G. McCaskie) was in a flooded field.

TAXONOMY. California specimens, including those from Poe Road 28 December 1993 (R. W. Dickerman; SDNHM 48750) and Obsidian Butte 22 September 1994 (P. Unitt; SDNHM 49110), are of the nominate subspecies, of the Palearctic, Alaska, and northeastern Canada. *A. i. morinella* (Linnaeus, 1766), of eastern North America, with deeper chestnut coloration on the scapulars and coverts, is a likely candidate to reach California. More work is needed to determine the subspecific identification of birds occurring at the Salton Sea, although on the basis of geography the vast majority are likely to be the nominate subspecies.

BLACK TURNSTONE
Arenaria melanocephala (Vigors, 1828)
Rare spring transient (late March to early June); casual fall transient (early July to early September); 2 winter records; 1 summer record. A rocky shoreline species endemic to the Pacific Coast, the Black

Turnstone is a casual migrant in the interior West, although it has proven to occur nearly annually at the Salton Sea, associated with the Ruddy Turnstone. Most records are from the Whitewater River delta and Salton City. The Black Turnstone was first noted at the sea when one was observed with four Ruddy Turnstones at the south end 17 May 1930 (Willett 1932). The species now averages at most three individuals per spring from 30 March (1979, Salton City; *AB* 33:805) to 7 June (1998, first-spring bird at Morton Bay; *FN* 52:503), with the majority of records from mid-April to mid-May. A bird in first-alternate plumage at the Whitewater River mouth 12 June 2000 (M. A. Patten) was exceptionally late. Most records involve lone individuals, but two birds have been seen together on many occasions (e.g., photographed at the Whitewater River mouth 31 May–4 June 1999; *NAB* 53:329), and there were three at the north end 23 April 1989 (*AB* 43:537) and at Obsidian Butte 28 May 1999 (*NAB* 53:329). Maximum counts are of seven at the Whitewater River delta 26 April (not May) 1994 (SBCM 54271–73; *FN* 48:341) and six there 21 April 1996 (*FN* 50:332).

This species has summered once at the Salton Sea, with at most five at Mullet Island 17 June–16 July 1993 (*AB* 47:1150). In contrast to the regularity of spring records, there are only seven fall records, four of adults in alternate plumage: one at the mouth of the New River 25 June 1989 (*AB* 43:1368), one at the south end 6 July 1996 (*FN* 50:996), two at the south end 12 July 1990 (*AB* 44:1186), and one at Red Hill 19 July 1997 (*FN* 51:1053). Adults in basic plumage photographed near Salton City 22 August 2001 (M. B. Stowe) and observed at the south end 28 August 1993 (*AB* 48:152) and at Obsidian Butte 1 August–8 September 2002 (G. McCaskie; B. Miller) provided the only other fall records. Additional fall reports have been published (*AB* 26:121, 27:919), but we do not consider them valid because of potential confusion with the Ruddy Turnstone (see Garrett and Dunn 1981). The Black Turnstone has wintered at the Salton Sea at least twice and probably thrice, including single birds at Obsidian Butte

2 February–8 April 1999 (*NAB* 53:329) and at Red Hill 26 and 27 February 1983 (*AB* 37:338). One or two birds at Salton City 12 March–22 April 1978 (*AB* 32:399, 1055) either wintered locally or were extremely early spring migrants.

ECOLOGY. Like the Ruddy, the Black Turnstone is generally found along barnacle beaches and on rock outcrops, but it sometimes forages on mudflats.

SURFBIRD
Calidris virgata (Gmelin, 1789)
Casual spring transient (mainly early April through early May); 1 fall record. The Surfbird is a scarce spring migrant through the Salton Sea though virtually unrecorded elsewhere in the interior Southwest. Like those of the two turnstones and the Red Knot, spring records of the Surfbird involve birds moving north from wintering grounds around the Gulf of California. Spring records extend from 9 April (1978, up to 10 at Salton City; see below) to 7 May (1993, *AB* 47:453), with a peak in late April. An exceptionally early spring migrant was observed at Salton City 18 March 1995 (*FN* 49:309). Typically lone individuals are found, but as many as ten (at most 5 per day) were seen at Salton City 9–22 April 1978 (*AB* 32:1055). Groups of five were noted at the mouth of the Whitewater River 23 April 1989 (*AB* 43:537) and at Desert Beach 25–29 April 1967 (*AFN* 21:540). An alternate-plumaged female was collected from the latter flock 29 April; both the skin (SDNHM 36097) and the complete skeleton (USNM 489303) were preserved. There are two mid-June records of adults in alternate plumage, of three (not 2) at the north end 8–15 June 1985 (*AB* 39:962) and one at the mouth of the New River 16 June 1990 (*AB* 44:1186), which we suspect were late spring stragglers. The only fall record is of an adult at Salton City 25 August 1985 (*AB* 40:158).

ECOLOGY. Salton City is a favored locale. Like the turnstones, the Surfbird is typically found foraging along barnacle beaches and rocky breakwaters. Spring migrants generally associate with flocks of Red Knots.

TAXONOMY. The Surfbird is frequently placed in the monotypic genus *Aphriza*, but morphologically (Jehl 1968) and genetically (R. E. Gill pers. comm.) this species is most similar to (and most closely related to) the Great Knot, *Calidris tenuirostris* (Horshield, 1821), so we merge *Aphriza* into *Calidris*.

RED KNOT
Calidris canutus (Linnaeus, 1758) subspp.
Fairly common spring transient (mid-March to early June); uncommon fall transient (late June to mid-November); casual in summer and winter. Although scarce elsewhere inland in the West, the Red Knot is a fairly common spring and uncommon fall transient through the Salton Sea region, with a few records for summer and winter. Spring records extend from 14 March (1978, Salton City; *AB* 32:1055) to 5 June (1983, mouth of the Whitewater River; *AB* 37:912), with the majority of birds passing through between mid-April and early May. This species is sometimes common in spring, but daily maxima are usually under 200 individuals. The highest count on record is of 582 around the Salton Sea 23 April 1989 (*AB* 43:437); about 300 were at the mouth of the Whitewater River 6 May 1972 (*AB* 26:905) and at Salton City 28 April 1984 (*AB* 38:960) and 21 April 1991 (*AB* 45:496). Rather than being late spring or early fall migrants, birds in basic or first-alternate plumage at the south end 7 June 1987 (*AB* 41:1487), 15 June 1996 (*FN* 50:996), and 16 June 1976 (*AB* 30:1003) probably summered, as did five at the mouth of the New River 25 June 1989 (*AB* 44:1186).

The Red Knot is scarcer in fall, the maximum number being 45 at the south end 14 July 1979 (*AB* 33:897). Fall records extend from 29 June (1985, 1 in alternate plumage at the mouth of the Whitewater River; *AB* 39:962) to 25 October (1971, north end; *AB* 26:120) and lack a distinct peak. Juveniles have been noted as early as 2 September (1991, 2 at Mecca; M. A. Patten). Migrants are exceptional after October. Individuals occurring in November, such as 2 at Salton City 3 November 1984 (G. McCaskie), up to 30 there 3–28 November 1985 (G. McCaskie), 6 there 11 November 1988 (M. A. Patten), 10 there 29 November 1986 (G. McCaskie), 4 at the south end 17 November 1984 (*AB* 39:102), 1 at the mouth of the New River 20 November 1976 (*AB* 31:223), and 20 around the sea 11–15 November 1999 (Shuford et al. 2000), were either exceptionally late migrants or, more likely, settled for the winter. Up to five in Salton City 26 December 1980–6 January 1981 (*AB* 35:335) were the first to winter at the Salton Sea. These birds were followed by 12 at Salton City 8 February 1982 (R. Higson), up to 5 there 11 November 1982– 12 March 1983 (*AB* 37:224, 338), 1 there 14 December 1993 (*FN* 48:248), and 1 near the mouth of the Alamo River 6 March 1998 (*FN* 52:257). Although it has been suggested that this species is now regular in winter at the Salton Sea (*AB* 41:328), it is not recorded annually in that season.

ECOLOGY. Knots are most frequent on barnacle beaches but also congregate on mudflats. All regional records are from the shoreline of the sea, adjacent backwaters and impoundments, and nearby irrigated agricultural fields (e.g., 20 in a flooded field near Red Hill 17 July 1993; G. McCaskie).

TAXONOMY. There has been a great deal of confusion over which subspecies of the Red Knot occurs in western North America. Likewise, the subspecies occurring in California has never been determined conclusively (Grinnell and Miller 1944; Unitt 1984). Five subspecies have been named: the nominate *C. c. canutus,* from the Taimyr Peninsula and western Siberia; *C. c. islandica* (Linnaeus, 1767), from Greenland and the northern Canadian Arctic; *C. c. rufa* (Wilson, 1813), from the southern Canadian Arctic; *C. c. rogersi* (Mathews, 1913), from Wrangel Island and Alaska; and *C. c. roselaari* Tomkovich, 1990, from the same region as *C. c. rogersi.* Ridgway (1919) recognized no subspecies, stating that "the few European specimens and one from China are not distinguishable from specimens from eastern North America." Almost all subsequent authors recognized at least *C. c. rufa,*

a large, long-billed, pale chestnut, white-bellied subspecies. Although recognized by Cramp and Simmons (1983), *C. c. islandica* is generally treated as a synonym of the large, long-billed, deep chestnut nominate subspecies (e.g., by Dement'ev et al. 1951; AOU 1957; Portenko 1972; Prater et al. 1977; Godfrey 1986). *C. c. islandica* differs only in its yellower upperparts in full alternate plumage (Hayman et al. 1986; Tomkovich 1992), a trait with much individual variation. Conover (1943) asserted that *C. c. rogersi* was not valid and merged it with *C. c. canutus,* a notion refuted by Portenko (1972). Previously, the American Ornithologists' Union followed Conover but noted that birds on the Pacific Coast "are regarded as intermediate between *C. c. canutus* and *C. c. rufa*" (AOU 1957:192). In many respects *C. c. rogersi* is intermediate, at least in the intensity of chestnut on the underparts in full alternate dress (Portenko 1972; Cramp and Simmons 1983; Hayman et al. 1986), but it has a shorter bill than either *C. c. canutus* or *C. c. rufa* (Prater et al. 1977; Cramp and Simmons 1983). The type locality of the recently described *C. c. roselaari* is Wrangel Island, a locale long treated as hosting *C. c. rogersi,* even from specimens (Portenko 1972). Because the latter was described from a migrant from Shanghai, however, its breeding range was only inferred.

Tomkovich (1992) provided a timely taxonomic revision that cleared this murky picture, although some questions remain. He considered *C. c. rogersi* the breeder on the Chukotski Peninsula of northeast Siberia, the principal wintering bird in Australia, and the migrant along the east Asian coast. This subspecies is smaller and whiter-bellied than the adjacent *C. c. roselaari,* which breeds on Wrangel Island and in northern Alaska and migrates along the Pacific coast of North America. Tomkovich suggested that the winter range of *C. c. roselaari* was the Gulf of Mexico, which seems unlikely; instead, its winter range is probably the Pacific coast of the Americas. He did not specify the winter range or migratory route of *C. c. rufa,* but it undoubtedly winters in the Gulf of Mexico and on the Atlantic coast of South America and migrates east of the Rocky Mountains. The nominate *C. c. canutus*, including *C. c. islandica*, migrates and winters in the Old World. Thus, virtually all California specimens are *C. c. roselaari*, and given its inferred wintering grounds, it is the only subspecies likely to occur regularly in the state. On the basis of mensural characters (Table 10) and the rich, dark rufous underparts of full adults, most birds from the Salton Sea can be classified as this subspecies. However, a male collected at the Whitewater River delta 20 April 1968 (SBCM 33019) is small (see Table 10) and white bellied, fitting *C. c. rogersi* (see Tomkovich 1992; Engelmoer and Roselaar 1998). *C. c. rufa* may occur in the Southwest on occasion, as does *Limnodromus griseus hendersoni*, a shorebird whose breeding range and migratory routes are somewhat similar.

SANDERLING

Calidris alba (Pallas, 1764)

Fairly common spring transient (mid-March to late May); uncommon fall transient (late July to late September); rare to uncommon in winter. A common bird of beaches along the Pacific coast, the Sanderling is typically a scarce migrant through the interior but fairly common to uncommon at the Salton Sea (it is rare but regular in winter). Like other coastal shorebirds that migrate through the Salton Sink, the Sanderling is much more common in spring than in fall. For example, it was deemed "fairly common" at the south end 17 May 1930 (Willett 1932; LACM 21468), and about 150 individuals at Salton City 22 April 1978, 28 April 1984, and 3 May 1997 (all G. McCaskie); 265 at the Salton Sea 21 April 1990 (*AB* 44:496); and 249 there 17 and 18 April 1999 (Shuford et al. 2000) are the largest spring tallies.

TABLE 10

Mensural Characters of Red Knots from the Salton Sea

SPECIMEN	LOCATION	BILL LENGTH (mm)	FLATTENED WING (mm)
SBCM 33019 (♂)	Whitewater River delta	30.2	153
SBCM 34087 (♂)	Whitewater River delta	33.0	160
SBCM 33017 (♂)	Whitewater River delta	33.4	160
SBCM 33015 (♂)	Whitewater River delta	36.0	161
C. c. roselaari (♂)	Wrangel Island	33.3–38.9	159.5–172.5
C. c. rufa (♂)	Eastern North America	33.8–37.5	153.5–176.0
C. c. rogersi (♂)	Chukotka/Anadyrland	28.9–33.5	150.0–166.0
SBCM 31972 (♀)	Whitewater River delta	35.1	165
SBCM 33018 (♀)	Whitewater River delta	36.3	170
SBCM 33016 (♀)	South end	36.5	168
LACM 102233 (♀)	Whitewater River delta	37.8	170
SBCM 33778 (♀)	Whitewater River delta	38.1	169
C. c. roselaari (♀)	Wrangel Island	34.7–38.3	166.0–175.5
C. c. rufa (♀)	Eastern North America	34.2–39.0	161.5–173.0
C. c. rogersi (♀)	Chukotka/Anadyrland	31.5–35.3	162.0–170.5

Source: Comparative data are from Tomkovich 1992.

Spring arrival dates are difficult to determine because of the presence of a small number of wintering birds (see below). Migrants appear to arrive in late March; for example, the no more than 30 at Salton City 27 November 1977–5 March 1978 (*AB* 32:399) had increased to 40 by 1 April 1978 (G. McCaskie). One in alternate plumage at Salton City 17 March 1991 (M. A. Patten) provided the earliest spring record of an apparent migrant. Transients peak in late April and early May, the latest being ten at the Whitewater River delta 4 June 1999 (G. McCaskie).

Fall records extend from 23 July (1989, 2 adults in alternate plumage at Salton City; G. McCaskie) to 28 September (1986, juvenile at the Whitewater River delta; G. McCaskie et al.), with high counts being 70 around the Salton Sea 14 September 1990 (W. D. Shuford et al.) and 40 adults at Salton City 5 September 1988 (M. A. Patten). Numbers fluctuate from year to year, but Sanderlings winter annually at Salton City and elsewhere along the western shore. Winter records extend from 31 October (1971, Salton City; G. McCaskie) into March (see above). Generally at most 15 individuals are found in that season, but at least 50 were present "all winter" in 1993–94 (*FN* 48:248), and an amazing 109 were along the western shore 1–9 December 1994 (*FN* 49: 197).

ECOLOGY. At all seasons the Sanderling frequents barnacle beaches, particularly around Salton City. Smaller numbers are noted on mudflats and around backwaters; some occasionally occur on breakwaters or rocks. This species tends to form pure flocks when on barnacle beaches on the western shore but mixes with other small *Calidris* sandpipers on mudflats, especially around the Whitewater River delta.

TAXONOMY. Engelmoer and Roselaar (1998) distinguished Nearctic from Palearctic birds as *C. a. rubidus* (Gmelin, 1789), differing in its longer bill and tarsus and more extensive grayish white fringes on the upperpart feathers in definitive alternate plumage. Mensural differences in their table 39 are inadequate on their own to support these subspecies at even the 75 percent level; we have been unable to evaluate alleged color differences. We thus leave the species monotypic, following Peters (1934:281).

SEMIPALMATED SANDPIPER
Calidris pusilla (Linnaeus, 1766)

Rare spring transient (late April to early June); casual fall transients (early July to mid-September). Owing to difficulty in field identification, the Semipalmated Sandpiper was largely undetected in California until the late 1970s. Careful examination of small *Calidris* sandpipers proved the Semipalmated to be a rare to uncommon fall migrant along the coast and through the Great Basin and the Mojave Desert and a rare but annual spring migrant through the Salton Sea (Luther et al. 1979; Luther 1980; Binford 1983, 1985). Indeed, the first California record of the species was of an adult collected near Niland 7 May 1960 (Cardiff 1961; SBCM 33032). Spring records involve birds in alternate plumage. In a typical spring one to three are found, but ten were noted 4–19 May 1991 (*AB* 45:496). Records extend from 21 April (1979, New River through 24 April; Binford 1983) to 6 June (1976, Salton Sea NWR; Luther et al. 1979), with a peak during the first half of May. Lone individuals are the rule, but high counts are of at most 6 at the south end 2–6 May 1984 (*AB* 38:960) and south of Wister 10 May 1992 (*AB* 46:480) and no more than 5 at Wister 11–19 May 1991 (*AB* 45:496).

The Semipalmated Sandpiper is a casual fall vagrant through the Salton Sea, a status at odds with its occurrence elsewhere in California and the West, where it is generally much more frequent in fall (Roberson 1980; Luther et al. 1983; Paulson 1993). The species may in fact be regular in that season, but concerted searching through large flocks of small *Calidris* has yielded only about 17 fall records from 28 June (2002, adult at Wister; *NAB* 56:486) to 14 September (1981, juvenile photographed at the south end; R. E. Webster), including a specimen of an adult female from Red Hill 30 June 1985 (R. Higson; SDNHM 49332). Most July records are of adults in alternate plumage, whereas those in August are of juveniles, including an early one at the south end 29 July 1995 (*FN* 49:980). Despite later sight reports (see Lehman 1994), the latest unquestionable record for California is of three juveniles at Harper Dry Lake in the central Mojave Desert 27 September–2 October 1991 (*AB* 46:149), two of which were collected (SBCM 53134, 53161). The identity of a purported Semipalmated Sandpiper in basic plumage at the south end 13 October 1989 (*AB* 44:161) is therefore tenuous.

ECOLOGY. Semipalmated Sandpipers are invariably found among flocks of Western Sandpipers congregated in shallow impoundments adjacent to the Salton Sea or at river or channel mouths at the edge of the sea. A few have been recorded in the Imperial Valley well away from the sea.

WESTERN SANDPIPER

Calidris mauri (Cabanis, 1856)

Abundant transient (early April to early June, late June to mid-October); common winter visitor; rare summer visitor. The Western Sandpiper is the most abundant shorebird at the Salton Sea, outnumbering even the resident Black-necked Stilt and American Avocet and the abundant migrant Long-billed Dowitcher (see Table 9). The first migrant adult Westerns of the fall arrive in alternative plumage as early as 20 June (1993, Salton City; G. McCaskie). Returning adults reach a peak in mid-July, just before the first juveniles appear at the end of July. Migrants as a whole continue through mid-October, with peak numbers from mid-July through August, and counts may be extremely high by early fall, such as 30,000 at the south end 16 July 1998 (M. A. Patten) and more than 34,000 around the sea in August 1999 (Shuford et al. 2000). Although numbers decrease through fall, they remain high through the winter, as many hundreds to a few thousand are present annually around the sea in that season (Shuford et al. 1999, 2000). Spring migration again brings massive numbers through the Salton Sink. During that season the Western Sandpiper outnumbers all other shorebird species in the region (see Table 9; Shuford et al. 1999). Migrants arrive in early April, with peak movement from mid-April to mid-May. As

with the Black-bellied Plover, the Marbled Godwit, and the Long-billed Dowitcher, the latest spring transients pass through just before the first fall migrants arrive. Birds in full alternate plumage, and thus presumably transients, have been noted as late as 7 June (1998, Whitewater River delta; M. A. Patten), although most depart by mid-May. This species is rare through the summer; an amazing count of about 60 at the mouth of the New River through June 1991 (*AB* 45:1161) aside, generally fewer than 10 are present in that season, with none found in some years.

ECOLOGY. More so than any other small *Calidris* sandpipers occurring in the region, the Western Sandpiper is a denizen of open mudflats, along the shoreline of the Salton Sea, in large freshwater impoundments, or in low water in bordering ponds and marshes. It avoids drainage and irrigation canals more than the Least Sandpiper does, though it still occurs in such settings. In addition to foraging on open mud or sand, the Western Sandpiper frequently forages in belly-deep water, behaving much like a miniature Stilt Sandpiper.

LITTLE STINT

Calidris minuta (Leisler, 1812)

One spring record. There are but six records of the Little Stint for California, an alternate-plumaged adult photographed at Wister 18 May 1991 (Patten, Finnegan, et al. 1995) being the only one for spring. This Eurasian species is a casual vagrant anywhere in North America. See the Red-necked Stint account in the Hypothetical List for a discussion of a stint specimen that may pertain to the Little.

LEAST SANDPIPER

Calidris minutilla (Vieillot, 1819)

Common transient and winter visitor (late June to mid-May); rare in summer. In the interior West the Least Sandpiper is generally outnumbered by the Western Sandpiper during migration periods, but it is the only widespread "peep" (small *Calidris* sandpiper) in winter. The Least is typically outnumbered by the Western by a factor of

100:1 or even 1,000:1 during peak migration periods in both spring (mid-April to mid-May) and fall (mid-July to mid-September), especially along the shoreline of the Salton Sea. The numbers are more balanced in the Mexicali and Imperial Valleys, with ratios of 1:1 or even 1:10 or 1:100 in favor of the Least Sandpiper. Winter numbers of these two species around the edge of the sea are fairly similar, with the Western tending to be slightly more numerous (Shuford et al. 1999, 2000); however, the Least is by far the most numerous wintering peep throughout the Coachella, Imperial, and Mexicali Valleys. Fall migrants return by 25 June (1988, 2 at the mouth of the New River; M. A. Patten, G. McCaskie), but substantial numbers do not start appearing until the end of July, coinciding with the appearance of the first juveniles, the earliest being 4 August 1984 (G. McCaskie). Fall movement peaks in August and September (e.g., 4,149 around the sea 24 August 1996; W. D. Shuford et al.), the majority of wintering birds having returned by then. Spring migration takes place from late March to the end of April, with peak counts reaching a few hundred per day but with a high count of 3,476 on 25 April 1992 (W. D. Shuford et al.). The last few trickle through in early May, with 21 May (1995, 5 near Wister; M. A. Patten) being the latest date, aside from an exceptionally late individual at the mouth of the Whitewater River 2 June 1990 (G. McCaskie). All peeps are rare in the Salton Sink during summer, with the Western more frequent than the Least in that season. Nonetheless, the latter is nearly annual, with records for two of every three years on average and a maximum of three at the south end 19 June 1977 (G. McCaskie).

ECOLOGY. Like other peeps in the region, the Least Sandpiper reaches peak abundance on mudflats, whether directly along the edge of the Salton Sea (especially around river mouths) or in adjacent large freshwater impoundments and flooded fields. However, in these areas it is overwhelmingly outnumbered by the Western Sandpiper. By contrast, the Least is far more likely than the Western to forage in wet fields away from the sea and narrow drainage ditches. Most Leasts forage on open mud or sand, seldom venturing into water deeper than the intertarsal joint. More importantly, however, the Least frequents freshwater habitats throughout the valleys, where it often greatly outnumbers the Western.

WHITE-RUMPED SANDPIPER
Calidris fuscicollis (Vieillot, 1819)

Four spring records (late May to mid-June). There have been four late spring records of the White-rumped Sandpiper for the Salton Sea, all of birds in alternate plumage from 30 May to 16 June. A female collected at the mouth of the Whitewater River 7 June 1969 furnished a first record for California (Dunn 1988; SDNHM 37201). Other records are of individuals photographed at Salton Sea NWR 16 June 1976 (Luther et al. 1979), observed at the mouth of the Whitewater River 30 May 1985 (Dunn 1988), and photographed south of Wister 30 and 31 May 1992 (Heindel and Patten 1996). The late spring pattern at the Salton Sea fits well with the rest of the West; for example, five of the other eight California records fall within the period 17 May–12 June. A fall report for the Salton Sea lacks convincing details (Pyle and McCaskie 1992).

BAIRD'S SANDPIPER
Calidris bairdii (Coues, 1861)

Rare fall transient (mainly mid-July to mid-October); 5 spring records (late April to early June). Baird's Sandpiper is an uncommon bird in the Pacific Coast states, although it migrates through California in fall in small numbers (with moderate numbers passing through east of the Sierra Nevada axis). In the Salton Sea region fall adults are recorded only casually, with records extending from 5 July (1989, near Wister; *AB* 45:1161) to 22 August (1981, near Niland; G. McCaskie). Juveniles are more frequent but still rare, with records extending from 29 July (1995, mouth of the New River; G. McCaskie) to 15 October (1982, mouth of the Whitewater River; G. McCaskie) and a peak from mid-August to mid-September. Small flocks of juveniles are occasionally noted, including 8 near Red Hill 14 August 1971 and at Salton City 11 September 1999 (both G. McCaskie) and an unparalleled 23 around the Salton Sea 19 August 1989 (W. D. Shuford et al.). A juvenile at the north end 20 November 1988 (*AB* 43:168) was exceptionally late for anywhere in the United States. This species is casual in California in spring, when the majority of the population is migrating through the Great Plains. There are five spring records for the Salton Sea, of one at Fish Springs (the south end) 27 April 1917 (Dawson 1921), two at the Whitewater River delta 28 April 1968 (*AFN* 22:575; SBCM M4093), one at Salton City 8 May 1982 (*AB* 36:894), at most four at the mouth of the Alamo River 15–20 May 1977 (*AB* 31:1047), and one at the south end 1 June 1986 (*AB* 40:524). The exact location of a Baird's Sandpiper reported "near the Salton Sea" 21 June 1954 (*AFN* 9:56) is unknown; the bird may have been outside the Salton Sink. Besides, it would establish the sole summer record for the Southwest, so we question the record's validity.

ECOLOGY. Baird's Sandpipers are most often found on mudflats or in flooded fields, although they are occasionally found on sandy beaches and even in dry fields. Most have been in the company of flocks of Western and Least Sandpipers.

PECTORAL SANDPIPER
Calidris melanotos (Vieillot, 1819)
Rare to occasionally uncommon fall transient (September–November); 3 spring records (mid-March to mid-May); 1–2 winter records. Even though the Pectoral Sandpiper is not a particularly rare migrant through California, it is seldom seen in the Salton Sink; none are recorded in most years. Fall records extend from 28 August (1996, near Calipatria; G. McCaskie) to 24 November (1990, Finney Lake; *AB* 45:151) and peak in late September. An extremely early adult was seen at the Whitewater River delta 30 June 1990 (*AB* 44:1186). Like those of many species, the numbers of migrants in the Salton Sink fluctuate, with sizable numbers recorded some years. In 1991, about 15 individuals were recorded from 22 September (3 juveniles near Mecca; M. A. Patten) to 26 October (4 juveniles 10 km southeast of Brawley; M. A. Patten). In 2000, double-digit counts were frequent in the

Imperial Valley from 31 August (2 juveniles near Niland; G. McCaskie) to 1 October (13 in a flooded field 8 km e. of Ramer Lake; M. A. Patten), with about 60 recorded all told, including a maximum of 17 individuals 6 km southwest of Niland 17 September (G. McCaskie). This species is a scarce spring migrant through California, rarer than Baird's Sandpiper. In the Salton Sink there are three spring records of the Pectoral Sandpiper, of single birds near Calipatria 17 March 1973 (*AB* 27:663) and at the south end 31 March 1979 (*AB* 33:805) and 18 May 1968 (*AFN* 22:575). One at the Imperial Warm Water Fish Hatchery 20 February–6 March 1982 (*AB* 36:331) represents the only winter record for California. One in a flooded pasture at Niland 22 December 1992 (*AB* 47:301) either was wintering locally or was an extremely late fall transient.

ECOLOGY. The Pectoral is as much a "grass-piper" as a sandpiper, frequenting flooded fields (particularly in the Salton Sink), pastures, ponds, and lake margins supporting dense grassy or weedy vegetation.

DUNLIN
Calidris alpina pacifica (Coues, 1861)
Common spring transient (early April to mid-May); uncommon fall transient and winter visitor (mid-September through April); casual in summer. Along with the Least Sandpiper, the Dunlin is the only *Calidris* likely to be found wintering in the interior West, although small numbers of the Western Sandpiper occur locally. All three species are numerous in winter at the Salton Sea, the two small species greatly outnumbering the Dunlin. Like most shorebirds in the Salton Sink, however, the Dunlin reaches its peak abundance during migration, and wintering birds cloud arrival dates in spring and departure dates in fall. Migrants in alternate plumage are northbound along the Pacific coast in April and May, so an alternate-plumaged bird at the mouth of the Whitewater River 5 April 1992 (M. A. Patten) appears to represent the earliest spring migrant. Peak movement takes place in late April and early May (e.g., 2,258 on 25 April 1992; W. D. Shuford et al.), the last coming through 20 May

(1995; G. McCaskie). An exceptionally late bird was at the north end 3 June 1980 (McKernan et al. 1984). Because the Dunlin molts into basic plumage on its breeding grounds prior to moving south, fall migrants arrive much later than virtually all other migrant shorebirds, with one at the north end 17 September 1980 (McKernan et al. 1984) and seven at Wister 19 September 1999 (B. Miller) being the earliest; the majority do not arrive until mid-October (McKernan et al. 1984) or early November (N. Warnock pers. comm.). Many of these migrants stay through the winter, when the species is uncommon (though locally fairly common around Red Hill and Obsidian Butte). Departure dates of wintering birds are not clear because spring migrants from farther south pass through as winter birds are leaving; still, it appears that the winter numbers begin to decrease by March (Shuford et al. 2000).

This species is casual in southern California in summer, with few valid records from June through August. At the Salton Sea there are a dozen records of birds in definitive or first-alternate plumages from 11 June (1978, mouth of the Whitewater River; *AB* 32:1208) to 3 September (1973, 3 at the mouth of the Whitewater River; G. McCaskie). Records from the first half of June (e.g., 11 June 1978, cited above, and 15 June 1968 at the mouth of the Whitewater River; *AFN* 22:648) may pertain to extremely late spring migrants, but other records likely involved summering birds. It is possible that birds in fresh alternate plumage in July (e.g., Bruchard Bay 2 July 1988; M. A. Patten) are fall transients that migrated before molting, but this hypothesis redefines what we know about the species' molt biology (see above).

ECOLOGY. The Dunlin frequents mudflats along the shoreline of the sea and adjacent freshwater and brackish impoundments. Small numbers occur in flooded fields, particularly in the Imperial Valley.

TAXONOMY. Geographic variation in the Dunlin is complex. Dunlins in western North America, including all specimens from California (but see below), are the large, long-billed *C. a. pacifica*

(MacLean and Holmes 1971), breeding in southern Alaska. *C. a. arcticola* (Todd, 1953), of northern Alaska, smaller and shorter-billed (Browning 1991; Engelmoer and Roselaar 1998), migrates along the Asian coast. *C. a. pacifica* is similar to *C. a. hudsonia* (Todd, 1953), of the Canadian Arctic and eastern North America, in size and structure but differs in alternate plumage in having a white band separating the black belly patch from the black breast streaks (Browning 1977b).

Genetic similarity between some birds from California and Washington and some from eastern Siberia (Wenink and Baker 1996) implies that either *C. a. sakhalina* (Vieillot, 1816), of Chukotka, or *C. a. kistchinski* Tomkovich, 1986, of Kamchatka, occasionally reaches the Pacific coast. One or both of these subspecies has occurred in the western Aleutians (Gibson and Kessel 1997), but they are otherwise unknown in the New World. The similarity in mitochondrial DNA may be an artifact of low sample size—both of individual Dunlins sampled and of base pairs sequenced—rather than an indication of genuine movement between these populations (see Wenink et al. 1993:97).

CURLEW SANDPIPER
Calidris ferruginea (Pontoppidan, 1763)
Three records, 2 in spring and 1 in fall. Of some 30 California records only a half-dozen Curlew Sandpipers have been found inland. Three of these records are from the Salton Sea: an alternate-plumaged bird at Salton City 27 and 28 April 1974 (Luther et al. 1979), a juvenile photographed at the mouth of the Whitewater River 13–16 October 1984 (Dunn 1988), and a bird in first-alternate plumage at the mouth of the Whitewater River 16–26 April 1994 (Howell and Pyle 1997; SBCM 54281). The vast majority of California's records are of fall migrants, so the two spring records are noteworthy.

STILT SANDPIPER
Calidris himantopus (Bonaparte, 1826)
Fairly common spring (mid-April to mid-May) and fall (July to mid-September) transient; uncommon winter visitor; 2 summer records; casual away from

FIGURE 43. The Salton Sea represents the northern extreme of the Stilt Sandpiper's winter range. Photograph by Kenneth Z. Kurland, January 2001.

the south end. Like the American Golden-Plover and Baird's Sandpiper, the Stilt Sandpiper (Fig. 43) migrates principally through the Great Plains. It is generally rare anywhere west of the Rocky Mountains, except at the south end of the Salton Sea and in the adjacent Imperial Valley, where it is fairly common in spring and fall and uncommon in winter. The species' occurrence in the region may be relatively recent. It was not recorded in double digits until March 1953 (*AFN* 7:235) and still appears to fly over northeastern Baja California (Patten et al. 2001) and Sonora (Russell and Monson 1998) before reaching its main wintering grounds in western Mexico and southward. The first birds of the fall appear at the south end in July, the earliest being an adult in alternate plumage at Wister 29 June (1991; *AB* 45:1161). Numbers usually build to 100–150 in August and September before tailing off by winter, when flocks of about 30–40 birds are the norm, although 130 were seen at the south end 20 February 1993 (*AB* 47: 301) and 164 were there in late January 1999 (Shuford et al. 2000). Numbers remain fairly stable through the winter but drop in early April as wintering birds move northward. An influx

of spring migrants arrives in mid-April, with peak passage in the first half of May. Concentrations at the south end of more than 400 on 9 May 1987 (*AB* 41:488) and 450 on 11 May 1991 (*AB* 45:496) give an indication of the number of birds that move through the region in spring. Most have departed by mid-May, one near the mouth of the New River 21 May 1967 (*AFN* 21:540) being the latest for the south end. An exceptionally late spring migrant was seen at the north end 31 May 1986 (see below). The only summer records are of individuals in basic plumage at the south end 13 June 1992 (*AB* 46:1178) and at Wister 21 June 1999 (*NAB* 53:433).

Oddly, this species is only a casual spring and fall transient at the north end, where spring records extend from 22 April (1978; *AB* 32:1055) to 31 May (1986; *AB* 40:524) and fall records extend from 23 July (1994, 2 birds; *FN* 48:989) to 16 September (1973, 3 birds; *AB* 28:108, SBCM 31962, M5260). It has twice been recorded in winter at the north end, with at most three near Oasis 16 December 1993–22 January 1994 (*FN* 48:248) and another there 2 January 1999 (*fide* J. Green). There is one record for Salton City, where two were photographed 27 November–

FIGURE 44. For a species of Eurasian origin, the Ruff is surprisingly frequent in the Salton Sink, with more than 25 records. Photograph by Kenneth Z. Kurland, February 1999, Obsidian Butte.

4 December 1993 (*FN* 48:248). Remarkable given the species' status in the Imperial Valley, a bird in basic plumage at Cerro Prieto 23 August 1997 (Patten et al. 2001) provided the only record for the Mexicali Valley and one of fewer than five for the Baja California peninsula (R. A. Erickson pers. comm.).

ECOLOGY. Because the Stilt Sandpiper feeds belly-deep in water, it prefers shallow impoundments bordering the sea, and it avoids mudflats. Small numbers also occur in flooded fields and shallow ponds in the Imperial Valley.

RUFF

Philomachus pugnax (Linnaeus, 1758)

Casual fall (mid-July to early November) and winter (early November through March) vagrant. A shorebird of the Palearctic, the Ruff is a rare annual fall and winter vagrant to California and elsewhere on the Pacific Coast (Roberson 1980; Paulson 1993). The approximately 25 Salton Sea area records fit this pattern of occurrence, with slightly more in fall than in winter. The roughly 20 fall records extend from 12 July (1999, adult female near Wister until 13 July; *NAB* 53:433) to 6 November (1989, juvenile near Red Hill; *AB*

44:162). Nine records are of adults from 12 July (cited above) to 20 October (1997, adult male 4 km east of Imperial from 23 August; *FN* 52: 126). Surprisingly, apart from a bird of unknown age or sex near Brawley 5 October 1995 (*FN* 50: 115), there are but three records of juveniles, the 6 November bird cited above and males near Brawley 23 August 1997 (*FN* 52:126) and 8 km east of Ramer Lake 1 October 2000 (*NAB* 55:103). The paucity of juveniles relative to adults is typical of fall shorebirds in the Salton Sink, even among regular migrants. Ten winter records (Fig. 44), some apparently involving returning individuals, extend from 4 November (1986, one of unknown sex at El Centro through the winter; *AB* 41:144) to 26 March (1996, female at Morton Bay from 12 March; *FN* 50:222). An adult female near Brawley 1 November 2002 (G. McCaskie) either was a late migrant or was wintering locally. Records involve lone individuals, save for a male and female together at El Centro 20 November 1988–11 February 1989 (*AB* 43: 168), the male staying until 4 March 1989 (*AB* 43:366). Reports from the south end (*FN* 49:196) and near Oasis (*FN* 48:248) apparently were not documented.

ECOLOGY. All Ruff records are from shallow impoundments near the edge of the Salton Sea or from shallow freshwater ponds or flooded fields in the Imperial Valley.

SHORT-BILLED DOWITCHER
Limnodromus griseus (Gmelin, 1789) subspp.
Common spring transient (early March to mid-May); fairly common fall transient (late June to early November); 3 summer records; unrecorded in winter. Because of the difficulty of identifying *Limnodromus* in the field, the true status of the Short-billed Dowitcher in the interior West has only recently been determined. In both spring and fall the Short-billed Dowitcher is an earlier migrant than its congener, spring birds appearing several weeks earlier than Long-billed Dowitchers and juveniles appearing in fall a month earlier. The Short-billed is an uncommon spring and fall transient through most of the West, with double-digit counts being expected maxima. As with many shorebirds, gulls, and other waterbirds, however, the Salton Sea provides a glaring exception. This species is a common to occasionally abundant spring transient from 4 March (2000, calling bird at the mouth of the Whitewater River; M. A. Patten) to 16 May (1994, male 13 km west-northwest of Calipatria; R. Higson, SDNHM 49099), with peak movement during the latter half of April (e.g., 3,413 tallied at the Salton Sea 23 April 1989; *AB* 43: 537). During spring it frequently outnumbers the Long-billed Dowitcher, especially around the north end and along the eastern and western shorelines of the sea. Lone birds at the south end 11 June 1977 and at Wister 10 June 1985 (with a damaged wing) and two in basic plumage at the mouth of the Whitewater River 8 June 1991 (all G. McCaskie), were presumably summering locally.

Like those of most shorebirds passing through the Salton Sea, numbers of the Short-billed Dowitcher are markedly lower in fall. Still, this species is a fairly common migrant from 25 June (1994, 25 at the south end; G. McCaskie) to 12 November (1999, ca. 8 at the Salton Sea Naval Test Base; W. D. Shuford) and, exceptionally, 28 November (1985, 2 identified by call at Salton City; G. McCaskie). Numbers peak from late July (when juveniles appear) through August, and most have passed through by mid-October. Although this species has been reported on various Salton Sea area Christmas Bird Counts (*AFN* 20:376; *AFN* 20:377; *AB* 29:580), we believe all such reports to be in error. There are no convincingly documented winter records anywhere in the interior Southwest, and the two at Salton City in 1985 (cited above) furnished the latest unassailable record for the interior of southern California.

ECOLOGY. Conventional dogma places the Short-billed Dowitcher in salt-water habitats, the Long-billed in freshwater habitats. As in many other cases, this dogma is only partially true. The Short-billed Dowitcher does reach peak abundance in the region along the shoreline of the Salton Sea and in saline or brackish backwaters, but it can also be numerous in freshwater marshes, ponds, and canals. Conversely, the Long-billed Dowitcher is more numerous in freshwater impoundments (e.g., as at Wister and Unit 1) but can be abundant along the shoreline of the sea. On the immediate shoreline the Short-billed Dowitcher prefers barnacle beaches, whereas the Long-billed Dowitcher prefers mudflats, but this tendency also has exceptions.

TAXONOMY. Virtually all California specimens are of the western subspecies, *L. g. caurinus* Pitelka, 1950, which has a largely whitish vent, heavy black barring on the flanks, and modest breast spotting. The smaller Great Plains subspecies, *L. g. hendersoni* Rowan, 1932, with redder underparts and little to no spotting on the center of the breast, has been taken in California (Whelan Lake, San Diego County, 24 September 1961; Unitt 1984), Baja California (Estero de Coyote 13 April 1927; SDNHM 11412), and Arizona (16 km north of Douglas 4 September 1964; Monson and Phillips 1981). Adults in full alternate plumage at the mouth of the Whitewater River 1 May 1993 and 24 April 1994 (both M. A. Patten) showed the characters of *L. g. hendersoni*, which is likely a rare annual migrant through eastern California.

LONG-BILLED DOWITCHER

Limnodromus scolopaceus (Say, 1823)

Abundant transient and winter visitor (late June to early June); uncommon summer visitor. There is but the briefest portion of the year when the Long-billed Dowitcher is not present in full force at the Salton Sea. Fall migrants arrive as early as 25 June (1989, 5 in alternate plumage at the south end; *AB* 43:1368), but the last of the spring migrants do not depart until late May or early June, with 8 June (1991, 3 in alternate plumage at the mouth of the New River; G. McCaskie) being the latest date. The majority occur from mid-July to mid-May, when the species typically numbers in the many thousands (Shuford et al. 2000). Numbers are higher during peak migration periods (early April to mid-May, late July to early November), especially in fall (e.g., 20,000 in the Imperial Valley 25 October 1998; M. A. Patten). Nonbreeding individuals summer at the Salton Sea each year, particularly in large impoundments around the south end (e.g., Wister, Rock Hill, Bruchard Bay), the highest summer count being 40 at the south end 20 June 1991 (G. McCaskie).

ECOLOGY. This dowitcher is most common in freshwater habitats, whether adjacent to the sea or scattered through the valleys. See the Short-billed Dowitcher account for details and caveats.

WILSON'S SNIPE

Gallinago delicata (Ord, 1825)

Fairly common transient and winter visitor (mid-August to early May). Being secretive, Wilson's Snipe is frequently overlooked, yet it is common across much of North America and a fairly common transient and winter visitor to the Salton Sea region. Most snipes occur from late September through mid-April, the earliest being 13 August (1983, Unit 1; G. McCaskie) and the latest, 2 May (1987, 2 at the Whitewater River delta; M. A. Patten). Numbers apparently vary from year to year (McCaskie 1970d), perhaps indicative of facultative movement from northerly climes during cold weather. Even so, with concerted effort double- or even triple-digit counts

can be had in most years, particularly in midwinter (Jurek 1974). There is no evidence of breeding in the region, but a female collected at Mecca 22 March 1911 "contained an egg the size of a small pea" (van Rossem 1911; UCLA 11293).

ECOLOGY. Wilson's Snipe is tied to freshwater habitats bordering the edge of the Salton Sea and is found through much of the Coachella, Imperial, and Mexicali Valleys. It is frequently encountered in irrigation ditches, where it can be surprisingly numerous, and along the edges of ponds and marshes. It sometimes occurs in wet fields supporting sufficient dense cover. Claims of the species' being "quite common on the mud flats at the edge of Salton Sea" (van Rossem 1911) are undoubtedly in error and likely pertain to the Long-billed Dowitcher.

TAXONOMY. Aside from the regular occurrence of the similar Common Snipe, *Gallinago gallinago* (Linnaeus, 1758), in Alaska (Gibson and Kessel 1997) and a single record for Labrador (Todd 1963), snipes in the *G. gallinago* complex in North America are Wilson's Snipe, only recently reaccorded species status (Banks et al. 2002). Wilson's Snipe differs from the Common Snipe, both the nominate subspecies, ranging throughout Eurasia, and *G. g. faeroeensis* (Brehm, 1831), of Iceland and islands north of Scotland, in having the underwing coverts and axillaries densely barred rather than clean white and the secondaries lacking an obvious white trailing edge. Wilson's Snipe is also more darkly marked above, with narrow, white feather fringes rather than wider, buffy ones, and it tends to have 16 rectrices, the outer ones narrower, rather than 14 in all, the outer ones wider, although rectrix count varies considerably and there is much overlap (*fide* R. J. Chandler).

WILSON'S PHALAROPE

Phalaropus tricolor (Vieillot, 1819)

Common transient (mid-March to mid-May, mid-June through October); rare winter visitor (November through mid-March). Wilson's Phalarope depends on large saline lakes in the interior West as staging areas during migration but, curiously, does not use the Salton Sea in that manner (Jehl 1988, 1994). Wilson's Phalarope is a common transient at the sea, but concentrations do not number in the hundreds of thousands, as at Mono Lake in eastern California and the Great Salt Lake in northern Utah (Jehl 1988). Instead, migratory flocks typically number in the hundreds, concentrations of more than 1,000 individuals being unusual. This species is equally numerous in spring and fall. At both seasons it is an early migrant. Spring transients move through from 12 March (1988, Salton City; M. A. Patten) to mid-May (e.g., south end 20 May 1995; G. McCaskie), with peak spring passage in late April and early May. Our ability to determine late dates for spring migrants (and early dates for fall migrants) is hindered by the presence of Wilson's Phalaropes in late May and early June, such as one at the north end 26 May 1980 (McKernan et al. 1984), a female in alternate plumage at the Whitewater River delta 27 May 2000 (M. A. Patten), and a male and female together at Salton City 1 June 2000 (G. McCaskie).

Wilson's Phalarope is the first migratory shorebird to return in fall, the earliest alternate-plumaged females being recorded by early June (e.g., 15 at Wister 3 June 2001; G. McCaskie). A flock of 350 at the south end 15 June 1985 (G. McCaskie) indicates how common this species can be by mid-June. The phalarope remains common through early July, when many other southbound shorebird species first appear. The first juvenile phalaropes are detected in mid-July (e.g., 3 at Mecca 13 July 1999; M. A. Patten). Peak passage is from mid-July through August, with some moving through in September and small numbers lingering through October. Although Wilson's Phalarope is a rare winter visitor to the region, it is recorded nearly annually. As a result, it is impossible to determine whether birds present in late October or November are late fall migrants or are attempting to winter locally. Up to ten at Obsidian Butte 24 October 1999–9 April 2000 (M. A. Patten, P. E. Lehman) furnished the date span for known wintering birds, although most winter records are from between mid-December and early February. Generally

lone individuals are recorded during that season, but, in addition to the up to ten noted above, four were at the south end 16 and 17 February 1975 (AB 29:742) and at most three were at Obsidian Butte 29 November 1998–28 February 1999 (NAB 53:209). Interestingly, numbers tend to be lower January–March, suggesting that some late fall birds are indeed migrants.

ECOLOGY. As along the coast, Wilson's Phalarope prefers freshwater ponds or still backwaters but avoids the Salton Sea itself. Some feed on mudflats along the edge of the sea, spinning as they forage as if they were swimming. The species is frequently encountered in flooded fields in the Imperial Valley, sometimes in flocks of more than 100 individuals.

RED-NECKED PHALAROPE
Phalaropus lobatus (Linnaeus, 1758)
Abundant fall transient (July through November); common spring transient (April to early June); casual in winter and summer. In the West in both spring and fall the Red-necked Phalarope migrates later in the season than does Wilson's Phalarope. Migrant Red-necked Phalaropes generally reach peak numbers two to three weeks later than Wilson's Phalaropes, and the last depart two to three weeks later. Spring migrants arrive as early as 5 April (1981, "Salton Sea"; AB 35:862), even 31 March (1990, male in alternate plumage at Obsidian Butte; G. McCaskie). Small numbers of presumed early spring migrants were found in coastal southern California during March 1995, so four at Morton Bay 18 March 1995 (FN 49:309) may have been migrants rather than wintering birds. Even so, most spring migrants do not arrive until late April, with passage peaking in the first half of May (e.g., 3,000 at the south end 11 May 2000; G. McCaskie). Because spring records extend into summer, it is difficult to determine dates for the latest spring transients, but obvious migrants are present at the end of May and into early June (e.g., 4 at Salton Sea NWR 7 June 1998; M. A. Patten). Stragglers have remained through the summer, but the species does not occur annually in that season. Birds in early June could well be late spring

migrants, but three at the mouth of the White-water River 15 June 1985 and near Oasis 18 June 1994 (both G. McCaskie) were almost certainly summering locally. A remarkable number remained through the summer of 1992 near Wister, with as many as 150 present 6 June–11 July (G. McCaskie, M. A. Patten).

Like that of Wilson's Phalarope, fall migration is more protracted. Unlike that species, the Red-necked Phalarope is more numerous in fall than in spring; numbers may build to impressive levels during peak movement. The first arrivals appear by 1 July (2000, mouth of the Whitewater River; G. McCaskie), but the species' numbers do not peak until late August and early September (e.g., 10,000 at Cerro Prieto 23 August 1997 and more than 7,500 there 4 September 2000; Patten et al. 2001). Large numbers are still moving through in October and even early November (e.g., 600 at Salton City and at the south end 2 November 1993; M. A. Patten). Because this species occasionally winters at the Salton Sea, fall departure dates are difficult to determine. Even so, 40 at the south end 11 November 1978 (Garrett and Dunn 1981) were undoubtedly migrants, and we believe one at the north end 27 November 1970 (Garrett and Dunn 1981) and four at Obsidian Butte 29 November 1998 (G. McCaskie) to be such. This species is casual in winter, averaging one record every three to four years from 1 December (1999, 2 at Obsidian Butte through 5 February 2000; NAB 54: 201) to 22 March (1959, south end from 1 March; Garrett and Dunn 1981), with up to 15 near Red Hill 8–17 December 2002 (fide G. McCaskie).

ECOLOGY. The Red-necked Phalarope is as likely as Wilson's to occur on freshwater ponds or still backwaters adjacent to the Salton Sea. Unlike Wilson's, however, it actively forages on the sea itself, sometimes in large flocks in protected bays, coves, and inlets.

RED PHALAROPE

Phalaropus fulicarius (Linnaeus, 1758)
Rare fall transient (late July through November); casual spring transient (mid-April through May); 4 summer records; 2 winter records. In contrast to the other two phalarope species, the Red Phalarope is more pelagic. It is thus rare in the interior of North America south of its breeding grounds. A few are detected annually in the interior Southwest during migration, particularly in fall. In the Salton Sink, spring records are few, averaging one every four to five years from 24 April (1994, first-spring male 5 km west of Niland; M. A. Patten, SDNHM 48887) to 30 May (1970, one in basic plumage at the mouth of the Whitewater River; AFN 24:644), but include a flock of four at the north end 10 May 1980 (AB 34:815). Three were reported without details 17 and 18 April 1999 (Shuford et al. 2000). One in basic plumage near Salton City 26 March 1995 (FN 49:309) was likely an exceptionally early spring transient given that this species is on the move along the southern California coast by mid-March. The Red Phalarope has been recorded four times in summer at the Salton Sea, with individuals at the Whitewater River delta 18 June 1977 (AB 31:1190) and 14 July–20 August 1973 (AB 27:919, 28:108), a female in alternate plumage at Wister 20 June 1999 (NAB 53:433), and an individual at the south end 24 June 1973 (AB 27:919).

Records of the Red Phalarope number many more in fall, when the species approaches annual occurrence. The first regional records are from fall, with females 3 km west of Niland 26 October 1940 (K. E. Stager; UMMZ 124413) and at Unit 1 on 6 September 1953 (Cardiff 1956; SBCM 33135). Fall records extend from 22 July (1972, north end; AB 26:905) to 4 December (1982, up to 25 at Red Hill from 9 November; AB 37:224). There is no obvious peak in occurrence, although most recent records are from October and November, when large flocks of Red-necked Phalaropes move through the Salton Sink. An incursion of Red Phalaropes into the desert Southwest in fall 1982 brought up to 35 birds to Red Hill, Unit 1, and Salton City 9 November– 4 December (AB 37:224). The only other record of a flock in fall was of up to five photographed at Salton City 11–23 November 1969 (AFN 24: 100). A bird at Salton City 11 January 1970 (E. A. Cardiff; SBCM 4446) was probably a wintering

TABLE 11
Approximate Mean Numbers of Gulls
Annually Wintering in the Salton Sea Region

SPECIES	MEAN NUMBER
Laughing Gull	1–3
Franklin's Gull	Vagrant
Little Gull	Vagrant
Bonaparte's Gull	500
Heermann's Gull	Vagrant
Mew Gull, *L. c. brachyrhynchus*	1–2
Ring-billed Gull	500,000
California Gull, mostly *L. c. californicus*	75,000
Herring Gull, *L. a. smithsonianus*	7,500
Thayer's Gull, *L. g. thayeri*	5–10
Lesser Black-backed Gull, *L. f. graellsii*	Vagrant
Slaty-backed Gull	Vagrant
Yellow-footed Gull	<10
Western Gull, mostly *L. o. wymani*	2–3
Glaucous-winged Gull	3–5
Glaucous Gull, *L. g. barrovianus*	<1
Black-legged Kittiwake	Vagrant

Source: Information compiled by Michael A. Patten, partially published in *FN* 52:257.

straggler from this small November incursion; one near Red Hill 8–17 December 2002 (G. McCaskie et al.) provided the only other winter record for the Salton Sink.

ECOLOGY. The vast majority of Red Phalaropes have been found in flocks of Red-necked Phalaropes, with which they share habitat preferences in the region.

LARIDAE • *Skuas, Gulls, Terns, and Skimmers*

More gulls winter at the Salton Sea than at any other locale in the interior West. Such species as the Herring Gull, generally rare in the interior Southwest, are common at the Salton Sea in winter (Table 11). Other coastal species, such as the Mew, Thayer's, the Western, the Glaucous-winged, and the Glaucous Gull, are essentially accidental vagrants elsewhere in the interior West but are regular in small numbers at the Salton Sea. Even though the numbers of the large gulls are high during midwinter, they reach a peak in late fall and early winter and again in late winter and early spring as thousands move to and from wintering grounds in the Gulf of California (e.g., see Pugesek et al. 1999). Total gull numbers are dominated by three species, the Ring-billed, the California, and the Herring, which together account for more than 98 percent of the gulls wintering in the region. Counts in summer are dominated by the first two species, which account for more than 92 percent of gulls, although both the Laughing and the Yellow-footed Gull occur in the hundreds to thousands by late summer.

In addition to being a haven for gulls, the Salton Sea is an important breeding site for terns and the Black Skimmer. The sea supports the only large nesting population (100–170 pairs) of the Gull-billed Tern in the western United States and four to five times that number of Black Skimmers. Although the Caspian Tern is a relatively recent (re)colonist, thousands nest at Mullet Island and elsewhere around the sea, as do a few Forster's Terns, albeit not annually. The California Gull began nesting at Obsidian Butte in 1996, extending its breeding range southward some 500 km from Mono Lake, and the Laughing Gull has nested sporadically over the past half-century. Migrant larids are perhaps less conspicuous at the Salton Sea, notable exceptions being Bonaparte's Gull and the Black Tern, both of which occur in the thousands during spring and fall. Affinities between the Salton Sea and the Gulf of California are evidenced by the regular fall passage of jaegers, the regular postbreeding occurrence of large numbers of Laughing and Yellow-footed Gulls (the only locale in the United States where the latter species occurs), and the regular breeding of various terns.

POMARINE JAEGER

Stercorarius pomarinus (Temminck, 1815)
Casual summer vagrant (late June through August); 4 spring records (May–early June); 1 or 2 fall records.
The Pomarine Jaeger is a pelagic species that is a casual vagrant in the interior Southwest. Salton Sea records are concentrated at the north end. Unlike the Parasitic and Long-tailed Jaegers, however, the Pomarine is encountered more frequently during summer than during migration. Indeed, it has become clear in the past decade that any jaeger on the Salton Sea in summer is most likely a Pomarine. There are eight, perhaps nine (see below) summer records from 30 June (1984, adult at the north end with up to 5 present by 14 July; *AB* 38:1062) to 24 August (1991, worn subadult female photographed from 3 August; *AB* 45:1161, SBCM 53595). Other summer records

are of an adult at Poe Road 10 July 1993 (*AB* 47:1150), one with an unidentified jaeger at Salton City 12 July 1990 (*AB* 44:1186), a subadult near Salton City 16 July 1995 (*FN* 49:980), an adult and a subadult at the Whitewater River delta 22 and 23 July 1995 (*FN* 49:980), an adult at the north end 3 August 1993 (*AB* 47:1150), and a subadult photographed at Morton Bay 13 August 1996 (*FN* 50:996). A subadult jaeger off the mouth of the Whitewater River 3 July 1994 (*FN* 48:989) was probably a Pomarine (M. A. Patten).

There are also five or six records of migrants, four in spring and one or two in fall. In spring, single adults were at Fig Lagoon 18 May 1996 (*FN* 50:333) and at the north end 29 May 1983 (*AB* 37:912), two adults were at the north end 4 June 1984 (*AB* 38:960), and a remarkable flock of 27 adults was at the mouth of the Whitewater River 4 May (not April) 1986 (*AB* 41:147). The flock approached from the south in tight formation, and upon reaching the north shore it broke into smaller flocks and settled on the water (*AB* 40:524). The only unequivocal fall record is of a fresh juvenile at Salton City 31 October 1992 (*AB* 47:149), because an adult at the north end 11 September 1977 (*AB* 32:258) may have been present since summer. The identification of immatures at the north end 8–12 September 1977 and possibly the same birds 24 September 1977 (*AB* 32:258) is questionable because young Pomarine Jaegers are virtually unrecorded in the interior Southwest in that season and September is the peak month for Parasitic Jaeger movement through the region (see below).

ECOLOGY. Apart from the bird at Fig Lagoon, all Pomarine Jaeger records are from the Salton Sea, generally on or very near the shoreline. This large species kleptoparasitizes California Gulls and Caspian Terns; the August 1991 bird was chasing Yellow-footed Gulls.

PARASITIC JAEGER

Stercorarius parasiticus (Linnaeus, 1758)
Rare fall transient (late August to early November); 2 summer records; 1 spring record. The Parasitic Jaeger is a rare but regular fall visitor to the

interior Southwest, principally to the Salton Sea. This pelagic species was virtually unknown in inland California until 1964, when an immature off the mouth of the Whitewater River 6 September was followed by eight immatures 18 September, two of which were collected 20 September (McCaskie and Cardiff 1965; SBCM 33172, 50901). This species has since proven to be annual in small numbers from 22 August (1992, north end; *AB* 47:149) to 28 November (1976, north end; *AB* 31:223), with the majority in September, smaller numbers until mid-October, and a few in early November. Two early winter records, of an immature at the south end 4 December 1982 (*AB* 37:338) and a jaeger "believed to be Parasitic" at the north end 17 December 1986 (*AB* 41:328), presumably involved extremely late fall migrants. The majority of Parasitic Jaeger records are from the north end, particularly around the Whitewater River delta, but there are many records from the mouths of the Alamo and New Rivers and near Salton City. More than 80 percent of fall records are of juveniles. Small concentrations have been noted, usually of three to seven birds (e.g., 6–7 at the north end 10–19 September 1976; *AB* 31:223). A group of 22 juveniles at the north end 16 September 1984 (*AB* 39:103) furnished the highest count, followed by 16 at the south end 10 September 1994 (*FN* 49:101). There are two definite summer records, of full-tailed adults at the Whitewater River delta 20 July 1985 (*AB* 39:962) and 1 July 2000 (*NAB* 54:423). Parasitics reported 13 July 1985 (*AB* 39:962) and 30 July 1984 (*AB* 38:1062) occurred prior to our determination that the Pomarine is the expected jaeger in summer at the Salton Sea (see above), so we consider those records tentative. The only spring record is of one or two adults at the north end 27–29 May 1984 (*AB* 38:960).

ECOLOGY. Virtually all regional Parasitic Jaeger records are from the Salton Sea itself, with the Whitewater River delta claiming the vast majority. A few occur on fresh water away from the Salton Sea (e.g., a juvenile at Cerro Prieto 4 September 1995; Patten et al. 2001). At the Salton Sea this species kleptoparasitizes Ring-billed and California Gulls and Caspian, Forster's, and Common Terns.

LONG-TAILED JAEGER
Stercorarius longicaudus pallescens Loppenthin, 1932
Casual fall transient (late August to early October); 1 spring record. The most pelagic of the jaegers, the Long-tailed is a casual fall transient in the interior West, including at the Salton Sea. There are about 17 records for the Salton Sea, the majority around the mouth of the Whitewater River. The Long-tailed Jaeger has been recorded as early as 25 (not 24) August (1974, juvenile at the mouth of the Whitewater River; Garrett and Dunn 1981, SDNHM 39393) and as late as 4 October (2002, adult photographed at the mouth of the Whitewater River; C. McGaugh). Additional records of juveniles extend from 3 September (1996, Bombay Beach; *FN* 51:120) to 26 September (2001, Salton City; G. McCaskie), including one found dead 5 km west of Niland 5 September 1985 (*AB* 40:159; SDNHM 45369). Adults were present at the Whitewater River mouth 10 and 11 September 1984 (*AB* 39:103) and 20 September 1980 (*AB* 35:226), and two were seen at Salton City 26 September 1993 (*AB* 48:152). A bird in its second fall at the mouth of the Whitewater River 3 October 1987 is noteworthy (*AB* 42:136; LACM 103840). A "full-tailed adult" flying north along the Whitewater River into the Coachella Valley 23 May 1987 (*AB* 41:488) is one of few spring records for the Southwest, including the open ocean off California.

ECOLOGY. The only regional Long-tailed Jaeger records away from the Salton Sea are of two birds on fresh water at Cerro Prieto, a juvenile 26 August–4 September 1995 and an adult 4 September 1995 (Patten et al. 2001). This small species has been observed at the sea kleptoparasitizing Ring-billed Gulls and Forster's, Common, and Black Terns. The gut of the 1987 specimen contained numerous Mozambique Tilapia (*Tilapia mossambica*).

TAXONOMY. Long-tailed Jaegers in the Nearctic are the pale *S. l. pallescens* (Manning 1964; Cramp and Simmons 1983). Only 6 percent of

FIGURE 45. The Salton Sink is the only region of the western United States where the Laughing Gull occurs regularly, primarily as a nonbreeding summer visitor, although it has bred. Photograph by Kenneth Z. Kurland, January 1999.

adults in the North American range of *S. l. pallescens* show the dark belly of typical *S. l. longicaudus* Vieillot, 1819, but intermediates do occur, especially in Greenland (Olsen and Larsson 1997:160).

LAUGHING GULL

Larus atricilla Linnaeus, 1758

Fairly common summer and fall visitor (mainly mid-June to early September); rare in winter and spring; has bred. In stark contrast to its abundance along the Atlantic coast and in the Gulf of Mexico, the Laughing Gull (Fig. 45) is a limited component of the avifauna of the western United States, the Salton Sea being the only locale where it occurs regularly. It formerly nested at the south end, beginning in 1928 (Miller and van Rossem 1929; MVZ 3723), with breeding birds arriving typically by mid-April and exceptionally as early as 3 April (1947; Garrett and Dunn 1981) or even 31 March (2002, 4 adults; E. Strauss). The species bred annually at the sea until 1957 (*AFN* 11:429). It did not nest in the region again until 1994, when a nest contain-

ing two chicks was discovered off Johnson Road 5 August (*FN* 48:989). It has since nested at the geothermal ponds at Cerro Prieto in June 1998 (2 pairs; Molina and Garrett 2001) and at Salton Sea NWR 27 May–15 July 1999 (Molina 2000) and 17 May–28 August 2000 (3 pairs; *NAB* 54: 423), and 7 April–July 2001 (8 pairs; K. C. Molina, *NAB* 55:482). Palacios and Mellink (1992) discovered this species breeding on Isla Montague, Baja California, at the mouth of the Colorado River.

Despite recent breeding, the Laughing Gull is now largely a postbreeding visitor to the region, from May through November. The species occasionally arrives in early April, an adult at the south end 7 April (1991; *AB* 45:496) being the earliest in recent decades. Adults at the south end 9 March 1996 (G. McCaskie), 13 March 1999 (*NAB* 53:329), and 18 March 1995 (*FN* 49:309) were either exceptionally early spring arrivals or lingering wintering birds. Small numbers are present in the Salton Sink every year by May, but the majority of Laughing Gulls occur from mid-June (e.g., 350 at the south end 15 June 1991; *AB*

45:1161) through early September. Peak counts are from July, when the first juveniles appear, and August, such as more than 800 (over 60% juveniles) in late July 1979 (*AB* 33:897), more than 750 around the south end 7 July 1990 (*AB* 44:1186), and 600 (mostly adults) in mid-August 1977 (*AB* 31:1190). Large numbers sometimes continue through October (e.g., 100 at the south end 27 October 1973; *AB* 28:108). The few lingering annually into late December typically depart by year's end, the latest by 14 January (1996; *FN* 50:222). Few have spent the entire winter, with one at the south end 7 December 1980–8 February 1981 (*AB* 35:336) providing the first record of an overwintering bird. Presumed wintering individuals have been recorded as late as 12 March 1983 (adult in basic plumage at the mouth of the New River; G. McCaskie) (see above for other March records). Generally only one or two birds stay all winter, but as many as 5 were at Finney Lake 2 January–February 1982 (*AB* 36: 331), and six adults were there during the winter of 1982–83 (*AB* 37:338).

ECOLOGY. Like all gulls in the region, the Laughing is commonly seen along the shore of the Salton Sea. It is also frequent, however, at freshwater lakes and ponds in the Imperial Valley, and large numbers are routine in flooded fields near the edge of the sea.

TAXONOMY. Cramp and Simmons (1983) recognized *L. a. megalopterus* (Bruch, 1855), of continental North America, and Miller and van Rossem (1929) ascribed Salton Sea specimens to it. Measurements of various populations of the Laughing Gull overlap broadly, however, so the species is generally regarded as monotypic (Ridgway 1919; Peters 1934; AOU 1957; Blake 1977).

FRANKLIN'S GULL

Larus pipixcan Wagler, 1831

Rare to uncommon spring transient (mainly April–May); rare summer visitor and fall transient (mainly August–October); 1 winter record. The highly migratory Franklin's Gull breeds on the northern prairies of central North America and at scattered locales through the West, including once on the Modoc Plateau of northeastern California (*AB* 43:1363). Small numbers regularly pass through the Salton Sink. The species is generally rare but sometimes uncommon in spring, when alternate-plumaged adults move through the region. Spring records extend from 1 April (1978, adult at Desert Shores; *AB* 32:1055) into June, peaking from late April through May. A remarkable 85 Franklin's Gulls on the Salton Sea 26 May 1997 (*FN* 51:928), including 59 second-summer birds together near Oasis (M. A. Patten), is the largest concentration recorded in California. Other large flocks were 19 at the Whitewater River delta 15 May 1983 (*AB* 37:913) and 12 there 2–8 June 1991 (*AB* 45:496), with 1–3 being a more typical daily count.

Summer records are difficult to determine, as some summering individuals overlap with late spring transients. The species appears to be rare in summer, most records involving birds in first-summer plumage, although a number of second-summer birds and a few adults have occurred. Definite summer records extend from 27 June (1981, 2 adults through 13 July; *AB* 35: 979) to 30–31 July (1983, first-summer bird at the north end; *AB* 37:1027) and include a specimen from the mouth of the Whitewater River 9 July 1978 (skeleton, SDNHM 40803). Generally no more than one or two birds are found during summer, but groups of up to six (1972, *AB* 26:906; 1983, *AB* 37:1027) and five (1980, *AB* 34:930) have occurred. Fall transients appear when some summering birds are still present. Fall migration is more protracted, the first (usually juveniles) appearing in late July, the last (usually first-winter birds) in mid-November. Dates extend from 16 July (1994, juvenile near Oasis; *AB* 48:989) to 26 November (1988, first-winter bird at North Shore from 20 November; *AB* 43:168). Twelve at the south end 23 October 1971 (*AB* 26:121) represents the maximum in fall. Other notable autumn counts are six at the south end 12 October 1975 (*AB* 30:127), five immatures at the north end 18 October 1964 (McCaskie and Cardiff 1965; SBCM M3614), and five at Red Hill 18–29 October 2000 (G. McCaskie). At least 30 were seen around the south end at various times 17 August–5 November 2000 (*NAB* 55:103). Seven reported near Calipatria 18 November 1966 (*AFN* 21:78) may have been Laughing Gulls, as field marks distinguishing the two species were not well known at the time. A Franklin's Gull at the south end 29 January 1989 (*AB* 43:366) is the only one recorded in winter in the eastern half of California.

ECOLOGY. Franklin's Gull is most often reported along the shore of the Salton Sea, particularly around river mouths and in extensive impoundments. Some have been found in flooded fields in the Imperial Valley. Franklin's seldom occurs in the company of Laughing Gulls but instead joins flocks of Ring-billed Gulls or terns.

LITTLE GULL

Larus minutus Pallas, 1776

Casual vagrant, with records year-round but mainly in spring and winter. The Little Gull is a casual visitor to California and the Salton Sea. At the sea there are nearly 20 records across all four seasons, predominantly in spring and winter. Seven spring records fall within the period 13 April (1996, adult at the north end through 21 April; McCaskie and San Miguel 1999) to 6 June (1999, first-summer bird photographed 6 km west of Niland beginning 23 May; *NAB* 53:329, Rogers and Jaramillo 2002). Additional first-summer birds were seen at the Whitewater River delta 18 April–9 May 1987 (Langham 1991) and near Salton City 8 May 1988 (Pyle and McCaskie 1992), and adults were present at Salton Sea NWR 7 May 1995 (Garrett and Singer 1998), Mecca 19 May 1995 (Garrett and Singer 1998), and the mouth of the Whitewater River 27 May 1996 (McCaskie and San Miguel 1999). There are three summer records of first-summer birds that molted before departing: one at the mouth of the Whitewater River 10 May–4 July 1992 (Patten, Finnegan, et al. 1995) and birds photographed 6 km west of Niland 31 May–24 August 1997 (Rottenborn and Morlan 2000) and 24 July–12 August 1995 (Garrett and Singer 1998). A first-summer bird at the mouth of the Whitewater River 6–12 June 1982 was probably attempting

to summer but died shortly after being discovered (skeleton, SDNHM 41796).

There are but two fall records. The first record for California was of an adult photographed near Mecca 16–21 November 1968 (cover of *AFN* 23[1]; Dunn 1988). An adult with a small flock of Bonaparte's Gulls at the mouth of the Whitewater River on 13 September 1988 (Pyle and McCaskie 1992) perhaps summered locally, although fall vagrants have arrived elsewhere in California even earlier (see Binford 1985; Dunn 1988). Winter records extend from 14 December (1993, first-winter bird photographed near Oasis through 8 January 1994; Erickson and Terrill 1996) to 21 April (1996, adult at the north end 2 March that molted into alternate plumage before departing; McCaskie and San Miguel 1999). Other winter records are of a first-winter bird at Salton City 17–31 January 1985 (Dunn 1988) and an adult and two first-winter birds at the Niland Boat Ramp beginning 24 January 1994 (Howell and Pyle 1997), with the adult and one immature staying through 18 February but the other immature found dead 12 February 1994 (LACM 107890). Additional reports from the south end (*AB* 27:663) and along the eastern shore (*FN* 52:126) are undocumented.

ECOLOGY. Little Gulls invariably are found with concentrations of Bonaparte's Gull. All occurrences have been along the edge of the Salton Sea at large impoundments, bays, or river mouths.

BONAPARTE'S GULL

Larus philadelphia (Ord, 1815)

Fairly common transient and winter visitor (October through May); uncommon summer visitor. The graceful Bonaparte's is the most common small gull across most of the West and is the small gull most likely encountered at the Salton Sea, where it is a fairly common transient and winter visitor. Records of migrants and wintering birds extend from 5 October (1980, north end; McKernan et al. 1984) to 31 May (1999, 5 at the mouth of the Whitewater River; M. A. Patten). Numbers are typically higher during migration, sometimes markedly so. Fall passage peaks in November (e.g., 1,500 at Salton City 25 November 1989; G. McCaskie), spring passage in late April to early May (e.g., 800 around the Salton Sea 3 May 1998; M. A. Patten). Winter numbers fluctuate from year to year but typically are in the low hundreds, with as many as 500 at Salton City in January and February 1998 (M. A. Patten) and 626 around the sea 11–15 November 1999 (Shuford et al. 2000) being exceptional. As with many waterbirds in the Salton Sink, differentiating late spring migrants from summering birds is nearly impossible. Small numbers summer each year, with a few dozen being the norm, although 250 were at the sea during June and July 1977 (Garrett and Dunn 1981), and 100 were at the north end during the summer of 1983 (AB 37:1027). Summering birds often molt, which they complete by mid-August; most disappear shortly thereafter.

ECOLOGY. Bonaparte's Gull favors still waters in inlets, impoundments, or backwaters, although some birds forage over the open sea. Water may be saline, brackish, or fresh, although the largest concentrations tend to be at the sea's shoreline, around either saline or brackish water.

HEERMANN'S GULL
Larus heermanni Cassin, 1852

Rare summer visitor (mid-April to early October); 4 winter records. The coastal Heermann's Gull is a familiar bird in the Gulf of California, where it is one of the most numerous breeding species. It is a casual visitor to the interior Southwest, with a handful of records for Arizona, southern Nevada, and the deserts of southeastern California. This species was first recorded at the Salton Sea when two immatures were found at the Whitewater River delta 3 July 1967 (AFN 21: 605; SBCM 33176, 33177). Since that time it has proven to be a rare annual visitor to the Salton Sea from mid-April to early October, though there are records from every month. Heermann's Gull is beginning to appear regularly in early spring, including pairs of adults, perhaps portending breeding in the Salton Sink. Even so, most records are probably of postbreeding birds or of

birds whose breeding attempts failed; virtually all records in April and May records have involved adults. Apart from long-staying birds discussed below, adults have been recorded as early as 7 April (1996, the north end; FN 50:333) and as late as 2 August (2001, Salton City from 1 August; G. McCaskie) and include a specimen of a male collected 22 km northwest of Westmorland 19 July 1991 (R. L. McKernan, SBCM 53546). An exceptionally early adult in full alternate plumage was at Obsidian Butte beginning 2 March 1999 (NAB 53:329); it stayed until 5 December 1999 (NAB 54:105), when it was in basic plumage. Presumably this same bird returned 3 (not 11) March–31 July 2000 (J. E. Pike; NAB 54:423), 13 (not 21) March–30 September 2001 (T. Benson; NAB 55:355), and 31 March–6 July 2002 (NAB 56:486). In 2001 and 2002 it was joined by a second bird 21 March (G. McCaskie).

After early July most records are of birds in first-alternate and juvenal plumages. Records of immatures extend from 3 July (1967, 2 at the Whitewater River delta; cited above) to 7 October (1995, juvenile at the north end from 30 September; FN 50:115), with a distinct peak from mid-July through August, the earliest juvenile at the mouth of the Alamo River 12 July 1969 (AFN 23:695), and a specimen of a first-winter female from 3 km southwest of Obsidian Butte 5 September 1987 (S. W. Cardiff; LSUMZ 156031). Individuals at the south end 2 November 1974 (AB 29:122) and Sheldon Reservoir 28 October 1996 (S. von Werlhof) were exceptionally late. There are four winter records, of a first-winter bird at the south end 22 November 1980–20 June 1981 (AB 35:226, 366, 863, 979), a first-winter bird at the mouth of the Alamo River 26 December 1996 (FN 51:643), a second-winter bird at Obsidian Butte 18–25 (not 24) January 1998 (FN 52: 257), and an adult at Obsidian Butte 24 January–9 May 1998 (FN 52:257, 391). Most records involve lone birds, but groups of two to three have been noted, four were at the mouth of the Whitewater River 15 and 16 July 2000 (NAB 54:423), and six were just outside the Salton Sink at Plaster City 14 July 1968 (AFN 22:649).

ECOLOGY. Most Heermann's Gull records are from the immediate vicinity of the Salton Sea, generally on barnacle beaches, jetties, and rock outcrops. There are scattered records throughout the Imperial Valley, however, from freshwater lakes, ponds, and flooded fields.

MEW GULL

Larus canus brachyrhynchus Richardson, 1831

Rare winter visitor (mid-November through April); casual in summer. Although it is fairly common along the Pacific coast, the Mew Gull is casual inland in the West. This status held true for many years at the Salton Sea, where the species was first recorded (a bird found dead) at the north end 15 January 1966 (*AFN* 20:459). The species has since proven to be regular in winter, however, with one to three records per year; as many as four have been found in a single winter, between Salton City and the south end 18 December 1984–3 February 1985 (*AB* 39:210). First-winter birds supply the majority of the records, but a fair number of adults and a few second-year birds have been recorded. Records extend from 11 November (2001, adult photographed at Red Hill; E. Strauss, G. Griffith) to 8 May (1971, first-summer bird at North Shore; *AB* 25:801), including a specimen of a first-winter bird from near Rock Hill 19 April 1969 (G. McCaskie; SDNHM 37202). Remarkably early was a juvenile at Salton City 13 September 1995 (*FN* 50:115). An adult at the mouth of the Whitewater River 24 May 1987 (*AB* 41:488) was extremely late. The Mew Gull has been recorded eight times in summer in the Salton Sink. An adult at the mouth of the Whitewater River 1 June–4 August 1984 (*AB* 38:1062) was the first recorded in summer; another was present there 6 June 1984 (D. L. Dittmann; SBCM 39393). Five other summer records are of first-summer birds at Morton Bay 6 June–22 September 1999 (*NAB* 53:432; S. Guers) and at the north end 29 June–16 July 1985 (*AB* 39:962) and of second-summer birds near Niland 28 August 1996 (*FN* 51:120), near Brawley 4–11 July 1998 (*FN* 52:503), at the mouth of the Whitewater River 6–17 July 1999 (*NAB*

53:432), and near Red Hill 2–23 June 2002 (*NAB* 56:486).

ECOLOGY. Winter records of this species are scattered across the region. Most are from the edge of the Salton Sea, with more in lagoons and backwaters than on the sea itself. The species has also been recorded in flooded fields and on small lakes and ponds in the Imperial Valley.

TAXONOMY. All Mew Gull records from the West, including the two specimens from the Salton Sea (cited above), pertain to the distinctive *L. c. brachyrhynchus*. This subspecies differs from the other three in having an extensive white swath separating the black primary tips from the dark-gray bases, especially across the sixth through ninth primaries. First-winter birds are also distinctive in being extensively brown below and having a mostly dark tail.

RING-BILLED GULL

Larus delawarensis Ord, 1815

Abundant winter visitor (mid-July to mid-May); uncommon nonbreeding summer visitor. The Ring-billed Gull is astoundingly abundant in the Salton Sea region during the winter, when hundreds of thousands (perhaps millions) are around the Salton Sea and in the adjacent Coachella, Imperial, and Mexicali Valleys. In that season it is not unusual to encounter tens of thousands feeding in single flooded or wet agricultural fields. Return and departure dates are muddied by the thousands of nonbreeding birds remaining annually through the summer. Even so, the earliest record of juveniles is 16 July (1988; G. McCaskie), indicating that fall migrants return by mid-July. Birds begin migrating northward in late March or early April; most have departed by mid-May. Although thousands are present each summer, the Ring-billed is typically outnumbered by the California Gull in that season.

ECOLOGY. The Ring-billed is the only gull in the region likely to be encountered in agricultural fields well away from the Salton Sea, accounting for more than 99 percent of all gulls in this habitat (M. A. Patten; G. McCaskie; Shuford et al. 2000). It is numerous along the shoreline of the Salton Sea, where it is often outnumbered in winter by the California Gull (particularly along the east and west shores) and at times by the Herring Gull (mainly around the south end). The Ring-billed is virtually always outnumbered by the California Gull in summer, regardless of habitat. Ring-billed Gulls forage and roost in flooded fields, dry agricultural fields (especially those that have been recently plowed), at freshwater reservoirs, lakes, and ponds, and on the sea.

CALIFORNIA GULL

Larus californicus Lawrence, 1854 subspp.

Common winter visitor (mid-July to mid-May); fairly common nonbreeding summer visitor; rare, local breeder. At the Salton Sea only the Ring-billed Gull is more numerous than the California Gull (see Table 11). Indeed, these two species are the only large gulls likely to be encountered through much of the interior West. In addition to being numerous in winter, many California Gulls wintering in Mexico migrate through the Salton Sea in March and April (Pugesek et al. 1999), a scenario likely true for many large gulls given their sizable concentrations in the northern Gulf of California in late winter and early spring (Howell 1999:56). Migrant California Gulls return to the Salton Sea by mid-July, the first juveniles (and thus the first definite migrants) appearing as early as 3 July (1994; G. McCaskie). Numbers build steadily, climbing to the tens or even hundreds of thousands by October. This species remains numerous through March, when outbound spring migrants trickle away, the last departing by mid-May. As with many species, departure dates are clouded by a substantial presence of summering birds. Thousands of nonbreeders remain at the Salton Sea each summer, when the California typically outnumbers the Ring-billed Gull, especially along the immediate shoreline. Amazingly, two pairs of California Gulls nested near Mecca during the summer of 1996 (FN 50:996). Although no young were fledged, this effort extended the breeding range about 500 km south of the nearest site, at Mono Lake in eastern California. The species has since bred annually at Obsidian Butte or at impoundments at nearby Rock Hill (Molina 2000), beginning with 22 pairs fledging young during July 1997 (FN 51:1053); some 40 pairs nested successfully in 1999 (Shuford et al. 2000).

ECOLOGY. Wintering California Gulls show a high affinity for the Salton Sea, making this species much scarcer than the Ring-billed Gull in flooded fields and similar habitats throughout the Coachella, Imperial, and Mexicali Valleys. Even so, it is sometimes common on large freshwater lakes and reservoirs in the Imperial Valley (e.g., Finney and Ramer Lakes and Fig Lagoon). Breeders occupy islands, such as a detached earthen berm at the north end and a small natural volcanic outcrop with sandy intrusions at the south end. Nesting took place amid a large colony of terns and skimmers at the north end but in a pure colony at the south end.

TAXONOMY. The California Gull was long regarded as monotypic, but Jehl (1987a) described the large, pale-mantled *L. c. albertaensis*, whose breeding range lies entirely east of the Rocky Mountains. Given that *L. c. californicus* is the breeder in the Great Basin, occurring as close as eastern California, breeders at the Salton Sea are undoubtedly the nominate subspecies. Wintering birds are more problematic, as the winter distribution of *L. c. albertaensis* is essentially unknown. Jehl (1987a) presumed it to be the same as that for the whole of the species, but this proposition seems unlikely. The species may exhibit a classic leapfrog pattern, with *L. c. albertaensis* wintering farther south. Only extensive collecting or examination of existing winter specimens can elucidate differences in the winter ranges of the subspecies. Both subspecies winter at the Salton Sea, with *L. c. californicus* predominating. Females collected from Red Hill 15 October 1999 (LACM 111004) and the Whitewater River delta 4 March 2000 (SDNHM 50397), 7 August 1971 (SBCM 33195), and 18 February 1987 (SBCM 52582) are *L. c. californicus*, as are a male taken from the latter locale 18 January 1992 (SBCM 53497) and a series of six in LSUMZ. The first female has an intermediate mantle color (Kodak gray score of 6) but a small bill (Table 12). Second-year males collected from Red Hill 22 December 1984 (DMNH 39143, 39144) are *L. c. californicus* and *L. c. albertaensis*, respectively (Fig. 46; see Table 12). A worn adult male taken from the mouth of the Whitewater River 12 June 1986 (LSUMZ 130768) is also *L. c. albertaensis* (see Table 12). Notably, an adult male collected from San Felipe, on the northern Gulf of California, 19 April 1926 (SDNHM 10466) is *L. c. californicus*. We consider field identifications tentative, although one such field study suggested that about 5 percent of California Gulls wintering at the Salton Sea were *L. c. albertaensis* (King 2000). On the basis of specimens, we conclude that the nominate subspecies predominates in the Salton Trough by a ratio of at least 5:1.

HERRING GULL

Larus argentatus smithsonianus Coues, 1862

Common winter visitor (mid-September to early May); rare summer visitor. Like that of so many other waterbirds, the Herring Gull's status at the Salton Sea is at odds with its status elsewhere in the interior West. In general this species is highly coastal, large numbers occurring only as far inland as the cismontane valleys. It is extremely rare across much of the desert Southwest, although it is rare but regular in small numbers during winter along the lower Colorado River (Rosenberg et al. 1991). By contrast, thousands regularly migrate through and winter at the Salton Sea, from 13 September (1992, juvenile at the mouth of the Whitewater River; M. A. Patten) to 12 May (1990, adult at Obsidian Butte; G. McCaskie), although differentiating between late spring migrants and birds attempting to summer is difficult. Most Herring Gulls occur between mid-October and mid-April, with 10,000–15,000 during the winter of 1997–98 (*FN* 52:256) being the largest count on record. The species has steadily increased in the region since the early 1980s, with average winter numbers around 7,500 individuals by the turn of the twenty-first century (see Table 11). It rarely remains through summer (*contra AFN* 7:37, a

report of 300 during August 1952) from 25 May (1996, at most 2 at the Whitewater River delta through 3 August; G. McCaskie) to 28 September (1986, 2 at the Whitewater River delta present from 12 July; M. A. Patten et al.). It is recorded annually, or nearly so, in that season, with normally only one to two individuals during any given summer.

ECOLOGY. Of the three large gulls common in winter (the others being the Ring-billed and the California), the Herring is the species most tied to the saline environment afforded by the Salton Sea. Most records are from the immediate shoreline, where birds roost and forage, although landfills (e.g., as at Brawley) and lakes in the Imperial Valley sometimes attract sizable numbers, and a very few occur in flooded fields.

TAXONOMY. Most of North America hosts only *L. a. smithsonianus*, among the most variable avian subspecies. All North American birds are large and tend to have a mostly dark tail in juvenal plumage, but there is much variation in the plumage pattern of birds in their first winter, with a seemingly bimodal distribution, each mode corresponding with an ocean. First-winter Herring Gulls on the Pacific coast are mostly dark brown with a dark tail, recalling a first-winter *L. occidentalis,* whereas first-winter Herring Gulls on the Atlantic coast are whiter below with distinct brown streaking and more white in the tail, recalling *L. a. argentatus/argenteus* or *L. fuscus* (M. A. Patten pers. obs.). All adults of *L. a. smithsonianus* have the mantle pale gray, the shade of a Ring-billed Gull's. Despite the geographic likelihood, there are no confirmed records for the coterminous United States of the Vega Gull, *L. a. vegae* Palmén, 1887, of Siberia and western Alaska. This subspecies differs from *L. a. smithsonianus* in its darker mantle (closer to that of a California Gull or a nominate Western Gull), glossier black wingtips, and patterning on the outer primaries. Even though *L. a. vegae* is sometimes treated as a distinct species (e.g., by Ridgway 1919:618; Kennerley et al. 1995), most authorities consider it conspecific with *Larus argentatus* (Dwight 1925; Peters 1934; Dement'ev et al. 1951:531; Cramp and Simmons 1983:836;

TABLE 12

Characterization of Subspecies of the California Gull *(Larus californicus)*

SPECIMEN	LOCATION	PALENESS (L)	BILL DEPTH (mm)	BILL LENGTH (mm)
Male *L. c. albertaensis*				
SDNHM 45795	Alberta; Beaverhill Lake	51.9	17.6	47.4
SDNHM 44134	Alberta; Beaverhill Lake	50.1	18.0	50.2
SDNHM 44136	Alberta; Frog Lake (topotype)	54.0	17.7	52.7
SDNHM 44133	Alberta; Beaverhill Lake	50.9	18.1	53.0
DMNH 39144	California; Salton Sea	50.8	16.8	50.2
LSUMZ 130768	California; Salton Sea	—	17.3	51.1
LSUMZ 152027	Louisiana; Cameron Parish	—	16.8	55.6
LSUMZ 162250	Louisiana; Cameron Parish	—	17.2	52.1
Male *L. c. californicus*				
SDNHM 45794	California; Mono Lake	43.4	16.7	46.8
SDNHM 21511	Oregon; Klamath Falls	48.5	17.1	45.2
SDNHM 21509	Oregon; Voltage	50.2	16.7	47.2
DMNH 39143	California; Salton Sea	46.6	16.3	41.8
SBCM 53497	California; Salton Sea	—	16.0	45.8
Female *L. c. albertaensis*				
SDNHM 44135	Alberta; Frog Lake (topotype)	56.0	16.6	46.8
Female *L. c. californicus*				
SDNHM 44123	California; Mono Lake	47.6	14.6	39.4
SDNHM 21512	Oregon; Voltage	48.1	13.6	41.6
SDNHM 42466	California; Stockton	47.6	14.5	41.6
SDNHM 42464	California; Stockton	45.7	15.0	42.9
LACM 111004	California; Salton Sea	—	14.5	39.6
SDNHM 50397	California; Salton Sea	44.7	15.3	39.9
SBCM 33195	California; Salton Sea	—	13.2	37.6
SBCM 52582	California; Salton Sea	—	14.8	40.2

Note: Paleness was determined using a Minolta CR-300 Chroma Meter; higher values of *L*, the metric used, indicate paler coloration. Bill depth was measured at the gonys (per Jehl 1987a), and bill length was measured as the exposed culmen. See Jehl (1987a) for additional mensural data.

FIGURE 46. Examples of both subspecies of the California Gull, collected at Red Hill in December 1984. Note the longer, heavier bill of *Larus californicus albertaensis* (bottom), the rarer of the two subspecies at the Salton Sea, comprising only about 10 percent of the winter population.

Sibley and Monroe 1990; Gibson and Kessel 1997; AOU 1998).

THAYER'S GULL

Larus glaucoides thayeri Brooks, 1915

Rare winter visitor (October through mid-May); 5 summer records. The Iceland Gull breeds throughout the Canadian arctic, with Thayer's Gull occupying the western half of that range. The majority of Thayer's Gull populations winter in central coastal California, with smaller numbers occurring south to northwestern Baja California. Thayer's Gull is a rare bird in the interior West but annual in small numbers at the Salton Sea, where it was first recorded near the mouth of the Alamo River 22 March 1969 (adult female; *AFN* 23:521, Devillers et al. 1971, SBCM 33192). This gull was considered casual in the region through the early 1980s (Garrett and Dunn 1981), but it is now clear that 5–10 occur in an average winter (see Table 11). Exceptionally, about 30 were at the Salton Sea during the winter of 1997–98 (*FN* 52:257), including as many as 15 at Salton City (M. A. Patten), and more than 15 were at the Salton Sea during the winter of 1982–83 (*AB* 37:338). The vast majority are birds in their first winter, although all identi-

fiable ages have been recorded. Records extend from 29 September (1985, juvenile at Salton City; *AB* 40:159) to 11 May (1985, first-summer bird at the mouth of the Whitewater River from 5 May; G. McCaskie), with one at the south end 20 May 1977 (*AB* 31:1047) and a first-summer bird at the mouth of the Whitewater River 25 May 1996 (G. McCaskie) being exceptionally late. The majority of records are from early November to mid-April (especially December–February). A first-summer bird at the mouth of the Whitewater River 23 April–6 August 2000 (*NAB* 54:423) is the only Thayer's known to have summered in California. Birds at that locale 30 April–11 June 1983 (*AB* 37:1027), 3 June 1984 (*AB* 38:1062), 7 June 1981 (*AB* 35:979), and 31 May–23 June 2002 (*NAB* 56:486) represent the only other regional records for late spring or summer.

ECOLOGY. Most regional records of Thayer's Gull are from the Salton Sea and the landfill at Brawley, with nearly all found among congregations of Herring Gulls. Thayer's can be found anywhere along the shoreline, with seemingly no preference for substrate, as it roosts on sand, barnacles, earthen berms, and rock outcrops.

TAXONOMY. Geographic variation in the Iceland Gull complex is complicated and partly clinal. North American breeders exhibit a reasonably smooth cline of darkening wingtip coloration from east to west (Weber 1981), so intermediates are commonplace. Currently (AOU 1998), blacker-winged breeders, from western Canada, are accorded full species status as the monotypic *Larus thayeri*. Grayer-winged breeders, from eastern Canada, are referred to as *L. g. kumlieni* Brewster, 1883, a subspecies of the Old World Iceland Gull, a white-winged species somewhat resembling a miniature *L. hyperboreus*. The basis for treating *L. g. kumlieni* and *L. thayeri* as separate species, however, rests almost entirely on Smith (1966), which has been discredited (Snell 1989, 1991). Furthermore, these two taxa freely interbreed on Southampton Island (Gaston and Decker 1985) and are increasingly treated as conspecific (e.g., by Godfrey 1986; Sibley and Monroe 1990:256; Gibson and Kessel 1997). There thus seems to be no sound biological basis for treating these two forms as distinct species. Because the clinal variation noted above appears to be smooth, with darkening wingtips from east to west, it is probable that *L. g. thayeri* is not even a valid subspecies, but more work is needed to determine subspecific limits. For now we follow Gibson and Kessel (1997) and Pittaway (1999) in maintaining the subspecies *L. g. glaucoides, L. g. kumlieni,* and *L. g. thayeri* but merge them into a single species. We suspect, however, that birds in the Old World will prove to be a species distinct from those in the New World, with neither exhibiting subspecific variation.

LESSER BLACK-BACKED GULL
Larus fuscus graellsii Brehm, 1857
Casual winter vagrant (late October to mid-March); 2 fall records. The Lesser Black-backed Gull is a European species. Records of this species on the East Coast have increased dramatically in recent years, as have records for the Gulf of Mexico and the Midwest (Post and Lewis 1995). This increase has brought at least a dozen to the Salton Sea in fall and winter. Aside from one in first-winter plumage in 1998, records have been of adults or fourth-winter birds, probably owing to the difficulty of field identification of birds in immature plumages. The first regional record, and only the second for California, was of an adult photographed at Red Hill 18 December 1984–5 January 1985 (Dunn 1988). Winter records extend from 25 October (1986, adult photographed at Rock Hill through 24 February 1987; Pyle and McCaskie 1992) to 17 March (2000, Salton Sea State Recreation Area from 25 January; *NAB* 54:221). An adult at the Whitewater River delta 14 February–8 March 1998 (Erickson and Hamilton 2001) also stayed late; this bird returned to the same locale 26 January–27 February 1999 (*NAB* 53:209), when it was photographed (Rogers and Jaramillo 2002), and again 10 January–6 February 2001 (CBRC 2001-073). Other winter records are of adults photographed at Red Hill 13 December 1986–24 February 1987 (Pyle and McCaskie 1992) and at the Salton Sea State Recreation Area 21 January–7 March 1998 (Erickson and Hamilton 2001) and adults observed at Obsidian Butte 19 January 1998 (Erickson and Hamilton 2001) and at Brawley 22–27 January 1996 (McCaskie and San Miguel 1999) and 30 December 1998–1 February 1999 (Erickson and Hamilton 2001). The bird at the state recreation area (see above) returned 25 January–17 March 2000, 26 January–13 March 2001 (CBRC 2001-039), and 11 November 2001–18 March 2002 (*NAB* 56:357; C. McGaugh et al.). A first-winter bird at Obsidian Butte 24 January–8 March 1998 is the only one in this plumage recorded in California (Erickson and Hamilton 2001). An adult or fourth-winter bird photographed at the Whitewater River delta 14 September–5 October 1986 (Langham 1991) and a fourth-winter bird at Red Hill and vicinity 17 September–21 November 1999 (CBRC 1999-176; *NAB* 54:105) provided the only two records for California of fall vagrants. The latter bird returned 30 September–13 October 2000 (CBRC 2000-123; *NAB* 55:103).

Five additional reports, from the mouth of the Whitewater River (Pyle and McCaskie 1992), Red Hill (Roberson 1993; Patten and Erickson 1994), the south end (Heindel and Garrett 1995), and Salton City (*FN* 52:257), lack adequate documentation. A confusing third-winter gull at Obsidian Butte 22 December 1998–7 March 1999 was initially thought to be a Slaty-backed Gull (CBRC 1999-061), but when it was refound in late January its legs were yellowish, eliminating that possibility. We believe it to have been either a large male Lesser Black-backed Gull or perhaps a Lesser Black-backed × Herring Gull hybrid or *L. fuscus tamyrensis*, a taxon unrecorded in North America. This bird returned 13 November–5 December 1999 in fourth-winter plumage (M. A. Patten et al.). An attempt to collect it failed, so its identity remains a mystery.

ECOLOGY. Lesser Black-backed Gulls have been found in large flocks of Herring Gulls along the shore of the Salton Sea, especially at inlets and river mouths. Two records are from the refuse dump in Brawley.

TAXONOMY. There are no specimens of the Lesser Black-backed Gull for California, but all adults have shown the characters of *L. f. graellsii*, which has a mantle roughly equivalent in shade and color to the dark slate gray of *L. occidentalis wymani*. Of the five subspecies of the Lesser Black-backed Gull, *L. f. graellsii* has nearly the palest mantle. This subspecies occurs in the British Isles and has recently colonized Iceland and adjacent continental Europe (Post and Lewis 1995). Although there are a few North American records of the darker *L. f. intermedius* Schiöler, 1922, of southern Scandinavia, the overwhelming majority of continental records, including the numerous specimens, are of *L. f. graellsii* (Post and Lewis 1995), one of them west to Alaska (Gibson and Kessel 1992; UAM 5708). Without a California specimen the paler *L. f. taimyrensis* Buturlin, 1911, of central Siberia, perhaps cannot be excluded, although it is extremely unlikely given the spread of *L. f. graellsii* across North America. *L. f. taimyrensis* is unrecorded (and unlikely) in the Western Hemisphere despite an alleged first-winter specimen from Alaska (Bailey 1948; DMNH 9786; see also Gibson and Kessel 1992, 1997). Together *L. f. taimyrensis* and *L. f. heuglini*, of northern Russia, are sometimes considered a full species, Heuglin's Gull (*L. heuglini*; e.g.,

Kennerley et al. 1995), but they are generally treated as subspecies of *L. fuscus* (Cramp and Simmons 1983:815; Sibley and Monroe 1990; AOU 1998). *L. f. heuglini* has a mantle color like that of *L. f. graellsii* and is only marginally distinct from that subspecies, differing only in its larger average size, its slightly later molt timing, and its tendency for a whiter head in basic plumage.

YELLOW-FOOTED GULL
Larus livens Dwight, 1919

Common postbreeding visitor (May through September); rare winter visitor (October through early March); casual in spring. The Salton Sea is the only location in the United States where one is likely to encounter the Yellow-footed Gull (Fig. 47). This species is an endemic breeder to the Gulf of California, where the total population numbers about 20,000 pairs (Anderson 1983), but a few thousand postbreeding wanderers annually move north to the Salton Sea. Its occurrence at the Salton Sea is a relatively recent phenomenon, an adult observed near Salton City 22 August 1965 providing the first record (Devillers et al. 1971; McCaskie 1983). By 1970 numbers had increased to nearly 50 individuals annually during summer and early fall (McCaskie 1983; Patten 1996). It is now a common postbreeding visitor, numbering in the hundreds to low thousands each summer and fall (Patten 1996). There is no evidence of nesting, but it is expected because adults have been observed performing mating displays and vocalizations (M. A. Patten).

The first postbreeding visitors arrive in May (e.g., 25 at Salton City 2 May 1992; *AB* 46:481), save for exceptionally early birds at Salton City 22 April 1973 (*AB* 27:820) and at the mouth of the Whitewater River 29 April 1967 (*AFN* 21:541). Numbers build quickly—for example, 150 were at the Salton Sea by 31 May 1992 (*AB* 46:1178)—with the majority arriving in June. Juveniles generally arrive in early July, two at Poe Road 22 June 2000 (G. McCaskie) being the earliest. The Yellow-footed Gull reaches its peak abundance in the Salton Sink in late July and August, such as more than 1,000 in early August 1990 (*AB* 44:1187), with counts in the thousands now

FIGURE 47. The Yellow-footed Gull, until 1965 unique to the Gulf of California, continues to increase at the Salton Sea. Photograph by Kenneth Z. Kurland, August 1999.

commonplace in that month. Numbers remain high into September but dwindle during that month; most birds depart by early October. Since the winter of 1997–98 small numbers have wintered annually. Only a few (3–5) individuals are noted most years; double-digit counts are unusual, but 10 were seen at the south end during the winter of 1997–98 (*FN* 52:257), 15 were counted there 21 December 1982 (*AB* 37:772), and 15 were at Salton City 26 February 1978 (*AB* 32:399).

ECOLOGY. Like the Western Gull, the Yellow-footed shows a distinct predilection for the shore of the Salton Sea. It favors barnacle beaches, jetties, earthen berms, and rock outcrops but usually avoids mudflats. Birds are occasionally noted foraging in flooded fields near the margin of the sea and, rarely, deep within the Imperial and Mexicali Valleys; those arriving in midsummer frequently depredate nesting tern colonies.

WESTERN GULL

Larus occidentalis Audubon, 1839 subspp.
Rare perennial visitor (but principally September through March). The Western Gull is the most common large gull along the Pacific coast of the coterminous United States, especially in southern California, but it is rarely encountered away from coastal influence. It is casual east of the mountains, except at the Salton Sea, where it is a rare perennial visitor whose occurrence has been increasing in frequency since the mid-1980s and especially since the mid-1990s. The first area record was of a third-winter bird at Salton Sea NWR 17 January–13 February 1965 (McCaskie and Cardiff 1965). A third-winter bird at the south end 29 March 1969 (*AFN* 23:521) was the only other one recorded through the 1970s (Garrett and Dunn 1981; McCaskie 1983). A few Western Gulls appeared in both 1983 and 1984, and then they recurred annually in small numbers until the mid-1990s, when they increased sharply. The species is now a rare visitor to the Salton Sea. Records are scattered across the entire year but are concentrated in winter from 3 September (1987, 2 juveniles at the White-water River delta; *AB* 42:136, LSUMZ 156022) to 19 March (1983, first-winter bird at Salton City from 5 March; *AB* 37:338). Early juveniles were photographed at Salton City 15 August 1986 (*AB*

44:144) and at the north end 17 August–7 September 1996 (*FN* 51:121), and two early adults were at Obsidian Butte 25 August 1999 (G. McCaskie). Similarly, one at Salton City 12–26 April 1986 (*AB* 40:524) was probably a late-wintering bird. Generally only one to three birds are encountered, but a concentration of at least five at Salton City 29 September–28 November 1985 (*AB* 40:159) was notable, especially for the mid-1980s. More recently, at least six were at the Salton Sea 28 January–8 March 1998 (*FN* 52:257). This species is scarcer, but also increasing, in spring and summer, from 2 May (1992, second-summer at Salton City; *AB* 46:481) into September (e.g., third-summer at the south end 10 August–7 September 1996; *FN* 52:121). In some cases birds remain through the summer, such as one at the south end 9 June–28 July 1990 (*AB* 44:1187) and up to three photographed at Morton Bay 8 June–24 July 1999 (*NAB* 53:433). Most summering birds are nonadults that perhaps wintered in the Gulf of California and failed to travel beyond the Salton Sea when migrating north in spring.

ECOLOGY. All area records of the Western Gull are from the shore of the Salton Sea, with none on fresh water even a modest distance away (in contrast to the Herring Gull and even the Yellow-footed Gull).

TAXONOMY. An adult female collected at Salton City 18 November 1984 (*AB* 39:103; UCSB 16829) belongs to the southern subspecies *L. o. wymani* Dickey and van Rossem, 1925, which breeds along the Pacific coast from central Baja California north to Point Conception in southern California. Most sight records pertain to birds showing characters of this subspecies, including pale irides and an extremely dark slate-gray mantle (the same shade as on a *L. f. graellsii* Lesser Black-backed Gull and quite similar to that of a Yellow-footed Gull). A third-winter bird at the south end 27 February 1983 (*AB* 37:338), a second-winter at Salton City 5 February 1984 (*AB* 38:357), a fourth-winter at Obsidian Butte 9 January 2000 (G. McCaskie, M. A. Patten), and a second-summer at the mouth of the Whitewater River 10 September 2000 (M. A. Patten) showed

characters of the nominate subspecies, which breeds from Point Conception north to Puget Sound. A few other wintering birds have appeared to belong to this subspecies, with its dusky irides and medium-dark gray mantle (slightly darker than a nominate California Gull's). There are no conclusive specimens for the Salton Sea, but this subspecies has been collected on the Colorado River, where a first-winter female (with scattered gray mantle feathers; T. R. Huels *in litt.*) was taken at Lake Havasu 12 December 1946 (Phillips et al. 1964; UA 4914).

GLAUCOUS-WINGED GULL
Larus glaucescens Naumann, 1840
Rare winter visitor (mid-October to mid-May); 3 summer records. The Glaucous-winged Gull is a coastal species breeding from Siberia through coastal Alaska to Puget Sound and wintering south to Baja California. It is casual in most of the interior West, but it is a rare annual winter visitor to the Salton Sea region and has been noted with increasing frequency elsewhere in the interior (Binford and Johnson 1995). Most Salton Sea records involve birds in their first winter, but adults have been noted frequently, and every age class has been recorded. In an average winter 3 to 5 are found on the Salton Sea (see Table 11), but there were 12 in the winter of 1997–98 (*FN* 52:257) and at least 10 in the winter of 1993–94 (*FN* 48:248). Records extend from 13 October (1987, south end; *AB* 42:136) to 8 May (1971, mouth of the Whitewater River; *AB* 25:801), with the vast majority from November to mid-April. A worn second-winter female at the mouth of the Whitewater River 21 May 1984 (SBCM 39394) and first-summer birds there 1 May–6 June 1982 (*AB* 36:1016), near Salton City 1 June 2000 (G. McCaskie), at the south end 17 June 1982 (*AB* 26:905), and at Wister 20 June 1981 (*AB* 35:979) were late and may have been attempting to summer. There are only four summer records, however, of at most three first-summer birds at the mouth of the Whitewater River late May–14 August 1984 (*AB* 38:1062; G. McCaskie), at most two first-summer birds there 22 June–22 July 2000 (G. McCaskie), and a heavily worn

first-summer bird at the mouth of the New River 15 July 1989 (*AB* 43:1368). A bird seen at the south end 30 September 2000 (H. King; W. J. Moramarco) presumably summered, although it may have been an early fall migrant.

Hybrids present an identification problem in nearly all large species of *Larus*. Several apparent hybrids of the Glaucous-winged Gull have been detected at the Salton Sea; most appeared to be hybrids with the nominate subspecies of *L. occidentalis,* a combination common in the Puget Sound region (Hoffman et al. 1978) and regular in winter south to southern California. Exceptions were adults at the south end 20 December 1977 (*AB* 32:884) and 11 November 1988 (G. McCaskie; M. A. Patten) suspected to be *L. glaucescens* × *L. argentatus* hybrids and a first-winter bird at Salton City 21 February 1998 judged to be a *L. glaucescens* × *L. hyperboreus* hybrid (M. A. Patten; M. T. Heindel).

ECOLOGY. The Glaucous-winged Gull may occur virtually anywhere along the shoreline of the Salton Sea, with Salton City, the Whitewater River delta, and the Red Hill–Rock Hill–Obsidian Butte area being favored locales.

GLAUCOUS GULL

Larus hyperboreus barrovianus Ridgway, 1886

Casual winter vagrant (early December to mid-April); 2 summer records. The Holarctic Glaucous Gull is a bird of the far north. It seldom reaches even coastal southern California and is casual in the interior West. The first Salton Sink record was of a first-winter male at Red Hill 15 February–22 March 1969 (Devillers et al. 1971; SBCM 33216). There have since been about 23 additional records, all but 4 in winter. During that season records extend from 9 December (1972, adult photographed at the south end; *AB* 27:663) to 15 April (1978, second-winter bird at Salton City from 26 February; *AB* 32:399, 1055), the majority being from late December to mid-March. One at the mouth of the Whitewater River 14 April–24 May 1987 (*AB* 41:488) and one to two there 11 May–2 June 1996 (*FN* 50:333) were probably late spring transients. Surprisingly, this species has summered twice at the mouth of

the Whitewater River, 14 May–13 August 1983 (*AB* 37:912, 1027) and 13 May–8 November 1986 (*AB* 40:1255, 41:144). All records have involved birds less than a year old, apart from the adult and second-winter bird cited above and adults at the mouth of the Alamo River 21–24 December 1989 (*AB* 44:329) and 8 February 1997 (*FN* 51:802). An apparent hybrid Glaucous × Herring Gull, often called "Nelson's Gull" (but see Jehl 1987b), was at Oasis 21 March 1999 (K. L. Garrett).

ECOLOGY. Save for a bird at the refuse dump at Brawley, all Glaucous Gulls in the region have been on the shore of the Salton Sea, most frequently at river mouths.

TAXONOMY. The geographic variation in this species is complex. Banks (1986a) recognized four subspecies, of which only the small, small-billed, darker-mantled *L. h. barrovianus,* of Alaska, has been recorded on the West Coast. The Salton Sea specimen is a large male (wing chord 490 mm, exposed culmen 60.8 mm), leading Devillers et al. (1971) to suggest that it was *L. h. hyperboreus* Gunnerus, 1767, a large, heavy-billed, dark-mantled form of the Palearctic. Banks (1986a:156) examined this specimen and concluded that it was "best considered *barrovianus* although its measurements overlap those of *leuceretes,*" referring to *L. h. leuceretes* Schleep, 1819, the large, heavy-billed, pale-mantled subspecies of eastern Canada and Greenland.

BLACK-LEGGED KITTIWAKE

Rissa tridactyla pollicaris Ridgway, 1884

Casual late fall and winter vagrant (late November through mid-February); 3 or 4 summer records. The Black-legged Kittiwake, a pelagic species, is rare inland in North America. There are nine records for the Salton Sea, the first of a first-summer male at the mouth of the Whitewater River on the anomalous date 22 July 1967 (*AFN* 21:605; SDNHM 36111). Despite the paucity of summer records elsewhere in California, there are two other records of birds summering in the Salton Sink, one of a first-summer bird at the mouth of the Whitewater River 15 June–3 August 1968 (*AFN* 22:649), the other of five "immatures" there 11 May 1969 (*AFN* 23:626), with one staying until 1 August 1969 (*AFN* 23:695). The timing of these records suggests that the birds were northbound migrants coming from the Gulf of California, where the species has wintered (Russell and Monson 1998). The other seven records, one of which involved two individuals, are of first-winter birds in fall and winter, including an exceptionally early one at the north end 16 September 1973 (*AB* 28:108) that may have summered locally. The remaining records better fit the timing of occurrence elsewhere in the interior of North America, with one at the south end 24 November 1967 (*AFN* 22:649); two photographed at Oasis 28 November 1980 (*AB* 35:226), with one remaining until 6 January 1981 (*AB* 35:336); one photographed at Unit 1 on 8 December 1996 (*FN* 51:802); one photographed at Obsidian Butte and Salton Sea NWR 17 December 1995–10 February 1996 (*FN* 50:224); one near Oasis 27 December 1974 (*AB* 29:580); and a first-winter bird near Mecca 30 December 2000 (*NAB* 55:228).

ECOLOGY. Some kittiwakes have been found along the shore of the Salton Sea, especially near the Whitewater River delta, whereas others have been seen at freshwater marsh ponds and along freshwater channels.

TAXONOMY. Black-legged Kittiwakes of the Pacific Ocean, *R. t. pollicaris*, are larger and have the primaries more extensively black than do nominate birds of the Atlantic.

SABINE'S GULL

Xema sabini (Sabine, 1819)

Rare fall transient (mid-September to mid-October); 3 spring records (May); 7 late spring and summer records (mid-June through August). Small numbers of the highly pelagic Sabine's Gull appear annually in fall across the interior of North America, implying a broad-front, normally nonstop migration. This handsome gull is a rare fall migrant through the Salton Sea region, where it is recorded once every two to three years; it is casual in spring and summer. Fall dates extend from 8 September (2001, juvenile at Obsidian Butte; G. McCaskie et al.) to 17 October (1980, ju-

venile at the north end; *AB* 35:226). First-summer birds at the mouth of the Whitewater River 7–9 August 1985 (*AB* 40:159) and at the mouth of the New River 8 August 1981 (*AB* 36:218) were presumably exceptionally early fall vagrants. Conversely, a juvenile photographed near Obsidian Butte 7 November–3 December 1998 (*NAB* 53:104; K. Sturm) provided one of the latest records for California. Most records involve lone individuals, but three were together at the north end 17–20 September 1981 (*AB* 36:218) and 3–16 October 1987 (*AB* 42:136), and a flock of as many as 8 juveniles was near Oasis 5–13 October 1991 (*AB* 46:149). The vast majority of fall records are of juveniles in the vicinity of the Whitewater River delta (e.g., LACM 104292, taken 3 October 1987).

There are but three records of spring transient Sabine's Gulls, all from May: an adult near Oasis 7 May 1994 (*FN* 48:341), two first-summer birds at the north end 29 May 1983 (*AB* 37:912), and a first-summer bird near Mecca 31 May 1992 (*AB* 46:481). A report of 24 distant adults off the mouth of the Whitewater River 18 May 1991 (*AB* 45:496) may have been correct but lacks documentation adequate for so unprecedented a record. There are seven records of birds in late spring and summer, from 9 June (1996, adult at the mouth of the Whitewater River through 15 June; *FN* 50:996) to 16 July (2000, first-summer bird at the mouth of the Whitewater River from 15 July; G. McCaskie, M. A. Patten). Other summer records are of adults near Red Hill 23 June 1968 (*AFN* 22:649), at the north end 24 June 1966 (*AFN* 20:600; first regional record) and 15 July 1977 (*AB* 31:1190), and near Mecca 1 July 2001 (*NAB* 55:482) and of a first-summer bird at the north end 4 July 1991 (*AB* 45:1161). Amazing was an adult dodging traffic on Interstate 8 near Seeley 27 June 1969 (not July, not 1968) (*AFN* 23:695; Garrett and Dunn 1981; G. McCaskie).

ECOLOGY. Nearly all area records involve gulls on the Salton Sea itself, which is not surprising for a pelagic species. The two exceptions are the adult near Seeley (cited above) and a juvenile at Sheldon Reservoir 3–5 October 1990 (*AB* 46:149).

TAXONOMY. This species is monotypic, as *X. s. woznesenskii* Portenko, 1939, is not diagnosably distinct (Vaurie 1965; Cramp and Simmons 1983).

GULL-BILLED TERN
Sterna nilotica vanrossemi Bancroft, 1929
Fairly common breeder (mid-March to mid-October); casual in winter (late November to early February).
For many decades the Salton Sea was the only known locality for the Gull-billed Tern in the western United States. Although the species now breeds at the south end of San Diego Bay on the coast of southern California, the Salton Sea remains its breeding stronghold in the West and the only locale in the interior West where the species has been found aside from two records from southwestern Arizona (Monson and Phillips 1981). This subtropical species was first discovered in the region in May 1927, when 500 pairs were found nesting at the south end (Pemberton 1927; MVZ 50668, 50669). Breeding numbers vary from year to year but have generally declined with rising water levels since the early 1930s (Grinnell and Miller 1944; Garrett and Dunn 1981), reaching a low of 17 pairs in 1976 (Remsen 1978). The population has since recovered, with 100–150 pairs breeding annually in the early 1990s (Parnell et al. 1995; *AB* 44:1187; Shuford et al. 2000) and 170 pairs nesting successfully during the summer of 1997 (*FN* 51:1054). In addition to populations around the Salton Sea, up to 200 pairs nest at Cerro Prieto (e.g., 191 nests there 19 May 2001; K. L. Garrett). Pairs form shortly after arrival, and nesting begins as early as mid-April; in 1949, for example, 500 were present with the first eggs 15 April (*AFN* 3:224). Nesting continues through June; for example, there was a nest near Kane Springs 9 June 1928 (MVZ 3734), there were 13 adults with 7 young at Salton City 21 June 1986, and there were 75 pairs with nests at the south end 28 June 1986 (*AB* 40:1255).

Gull-billed Terns arrive early in spring, generally reaching the Salton Sea by mid-March, two at the north end 11 March 1984 (*AB* 38:961) being the earliest, apart from an exceptionally early

individual 5 March 1995 (K. C. Molina). Numbers build quickly, with a few hundred birds present by April. Most depart by September, the latest records being of six at the south end 12 October 1975 (G. McCaskie) and one at Obsidian Butte 8 September–12 October 2001 (G. McCaskie; M. Pollock). This species is extremely rare in winter, with eight records from 21 November (1979, north end; *AB* 34:201) to 1 February (1986, near Brawley; *AB* 40:335). The other six records are of individuals at the north end 27 November 1977 (*AB* 32:258) and 1 December 1979 (*AB* 34:306), one at Wister 18 December 2001 (K. L. Garrett), four on the Salton Sea (south) Christmas Bird Count 28 December 1978 (*AB* 33:666), one shot by a waterfowl hunter near Obsidian Butte 31 December 1971 (*AB* 26: 655; WFVZ 22708), and two at the north end 20 January 1979 (*AB* 33:314). We question a late November report of ten at Salton Sea NWR (*AFN* 13:66).

ECOLOGY. Nesting Gull-billed Terns occupy protected spits, berms, and islets consisting of sand and barnacles (Parnell et al. 1995). They forage at freshwater ponds and impoundments adjacent to the Salton Sea and over flooded fields, where they glean food from the surface (unlike most other terns, they do not dive). Some hawk insects over Alfalfa and other agricultural fields (Abbott 1940), particularly if the fields are at least partly flooded. Breeding terns require shallow water near nest sites, where they soak their bellies so that they can cool their eggs (Grant 1978).

TAXONOMY. The Salton Sea is the type locality of *S. n. vanrossemi*, the type being an adult male from the south end 21 May 1928 (Bancroft 1929; UCLA 22838). It differs from the nominate subspecies, of the Old World, in its shorter tail and flatter gonys and from *S. n. aranea* Wilson, 1814, of eastern North America, in its much larger size.

CASPIAN TERN
Sterna caspia Pallas, 1770
Common breeding visitor (mid-April through October); uncommon to irregularly fairly common

winter visitor. Caspian Tern populations at the Salton Sea endured a roller-coaster ride over the past eight decades. In 1927 the species was discovered nesting at the south end (Pemberton 1927), where it maintained a steady, albeit small (fewer than 50 pairs), breeding population until the early 1950s (Remsen 1978). Thereafter it was mainly a common spring and abundant fall transient through the Salton Sea, remaining commonly through summer as a nonbreeder and uncommonly through winter (Garrett and Dunn 1981). Breeding resumed at the Salton Sea in the early 1990s, 30 nests on Mullet Island 25 June 1992 (*AB* 46:1178) being the first in more than 30 years. As elsewhere in the West (Wires and Cuthbert 2000), breeding numbers have increased dramatically since, with 60 pairs nesting in summer 1993 (*AB* 47:1150), 120 pairs successfully fledging young at the south end during the summer of 1994 (*FN* 48:989), 1,000 pairs nesting at Mullet Island during the summer of 1996 (*FN* 50:996), and more than 1,200 pairs nesting at the Salton Sea during the summer of 1997 (*FN* 51:1054). The number of breeding pairs had dropped to 211 by 1999 (Shuford et al. 2000). The first breeders arrive by 13 April (1940, south end; Abbott 1940). Nesting begins in May (e.g., near Mecca in 1998; M. A. Patten), and juveniles fledge into mid-September. Most Caspians depart by the end of October, but some remain through winter. The species is generally uncommon in that season, with counts seldom exceeding 20 birds, but it can be fairly common at times; for example, there were 91 at the south end 26 December 1981 (*AB* 36:758) and 75 around the sea 21 February 1998 (M. A. Patten). Winter numbers are probably increasing in concert with breeding numbers.

ECOLOGY. Caspian Terns nest on exposed sandy substrates on islands or protected dikes and berms. Mullet Island supports by far the largest number of breeders. The species forages anywhere along the shoreline, along rivers and wide irrigation channels, and in ponds, lakes, and reservoirs in the valleys. It feeds mainly on small fish captured from the surface or via shallow dives. The largest roosting flocks congregate on mudflats at the river mouths, on exposed mud or sand in shallow impoundments, and in flooded fields adjacent to the sea. When roosting, the Caspian Tern often forms large mixed-species flocks with Black Skimmers and various species of gulls, although it tends to maintain pure subflocks within these aggregations.

ROYAL TERN
Sterna maxima maxima Boddaert, 1783

Casual summer vagrant (mid-May to mid-July). The normally oceanic Royal Tern has been recorded only five to six times inland in the West, all at the Salton Sink in midsummer. Three records are from the mouth of the Whitewater River, where birds in basic plumage were photographed 20 July 1990 (*AB* 44:1186) and observed 4 July 1991 (*AB* 45:1162) and 23 June–1 July 2001 (*NAB* 55:482). A first-summer bird was seen at the mouth of the New River 2 July 1995 (*FN* 49: 980), and two were at Cerro Prieto 10 May 1997 (Patten et al. 2001). We have been unable to examine photographs of a reported adult in alternate plumage at the mouth of the Whitewater River 12 June 1986 (*AB* 40:1255). The record is likely sound, but the plumage is anomalous for the date (see below).

TAXONOMY. The nominate subspecies is the only one occurring in North America, including birds nesting in coastal southern California and in the Gulf of California. Breeders in western North America retain definitive alternate plumage for only a short time, the molt into basic plumage being complete by April. Breeders on the Gulf of Mexico and on the East Coast apparently retain alternate plumage into June, leading to speculation that the bird reported in 1986, if correctly identified, reached the Salton Sea from farther east (*AB* 40:1255).

ELEGANT TERN
Sterna elegans Gambel, 1848

Casual summer vagrant (late April to late August); has bred. The Elegant Tern is a locally common breeder in coastal southern California (Collins et al. 1991) and the Gulf of California. It is a casual vagrant in the interior Southwest, where all but a handful of records are of adults at the Salton Sea. The species has been recorded about 20 times in the region, all since an adult was photographed at the mouth of the Whitewater River 20 May–1 June 1985 (AB 39:350). Since the mid-1990s the Elegant Tern has been recorded annually at the sea. We believe that birds in alternate plumage are spring migrants who overshot breeding colonies in the northern Gulf of California but that those in basic plumage are postbreeding birds who wandered north. Records extend from 24 April (1994, 2 at the mouth of the Whitewater River until 25 April; FN 48:341) to 24 August (1997, south end; FN 52:126), with most from mid-May (near Mecca 8 May 1999; NAB 53:330) to mid-July (photographed at Obsidian Butte 20 July 1990; AB 44:1187). The Whitewater River delta has proven especially attractive to the species. A juvenile at Red Hill 2 July 1999 (NAB 53:432) and a first-summer bird with two adults at Salton City 1–8 August 2001 (NAB 55:483) provided the only records of nonadults for the Salton Sea. Up to two adult Elegant or Royal Terns at Salton Sea NWR 6–15 June 2000 were probably the former species (NAB 54:423). An adult in alternate plumage at Cerro Prieto 19 May 2001 (K. L. Garrett) furnished the first record for the Salton Sink away from the Salton Sea. Astoundingly, this record was followed by a pair on a nest at Cerro Prieto 27 April 2002 (NAB 56:940), although the birds could not be found on searches later in the season.

ECOLOGY. Most Elegant Terns recorded at the Salton Sea have been at river mouths, either roosting on mudflats or foraging near the mouths. A few individuals have occurred elsewhere along the shoreline. The Cerro Prieto birds frequented geothermal ponds.

COMMON TERN
Sterna hirundo hirundo Linnaeus, 1758
Common fall transient (mid-July to mid-November); uncommon spring transient (late April to mid-June); rare in summer. In stark contrast to virtually all other locations in the interior West, the Salton

Sea regularly hosts thousands of Common Terns during the fall migration. Indeed, it is not unusual for this species to be the most abundant tern at the sea in September. Fall migrants appear by mid-July (e.g., 15 at the south end 16 July 1988; M. A. Patten), with numbers building to the low hundreds by August (e.g., 350 at the Whitewater River delta 22 August 1981; G. McCaskie), before reaching a peak in mid-September, such as 715 at the mouth of the Whitewater River 16 September 1976 (AB 31:223), more than 1,000 there 20–23 September 1980 (AB 35:227), more than 1,500 there 20 September 1981 (G. McCaskie), and 2,000 around the sea 26 September 2001 (G. McCaskie). Numbers decline through October (e.g., 350 at the Whitewater River delta 13 October 1984; G. McCaskie), the last birds normally moving through by mid-November (e.g., 30 at the north end 13 November 1979; Garrett and Dunn 1981). Exceptionally late fall migrants were at the mouth of the Whitewater River 6 December 1978 (Garrett and Dunn 1981) and 14 December 1980 (2; AB 35:336). A December report for the north end (AB 37:771) lacks convincing documentation.

Spring migration is far less dramatic, with generally only small numbers passing through from 18 April (1973; AB 27:820) to mid-June (e.g., 3 at an unspecified location 18 June 1966; AFN 20:600), although 50 were tallied at the north end 13 May 1967 (Garrett and Dunn 1981). Birds occurring after mid-June are difficult to classify; they may be late spring transients (e.g., an adult in alternate plumage at Red Hill 22 June 2000; G. McCaskie), nonbreeding summering individuals (e.g., a first-summer bird at the mouth of the Whitewater River 15 July 1989; M. A. Patten), or perhaps exceptionally early fall transients. Generally only one to three birds are present in summer, but 15 individuals were at the south end 19 June 1976 and at the mouth of the Whitewater River 27 June 1964 and 2 July 1977 (all G. McCaskie).

ECOLOGY. All records of the Common Tern are from the shoreline of the Salton Sea, where birds forage over the sea and roost on mudflats, barnacle beaches, and exposed earth around backwaters. The largest concentrations are around the mouths of rivers, especially the Whitewater.

TAXONOMY. Common Terns across North America, Europe, North Africa, and much of Asia are the nominate, *S. h. hirundo,* including the seven Salton Sea specimens (HSU 4156 and 6 at SBCM). There is no physical evidence that *S. h. longipennis* Nordmann, 1835, of Siberia and northern China, has occurred in North America away from westernmost Alaska. In breeding condition it differs from the nominate subspecies in its black bill, browner legs, and grayer body.

ARCTIC TERN
Sterna paradisaea Pontoppidan, 1763
Casual spring transient (June); 2 summer records (mid-July); 2 fall records. The highly pelagic Arctic Tern is extremely rare in the interior of North America. At the Salton Sea it is a casual spring transient, with 14 records involving 22 individuals, averaging one every two to three years. All spring transients have occurred in June (especially the first half), from 1 June (1986 and 2000, mouth of the Whitewater River, the latter staying until 9 June; *AB* 40:524, *NAB* 54:423) to 22 June (1991 and 2000, 2 adults at Poe Road and a first-summer bird near Salton City; *AB* 45:1162, *NAB* 54:423). Additional records are of adults at the mouth of the Alamo River 2 June 1990 (*AB* 44: 497), the mouth of the Whitewater River 3 June 1987 (*AB* 41:488) and 9 June 1984 (*AB* 38:961), and the south end 4 June 1978 (*AB* 32:1055), 10 June 1979 (*AB* 33:806), and 20 (not 27) June 1993 (*AB* 47:1150; G. McCaskie). A first-summer bird was at Salton City 2 June 1993 (*AB* 47:453). In addition to the two at Poe Road cited above, multiple birds have been encountered on three occasions, with an adult and a first-summer bird at the mouth of the Whitewater River 9 June 1990 (*AB* 44:497), five birds at the south end and Salton City 10 June 1979 (*AB* 33:806), and three at Salton City 13 June 1976 (*AB* 30:1004). Both summer records are from 1990 at the mouth of the Alamo River, where there was an adult 7 July and a first-summer bird 14 July (both

AB 44:1187). These birds were possibly extremely late spring vagrants. The two fall records involved adults at the mouth of the Whitewater River 3 October 1986 (*AB* 41:144) and 30 September (not 1 October) 1995 (*FN* 50:115; M. A. Patten).

ECOLOGY. All regional Arctic Terns have been recorded over the Salton Sea. Most were flying, but a few were roosting on mudflats. Most have been with flocks of other small *Sterna* species, usually the Common Tern.

FORSTER'S TERN
Sterna forsteri Nuttall, 1834
Common summer visitor (late March to mid-November), with some breeding; fairly common winter visitor. Like many of the breeding larids at the Salton Sea, Forster's Tern has an eccentric history. It frequented the sea throughout the year, especially during migration, for many years before it began to breed there in 1970, when two nests were discovered at the mouth of the New River 17 May (*AFN* 24:717). The species has nested irregularly around the sea ever since, the breeding population climbing to 20 pairs in 1972 (*AB* 26:906) and a maximum of 200 pairs in 1978 (north end; *AB* 32:1208). Breeding is not annual, however, and only a few pairs attempt to nest in most years, mainly between Wister and the mouth of the New River; the species also breeds in small numbers at Cerro Prieto (Molina and Garrett 2001). A group of 20 pairs near the mouth of the New River 7 July 1990 (G. McCaskie) has not been exceeded since. Egg dates range from 16 May through July, with fledging as late as the end of August (Grant 1982). Because Forster's Tern remains at the sea fairly commonly through the winter, arrival and departure dates cannot be determined. Regular counts show that breeders arrive in late March and mostly depart by mid-November. The species is least numerous at the sea from mid-January to mid-March.

ECOLOGY. Breeding Forster's Terns occupy sites used by the Gull-billed Tern, the Caspian Tern, and the Black Skimmer, often forming mixed colonies, although Forster's is virtually always greatly outnumbered by the other three species. It forages on the sea (especially along the shallow periphery), on freshwater lakes, ponds, and reservoirs, and along the rivers and the larger irrigation canals with open water. It mostly feeds on small fish (e.g., *Gambusia*) and large invertebrates. Some feed in agricultural drains in the Imperial and Mexicali Valleys well away from the sea (E. Mellink; M. A. Patten).

TAXONOMY. Forster's Tern is generally considered monotypic (Peters 1934; AOU 1957), but there may be a basis for recognizing the smaller, paler-mantled *S. f. litoricola* Oberholser, 1938, of eastern North America and the Gulf of Mexico (McNicholl et al. 2001). A quantitative study is needed to determine the subspecies' validity. If it were judged valid, Forster's Terns in the Salton Sink would be the nominate subspecies.

LEAST TERN

Sterna antillarum (Lesson, 1847) subsp.?

Rare spring and summer visitor (mid-April to mid-September). The Least Tern is an annual rare spring and early summer visitor to the interior West, where the vast majority of records are from the Salton Sea. It has been recorded as early as 15 April (1978, adult at Unit 1; *AB* 32:1055) and as late as 13 September (1976, south end; *AB* 31:223), with the majority of records for May and June. There are only six records after July. In addition to the September record cited above, late for anywhere in California, two remained (from 9 June) at the mouth of the Whitewater River until 5 August 2000 (*NAB* 54:423), one was there 17 August 1974 (*AB* 29:122), one was at the south end 12 August 1971 (*AB* 25:906), and two were there 31 August 1985 (*AB* 40:159). Juvenile Least Terns at Red Hill 14 July 2001 (*NAB* 55:483), at Wister 27 July 2002 (*NAB* 56:486), north of Salton City 17 August 2000 (*NAB* 54:423), and at Fig Lagoon 18 August 2001 (P. Saraceni) are the only ones of that age ever found in the Salton Sink. Most regional records of the Least Tern involve one or two birds, apparent pairs being a surprisingly frequent occurrence. Groups of three were observed at the north end 21 May 1980

(*AB* 34:816), 6–7 June 1981 (*AB* 35:979), 19 June 1976 (*AB* 30:1004), and 27 June 1964 (McCaskie and Cardiff 1965) and at the south end 2 June 1991 (*AB* 45:1162), and six were together near Oasis 11 June 1998 (*FN* 52:503). Interestingly, adults are encountered more frequently than first-summer birds, in about a 3:1 ratio.

ECOLOGY. The Least Tern most often occurs on mudflats near the deltas of the three rivers, where it often forages in fresh water in the rivers or nearby ponds. One at Ramer Lake 3 May 1997 (*FN* 51:928) and two at Fig Lagoon 31 May 1997 (*FN* 51:928) were well away from the Salton Sea. The increase in records over the past decade, coupled with the prevalence of apparent pairs of adults, may portend breeding at the Salton Sea (see *NAB* 56:486).

TAXONOMY. Subspecies taxonomy of the Least Tern in western Mexico is in need of revision (Patten and Erickson 1996). According to current taxonomy birds in the northern Gulf of California, presumably the source of birds at the Salton Sea, should be *S. a. mexicana* van Rossem and Hachisuka, 1937, a slightly smaller, darker form breeding in Sonora and Sinaloa. However, birds breeding in northeastern Baja California have been treated as *S. a. browni* Mearns, 1916 (Palacios and Mellink 1996), the Pacific Coast subspecies that differs from other United States forms in having a paler nape (Johnson et al. 1998). Only *S. a. browni* has been taken in California, but without regional specimens and a taxonomic revision the subspecies of the Least Tern occurring at the Salton Sea will remain unknown.

BLACK TERN

Chlidonias niger surinamensis (Gmelin, 1789)

Common transient (mid-April through May, August to mid-November); uncommon summer visitor; casual in winter (late November to early February). Contrary to earlier claims (e.g., *AFN* 13:401 and Cogswell 1977), there is no evidence that the Black Tern has ever bred at the Salton Sea (or anywhere else in the Southwest). Still, the Salton Sea is one the species' most important migratory stopover sites in western North America (Shuford et al. 2000). The Black Tern is a common to irregularly abundant (mainly in fall) transient and an uncommon to irregularly fairly common nonbreeding summer visitor in the region. Numbers during peak migration periods dwarf numbers elsewhere in the Southwest. Spring migrants have arrived as early as 10 April (1962; Garrett and Dunn 1981), and fall migrants have lingered as late as 11 November (1967 and 1988, 3 and 4 near Red Hill; *AFN* 22:90, M. A. Patten). Because dozens to hundreds of non-breeders summer each year, departure dates of spring migrants and arrival dates of fall migrants are clouded. Peak spring passage takes place from late April to mid-May (e.g., 2,000 on 3 May 1998; M. A. Patten), peak fall passage during the first half of August (e.g., 15,000 on 6 August 1977; G. McCaskie). Juveniles have been detected as early as 2 August (1993; G. McCaskie), so fall migration begins no later than late July or early August, and most have passed through by mid-October. Summer numbers reach their nadir from mid-June to mid-July. The Black Tern is casual in winter, from 28 November (1982, south end through 21 December; *AB* 37:338) to 2 February (1987, near El Centro; *AB* 41:330), with one recorded every five-plus years on average.

ECOLOGY. Most Black Terns forage directly over the Salton Sea and roost on exposed mud or sand spits around shallow backwaters adjacent to the sea. This species also commonly forages and roosts in flooded fields, particularly in the Imperial Valley. Flocks are occasionally noted foraging over dry fields, with birds flying about in nighthawk-like fashion as they capture flying insects.

TAXONOMY. Geographic variation is minimal, the North American *C. n. surinamensis* differing from the nominate subspecies, of the Old World, only in its shorter wings and longer legs.

BLACK SKIMMER

Rynchops niger niger Linnaeus, 1758

Fairly common breeder (early April to December); 4 winter records. In the big picture the Black Skimmer (Fig. 48) is a rare bird in the West, but in the past 30 years this species has undergone a significant range expansion into California (Collins and Garrett 1996), making it a regular component of the state's avifauna. The skimmer was first recorded at the Salton Sea 3 July 1968, when five were observed at the mouth of the Whitewater River (McCaskie and Suffel 1971; LACM 76788). The species was recorded on occasion over the next three years, accumulating five records by the summer of 1971 (including SBCM 33298). In 1972 it invaded (Winter 1973; Winter and McCaskie 1975) and began nesting regularly (McCaskie et al. 1974; Grant and Hogg 1976). By 1977 there were as many as 500 birds with 100 nests, although the breeding population crashed to near zero shortly thereafter (Collins and Garrett 1996; Molina 1996). By the mid-1980s the breeding population had experienced a huge upsurge, with 500 pairs breeding by 1987 and generally 100–400 pairs breeding annually since (Molina 1996; Shuford et al. 2000; *NAB* 55:483). Breeding colonies have been found near the Whitewater River delta, at various locales around the south end, at Salton City, at Ramer Lake, and at Cerro Prieto and range in size from ten to hundreds of pairs (Molina 1996; K. L. Garrett).

Breeders arrive in early April, ten at the mouth of the Whitewater River 1 April 1978 (G. McCaskie) being the earliest. Nesting commences in May; the first eggs are laid in mid-May (the earliest being 15 May; Grant 1982), and some pairs are on eggs through September (the latest being 27 September; Grant 1982). Hatching occurs from mid-June into September, and fledglings appear from late July through October (Molina 1996). Most depart the sea by mid-October, but sometimes dozens remain into November, and a few linger into early December or even the end of December, such as two at the mouth of the Alamo River 28 December 1995 (*FN* 50:552), two at the mouth of the Whitewater River 21–26 December 1980 (*AB* 35:727; Garrett and Dunn 1981), and several there 5 January 2002 (*fide* H. King). There are only four records of birds overwintering, however, with one at the mouth of the Alamo River 27 December 1980–8 February 1981 (*AB* 35:336), one or two at Red Hill 13 December 1986–24 January 1987 (*AB* 41:330),

FIGURE 48. The Black Skimmer's colonization of the Salton Sink was part of an aggressive northward range expansion on the Pacific Coast of North America. Photograph by Kenneth Z. Kurland, June 2000.

eight at the mouth of the Whitewater River 2 March 1996 (G. McCaskie), and one at Red Hill 12 February 2000 (*NAB* 54:221).

ECOLOGY. Like the Gull-billed Tern, breeding skimmers require shallow water near nest sites in order to soak their bellies to aid in cooling their eggs (Grant 1978). Most nest within or adjacent to Gull-billed and Caspian Tern colonies. The species requires calm, shallow water, fresh or saline, for foraging.

TAXONOMY. Skimmers north of Panama are the small nominate subspecies, further distinguished from the two South American subspecies by the broad white edges to the secondary tips and outer rectrices.

ALCIDAE • *Auks, Murres, and Puffins*

ANCIENT MURRELET

Synthliboramphus antiquus (Gmelin, 1789)

Two spring records. The only species of the highly pelagic auk family to be found at the Salton Sea is the Ancient Murrelet. Birds in alternate plumage at the mouth of the Whitewater River 16 June 1984 (*AB* 38:1062) and 23 May 1987 (*AB* 41:488) were presumably moving north from the Gulf of

California. Munyer (1965) and Verbeek (1966) summarized inland occurrences of this species, citing 25 records from Canada and the United States through the early 1960s. Neither author mentioned an undated specimen from nearby Palm Springs, in the northern Coachella Valley (Garrett and Dunn 1981). The collection date is unknown, and details about the specimen, including its whereabouts, were never published, so the record was treated as tentative by Erickson et al. (1995). The female specimen (SBCM 30797) appears to be a fall immature (M. A. Patten). Its accession number suggests that the specimen was collected in the early 1960s.

COLUMBIFORMES • *Pigeons and Doves*

Although no native pigeons are regular in the region, doves are common and conspicuous. The White-winged Dove, the Mourning Dove, and the Common Ground-Dove all breed commonly, with the first two species being locally abundant. Furthermore, the Inca Dove breeds in the suburban portions of the cities and towns in the Imperial Valley and the northern Mexicali Valley.

COLUMBIDAE • Pigeons and Doves

BAND-TAILED PIGEON
Columba fasciata [monilis] Vigors, 1839
Four records, 2 in fall and 1 each in spring and winter. The Band-tailed Pigeon is an irregular wanderer into lowlands from oak woodlands of the foothills and mountains. It has been recorded four times in the Salton Sink. One was shot by a hunter 12 km northeast of Calexico 4 October 1941 (Neff 1947). Subsequent records are of individuals at the south end 9 May 1974 (*AB* 28:852) and 12–13 August 1974 (*AB* 29:122) and an immature at Brawley 24 February 1990 (*AB* 44:329).

TAXONOMY. All California specimens are the smaller, darker *C. f. monilis,* of the Pacific Coast. The subspecies occurring in the Salton Sea region, however, is unknown (the specimen was not preserved). The larger, paler *C. f. fasciata* Say, 1823, of the southern Rocky Mountains and mainland Mexico, is possible, as it accounts for all Arizona records and has been taken as far west as Yuma, on the lower Colorado River (Phillips et al. 1964:41).

WHITE-WINGED DOVE
Zenaida asiatica mearnsi Ridgway, 1915
Common breeder (late March to early September); rare but increasing in winter (mid-October to mid-March). It appears that the White-winged Dove underwent a major westward expansion of its range sometime shortly after the turn of the twentieth century (Phillips 1968; Rosenberg et al. 1991). After it colonized the Colorado River, Ricketts (1928) observed that "sometimes a few drift into the Imperial Valley," although breeding there was noted as early as 1920 (pair at Brawley 4 May–12 June; Fortiner 1920a). This species is now widespread and common throughout the Sonoran Desert and a common breeding visitor to the Salton Sea area. The first of the spring arrive in April, sometimes as early as 22 March (1976; *AB* 30:886). Most depart by the onset of the dove-hunting season, 1 September, but a few linger into late fall, exceptionally to 5 November (1939, Calexico; D. Dunann, SDNHM 18088),

8 November (2000, Brawley; G. McCaskie), and 16 November (1996, Westmorland; D. Chappel, SBMNH 6637). Breeding takes place from early May through July.

The White-winged Dove is resident not far west and southwest of the Salton Sea, in the Anza-Borrego Desert (Unitt 1984) and in northern Baja California (e.g., Meling Ranch; M. A. Patten). It may soon become a resident of the Salton Sink. Two observed near Brawley 22 February 1963 (*AFN* 17:359) were the first to be recorded in the Salton Sink during winter. The species did not winter again until 1984, when one was noted at the north end 21 January (*AB* 38:358). These records were followed by individuals at the north end 4 January 1977 (*FN* 51:638) and at El Centro 8 February 1997 (*FN* 51:802). Since 1998–99 the White-winged Dove has been recorded annually in winter from 10 October (2002, 2–10 at Calipatria to at least 24 January 2003; G. McCaskie) to 21 March (2001, up to 4 at Calipatria from 19 December 2000; A. Howe, V. Howe).

ECOLOGY. This dove nests in any densely foliaged tree within commuting distance of drinking water and foraging habitat (weedy fields and suburban gardens); it even shows a preference for Saltcedar in some parts of the Sonoran Desert (Rosenberg et al. 1991). Hundreds routinely congregate in the southern Coachella Valley around canals lining citrus orchards and Date Palm groves.

TAXONOMY. Birds in the southwestern United States, Baja California, and Sonora belong to the mid-sized, pale subspecies *Z. a. mearnsi,* of which *Z. a. clara* van Rossem, 1947, is a synonym (Saunders 1968).

MOURNING DOVE
Zenaida macroura marginella (Woodhouse, 1852)
Abundant breeding resident. Few North American species are more conspicuous or more common than the Mourning Dove. Its status in the Salton Sea region, as an abundant breeding resident, is no different. It is migratory, with higher numbers in winter than in summer, but its general year-round abundance prohibits an accurate assessment of seasonal variation in num-

bers. It breeds mainly in spring and summer, beginning in mid-February and continuing into August, with a peak in May and June (Fortiner 1921). However, nests have been recorded over much of the year, with dates ranging from 18 January to 23 September (Fortiner 1921).

ECOLOGY. Breeding Mourning Doves have rather catholic tastes, using nest sites ranging from mature Fremont Cottonwood and scrubby Iodine Bush to snags and bare ground. The region's network of irrigation canals assures that this dove is nowhere far from the drinking water it needs to survive the summer. The species is mainly a terrestrial feeder. Fallow and grain fields act as concentrating points, with seeds from cultivated plants making up more than 50 percent this dove's diet (Browning 1962). In the 1950s and early 1960s principal foods were cultivated Barley seeds in winter and Common Flax seeds in summer (Browning 1962); Milo was also a prevalent crop food. Other principal foods were seeds from nonnative weedy plants, especially Lamb's Quarters, Dwarf Canary Grass, and Barnyard Grass (Browning 1962). Changes in local crops have undoubtedly led to changes in the species' diet.

TAXONOMY. The palest of the five subspecies is *Z. m. marginella*, of western North America. This subspecies also lacks the buff head of various insular forms in Mexico and is slightly larger than *Z. m. carolinensis* (Linnaeus, 1766), of eastern North America (Rand and Traylor 1950; Pyle 1997).

INCA DOVE
Columbina inca (Lesson, 1847)
Uncommon to fairly common breeding resident in the Imperial Valley; casual elsewhere. Like many columbids, the Inca Dove (Fig. 49) has expanded its range in the past century (Phillips 1968). The species is now an uncommon to fairly common (and increasing) breeding resident in the Imperial Valley, especially around Calexico, where at least 25 were first detected 4 February 1984 (*AB* 38:358). Further searching uncovered substantial populations around Mexicali, Calexico, and El Centro (Patten et al. 1993, 2001). By 1990 the

Inca Dove was spreading northward, as implied by at most seven at Brawley 21 January–28 April 1990 (*AB* 44:329). This species is now established in small numbers in Brawley, Calipatria, and Niland. In the Imperial Valley it has been noted north to Wister (e.g., 23–24 August 1997; G. McCaskie) and west to Westmorland (e.g., 20 October 1997; G. McCaskie), Mount Signal (e.g., 25 September 1994; M. A. Patten), and Seeley (e.g., 19 July 1997, 27 May 2000; G. McCaskie, M. A. Patten). Nesting data are few: a pair with three recently fledged juveniles was at Calipatria 24 June 1995 and a nest was there, about 3 m up a Chinaberry, during the spring of 2000 (G. McCaskie). The Inca Dove's status north of the Imperial Valley is tenuous. Aside from individuals at Salton City 3 September 1996 (G. McCaskie), the north end 2 January 1993 (*AB* 47:968), and the Salton Sea State Recreation Area 2 October 2000 (P. D. Jorgensen), it is unknown north of the valley. If correct, a report of two at Palm Springs in mid-February 1928 (Schneider 1928; cf. Grinnell and Miller 1944:568), the first for California, suggests that small numbers have moved through the Salton Sink for some time. We believe reports of "several" near Indio in September–October 1955 (*AFN* 10:56), two at Salton Sea NWR in April 1962 (*AFN* 16:448), and individuals at the latter locale on "two occasions" during the summer of 1962 (*AFN* 16:508) to be in error.

ECOLOGY. The Inca Dove occurs only in artificial habitats such as suburban neighborhoods, parks, and ranch yards. Even at Wister it prefers the lawns around the few homes to the nearby mesquite thickets or Saltcedar tangles. Like the Mourning Dove and the Common Ground-Dove, the Inca Dove forages almost exclusively on the ground.

COMMON GROUND-DOVE
Columbina passerina pallescens (Baird, 1859)
Common breeding resident. In much of California the diminutive Common Ground-Dove is closely linked with agriculture, especially citrus orchards. Settlement and cultivation of the Salton Sink enabled the species to increase greatly, if

FIGURE 49. Found only around human dwellings, the Inca Dove is spreading northwest. It first colonized the Imperial Valley in 1984 and 15 years later was becoming fairly common there. Photograph by Kenneth Z. Kurland, December 2000.

not to colonize the region outright. In either case the Common Ground-Dove now reaches its peak abundance in California around the Salton Sea. Indeed, the region is one of the few in the United States where one may see dozens in a single day without too much effort. The species is most common around the Whitewater, Alamo, and New Rivers, Wister, Fig Lagoon, and Finney and Ramer Lakes, as well as in Brawley and El Centro, but it may be encountered wherever there are trees. Breeding is concentrated in spring, with most activity from March into June (e.g., a nest with 2 eggs west of Niland 22 April 1962; SBCM E3376). However, nests have been found in nearly every month of the year (Fortiner 1920b, 1921).

ECOLOGY. As its name suggests, the Common Ground-Dove forages mainly on the ground. It is most numerous at open edges of riparian habitats (rivers, marshlands, lake edges) and in well-wooded suburban neighborhoods, parks, and ranch yards. Orchards, especially citrus, are heavily used in the southern Coachella Valley. This dove generally forages in shaded areas. It nests in dense thickets, frequently using Saltcedar.

TAXONOMY. Paralleling the Mourning Dove, ground-doves of western North America, *C. p. pallescens,* are paler than those in the East but are otherwise similar.

RUDDY GROUND-DOVE
Columbina talpacoti eluta (Bangs, 1901)

Two winter records. Beginning in 1981 (Rosenberg and Witzeman 1998) the Ruddy Ground-Dove has staged a minor invasion into the Southwest. In the ensuing two decades the number of records for Arizona climbed to over 100, the number for California reached about 80, several were recorded as far north as southern Nevada, and records for northern Sonora increased markedly (Russell and Monson 1998). The vast majority have occurred in fall and winter, typically from September to early April. The only acceptable records for the Salton Sink, of up to two males photographed 3 km southeast of El Centro 27 January–15 March 2002 (K. Z. Kurland; CBRC 2002-033) and of up to 6 photographed at Calipatria 8 December 2002+ (G. McCaskie et al.; CBRC 2002-2000), fit this seasonal pattern of occurrence. Ruddy Ground-Doves reported at

Wister and Ramer Lake were not documented well enough to establish identification (see Patten, Finnegan, et al. 1995; CBRC 2000-072).

TAXONOMY. All Ruddy Ground-Doves recorded in the Southwest have shown characters of the expected *C. t. eluta,* of western Mexico (Dunn and Garrett 1990). The lone specimen from California, of a male found dead 22 km north of Blythe, Riverside County, 26 November 1992 (LACM 107326), is *C. t. eluta,* grayer and with the rufous paler than subspecies farther east and south.

CUCULIFORMES • *Cuckoos and Allies*

Aside from the Greater Roadrunner, no other cuckoos occur regularly at the Salton Sea. This was not always the case, as the Yellow-billed Cuckoo, a state-listed endangered species whose populations have plummeted throughout California and the West (Laymon and Halterman 1987), likely occurred regularly as a migrant. Roadrunners, at least, remain a conspicuous resident; they are probably as common and as easily observed around the Salton Sea as they are anywhere in their range.

CUCULIDAE • *Cuckoos, Roadrunners, and Anis*

YELLOW-BILLED CUCKOO
Coccyzus americanus (Linnaeus, 1758)
Six summer records (late June to mid-July). The Yellow-billed Cuckoo has undergone a catastrophic decline throughout western North America. It formerly occurred as a breeder along the Río Hardy a short distance southeast of Cerro Prieto (outside the Salton Sink). The presence of two calling birds at Murguia, on the Río Hardy 9 July 1995 (Patten et al. 2001) suggests that a small population may persist there. Within the Salton Sink there are but six records, all of summer stragglers, although this species is such a late migrant in spring that records could pertain to transients. Individuals were at Red Hill 22 June 2000 (*NAB* 54:423), Finney Lake 23 June 1999 (*NAB* 53:432), 5 July 1974 (*AB* 28:949), and 7 July 2002 (*NAB* 56:486), Wister 10 July 1977

(*AB* 31:1190), and Salton Sea NWR 13 July 2002 (*NAB* 56:486).

ECOLOGY. Yellow-billed Cuckoos breeding in the West occupy gallery riparian forest dominated by Fremont Cottonwood and willows. The massive incursion of Saltcedar along all of the region's river courses has severely restricted this habitat and no doubt precluded cuckoos from colonizing. (This species was never detected even in cottonwood riparian forest along the historical New and Whitewater Rivers.)

TAXONOMY. This cuckoo is monotypic. A larger western subspecies, *C. a. occidentalis* Ridgway, 1887, was widely recognized (e.g., Peters 1940; AOU 1957) and recently championed (Franzreb and Laymon 1993). Numerous authors have questioned the validity of this subspecies, most recently Banks (1988, 1990). On average, cuckoos in western North America are significantly larger in wing chord, tail, bill length, and bill depth, but there is complete to near complete overlap in every mensural character (see Franzreb and Laymon 1993:table 1). Therefore, although western birds average larger, they are not distinguishable from eastern birds at the 75 percent level. A genetic study found little divergence in mitochondrial DNA (Fleischer 2001). California specimens are exclusively of breeders, thus pertaining to the western population. It is highly probable that many September and October records from coastal California are of vagrants from the eastern population, but even a specimen may be uninformative given the extensive overlap in measurements. Records from the Salton Sea are from a time of year when birds from the western population are most likely.

GREATER ROADRUNNER
Geococcyx californianus (Lesson, 1829)
Fairly common breeding resident. The Greater Roadrunner (Fig. 50), that famous denizen of the desert Southwest, is nowhere common, but it is probably more common in the Salton Sea region than just about anywhere else. Even without a focused effort a typical day can yield more

FIGURE 50. Symbolic of the Southwest, the Greater Roadrunner remains fairly common in the Salton Sink even where only small patches of native shrubs survive. Photograph by Kenneth Z. Kurland, December 1999.

than ten roadrunners, a remarkable number for this somewhat secretive predator. It reaches its peak abundance in the semi-native habitat remaining around the rivers, lakes, and marshes and bordering the sea's western and eastern edges. Breeding can begin quite early, even in late January, although more typically this species nests in late March and April (e.g., at Mecca 8 April 1916; MVZ 2858), as the weather must be warm enough for reptiles to be active so that young may be fed.

ECOLOGY. Well known for its ability to capture rattlesnakes (*Crotalus* spp.) and fast-moving lizards (e.g., *Cnemidophorus* sp.), the roadrunner also consumes more modest prey. One collected at Mecca 20 March 1908 had recently ingested numerous Orthoptera and Coleoptera and a *Phrynosoma* horned lizard. As is well known, the roadrunner forages almost exclusively on the ground, even overturning stones in search of prey (Jaeger 1947a). It nests in dense shrubs or low trees, frequently ones protected by thorns (e.g., mesquite), but also in dense saltbush. It is

most commonly seen in desert scrub but also frequents orchards and thickets bordering rivers, lakes, and irrigation canals. It is less frequent in agricultural habitats.

TAXONOMY. Roadrunners in Texas, which on average are slightly smaller than birds in California and Arizona, have been named *G. c. dromicus* Oberholser, 1974. However, there is extensive overlap in measurements, so this subspecies generally is not recognized (Rea 1983; Browning 1990).

GROOVE-BILLED ANI
Crotophaga sulcirostris Swainson, 1827

One fall record. Unitt observed a Groove-billed Ani near Seeley 25 October 1986 (Langham 1991), only the third recorded in California. This species remains extremely scarce in California, although there have been nearly ten additional records since 1986, all from late fall and winter (Mlodinow and Karlson 1999).

TAXONOMY. *C. s. pallidula* Bangs and Penard, 1921, of southern Baja California, was described as having duller green edges and grayer centers throughout the contour feathers relative to the nominate subspecies. The AOU (1957) included western Mexico within the range of *C. s. pallidula*, but Peters (1940) and Friedmann et al. (1950) followed Bangs and Penard (1921) in restricting it to the Cape District of Baja California; they considered it extinct. If the restricted treatment is followed, then all California records refer to the nominate subspecies. However, the single specimen at SDNHM from southern Baja California does not differ in color or measurements from comparable specimens from mainland Mexico. We thus infer that Greenway (1958: 365) was correct in suspecting that the supposed paleness was an artifact of fading and that the species is monotypic.

STRIGIFORMES • Owls

As a general rule of thumb, the only species of owls that are prevalent in the arid Southwest are the Barn Owl, the Great Horned Owl, and the Burrowing Owl. True to form, the Salton Sink supports good numbers of only these three species of owls. Because the first two species are nocturnal they are encountered far less often, but each is an uncommon to fairly common breeding resident in both settled areas and natural habitats with large trees. The Imperial Valley is the stronghold for the Burrowing Owl in California; it is the only locale in the state where the species remains common. Likewise, adjacent northeastern Baja California is the species' stronghold on that entire peninsula (Palacios et al. 2000). The Short-eared Owl, small numbers of which winter annually in the Imperial Valley, is the only other owl that occurs regularly in the Salton Sink.

TYTONIDAE • Barn Owls

BARN OWL

Tyto alba pratincola (Bonaparte, 1838)

Fairly common breeding resident; probably more numerous in winter. The Barn Owl is common over much of temperate North America, but it can be uncommon or even rare in portions of the desert Southwest (see Garrett and Dunn 1981). However, it is a fairly common, albeit seldom seen, breeding resident in the Salton Sink, with probably some winter influx (most records are between October and April). Breeding presumably takes place from February through July, yet definite breeding records for the region are few (as a result of difficulty of detection, not true rarity).

ECOLOGY. This species is especially prevalent in palm groves and in residential areas supporting large trees, although it also roosts in mesquite thickets (van Rossem 1911) and in large Fremont Cottonwood, Goodding's Black Willow, and Athel. It readily occupies disturbed habitats, often being especially abundant in agricultural areas with scattered trees (Garrett and Dunn 1981). It nests in palms, in buildings, and in nooks in stacks of hay bales. Barrows (1989) reported that the Barn Owl feeds principally on Botta's Pocket Gopher (*Thomomys bottae*), heteromyids, and the House Mouse (*Mus musculus*) in the Coachella Valley and vicinity.

TAXONOMY. This cosmopolitan species is highly variable, with 34–40 described subspecies, 23–25 of them island endemics (Voous 1988). The North American subspecies *T. a. pratincola* is the largest. It is also one of the whitest, although there is substantial individual and sexual variation in the extent of buff.

STRIGIDAE • Typical Owls

FLAMMULATED OWL

Otus flammeolus (Kaup, 1852) subsp.?

One fall record. Unlike most owls, the Flammulated Owl is highly migratory (Phillips 1942). Both spring and fall migrants have been found away from breeding areas in California, with more inland than coastal records (Collins et al. 1986). The only record for the Salton Sea region is of a fall migrant observed near Westmorland 4 October 1977 (*AB* 32:259). A Western Screech-Owl reported at the south end 6 September 1977 (*AB* 32:259) may actually have been a Flammulated Owl, given the locale and date.

TAXONOMY. Either *O. f. idahoensis* Merriam, 1892, a finely marked, modestly reddish gray subspecies of montane California and the Pacific Northwest, or *O. f. frontalis* Hekstra, 1982, a heavily marked, mainly gray subspecies from the Great Basin and Rocky Mountains, could migrate through southeastern California (see Browning 1990; Marshall 1997).

WESTERN SCREECH-OWL

Otus kennicottii yumanensis Miller and Miller, 1951
Rare, local breeding resident. The Western Screech-Owl was likely more common in the region during the first half of the twentieth century (van Rossem 1911; Clary 1933; Grinnell and Miller 1944), particularly when ranches and natural oases supported stands of large Fremont cottonwood. It was believed to be extirpated as a breeder by the late 1940s, with no recorded nesting after 1948 in either the Imperial or the Coachella Valley (Garrett and Dunn 1981). However, this species persists, and undoubtedly still nests, in both valleys. Aside from a female near Mecca 3 September 1987 (WFVZ 43346), recent records for the Coachella Valley are for winter at Coachella and Mecca on the Salton Sea (north) Christmas Bird Count (e.g., *AB* 45:994, 47:968; *FN* 50:848), but because this species undertakes little, if any, seasonal movement, these birds are almost certainly resident. Likewise, this species persists in Brawley, including calling birds in spring (e.g., 28 April 1990; B. E. Daniels). It was recently detected in the El Centro area (e.g., 1 bird 3 km southeast of El Centro for 5 weeks beginning 7 February 1996; K. Z. Kurland) and in Mexicali (Patten et al. 2001). It occurred formerly at Finney Lake (e.g., 4 on 4 April 1985; *AB* 39:350) and may still occur in the larger stands of trees both there and at nearby Ramer Lake. One reported at the south end 6 September 1977 (*AB* 32:259) was away from areas of known occurrence and thus may have been a migrant Flammulated Owl, although the presence of a calling Western Screech-Owl at Salton Sea NWR 6 March–10 April 1994 (K. C. Molina; K. L. Garrett) implies some local movement.

ECOLOGY. As noted above, this small owl is generally associated with stands of large trees, particularly Fremont Cottonwood. However, in residential Brawley, for example, it now occupies gum and other nonnative trees (M. A. Patten); a bird in Mexicali 4 September 2000 was in a Mexican Fan Palm (M. A. Patten). The Western Screech-owl also occurs in dense stands of mature mesquite, with large trees providing suitable nest sites.

TAXONOMY. Birds throughout the Colorado Desert, including the Coachella Valley (e.g., WFVZ 43346), are the small, pale *O. k. yumanensis*, with pinkish-gray ground color and fine, indistinct barring; note that *O. l. gilmani* Swarth, 1910, is a "superfluous" name that has been applied to part of this same population (Marshall 1967:33).

GREAT HORNED OWL

Bubo virginianus pallescens Stone, 1897
Uncommon breeding resident. The Great Horned Owl ranks with the Barn Owl as one of the two most widespread owls in the New World. For a bird of prey, it is relatively common throughout much of its range, although it tends to be less numerous in deserts. In the Salton Sea it is uncommon year-round, with most records from fall and winter (November through mid-March), although fall dispersants appear as early as August, for example, at the Whitewater River channel 4–25 August 1990 (G. McCaskie; K. A. Radamaker) and Finney Lake 10 August 1986 (*AB* 41:145). Even though it was "believed extirpated as a breeder" in the mid-1980s (*AB* 41:145), the species nests annually in small numbers. Nesting begins early; for example, it was nesting near Indio 22 March 1942 (C. A. Harwell), a nest at Brawley already had three young 8 April 1999 (B. Miller), and at least one young in a Brawley nest had hatched by 6 February 2000 (B. Miller). Young fledge from mid-June to mid-July. Favored breeding sites include Brawley, Wister, Westmorland, El Centro, and various locales in the southern Coachella Valley.

ECOLOGY. Great Horned Owls have nested in Fremont Cottonwood, in mesquite thickets, and in nonnative fan and other palms. Birds may also occasionally nest in other nonnative trees in the region, especially gum. Virtually any substantial tree serves as a roosting site. The Great Horned Owl takes larger prey than does the Barn Owl, including Audubon's Cottontail (*Sylvilagus auduboni*) and woodrats (*Neotoma* spp.), but likewise readily occupies both natural and disturbed sites (Barrows 1989).

TAXONOMY. Regional specimens are the pale, lightly barred *B. v. pallescens*, largely white and black below with unmarked feet and occurring from the San Joaquin Valley and the southern Great Basin south through the Sonoran Desert. There is no evidence of migratory subspecies from the north reaching the Salton Sink in winter, even though all area specimens were collected from 15 November (1984, 3 km southeast of Oasis; SDNHM 44326) to 3 April (1909, Silsbee; MVZ 8043).

ELF OWL
Micrathene whitneyi whitneyi (Cooper, 1861)
One fall record. The Elf Owl is nearing extirpation as a breeder in California (Halterman et al. 1989). The few pairs known to persist are along the lower Colorado River at Picacho State Recreation Area (R. L. McKernan pers. comm.). It is thus surprising that the only record for the Salton Sea of this tiny migratory owl is recent, of a presumed fall migrant photographed near Calipatria 21 September 1995 (*FN* 50:115).

TAXONOMY. California records pertain to the nominate subspecies, the darkest and brownest of the four named taxa (Voous 1988:170).

BURROWING OWL
Athene cunicularia hypugaea (Bonaparte, 1825)
Common breeder (March to September); uncommon to fairly common winter visitor. Although the Burrowing Owl (Fig. 51) is declining precipitously in much of California and the West, it remains fairly common in the Imperial and Mexicali Valleys. Indeed, this species is more common at the Salton Sea than anywhere else in California; nearly 6,500 individuals, nearly 70 percent of the entire California population, occur in the Imperial Valley (Shuford et al. 1999). Curiously, it is seldom recorded in the Coachella Valley,

where it has always been uncommon (van Rossem 1911); there are no known extant populations there, although at least two pairs were near Mecca during the summers of 1999 and 2000 (M. A. Patten). This owl is a common breeding visitor to the Imperial and Mexicali Valleys from early March to late September, most adults nesting from late March to June (e.g., Silsbee 5 April 1909; MVZ 94). Numbers peak in late summer, when fledglings begin to congregate around burrow entrances. Numbers are lower in winter, when birds migrate southward.

ECOLOGY. As its name suggests, this owl nests primarily in burrows dug into the sides of earthen irrigation ditches, especially those with regularly flowing water (Coulombe 1971). Occupied areas are invariably treeless and relatively flat save for the banks of ditches, canals, or berms. Burrows are generally dug by Round-tailed Ground Squirrels (*Spermophilus tereticaudus*), then modified by the owls. The Burrowing Owl will also occupy pipes and small culverts with similar surrounding habitat. It can frequently be seen perched atop the banks of ditches or on nearby poles and hay bales and foraging over open fields, particularly fields with dried, weedy vegetation. Unlike its larger cousins, the Barn and the Great Horned, this small owl feeds mainly on insects and small heteromyids (Barrows 1989).

TAXONOMY. Birds of continental North America outside of Florida are of the larger, paler subspecies *A. c. hypugaea* (Clark 1997).

LONG-EARED OWL
Asio otus wilsonianus Lesson, 1830
Casual winter visitor (late October to mid-March). The Long-eared Owl has experienced a steep decline throughout southern California (Bloom 1994) and elsewhere in the West. It was probably formerly more numerous as a wintering bird in the Salton Sea region, but there are few definite records, from 26 October (1986, near Seeley; C. A. Marantz) to 17 March (1956, 8 km northwest of Westmorland; SBCM 33469). The only other records are of single birds 6 km northeast of Calipatria 16 November 1962 (SBCM

FIGURE 51. The Imperial
Valley is the last region of
California where the rapidly
declining Burrowing Owl
remains common, but the
lining of irrigation canals could
threaten it even there. Photograph by Kenneth Z. Kurland,
July 1999.

33470), at the south end 18 December 1979 (*AB* 34:662), at the north end 2 January 2000 (*fide* C. McGaugh), along the New River near West-morland 6 January 1973 (*AB* 27:663), and near the mouth of the New River 20 November 1976 (*AB* 31:223). The Long-eared Owl is more fre-quent in undisturbed desert just outside the Salton Sink, for example, the Anza-Borrego Desert and the Chocolate Mountains (Massey 1998; R. L. McKernan *in litt.*), than it is on the heavily modified floor of the sink itself.

ECOLOGY. These owls have been noted in dense stands of large Athel, a favored roosting tree west of the region in the Anza-Borrego Desert (Massey 1998).

TAXONOMY. The allegedly paler western subspecies, *A. o. tuftsi* Godfrey, 1947, is best con-sidered a synonym of the widespread New World subspecies, *A. o. wilsonianus* (Rea 1983; Unitt 1984). North American birds differ from birds of the Old World in being more heavily barred below and having a contrasting orange facial disk.

SHORT-EARED OWL
Asio flammeus flammeus (Pontoppidan, 1763)
Rare winter visitor (mid-October to mid-March) to the Imperial Valley; 1 fall record. The partially crepuscular Short-eared Owl is a scarce winter visitor to the cultivated portions of the desert Southwest. Except for one at the north end 2 Jan-uary 2000 (*fide* C. McGaugh), in the Salton Sea region it has been recorded only in the Imperial Valley. Small numbers (probably no more than

20) winter annually in fallow agricultural fields from 11 October (2001, Rock Hill; R. J. Norton) to 24 March (1942, Wister; C. A. Harwell). It was apparently more numerous formerly, given that it was "next to the Burrowing Owl, the most common Raptore [sic]" (van Rossem 1911). However, eight at the south end 30 January 1970 (Garrett and Dunn 1981) is the maximum recent count, although this owl's secretive nature perhaps masks its true abundance. One at the south end 7 August 1971 (AB 25:907) was most likely an extremely early fall migrant, but it may have summered locally.

ECOLOGY. This species is partial to fields overgrown with tall grasses and Russian Thistle, often with some moisture. Most recent occurrences have been in the vast expanses of fields between Red Hill and Brawley, but some have been observed in similar habitat southeast of Brawley. Oddly, one was roosting in Iodine Bush on Obsidian Butte from 21 December 1999 into mid-February 2000 (M. A. Patten et al.).

TAXONOMY. Birds in continental North America are of the nominate subspecies, which is also widespread in the Old World.

NORTHERN SAW-WHET OWL
Aegolius acadicus acadicus (Gmelin, 1788)
Three records, 2 in winter and 1 in fall. In California the Northern Saw-whet Owl is a mountain species that wanders into the lowlands during some winters, perhaps in search of a better food supply. Three have been recorded in the Salton Sea region during fall and winter, the first two being north of Westmorland 3 February 1950 (Cardiff and Cardiff 1951; SBCM 39146) and at the south end 18 October 1969 (AFN 24: 100a). The accumulation of its pellets suggested that one that wintered at Mecca Beach 23 January–11 March 1971 (AB 25:629) had been there for several months prior to its discovery (Garrett and Dunn 1981). Just outside the region, one was at Regina, at the western base of the Chocolate Mountains, 4 February 1978 (AB 32:400).

TAXONOMY. All California records pertain to the widespread nominate subspecies, as *A. a.*

brooksi (Fleming, 1916) is unrecorded away from the Queen Charlotte Islands, off British Columbia (Sealy 1998).

CAPRIMULGIFORMES •
Nightjars, Oilbirds, and Allies

Were it not for the Lesser Nighthawk, nightjars would be virtually unknown in the Salton Sink. The nighthawk, though, is abundant. Hundreds litter skies on warm summer nights as they flutter erratically in search of flying insects. The Common Poorwill may well be an annual migrant through the region, but its retiring, nocturnal habits make detection difficult, so it has been recorded but a few times.

CAPRIMULGIDAE • Nightjars

LESSER NIGHTHAWK
Chordeiles acutipennis texensis Lawrence, 1857
Abundant breeder (mid-March through October); rare winter visitor (mid-November to mid-March).
Nightjars tend to be difficult to observe, but for two reasons the Lesser Nighthawk (Fig. 52) is an exception in the Salton Sea region. First, unlike most species of the Caprimulgiformes, nighthawks in general tend to be crepuscular. In the Salton Sink, Lesser Nighthawks are frequently observed well after sunrise and often fill the sky around sunset. Second, although common throughout the Southwest, this species is unusually abundant around the Salton Sea; one could easily tally a few hundred in the short time between sunset and nightfall. Despite the relatively mild winter climate, the Lesser Nighthawk is largely a breeding visitor to the Salton Sea, arriving in mid- to late March, with the earliest 18 March (1972, 1989, 1995; AB 26:808, G. McCaskie). Breeding begins early, with the earliest egg date being 18 April (Grant 1982). Nesting takes place mostly from early May to mid-July but continues well into the summer (e.g., a nest with 3 young photographed at the Whitewater River delta 3 August 1991; W. J. Moramarco). Most Lesser Nighthawks have departed by late September, and the latest date is 25 October (1996, El Centro; P. Unitt).

FIGURE 52. The Lesser Nighthawk is the only common nightjar occurring in the Salton Sink. Hundreds hawk insects during the dawn and dusk of summer but roost placidly during the day. Photograph by Michael A. Patten, June 1997, Wister.

Beginning with one 5 km northwest of Calexico 23 January 1922 (Howell 1922b) this species has been recorded over a dozen times in winter in the Salton Sea region, particularly in the Imperial Valley. Winter records have averaged about one every two winters since 1990, suggesting that a few winter in this valley every year and simply go undetected. Five or six were just east of the region, at Regina, 3 February–5 March 1978 (*AB* 32:400). The date span for winter records was provided by birds roosting at Finney Lake, from 8 November+ (2002; B. Miller) to 23 March (2000, from 18 November 1999; *NAB* 54:221).

ECOLOGY. Breeding nighthawks occupy xeric desert scrub, usually far from water (Grant 1982) and typically dominated by saltbush, but they occasionally occur in damper areas dominated by Iodine Bush and Arrowweed. Nests are often at least partly shaded (Grant 1982). Nighthawks frequently roost in Athel, and aligning with the branch, as well as on the ground. They forage over just about all habitats in the area but are most numerous over desert scrub near water (either fresh water or the Salton Sea).

TAXONOMY. Birds from central Mexico northward are of the large, pale *C. a. texensis,* of which *C. a. inferior* Oberholser, 1914, is a synonym (Dickerman 1985).

COMMON POORWILL
Phalaenoptilus nuttallii nuttallii (Audubon, 1844)
Rare transient (April to mid-May, October); probably annual in small numbers but few definite records. The secretive, nocturnal Common Poorwill is widespread throughout the West. It occurs in desert ranges on either side of the Salton Sink but is only a rare migrant in the region, although it is probably regular in small numbers and recorded infrequently simply because of its habits. There are only 14 definite records, evenly split between spring and fall. One near El Centro 28 March 2002 (K. Z. Kurland) was exceptionally early. Otherwise, spring records extend from 11 April (1908, Mecca; W. P. Taylor, MVZ 798) to 15 May (1999, male 5 km northwest of Frink; M. A. Patten), with other records from Finney Lake 12 April 1987 (G. McCaskie), Sheldon Reservoir 15 April 1990 (S. von Werlhof), Cerro Prieto 20 April 1905 (USNM 255023), and 3 km west of Westmorland 5 May 1996 (2 birds, 1 a male collected; R. Higson, P. Unitt, SDNHM 49499). Fall records extend from 1 October (1989, female 7 km northwest of Imperial; R. Higson, SDNHM 47667) to 3 November (1974, near Calipatria; *AB* 29:122), with others being of females north-northeast of Westmorland 3 October 1948 (Cardiff 1956; SBCM 33490), north of Westmorland 23 October 1947 (Cardiff 1956;

SBCM 33489), and at Unit 1 on 13 October 1999 (LACM 111006) and birds 9 km north of Niland 24 October 1976 (P. F. Springer) and at Niland 25 October 2000 (G. McCaskie). One reported at the north end 23 December 1978 was supported by "skimpy details" (*AB* 33:666). An alleged nest on the "east edge of the sea" 11 May 1940 (Abbott 1940) probably pertained to the Lesser Nighthawk, given the locale.

ECOLOGY. Migrant poorwills are found in areas with shrubby vegetation, whether desert scrub, thickets of Saltcedar, or untended edges of suburbia.

TAXONOMY. The nominate subspecies, large with the upperparts boldly patterned in silver-gray and black, breeds widely in the Great Basin and the Mojave Desert, south through the Anza-Borrego Desert, west of the Salton Sea (SDNHM specimens). Its range is thus more widespread than mapped by Grinnell and Miller (1944), and that of the smaller, more delicately marked, pinkish buff *P. n. hueyi* Dickey, 1928, occurring in the lower Colorado River valley and the Chocolate Mountains (R. L. McKernan), is correspondingly more restricted. *P. n. hueyi* perhaps occurs in the rocky hills east of the Salton Sink, but it has been insufficiently ascribed to "somewhere in the [Colorado] delta, below Mexicali, the last of March" 1917 (Grinnell 1928:130). Only two specimens from west of the Salton Sea (SDNHM 17937, 40974) appear intermediate between *P. n. nuttallii* and *P. n. hueyi;* the rest are typical of the former. The nominate subspecies is migratory, passing through the range of *P. n. hueyi* (Rea 1983). It is thus not surprising that the six regional specimens are of the nominate subspecies, although the May specimen is a little darker brown and more heavily marked than usual, tending toward *P. n. californicus* Ridgway, 1887, of cismontane California.

WHIP-POOR-WILL

Caprimulgus vociferus arizonae (Brewster, 1881)

One fall record. The only record of the Whip-poor-will for the Salton Sea is also one of few records of a migrant for California: a male was

roosting in mesquite at Salton Sea NWR 23 August 1975 (*AB* 30:127; SBCM 30033).

TAXONOMY. The specimen is apparently *C. v. arizonae* (Brewster, 1881), the subspecies that summers (and probably breeds, although there is yet no direct evidence of nesting) at a half-dozen locales in the southern California mountains (Jones 1971; Garrett and Dunn 1981). This subspecies is larger and browner than the nominate, although morphological differences are minimal and perhaps not consistent (Hubbard and Crossin 1974; Cleere 1998:207). Nonetheless, on the basis of differences in vocalizations the burry-voiced *C. v. arizonae* is sometimes treated as specifically distinct from the clear-voiced nominate subspecies (e.g., Howell and Webb 1995).

APODIFORMES • *Swifts and Hummingbirds*

Both swifts and hummingbirds are relatively poorly represented at the Salton Sea, although the Black-chinned, Anna's, and Costa's Hummingbirds breed in suburban settings in the Coachella, Imperial, and Mexicali Valleys. Furthermore, one of the great spring migration spectacles in this region is the occasional concentration of thousands of Vaux's Swifts at the north end. The White-throated Swift commonly forages over the southern Coachella Valley, where it is frequently encountered during winter and spring.

APODIDAE • *Swifts*

BLACK SWIFT

Cypseloides niger borealis Kennerly, 1857

Casual spring transient (early May to early June); 1 record each for summer and fall. Given that this species is a rare, local breeder in California, particularly in the mountains of southern California (Foerster and Collins 1990), it is not surprising that the Black Swift is only a casual spring migrant through the Salton Sink. It has been detected only six times in that season, from 3 May (1981, 3 at Finney Lake; *AB* 35:864) to 9 June (1990, near Imperial; K. L. Garrett). Other records are of one at Finney Lake 4 May 1969 (*AFN* 23:626), one at the south end 7 May 1988 (*AB* 42:482), two at Finney Lake 12 May 1998 (H. Detwiler), and one at the north end 25 May 1987 (*AB* 41:488). Records of ten at Desert Hot Springs 22 May 1980 (*AB* 34:816), single birds at San Gorgonio Pass 15 May 1978 and 19 May 1962 (Garrett and Dunn 1981), and two at Snow Creek Village 11 May 1982 (*AB* 36:964) suggest that small numbers may pass through the region in spring on a regular basis. One at the north end 30 September 1987 (*AB* 42:136) furnished the only record of a fall migrant. A specimen from Salton Sea NWR 29 July 1995 (*FN* 49:982; SDNHM 49237) was found a month before the beginning of the fall migration (Garrett and Dunn 1981), but the bird was partly mummified, so it was presumably passing through in late spring.

TAXONOMY. The only subspecies occurring in the United States, and the only one recorded in California, is *C. n. borealis,* a form larger and whiter on the underparts (females only) than other subspecies. Unfortunately, the poor condition of the 1995 specimen (only a wing and the skeleton could be preserved) prohibited positive identification to subspecies.

CHIMNEY SWIFT

Chaetura pelagica (Linnaeus, 1758)

One or 2 summer records (June). The Chimney Swift is a rare migrant through California and a summer visitor there, particularly along the southern coast (Devillers 1970a). It has proven to be the expected *Chaetura* in southern California during the summer. The only unequivocal record for the Salton Sea was of a bird studied at close range as it clung to a utility pole near the mouth of the New River 25 June 1994 (*FN* 48:989). A *Chaetura* swift at the mouth of the Whitewater River 11 June 1992 (*AB* 46:1179) most likely belonged to this species given the date and location. The underparts (including the throat) of that individual were uniformly dark, but the upperparts were not seen well, so the identification was treated as tentative (M. A. Patten). Even

so, a Vaux's Swift anywhere in lowland southern California in mid-June would be unprecedented (Garrett and Dunn 1981). The identity of a Chimney Swift reported at the north end 20 May 1972 (*AB* 26:809), a date when Vaux's Swifts commonly move through the area, is questionable.

VAUX'S SWIFT

Chaetura vauxi vauxi (Townsend, 1839)

Common spring transient (mid-April to mid-May); rare fall transient (September to mid-October). Large numbers of Vaux's Swifts move through the Salton Sea region each spring, often in mixed flocks with migrant swallows, as they return to breeding sites in the Sierra Nevada and the Pacific Northwest. Spring dates extend from 10 April (1994, 5 near Oasis; G. McCaskie) to 23 May (1999, Mecca; M. A. Patten). During peak movements in the first half of May numbers at the north end are often in the hundreds and occasionally in the low thousands. Mirroring its status elsewhere in the California desert, this species is much rarer during fall migration, with records from 1 September (1996, south end; P. Unitt) to 12 October (1975, south end; G. McCaskie); most are from the first half of September. Numbers in that season seldom reach double digits, and lone individuals are frequent, but 75 were observed at the north end 20 September 1981 (G. McCaskie). Vaux's Swift is generally rare in winter anywhere in California, although it can be numerous in coastal San Diego County. It is the only *Chaetura* swift recorded in North America in that season. It has been recorded once in winter at the Salton Sea, with two at the south end 12 February 1969 (*AFN* 23:522).

ECOLOGY. The vast majority of Vaux's Swifts are noted over water, ranging from the Salton Sea to small irrigation canals and ponds. Roosting sites are unknown in the region, but this species uses chimneys and tree trunks elsewhere in southern California.

TAXONOMY. California records pertain to the pale nominate subspecies, a highly migratory form and the only one breeding north of Mexico.

WHITE-THROATED SWIFT

Aeronautes saxatalis saxatalis (Woodhouse, 1853)

Fairly common nonbreeding perennial visitor; most numerous in winter (mid-December to mid-May). Grinnell and Miller (1944) opined that "hardly a cubic yard of atmosphere anywhere south of latitude 38° and within a mile of the ground surface can have escaped traversement by one or more White-throated Swifts in any ten-year period." This description captures this species' aerial domain well, as does their comment that "possibly the daily cruising radius of this bird is greater than in any other species, even the California Condor [*Gymnogyps californianus*]." The great distances through which this species moves while foraging, as well as its tendency to forage out of sight under high clouds or clear skies, makes assessing its status remarkably difficult. Bordering the Salton Sink, the White-throated Swift breeds in both the Santa Rosa Mountains, to the west, and the Mecca Hills, to the east, the latter hosting winter roosts of about 1,000 birds (R. L. McKernan). In the sink, therefore, this swift is most numerous and most regularly seen foraging over the Coachella Valley. It may be encountered throughout the year, but peak counts are during winter, and it is seldom noted from mid-June through October except during cloudy conditions. This species' status in the Imperial and Mexicali Valleys is far different: there are no summer records. Instead, it is a winter visitor, with records extending from 20 November (2000; G. McCaskie) to 22 May (1994; M. A. Patten), with one at Salton Sea NWR 2 June 2002 (G. McCaskie) being exceptionally late. Wintering birds may arrive earlier in the season, but clear skies prevail in fall well into November, making detecting swifts nearly impossible.

ECOLOGY. As noted above, this aerial feeder is seldom recorded under clear skies, even in the Coachella Valley. Thus, most records for the region are from late fall through spring, corresponding to the period when winter storms generate low clouds. Like Vaux's Swift, the White-throated Swift is typically encountered in large flocks. It often forms large mixed-species flocks with Vaux's Swift and various swallows during spring migration (March through May), particularly in the Coachella Valley.

TAXONOMY. The allegedly larger *A. s. sclateri* Rogers, 1939, is a synonym of the nominate subspecies (Behle 1973). Thus, all White-throated Swifts north of Central America are *A. s. saxatalis*, differing from *A. s. nigrior* Dickey and van Rossem, 1928, of El Salvador and Guatemala, in their larger size, paler dorsum, and more extensively white underparts.

TROCHILIDAE • *Hummingbirds*

The hummingbird fauna of the Salton Sink is depauperate in comparison with that elsewhere in the Southwest. Three species breed uncommonly to fairly commonly in ornamental vegetation having sufficient nectar-producing flowers and shrubs, particularly bottlebrush, gum, and Cape Honeysuckle, and around citrus orchards in the Coachella Valley. Only Costa's Hummingbird breeds in native habitats in the region, where it is restricted to flowering shrubs in washes. The remaining three species are uncommon to casual migrants.

BLACK-CHINNED HUMMINGBIRD

Archilochus alexandri (Bourcier and Mulsant, 1846)

Uncommon migrant and local breeder (mid-March to late September). The Black-chinned Hummingbird is widespread and common in much of the Southwest and breeds along the lower Colorado River and locally in the Mojave Desert. In the Salton Sea region, though, it is generally uncommon and occurs mainly as a migrant en route to and from cismontane California. Records extend from 14 March (1992, 8 km southwest of Westmorland; S. von Werlhof) to 25 September (2000, 3 km southeast of El Centro; K. Z. Kurland). More hummingbirds are noted during April and August, but some remain through the summer in cities and towns such as Indio, Coachella, and Mexicali (an example of the last is from 11 km east of Cerro Prieto 2 June 1928; C. C. Lamb, MVZ 52918). It is almost common in summer in the lusher areas of Brawley and El Centro (e.g., more than 15 in one yard 3 km southeast of El Centro 15 July 2000; K. Z. Kurland). The Black-chinned Hummingbird is al-most unknown in California after late September.

ECOLOGY. Regional breeding in this species is confined to suburban settings with suitable nesting trees. In native habitats this humming-bird relies on the California Sycamore and other trees to supply sufficient fuzz for its nest. Aside from a few planted in Brawley and El Centro, these sycamores are absent from the region. The Black-chinned Hummingbird will use other trees, however, such as planted maple and mul-berry (e.g., a nest in the latter at El Centro 16 May 1999; J. C. Burger). It uses Fremont Cottonwood also, but seemingly only in suburban areas, as breeding hummingbirds in natural habitats are invariably Costa's.

ANNA'S HUMMINGBIRD

Calypte anna (Lesson, 1827)

Uncommon, local resident and breeder; numbers augmented in winter (mid-August to mid-April). Anna's Hummingbird is principally a bird of the Pacific Coast. Its status in the Salton Sink has changed greatly in the past two decades, largely as a result of the maturation of exotic trees in the older parts of the larger towns. The trend has been duplicated in the lower Colorado River valley (Rosenberg et al. 1991) and in many other locales in western North America, from Oregon to northwestern Mexico (Zimmerman 1973). Formerly this species was a fairly common winter visitor to the Salton Sink from late Sep-tember to early April (Garrett and Dunn 1981). It was known to breed at the Brock Research Center, east of the Salton Sink, but it only rarely summered in the region and was not known to breed. There is still an influx of birds in winter from 16 August (1986, Brawley; B. E. Daniels) to 24 April (1988, Brawley; M. A. Patten), but sometime in the 1990s it colonized the region. Small numbers now summer annually there, as well as at El Centro (K. Z. Kurland) and pre-sumably at Brawley, in parts of the southern Coachella Valley, and probably in Calexico and Mexicali. This species appears to be established best in Brawley, Coachella, and Thermal, but even at those locales daily counts seldom exceed five individuals (e.g., 6 at Brawley 21 July 1999; C. A. Marantz).

ECOLOGY. In the Coachella and Imperial Val-leys this species is strongly tied to human set-tlements. Wintering birds frequent wooded sub-urban neighborhoods, parks, and ranch yards, particularly those with flowering gum or bottle-brush. Summering birds are found only in the most heavily vegetated portions of Brawley, El Centro, and the Coachella Valley. In the last it also frequents citrus orchards to some degree.

COSTA'S HUMMINGBIRD

Calypte costae (Bourcier, 1839)

Fairly common breeding resident. A characteristic bird of the Sonoran Desert, Costa's Humming-bird is a fairly common resident in the Salton Sink and the region's most numerous humming-bird. There is some seasonal movement, so that numbers peak from mid-February through April with the return of birds that have wintered to the south (Baltosser 1989). When ornamental flower-ing shrubs are present, numbers remain high through midsummer (e.g., 15 at Brawley 21 July 1999; C. A. Marantz), but they are somewhat lower in midwinter. Breeding takes place from early February (e.g., female on a nest at Brawley 5 February 1984; G. McCaskie) into April (e.g., nest with eggs at Mecca 2 April 1908; W. P. Tay-lor, MVZ 65), with most juveniles fledging by late April or early May.

ECOLOGY. Irrigation of the Coachella, Impe-rial, and Mexicali Valleys affected the status of hummingbirds in the Salton Sink profoundly. Both Black-chinned and Anna's Hummingbirds were largely or completely absent before settle-ment provided habitat that lured them to the region. For Costa's Hummingbird, however, irrigation probably has been a mixed blessing. The species was likely a breeding visitor, occur-ring in winter and spring only in washes fring-ing the Salton Sink. Agriculture removed much of its habitat, but it has since occupied suburban neighborhoods and ranch yards, where it re-mains year-round and presumably competes with Anna's and Black-chinned Hummingbirds. As for the other species, nonnative flowers (gum, bottlebrush, citrus trees) are primary attractants, with Tree Tobacco being a chief culprit through much of this hummingbird's expanding range (Baltosser 1989).

CALLIOPE HUMMINGBIRD

Stellula calliope (Gould, 1847)

Rare spring transient (late March to mid-May). A somewhat uncommon and local montane species of the West, the Calliope Hummingbird is a rare migrant through the lowlands of California, primarily in spring. The only records for the Salton Sea region are from the Imperial Valley in that season. It may be annual in small numbers (no more than 10 individuals), but records are few, from 27 March (1965, male at Salton Sea NWR; G. McCaskie) to 13 May (1972, south end; G. McCaskie). The majority are from the period from the second week of April to the first week of May, including specimens of females collected 11 km northwest of Niland 7 May 1961 (E. A. Cardiff; SBCM M3230) and at the Highline Canal east of Calipatria 7 May 1996 (G. L. Braden; SBCM 54866).

ECOLOGY. Like other migrant hummingbirds in the region, this species has been found only at flowering ornamentals, especially bottlebrush.

TAXONOMY. There are far too many genera of hummingbirds, with most defined by sexually selected characters of the males. Morphological features defining the monotypic genus *Stellula* are minor. Apart from the striped gorget, all plumage characters, especially the rectrix pattern, fit *Selasphorus*. We recommend merger of *Stellula* (and *Atthis*, of Mexico and Central America) into *Selasphorus*.

RUFOUS HUMMINGBIRD

Selasphorus rufus (Gmelin, 1788)

Uncommon spring transient (late February to early May); rare fall transient (mid-July to late September). The Rufous Hummingbird may be encountered in large numbers during migration in the Southwest but is a scarce migrant in the deserts of southeastern California. Its status is clouded by identification difficulties, the extremely similar Allen's Hummingbird not being distinguishable in the field under most conditions. Even adult males can pose problems (McKenzie and Robbins 1999). Nevertheless, given that virtually all adult males seen in the region are obviously Rufous Hummingbirds with full orange-rufous backs and that the Rufous is by far the more expected species anywhere in the desert Southwest (Phillips 1975a), we summarize records of all *Selasphorus* hummingbirds under this species. With that caveat, spring records extend from 19 February (2000, ca. 5 at Heber; D. S. Cooper) to 7 May (1988, 2 at the south end; G. McCaskie), with a peak in mid-April (e.g., 18 at Brawley 24 April 1988; M. A. Patten). This migration window is mirrored on the coastal slope of southern California, where the Rufous Hummingbird is a later migrant than Allen's by three to four weeks. Outside the region but not far southwest of the Mexicali Valley an adult male was collected in the Sierra Cucapah 24 February 1905 (Stone and Rhoads 1905; ANSP 48293). Fall records are few, with all from 10 July 1997 to 30 September 2000 (both 3 km southeast of El Centro; K. Z. Kurland), including a specimen from 3 km west of Westmorland 14 September 1991 (R. Higson; SDNHM 47762). A "bedraggled" bird at bottlebrush and Cape Honeysuckle at Brawley 29 October 2000 (P. A. Ginsburg) was exceptionally late.

ECOLOGY. This species has been encountered only in orchards, gardens, and residential neighborhoods in the region.

ALLEN'S HUMMINGBIRD

Selasphorus sasin sasin (Lesson, 1829)

Six records, 4 in spring (February) and 2 in fall (mid-July). Owing to the difficulty of distinguishing Allen's Hummingbird from the similar Rufous Hummingbird in the field, as well as its rarity, the former has been reliably reported only four times in the Salton Sink. All six records involved adult males feeding on ornamental flowers. Fall migrants occurred at Brawley 12 and 14 July 1984 (*AB* 38:1062). This species may occur regularly in small numbers at this time of year in heavily vegetated residential areas. Nevertheless, the vast majority of *Selasphorus* humming-

FIGURE 53. Allen's Hummingbird largely bypasses the Salton
Sink on both its spring and fall
migrations. This photograph,
from Kenneth Kurland's yard,
3 km southeast of El Centro,
supports one of only six records
considered valid. Photograph by
Kenneth Z. Kurland, 22 February 2000.

birds passing through the Salton Sink are of in-
determinate species, and the Rufous Hummingbird is undoubtedly the more frequent
(Phillips 1975a), probably by several orders of
magnitude. Although Allen's Hummingbird reportedly migrates through the California desert
in spring (Phillips 1975a), an adult male photographed 3 km southeast of El Centro 22–24 February 2000 (Fig. 53, *NAB* 54:221), another seen
at the same locale 15–19 February 2001 (*NAB*
55:228), and two there 1–18 February 2002 (K. Z.
Kurland) furnished the only records in that season for the Salton Sink. The recency of spring
records suggests that the species may be more
numerous in the region.

TAXONOMY. Even though there are no specimens, all three records must be of the highly
migratory nominate subspecies, as the longer-
billed *S. s. sedentarius* Grinnell, 1929, is resident on the Channel Islands and the adjacent
mainland and has limited, if any, seasonal
movements.

CORACIIFORMES • *Rollers, Motmots, Kingfishers, and Allies*

The colorful Coraciiformes reach peak diversity
in tropical and subtropical regions around the
world. Despite their being the dominant small
landbirds in North America in the early Tertiary,
only one species, the Belted Kingfisher, is widespread today.

ALCEDINIDAE • *Kingfishers*

BELTED KINGFISHER
Ceryle alcyon (Linnaeus, 1758)
Fairly common winter visitor (August to mid-May).
The Belted Kingfisher is widespread and fairly
common throughout North America, although
it is declining as a breeder in California and elsewhere in the West. It is a fairly common winter
visitor throughout the Salton Sea region, where
records extend from 31 July (1999, south end;
G. McCaskie) to 14 May (1988, north end, and
2000, Fig Lagoon; G. McCaskie), with most

present from late August to mid-April. One at the mouth of the Whitewater River 2 July 1988 (G. McCaskie) either was an exceptionally early fall transient or had summered locally.

ECOLOGY. Not surprisingly, given its dependence on fish, this species is tied to fresh water in the region, although it is not finicky about its source or size. The Belted Kingfisher can be found near sources ranging from large freshwater lakes (e.g., Finney and Ramer Lakes, Fig Lagoon) to small irrigation ditches. It only rarely takes fish directly from the Salton Sea, and then only from the edge near influxes of fresh water. It most frequently perches on utility wires and Saltcedar snags.

TAXONOMY. This species is generally regarded as monotypic (see Phillips 1962) because *C. a. caurina* Grinnell, 1910, is weakly differentiated at best (Rand and Traylor 1950; Mayr and Short 1970).

PICIFORMES • *Puffbirds, Jacamars, Toucans, Woodpeckers, and Allies*

Not surprisingly, the paucity of true woodland or forest habitat in the Salton Sea region has limited the number of woodpeckers there. Only the Ladder-backed Woodpecker, common in the deserts of the Southwest, is widespread. However, large trees now planted around ranches and in suburban neighborhoods lured Gila Woodpeckers to the Imperial Valley in the 1920s and support a good number of wintering Northern Flickers and a small number of wintering Red-naped Sapsuckers. The nine other species of woodpeckers recorded in the region are casual visitors at best, with several represented by fewer than ten records.

PICIDAE • *Woodpeckers and Allies*

LEWIS'S WOODPECKER
Melanerpes lewis (Gray, 1849)
Casual, irregular winter visitor (late September to early May). Lewis's Woodpecker, generally uncommon over most of its range, is famous for irregular and unpredictable wanderings away from

its usual haunts of oak savannah and open pine woodlands in the intermountain West (Ryser 1985). The first regional records (Cardiff 1956) were of males secured at Heise Springs 24 April 1948 (SBCM M565) and 1 October 1948 (SBCM 673). The species has since proven to be a casual and highly irregular winter visitor to the area from 23 September (1989, near Westmorland; G. McCaskie) to 5 May (1990, near the mouth of the New River; *AB* 44:497), with the majority recorded from mid-December to mid-March. There are seldom more than two birds in a given winter, but as many as six were observed at three "widely-scattered localities around the south end" 22 December 1986 (*AB* 41:330), six were together in Brawley during January and February 1990 (*AB* 44:330), and up to seven reached the Imperial Valley during the winter of 1996–97 (*FN* 51:643, 800).

ECOLOGY. Lewis's Woodpeckers are invariably attracted to orchards or open areas with large trees, from Fremont Cottonwood and Athel to the Mexican Fan Palm and the Date Palm. This woodpecker frequently flycatches from prominent perches on utility poles and tall trees.

RED-HEADED WOODPECKER
Melanerpes erythrocephalus (Linnaeus, 1758)
One summer record. The second California record of the Red-headed Woodpecker was of an adult photographed in eucalyptus at Wister 17 July–22 August 1971 (Cardiff and Driscoll 1972; Binford 1985).

TAXONOMY. The species is monotypic, the described subspecies *M. e. caurinus* Brodkorb, 1935, being insufficiently distinct (Short 1982).

ACORN WOODPECKER
Melanerpes formicivorus (Swainson, 1827) subsp.?
Six records, 3 in winter, 2 in fall, and 1 in summer. As suggested by its common name, the Acorn Woodpecker is normally a denizen of oak woodlands, though, like many woodpeckers, it is prone to wander, particularly in fall and winter. These seasons account for five of the only six records for the Salton Sea, all from the Imperial Valley:

FIGURE 54. Many miles from the nearest oak tree, this Acorn Woodpecker at the Wister Unit of the Imperial Wildlife Area was one of only five ever recorded in the Salton Sink. Photograph by Kenneth Z. Kurland, 17 July 1999.

near Westmorland 4 November 1973 (*AB* 28:108) and 4 October 1977 (*AB* 32:259), at El Centro 22–24 December 1986 (*AB* 41:330) and 19 January–28 February 2002 (*NAB* 56:224), and a male photographed at Calipatria 21 December 1999–5 February 2000 (*NAB* 54:221). A female photo-graphed at Wister 17–18 July 1999 (Fig. 54, *NAB* 53:432) was much earlier than most fall wanderers elsewhere in the desert Southwest (see Rosenberg et al. 1991).

TAXONOMY. Without a specimen a subspecific determination cannot be made. Either the large, heavy-billed *M. f. bairdi* Ridgway, 1881, resident through much of cismontane California, or the small, thin-billed *M. f. aculeatus* Mearns, 1890, resident in the interior Southwest and collected in southeastern California (Miller and Stebbins 1964:107), could account for the records.

GILA WOODPECKER

Melanerpes uropygialis uropygialis (Baird, 1854)

Uncommon and local breeding resident. Widespread settlement of the Imperial Valley brought with it the planting of many shade trees. The maturation of palms, Fremont Cottonwoods, and other large trees lured Gila Woodpeckers into

the area (van Rossem 1933a; Phillips 1968), presumably from the south, as the species was already known from the Río Hardy drainage south of the Mexicali Valley (Stone and Rhoads 1905; ANSP 48289, 48291). Hoffmann (1927) reported several near Holtville in 1927 as the first in the Salton Sink, but the species was taken along the Río Alamo 13 km east of Mexicali 21 and 23 April 1894 (E. A. Mearns; USNM 134022, 134023). By the early 1930s the species was established in the Imperial Valley, especially around El Centro, Calipatria, and Brawley (van Rossem 1933a; SDNHM 22624). The Gila Woodpecker is now a locally common breeding resident in the Imperial Valley, particularly at Brawley and El Centro, and in the Mexicali Valley. April and May are principal months for breeding (e.g., nest with 5 eggs from the Imperial Valley 24 April 1948; SBCM/WCH 8843). The species persisted into the late 1940s as far north as Calipatria (*AFN* 3:144) and Westmorland (3 specimens from 8 October 1949 to 2 February 1950; PSM 4129, 7305, 7306) but now generally does not occur north of Brawley. It occasionally wanders north to Calipatria (e.g., 18 December 1986; *AB* 41:1281) and Wister (e.g., 28 April 2000; G. McCaskie) and west to Seeley (e.g., 14 June 1997; G. McCaskie).

Despite apparently suitable habitat in the southern Coachella Valley, it has been recorded there only three times, 1 September 1963 (Garrett and Dunn 1981), 3 September 1984 (*AB* 39:103), and 24 May 1987 (*AB* 41:489).

ECOLOGY. This woodpecker favors palms, especially the Mexican Fan Palm, and large broadleaf trees such as Fremont Cottonwood, figs, and gums. It most often nests in palms, although it sometimes nests in utility poles or in large exotic trees (e.g., sycamores or gums).

TAXONOMY. Only the nominate subspecies occurs in the southwestern United States and northern Sonora (Rea 1983). The supposedly slightly paler, whiter-backed *M. u. albescens* van Rossem, 1942, of the Colorado River, is a synonym of *M. u. uropygialis* (Phillips et al. 1964; Short 1982). The two subspecies endemic to Baja California are either distinctly smaller or distinctly darker than the nominate subspecies.

WILLIAMSON'S SAPSUCKER
Sphyrapicus thyroideus (Cassin, 1852)
Two records. Williamson's Sapsucker is a high-mountain bird that rarely wanders to the lowlands of southern California, east and west of the mountains. The only regional records are of females 6.5 km south of Mecca 7 (not 5) October 1955 (*AFN* 10:58; WFVZ 1507) and at Brawley 18–19 January 1999 (*NAB* 53:209).

TAXONOMY. Two subspecies, *S. t. nataliae* (Malherbe, 1854), of the Rocky Mountains and the intermountain West, and *S. t. thyroideus,* of the Pacific slope, are commonly recognized on the basis of the former's smaller bill (Swarth 1917; Cowan 1938; AOU 1957; Phillips et al. 1964; Short 1982; Godfrey 1986). However, 16 specimens from Arizona and 16 from the Pacific slope in SDNHM hardly differ in bill size (mean lengths 20.3 mm, $s = 1.2$, and 21.8 mm, $s = 0.8$, and mean depths 6.9 mm, $s = 0.4$, and 7.1 mm, $s = 0.4$, respectively), far less than the minimum acceptable for a subspecies ($D_{ij} < -1.3$ in all cases). We therefore concur with Browning and Cross (1999) in regarding Williamson's Sapsucker as monotypic.

YELLOW-BELLIED SAPSUCKER
Sphyrapicus varius (Linnaeus, 1766)
Seven winter records (mid-November to early February). The Yellow-bellied Sapsucker is a rare but regular fall and winter vagrant to California from the eastern United States (Devillers 1970b), recorded seven times in the Salton Sink. A first-spring female 5 km west of Niland 1 February 1953 (E. A. Cardiff; SBCM 33690) was one of the first recorded in California. Records extend from 14 November (2002, juvenile at Brawley to at least 31 December; G. McCaskie) to 9 February (2002, at Mecca from 13 January; C.-T. Lee). Other records are of juveniles at Wister 22–30 December 1998 (K. L. Garrett et al.) and Finney Lake 7 February 1982 (*AB* 36:331) and of adults at Wister 16 November 2002+ (T. Benson) and at Finney Lake 8 February 1985 (*AB* 39:210).

ECOLOGY. Large Athel trees have hosted all of the Yellow-bellied Sapsuckers; this tree attracts a disproportionate number of sapsuckers occurring in the region.

TAXONOMY. This species is monotypic, the allegedly darker *S. v. appalachiensis* Ganier, 1954, being a synonym (Short 1982:176).

RED-NAPED SAPSUCKER
Sphyrapicus nuchalis Baird, 1858
Rare winter visitor (mid-September to mid-May). A species of the Great Basin, the Red-naped Sapsucker is the only sapsucker likely to be encountered in the desert Southwest, where many disperse in fall and winter. Even so, it is rare in the Salton Sea region, with generally only two to four recorded in any given winter. Records extend from 18 September (1997, 3 km southeast of El Centro; K. Z. Kurland) to 16 May (1953, female 5 km west of Niland; E. A. Cardiff, SBCM 2265), with most from late October to mid-March. This species is much more numerous in winter in the lower Colorado River valley (Rosenberg et al. 1991), a short distance east of the Salton Sink.

ECOLOGY. Any suitable large tree will suffice for this sapsucker, but it favors Fremont Cottonwood and nonnative Athel, pine, and gum.

RED-BREASTED SAPSUCKER

Sphyrapicus ruber daggetti Grinnell, 1901

Casual winter visitor (late October through Febru-ary). Although a fairly common breeder through-out the Sierra Nevada and the mountains of southern California, the Red-breasted Sapsucker is surprisingly rare in the desert Southwest, with only a few valid records for Arizona (Monson and Phillips 1981; Rosenberg and Witzeman 1998). It appears that most of the down-slope move-ment of this species is toward the Pacific Ocean. There are only ten well-documented records for the Salton Sea region, ranging from 24 October (2000, photographed at Wister through at least 4 December; G. McCaskie) to 27 February (1999, Brawley from 12 December 1998; G. McCaskie). Other records are from Calipatria 19 December 1985 (*AB* 40:1008), El Centro 26 December 1986 (*AB* 41:330), Niland 28 December 1995 (*FN* 50:852), the north end 29 December 1971 (*AB* 26:521) and 29 December 1990 (*AB* 45:994), the south end 13 January 1973 (G. McCaskie), and Wister 21 February 1998 (G. McCaskie; M. A. Patten) and 25 February 1999 (G. McCaskie).

ECOLOGY. All regional records of this species are from nonnative trees, particularly Athel and pines.

TAXONOMY. There are no specimens for the region. However, all birds carefully studied in the field have had the characters of *S. r. daggetti*, the duller, more heavily marked subspecies that breeds from the southern Cascades through the Sierra Nevada and south through the mountains of southern California. Furthermore, this sub-species accounts for virtually all lowland records in coastal southern California (Devillers 1970b; Garrett and Dunn 1981). The redder, less marked *S. r. ruber* occurs only rarely in California, with most records from the north coast. Several sap-suckers with largely red heads were considered to be hybrid Red-breasted × Red-naped Sap-suckers by some observers (e.g., the Niland bird; C. A. Marantz pers. comm.). However, fairly ex-tensive black markings on the auriculars appear to be within the range of normal variation of *S. r. daggetti* (Johnson and Johnson 1985).

LADDER-BACKED WOODPECKER

Picoides scalaris cactophilus (Oberholser, 1911)

Uncommon breeding resident. Widespread in the desert Southwest and the thorn forests of Mex-ico, the Ladder-backed Woodpecker is the only woodpecker that breeds widely in the Salton Sink. It does not occur in large numbers any-where in the region, but it can be found some-what consistently at any location with native trees. Breeding occurs from March to July over most of its range (Short 1982), including in the Salton Sink.

ECOLOGY. This species is most common in stands of mesquite (see van Rossem 1911) but also occurs in Fremont Cottonwood and Good-ding's Black Willow. It generally avoids pure stands of Saltcedar and only seldom uses gum and other large nonnative trees. Indeed, this species' scarcity in the Salton Sink is undoubt-edly largely the result of the dearth of large na-tive trees.

TAXONOMY. Only the short-tailed, short-billed *P. s. cactophilus* occurs in the southwestern United States, as various other named forms are insufficiently distinct to warrant recognition (Short 1968). *P. s. cactophilus* further differs from other subspecies in having more extensive white barring dorsally and in being larger overall.

NUTTALL'S WOODPECKER

Picoides nuttallii (Gambel, 1843)

One fall record. The only Nuttall's Woodpecker recorded in the Salton Sink is one of few ever re-corded outside of the species' normal range and habitat (oak and riparian woodlands). A calling bird carefully identified in a stand of Fremont Cottonwoods near Westmorland 3 September 1973 (*AB* 28:108) was considered to be a juvenile wandering during its first fall (J. L. Dunn).

DOWNY WOODPECKER

Picoides pubescens (Linnaeus, 1766) subsp.?

One spring record. There is but one record of the Downy Woodpecker for the Salton Sea, of a bird observed near Mecca 4 April 1977 (*AB* 31:1048). This species rarely wanders into deserts of southeastern California, with the Mecca bird being the southernmost on record.

TAXONOMY. The Salton Sea bird was presumably the cismontane *P. p. turati* (Malherbe, 1860), a small form with smoke-gray underparts recorded in Whitewater Canyon and Mission Creek, just north of the Coachella Valley. Nevertheless, there are vagrant records for southern California of the Great Basin subspecies *P. p. leucurus* (Hartlaub, 1852), a larger form with white underparts (see Grinnell and Miller 1944). Although on the basis of geography the former subspecies is more likely, the latter subspecies cannot be excluded without a specimen.

WHITE-HEADED WOODPECKER

Picoides albolarvatus (Cassin, 1850) subsp.?

One fall record. The White-headed Woodpecker is a high-mountain species that rarely wanders into the lowlands, so it is not surprising that there is but a single record for the area. One near Indio 28 November 1955 (*AFN* 10:58) occurred during a winter that also brought a Williamson's Sapsucker and Mountain Chickadees to the region. Another White-headed Woodpecker wintered in nearby Palm Springs from late November 1915 to 25 February 1916 (Jaeger 1947b).

TAXONOMY. The larger-billed subspecies breeding in the Transverse and the Peninsular Ranges, *P. a. gravirostris* (Grinnell, 1902), is presumably responsible for the Indio record. However, the nominate subspecies, breeding in the Sierra Nevada, cannot be excluded without a specimen, particularly in light of various vagrants from the Mojave Desert of California (Garrett et al. 1996) that were presumably the nominate on the basis of geography.

NORTHERN FLICKER

Colaptes auratus (Linnaeus, 1758) subspp.

Fairly common transient and winter visitor (late September to mid-April). The Northern Flicker is the most abundant woodpecker in the Salton Sea region during the winter, when breeders from the north, especially from the Great Basin, descend into the lowland Southwest. Then flickers out-

number Ladder-backed Woodpeckers in native habitat and Gila Woodpeckers in towns in the Imperial Valley. Records extend from 22 September (2000, 3 km southeast of El Centro; G. McCaskie) to 13 April (1942, Coachella Valley; C. A. Harwell).

ECOLOGY. Flickers occupy virtually any large tree, native or not. Because most large trees are now concentrated in settled areas, such as suburban neighborhoods, cemeteries, and ranch yards, this species is particularly common in those areas. It also occurs in stands of mature Athel and along the river courses where there are suitably large Goodding's Black Willow and Fremont Cottonwood. Only in dense stands of mesquite does the Ladder-backed Woodpecker dominate numerically.

TAXONOMY. The predominant wintering subspecies from the coast at San Diego east through central Arizona is *C. a. canescens* Brodkorb, 1935, identified by its relatively pale gray-brown back and contrasting grayer nape and crown (Rea 1983). It breeds in the interior from the Sierra Nevada across the Great Basin. Its mixing in winter with the coastal subspecies *C. a. collaris* Vigors, 1829, presumably accounts for its often being unrecognized (e.g., AOU 1957). Despite its prevalence in the region, *C. a. canescens* accounts for only 10 of 17 specimens from the Salton Sink. Four (SBCM 33623, SDNHM 43461 and 49358, WFVZ 1523) are of *C. a. collaris*, with crowns and napes uniformly dark brown and backs darker and browner. Some migration to the Salton Sink of this subspecies, which breeds in the Coast Ranges of California, is expected given that it constituted 20 percent of 60 wintering flickers Rea (1983) collected in central Arizona. A male from 12 km northwest of Imperial 4 October 1987 (P. Unitt; SDNHM 45102) is a striking example of *C. a. cafer* (Gmelin, 1788), from the Pacific Northwest, with deep maroon-chocolate upperparts, including the crown, and extensively vinaceous underparts. It is conspicuously darker above than any other southern California specimen at the San Diego Natural History Museum. It appears to represent the southernmost record, by a considerable distance,

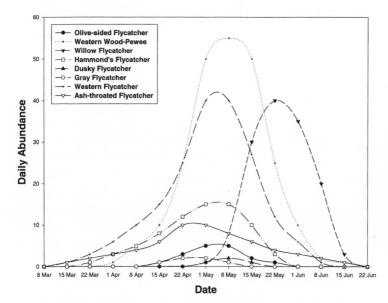

FIGURE 55. General peak abundance and timing of spring transient tyrant flycatchers (Tyrannidae) through the Salton Sink. The two common *Empidonax* flycatchers, the Western and the Willow, only slightly overlap in timing. Data are based on specimens and Michael A. Patten's field notes.

of *C. a. cafer,* as Grinnell and Miller (1944) did not report it south of Berkeley and Monson and Phillips (1981) did not report it from Arizona. Though not highly migratory, it apparently disperses long distances on occasion, with two specimens from Idaho (Burleigh 1972) and three from Utah (Behle 1985), including one from the latter near Kanab, in southwestern Utah.

Flickers of the Yellow-shafted forms of the East are rare, regular fall (mainly) and winter visitors to the West. Salton Sink records extend from 22 October (1964, male at Salton Sea NWR; *AFN* 19:79) to 30 March (1969, Niland; *AFN* 23: 626), average one every other year, and include a male specimen collected from 5 km north of Westmorland 16 January 1971 (E. A. Cardiff; SBCM 33616). All Yellow-shafted Flickers taken in California belong to the large, highly migratory northern subspecies *C. a. luteus* Bangs, 1898, which includes *C. a. borealis* Ridgway, 1911, as a synonym (Phillips et al. 1964; Short 1982). However, a female taken 6.5 km south of Mecca 7 October 1955 (E. N. Harrison; WFVZ 1524) is clearly a hybrid *C. a. luteus* × *C. a. canescens,* with a gray crown and face and distinctly orange remiges and rectrices. Howell's (1922b) report of a flicker with yellow "shafts" at or near Calex-

ico 22 January 1922 lacks details but was inferred by Grinnell and Miller (1944:230) to represent the Gilded Flicker, a species unknown in the Salton Sink (see the Hypothetical List).

PASSERIFORMES • *Passerines*

The Salton Sink is an important migration corridor for landbirds, particularly in spring. Numerous Neotropical migrants winter in western Mexico, and in spring many of these migrants follow the coast northward. Upon reaching the head of the Gulf of California, they follow two principal pathways, the low-lying basins of the lower Colorado River and the Salton Sink (Fig. 55). Fall migration through the region is far less dramatic, with all species occurring in lower numbers and/or lower concentrations (see, e.g., Guers and Flannery 2000). This pattern generally holds throughout the Southwest, as landbird migration is much more protracted in fall than in spring. The fall pattern, however, belies the harsh conditions of the Salton Sink. Many migratory landbirds avoid the region below sea level, perhaps because of high temperatures and resultant low food levels, instead following montane routes and overflying the desert floor.

As a group the tyrant flycatchers are a prevalent and complex component of the Salton Sink's avifauna. Six species breed, but only the Western Kingbird is common and widespread. The Black Phoebe, Say's Phoebe, and the Ash-throated Flycatcher are local, and the Vermilion and Brown-crested Flycatchers have been extirpated. The Black and Say's Phoebes are the only common winter visitors, although a small number of Vermilion Flycatchers winter annually, mainly in the Imperial and Mexicali Valleys, and one or two Gray Flycatchers are present in a typical winter. All other regularly occurring species are transients, most of which move through in spring, each at a slightly different time and in different numbers (see Fig. 55).

OLIVE-SIDED FLYCATCHER

Contopus cooperi (Nuttall, 1831) subspp.

Uncommon spring transient (late April to early June); casual fall transient (September). A denizen of boreal forests, both to the far north and at higher elevations, the Olive-sided Flycatcher is an uncommon migrant through most of the Southwest, where it is decidedly more common in spring than in fall. Spring records for the Salton Sea region extend from 22 April (1908, Mecca; W. P. Taylor, MVZ 891) to 2 June (2002, Wister; *NAB* 56:357). Just southeast of the region a female 11 km east of Cerro Prieto 14 June 1928 (Grinnell 1928; MVZ 52925) was exceptionally late, as was a bird near El Centro 9 June 2002 (*NAB* 56: 357). On average, one encounters this species on one of every two visits during May, but with peak counts of only two or three birds. The Olive-sided Flycatcher is an uncommon to rare fall migrant through the desert Southwest, and it is virtually unrecorded in the Salton Sink in that season (*contra* Garrett and Dunn 1981:55). Individuals 3 km southeast of El Centro 29 August–14 September 2002 (G. McCaskie), at Mexicali 1 September 2000 (Patten et al. 2001), Wister 8 September 2001 (P. A. Ginsburg), and the south end 10 September 1994 and 21 September 1996 (both G. McCaskie) provided the only five fall records.

ECOLOGY. This species frequents exposed snags at or near the crowns of tall trees. Like all pewees, and unlike *Empidonax* flycatchers, the Olive-sided Flycatcher typically returns to the same perch following a sally.

TAXONOMY. Despite average differences in measurements between populations in eastern and western North America (Oberholser 1974: 989), the Olive-sided Flycatcher generally has been regarded as monotypic since Wetmore's (1939) analysis (AOU 1957; Mengel 1963; Phillips et al. 1964; Traylor 1979a; Unitt 1984). However, birds breeding in southern California and northern Baja California, *C. c. majorinus* Bangs and Penard, 1921, are diagnosably larger (Todd 1963:487; Rea 1983; *contra* Monson and Phillips 1981). Bill and tarsus measurements broadly overlap, but in unworn males the wing chord measures at least 111 mm, and the tail, at least 74 mm, and in unworn females the wing chord measures at least 104 mm, and the tail at least 70 mm (Todd 1963:487–88). *C. c. majorinus* reportedly tends to be more extensively dark ventrally, but the difference is slight.

C. c. cooperi is the predominant subspecies migrating through southeastern California in spring (cf. Rea 1983:190). Three of five study skins from the Salton Sink are of *C. c. cooperi,* of a female found 3 km southwest of Niland 18 May 1962 (SBCM 36022; wing 101.7 mm, tail 66.6 mm) and males found 7 km northwest of Imperial 30 April 1988 (SDNHM 45265; wing 105.0 mm, tail 68.2 mm) and 12 km northwest of Calipatria 13 May 1937 (MVZ 71510; wing 103.7 mm, tail 70.5 mm). Likewise, two spring migrants from just east of the Salton Sink, a male from the Chocolate Mountains 3 May 1990 (SBCM 52485; wing 106.7 mm, tail 69.0 mm) and a female 3 km southeast of Palo Verde 10 May 1959 (SBCM 36021; wing 99.0 mm, tail 63.2 mm), are *C. c. cooperi,* as is the late female cited above from near Cerro Prieto (wing 98.9 mm, tail 66.5 mm). As expected on the basis of its much smaller range, *C. c. majorinus* is less frequent, with only two records: females were at Mecca 29 April 1908 (W. P. Taylor, MVZ 1088; wing 105.6 mm, tail 72.3 mm) and

11 km northwest of Imperial 4 May 1996 (P. Unitt, SDNHM 49532; wing 106.1 mm, tail 69.8 mm).

GREATER PEWEE

Contopus pertinax pallidiventris Chapman, 1897

One fall record. The only record of the Greater Pewee for the Salton Sink was the first for California. One collected near the mouth of the Alamo River 4 October 1952 (Cardiff and Cardiff 1953b; Roberson 1993) is now a life mount on public display at the San Bernardino County Museum (SBCM M1907). The Greater Pewee has been recorded just east of the region at Brock Research Center 29 September 1965 (Dunn 1988; LACM 60645), 28 October 1972–17 February 1973 (Winter and McCaskie 1975), and 24 December 1998–27 February 1999 (Erickson and Hamilton 2001; photograph in *NAB* 53:209). Two spring reports from the Salton Sink were insufficiently documented to establish California's first records in that season (Garrett and Singer 1998; Rottenborn and Morlan 2000).

TAXONOMY. California specimens are the gray, pale *C. p. pallidiventris*. This subspecies is the northernmost, breeding north to southeastern Arizona.

WESTERN WOOD-PEWEE

Contopus sordidulus Sclater, 1859 subspp.

Common spring transient (mainly late April to early June); fairly common fall transient (August to mid-October). The Western Wood-Pewee is one of the most common migrant passerines in the deserts of southern California. It generally does not arrive in spring until the second half of April, but there are several earlier records for the Salton Sea region (*AB* 37:911, 45:497), one for the exceptionally early date of 3 April (1991, 7 km northwest of Imperial; R. Higson, SDNHM 47305). Spring records extend to 16 June (1990, Salton Sea NWR; M. A. Patten), with peak passage during the first half of May (e.g., 30 at the south end 7 May 1988; G. McCaskie). The Western Wood-Pewee is rarer in fall, when records extend from 1 August (2002, Wister; G. McCaskie) to 13 October (2000, Wister; G. McCaskie). Most are from late August to mid-September, and daily counts seldom exceed two individuals.

ECOLOGY. This species prefers tall trees, native or nonnative, often perching conspicuously on exposed snags at or, more typically, below the crown. Unlike most *Empidonax* species, the pewees forage boldly. Extensive sallies are punctuated by deep aerial dips, with the bird often returning to the same perch.

TAXONOMY. Two subspecies are generally recognized north of the Mexican border, the dark *C. s. saturatus* Bishop, 1900, breeding in the Pacific Northwest, and the paler *C. s. veliei* Coues, 1866, breeding elsewhere in the species' North American range. The names *C. s. siccicola* and *C. s. amplus,* proposed by Burleigh (1960), were synonymized by Mayr and Short (1970), Browning (1977a), Rea (1983), and Behle (1985). Johnson (1965) synonymized even *C. s. saturatus,* though subsequent authors have continued to recognize it. Both *C. s. saturatus* and *C. s. veliei* should be expected to pass through the Salton Sink during migration, but our distinguishing them is hampered by a lack of material from the breeding range of *C. s. saturatus.* Six spring specimens from the Salton Sink, as dark as or darker than two from northwestern Oregon, were taken from 30 April (1989; SDNHM 47421) to 10 June (1989; SDNHM 47533). Three paler specimens were taken from 3 April (1989; SDNHM 47305) to 17 May (1989; SDNHM 47506). The apparent pattern of the more northern population's passing through later in the spring is exemplified by many other species. We cannot classify four fall juveniles because of a lack of juveniles known to be *C. s. saturatus.* Nevertheless, two earlier specimens, collected 25 August 1984 (SDNHM 43271) and 6 September 1992 (SDNHM 1992), are relatively pale, whereas two later ones, from 23 September 1989 (SDNHM 47608) and 24 September 1988 (SDNHM 45333), are conspicuously darker, especially on the crown.

ALDER FLYCATCHER

Empidonax alnorum Brewster, 1895

One fall record. Because of the tremendous difficulty involved in identifying the sister species in the Traill's Flycatcher complex (*E. traillii / alnorum*), little is known about patterns of vagrancy of the Alder Flycatcher in western North America. California has but four acceptable records of the Alder (Patten, Finnegan, et al. 1995; P. Unitt), although it may well occur regularly. The single record for the Salton Sea region is of a juvenile male 2 km west of Westmorland 28 September 1991 (R. Higson; SDNHM 47934).

WILLOW FLYCATCHER

Empidonax traillii (Audubon, 1828) subspp.

Common spring (mid-May through June) and fall (August to early October) transient. Among the species of *Empidonax,* the Willow and Western Flycatchers are easily the most common migrants in the Salton Sea region (see Fig. 55), mirroring their status as migrants elsewhere in the Southwest. The Willow Flycatcher is slightly more numerous in spring than in fall, with records extending from 5 May (1999, southwest of El Centro; B. Miller) to 23 June (1998, Salton Sea State Recreation Area; *FN* 52:504). It is a late migrant everywhere in the West, so despite a few early and mid-May records, most movement takes place during the last week of May and the first half of June. For example, dates for about 20 specimens range from 12 May (1937, 12 km northwest of Calipatria; J. M. Linsdale, MVZ 71491) to 19 June (1989, 7 km northwest of Imperial; R. Higson, SDNHM 47319), with all but the earliest from 20 May and later. One reported at the Salton Sea 23 April 1972 (*AB* 26:808) is undoubtedly in error as this species is virtually unknown in California before the first week of May. Fall records extend from 6 August (1994, near the mouth of the Alamo River; G. McCaskie) to 9 October (1977, 2 at the south end; G. McCaskie), with most birds passing through from late August to mid-September. Adults have been recorded as late as 27 August (1989, 7 km northwest of Imperial; R. Higson, SDNHM 47661); all later migrants are juveniles.

ECOLOGY. Migrant Willow Flycatchers can be found in any shrub or tree, native or not, throughout the region, but they prefer mesic areas. Thus, they are likely to be seen in scrubby riparian growth along ditches, around pools, and along the edges of ponds and lakes, as well as in well-watered yards and parks in residential areas.

TAXONOMY. Almost all Willow Flycatchers recorded in the region, spring or fall (MVZ, SBCM, and SDNHM specimens), are of the dark-backed, dark-crowned northwestern subspecies *E. t. brewsteri* Oberholser, 1918. *E. t. brewsteri* breeds as close as the Sierra Nevada but now occurs primarily in the Pacific Northwest. It migrates commonly through the breeding range of the pale southwestern *E. t. extimus* Phillips, 1948 (Unitt 1987). A specimen from 7 km northwest of Imperial 19 June 1989 (R. Higson; SDNHM 47319), still a late spring migrant on that date, is the only evidence for *E. t. extimus* occurring in migration away from its breeding range anywhere in the United States, claims to the contrary by Yong and Finch (1997) being unsupported. Like the Gray Vireo, *E. t. extimus* apparently does not normally stop on migration between its nesting sites and the international border. Even though the Colorado River was the original core of its range, there is no basis for supposing that this endangered subspecies (United States Fish and Wildlife Service in 1995) ever nested in the Salton Sink, although it formerly bred along the Río Hardy not far from Cerro Prieto (Patten et al. 2001).

LEAST FLYCATCHER

Empidonax minimus (Baird and Baird, 1843)

Two records. The Least Flycatcher is a rare annual fall and winter vagrant to California; it has bred in the state once, occurs sparsely in spring, and is scarce away from coastal locales. It has been recorded only twice in the Salton Sink. An immature male was collected in mesquite 10.5 km southwest of Brawley 31 December 1986 (*AB* 41:330; SDNHM 44572), and one was photographed 3 km southeast of El Centro 20 September 2002 (K. Z. Kurland). Of note is a record just east of the region at Brock Research Center 28 November 1978–17 February 1979 (*AB* 33:216, 315).

HAMMOND'S FLYCATCHER

Empidonax hammondii (Xántus, 1858)

Uncommon (occasionally fairly common) spring transient (late March to mid-May); 5 fall records (October); 1 winter record. Elsewhere in the California desert Hammond's Flycatcher is a fairly common transient, but it is much less common in the Salton Sea region. However, this species shows the same pattern in the region as it does elsewhere in the desert in that it is more common in spring than in fall, with occasional large concentrations (e.g., 20 at El Centro 5 May 1990; M. A. Patten). Spring records extend from 25 March (1989, 13 km south of Seeley; R. Higson, SDNHM 47199) to 22 May (1994, Finney Lake; M. A. Patten). Hammond's Flycatcher is an uncommon fall migrant through the California deserts. Surprisingly, there are but five fall records for the Salton Sink, from 28 September (1996, Salton Sea NWR; S. von Werlhof) to 18 October (1964, Whitewater River delta; E. A. Cardiff, SBCM 36067), the others being from Westmorland 2 October 1948 (E. A. Cardiff; SBCM 36048), 3 km north-northwest of Seeley 6 October 1986 (P. Unitt; SDNHM 44435) and the south end 9 October 1994 (M. A. Patten). A first-spring male at Thermal 19 March 1921 (FMNH 140923), a date plausible for an early spring migrant, was in heavy prealternate molt and thus must have wintered locally (Johnson 1970).

ECOLOGY. Migrant Hammond's Flycatchers are most often noted in Athel, Fremont Cottonwood, and mesquite, but they can be found in almost any broadleaf or coniferous tree, native or not.

GRAY FLYCATCHER

Empidonax wrightii Baird, 1858

Uncommon transient (mid-April to mid-May, late August to early October); rare winter visitor (mid-November to mid-March). Unlike other species of *Empidonax* in the Salton Sea region, the Gray Flycatcher is almost as frequent a winter visitor as it is a transient. In fact, the first regional records are of two at Mecca 5 January 1911 (van Rossem 1911). Winter records extend from 10 No-

vember (2001, 1 returning to Wister for its third winter; G. McCaskie et al.) to 23 March (1911, Mecca; van Rossem 1911) and average about one a year. This species is more numerous and regular in winter in desert scrub surrounding the Salton Sink, particularly east of the Imperial Valley and west toward the Anza-Borrego Desert. This species is an uncommon migrant in spring from 22 April (1956, northwest of Westmorland; E. A. Cardiff, SBCM 36079) to 14 May (1954, northwest of Westmorland; E. A. Cardiff, SBCM M1411) and in fall from 31 August (1986, 3 km north-northwest of Seeley; P. Unitt, SDNHM 44401) to 6 October (1986, 3 km north-northwest of Seeley; P. Unitt, SDNHM 44615).

ECOLOGY. Migrant Gray Flycatchers are found wherever there are trees. Wintering birds are most often found in natural habitat, particularly mesquite thickets and washes lined with Blue Palo Verde.

DUSKY FLYCATCHER

Empidonax oberholseri Phillips, 1939

Rare spring transient (mid-April to mid-May); 1 or 2 fall records. Even though the Dusky Flycatcher is a fairly common breeder in most of the higher mountain ranges in southern California and an uncommon migrant in the Mojave Desert of California, it is seldom recorded in the Salton Sea region. Records extend from 21 April (1962, north of Niland; E. A. Cardiff, SBCM 36076) to 16 May (1999, Sheldon Reservoir; M. A. Patten). All are from the Imperial Valley, save for two specimens from Mecca, a male collected 26 April 1908 (W. P. Taylor; MVZ 1047) and a female taken 29 April 1908 (C. H. Richardson Jr.; MVZ 1048). An extremely early spring migrant was taken just west of the Salton Sink about 6 km northwest of Ocotillo 5 April 1933 (I. N. Gabrielson; USNM 590462); this bird probably wintered locally. The Dusky Flycatcher is extremely rare in fall in the California desert. The one unequivocal fall record for the Salton Sink is of a juvenile female 12 km northwest of Imperial 15 September 1991 (R. Higson; SDNHM 47776). An *Empidonax* flycatcher suspected to belong to this species was at the south end 27 October 1973 (J. L. Dunn; G. McCaskie).

ECOLOGY. All Dusky Flycatcher records are from scattered patches of trees, principally ornamental. The bird at Sheldon Reservoir was in a patch of native Blue Palo Verde.

TAXONOMY. This species is monotypic, *E. o. spodius* Oberholser, 1974, being a synonym (Browning 1974).

WESTERN FLYCATCHER

Empidonax difficilis difficilis (Baird, 1858)

Fairly common spring transient (mid-March to early June); uncommon fall transient (mid-August to mid-October); 2 winter records. The Western Flycatcher is the dominant migrant *Empidonax* throughout California and much of the Southwest. In the Salton Sink its spring migration is long, extending from 15 March (1999, 3 km southeast of El Centro; K. Z. Kurland) to 12 June (1989, 7 km northwest of Imperial; R. Higson, SDNHM 47384), with a peak from late April through mid-May (e.g., 30 in the Imperial Valley 16 May 1999; M. A. Patten). Numbers are much lower in fall, when records extend from 6 August (2002, 3 km southeast of El Centro; K. Z. Kurland) to 24 October (2002, Wister; M. San Miguel), with a peak from late August through mid-September. Summer stragglers were near the mouth of the New River 1 July 2000 (G. McCaskie) and at Salton Sea NWR 6 July 1996 (G. McCaskie). The Western Flycatcher has twice wintered in the Imperial Valley, at Wister 14 December 1999–1 March 2000 (*NAB* 54:221) and 10 km east of Calipatria 11–20 December 1999 (*NAB* 54:221).

ECOLOGY. Migrant Western Flycatchers use just about any tree or shrub in the region but seem partial to Athel and various broadleaf ornamentals planted throughout the area (including orchard trees).

TAXONOMY. The systematics and taxonomy of the Western Flycatcher (*Empidonax difficilis/occidentalis*) complex continue to be debated hotly (Johnson 1980, 1994a; Phillips 1986:xxviii, 1991:l, 1994a; Johnson and Marten 1988; Howell and Cannings 1992; Gilligan et al. 1994;

Howell 1999:6). At present two species are recognized (AOU 1998). However, the picture painted by Johnson (1980) in his impressive monograph on the systematics of this group is not so clear. First, the range of the Pacific-slope Flycatcher (*E. d. difficilis*) extends much farther east (see below) than Johnson showed, suggesting either a far greater area of sympatry or a broader hybrid zone. Second, contact calls reported as diagnostic by Johnson (1980) apparently are not always so (Campbell et al. 1997:78; J. Morlan *in litt.*) since male Pacific-slope Flycatchers can utter either contact call. Worse yet, not even the full song, but only the third note, is diagnostic! The Pacific-slope Flycatcher breeds from montane Baja California and southern California in and west of the Peninsular Ranges and the Sierra Nevada. Its range extends north throughout the Pacific Northwest, then east to western Montana (K. L. Garrett; C. A. Marantz pers. comm.) and probably southeastern Oregon (Gilligan et al. 1994).

The Cordilleran Flycatcher, *E. d. hellmayri* Brodkorb, 1935, breeds in the Great Basin ranges as far west as the Modoc Plateau, of northeastern California, and the various "sky islands" of western and southern Arizona. This taxon has also been attributed to Clark Mountain (Johnson 1995a), although the basis for this claim is unknown. There are a few reports from eastern California of Western Flycatchers giving the full song of the Cordilleran Flycatcher, such as from the White Mountains (*AB* 44:497) and the east slope of the Sierra Nevada in the Owens Valley (T. Heindel and J. Heindel, pers. comm.). Thus, the Cordilleran Flycatcher may be a scarce migrant through the California deserts, although there are no records from there. If properly sexed, a female from Seven Wells, on the Río Alamo about 40 km east-southeast of Mexicali, 16 April 1894 (F. X. Holzner; USNM 133714) is *E. d. hellmayri* (wing chord 68.5 mm, tail 59.2 mm). However, this subspecies does not arrive on its breeding grounds in southern Arizona until 7 May (Phillips et al. 1964:88), so we suspect that the specimen is missexed; its measurements fit a male *E. d. difficilis*. Despite a claim from Arizona

(Monson and Phillips 1981), the larger, duller, grayer *E. d. insulicola* Oberholser, 1897, an endemic breeder of the Channel Islands (Johnson 1980), is unrecorded on the California mainland. Thus, regardless of the species-level taxonomy followed, based on contact calls or specimen measurements all records of the Western Flycatcher for the Salton Sink pertain to *E. d. difficilis*.

BLACK PHOEBE
Sayornis nigricans semiatra (Vigors, 1839)
Common winter visitor (mid-August through March); fairly common breeder. The Black Phoebe is conspicuous and common at lower elevations throughout the Southwest wherever there is water. In the Salton Sink this species is mainly a winter visitor from mid-August through March. Exact arrival and departure dates are impossible to determine because of the ever increasing breeding population. Breeders were unknown from the area until Cardiff (1950) discovered the species nesting in the Imperial Valley. There is now a substantial breeding population, although count data are lacking. Breeding often occurs early in spring; for example, a pair at Unit 1 had already fledged two young by 18 April (1992; M. A. Patten). Some pairs produce second broods or simply breed later (e.g., a fresh juvenile at the Whitewater River 17 July 1999; M. A. Patten). Although the Black Phoebe greatly outnumbers Say's Phoebe as a breeder in the region, winter counts are often roughly equal (i.e., there is a larger winter influx of Say's relative to the Black).

ECOLOGY. Black Phoebes are seldom found far from water, preferring running or still fresh water. Breeders require suitable vertical faces, generally shaded with a tight overhang, and a supply of mud to construct their nests. Concrete-lined drainage culverts provide a favored nesting substrate, particularly if they are located over slow-flowing water.

TAXONOMY. Black Phoebes in the United States (and much of Mexico) are of the northern subspecies *S. n. semiatra,* characterized by its wholly white undertail coverts.

EASTERN PHOEBE
Sayornis phoebe (Latham, 1790)
Casual winter vagrant (late November through mid-March); 2 fall records. The Eastern Phoebe is a rare fall and winter vagrant to California from the eastern United States. Following a record of one at Wister 11 January–23 February 1952 (*AFN* 6:214), about 17 Salton Sea records have fit this pattern of occurrence, although only 2 are from fall. Winter records extend from 26 November (1970, Ramer Lake; *AB* 25:109) to 19 March (1988, south end from 23 January; *AB* 42:322) and include an adult female collected along the Alamo River 7.5 km northwest of Calipatria 18 December 1988 (R. Higson; SDNHM 46504) and a bird photographed near Brawley 15 January–6 March 2000 (*NAB* 54:221). Fall records are of individuals at Brawley 12–14 October 1986 (*AB* 41:145) and 4.5 km south of Westmorland 31 October 1970 (*AB* 25:109; SBCM 36073).

ECOLOGY. This species occurs in mesic habitats similar to those used by the Black Phoebe, including river courses and edges of canals and ponds with suitable perches (whether shrubs, low trees, low posts, or fence lines).

SAY'S PHOEBE
Sayornis saya (Bonaparte, 1825) subspp.
Common winter visitor (mid-September to early April); uncommon breeder. Say's Phoebe is common in winter in agricultural areas of the Coachella, Imperial, and Mexicali Valleys, as it is throughout the Sonoran Desert. Small numbers of breeding residents cloud the determination of migration dates, but wintering birds occur largely from mid-September (e.g., 6 in the Imperial Valley 15 September 2000; G. McCaskie) to early April. Firm departure dates are unknown because worn spring specimens cannot be identified to subspecies. Birds have been recorded as late as 6 May (1909, 9 km west of Imperial; F. Stephens, MVZ 8195) away from (currently) known breeding locales. Say's Phoebe is an uncommon breeder at the Salton Sea, preferring the arid country at the edges of the region over the irrigated valley floors. Salton City, Brawley, and the southern Coachella Valley are

principal sites. It may breed at San Sebastian Marsh, where the species was noted 25–31 May 1997 (R. P. Henderson). Nesting begins early; for example, a pair was nesting west of Westmorland by February (R. Higson), and a pair was feeding young at Indio 12 March 1942 (C. A. Harwell). Birds fledge by June or early July. Numbers are seemingly lower from June through August, suggesting that postbreeding dispersal carries individuals away from the Salton Sink, a phenomenon noted elsewhere in the Sonoran Desert (Monson and Phillips 1981; Rea 1983).

ECOLOGY. This species readily occupies agricultural habitats, using substrata from power lines to lower posts as perches from which to sally for flying insects. Birds frequently occur in open desert scrub, as this species is far less tied to water than is the Black Phoebe. Breeders, however, are more restricted, with all documented nesting on buildings or other structures, typically under eaves or in similar shaded areas 2–3 m above the ground.

TAXONOMY. Most Say's Phoebes occurring in the region are winter visitors from the north of the darker, browner nominate subspecies, of which *S. s. yukonensis* Bishop, 1900, is a synonym (Browning 1976). The local breeding population, however, is evidently the paler, grayer *S. s. quiescens* Grinnell, 1926. Whereas there are no specimens indisputably representing the breeding population, two juveniles from Laguna Hanson in the Sierra Juarez, a short distance to the southwest (SDNHM 9529, 31854), are clearly *S. s. quiescens,* as is the breeding population along the lower Colorado River (Rea 1983). Furthermore, specimens from 12 km northwest of Imperial 4 October 1987 (SDNHM 45100) and from the "Salton Sea" 14 February 1940 (SDNHM 23172) are of *S. s. quiescens,* as are a few others from just outside the region (e.g., Palm Springs, 14 January 1980; SDNHM 610). We suspect that this subspecies disperses short distances from desert habitats where it breeds into the agricultural areas covering most of the region (cf. Willett 1934).

VERMILION FLYCATCHER
Pyrocephalus rubinus flammeus van Rossem, 1934
Rare winter visitor (mid-September through March); rare breeder in the Imperial Valley; formerly a fairly common breeder in the Coachella Valley. The Vermilion Flycatcher has undergone a significant and interesting range shift during the past five decades (Phillips 1968). It was formerly a widespread and fairly common breeder in the Sonoran Desert, including the Imperial and Coachella Valleys (Grinnell and Miller 1944; Garrett and Dunn 1981), and along the lower Colorado River (Grinnell 1914) but absent from elevations greater than 150 m (Grinnell and Miller 1944). It was particularly numerous in the Coachella Valley (van Rossem 1911; Hanna 1929, 1936; Stevenson 1932), with "over a dozen within a few hours on several occasions" (Hanna 1935). The species has now been virtually extirpated as a breeder in the Sonoran Desert of California (Garrett and Dunn 1981; Rosenberg et al. 1991), but it has colonized a dozen oases in the Mojave Desert, all of which are at elevations well above 150 m.

This gaudy flycatcher is mainly a summer visitor to the southwestern United States, although a few winter in the Sonoran Desert and in cismontane southern California. Its status in the Coachella Valley through the 1950s and in the Imperial Valley through the early 1920s reflected this pattern. At that time birds arrived on their breeding grounds by mid-February and typically departed by late September, with an adult male 5 km north of Calexico 14 October 1921 (UCLA 5952) being the latest. Nesting took place as early as 3 March (1935, nest with 2 eggs; Hanna 1935), with fledglings by 25 March (1934; Hanna 1935). Breeding continued through April, as evidenced by nests with 3 eggs 15 April 1928 (Hanna 1929) and 16 April 1933 (Thermal; PSM 15309) and a female carrying eggs collected west of Coachella 15 April 1908 (Stevenson 1932; MVZ 828). Breeding activity presumably persisted into early July. This species no longer breeds in the Coachella Valley, but 9 pairs bred on the eastern fringe of the Imperial Valley, along the Highline Canal north of Holtville, in

1984 and 1985, and 11 pairs bred there in 1995 and 1996. An adult male 10 km east-northeast of Calipatria 21–22 April 1999 (B. Mulrooney) hints at persistent breeding along that canal. It may also breed at the golf course along the New River in Mexicali, where up to four were noted 6–20 August 1995 (Patten et al. 2001).

The Vermilion Flycatcher winters regularly in small and declining numbers in the Imperial and Mexicali Valleys, from 7 September (1999, female at Brawley; B. Mulrooney) to 21 April (2001, 2 at Wister; S. Koonce). It occasionally winters in the Coachella Valley, dating from records from Mecca 21 December 1909 (MCZ 318609) and 3 January 1911 (van Rossem 1911) and from Thermal 27 January 1918 (UCLA 11966). This change in status, from chiefly a breeder to chiefly a winter visitor, has been documented in other parts of its Sonoran Desert range (Rea 1983:191).

ECOLOGY. When it formerly bred, this species was associated with low-lying, open riparian areas and ranch lands with surface water and dominated by Honey Mesquite, Goodding's Black Willow, and Fremont Cottonwood. Wintering birds occupy cemeteries, ranch lands, yards, and cattle pastures with trees.

TAXONOMY. California birds are of the western Mexico subspecies *P. r. flammeus,* the palest subspecies; indeed, its type locality is Brawley. It differs from *P. r. mexicanus* Sclater, 1859, of Texas south through eastern Mexico, in being paler and grayer above, with the adult male's underparts mottled orange-red rather than unmottled deep red.

DUSKY-CAPPED FLYCATCHER
Myiarchus tuberculifer olivascens Ridgway, 1884
Three winter records. There are three Salton Sink records of the Dusky-capped Flycatcher, a rare fall and winter vagrant to California. Calling birds were observed along the Alamo River 7 km northwest of Calipatria 22–30 December 1987 (Pyle and McCaskie 1992), photographed at Finney Lake 12 January–28 March 1998 (Erickson and Hamilton 2001), and observed along the

New River 7 km southwest of Brawley 23 March 2000 (*NAB* 54:221). These records fit nicely within the temporal pattern of occurrence elsewhere in California, but are geographical outliers: the vast majority of records (more than 90% of more than 50) are from the coastal slope. Prior claims for the region (California Department of Fish and Game 1979) are baseless.

TAXONOMY. The only specimen for California, from Furnace Creek Ranch in Death Valley, Inyo County, 23 November 1968 (LACM 66519), is the geographically expected northwestern subspecies *M. t. olivascens* (Suffel 1970), which presumably accounts for all other California records.

ASH-THROATED FLYCATCHER
Myiarchus cinerascens Lawrence, 1851
Fairly common transient (mid-March to mid-May, mid-July to mid-October); uncommon to rare, local breeder in the Coachella Valley; rare winter visitor. Despite breeding commonly through a wide range of habitats in the West, the Ash-throated Flycatcher nests regularly only in a limited area around the Salton Sea (the Coachella Valley, where it is local and uncommon to rare). The species may nest sporadically in the Imperial Valley, where three to four summered in 2000 (G. McCaskie): one occupied mature Fremont Cottonwood riparian woodland mixed with mesquite along the New River about 4 km east of Fig Lagoon 1–6 June, at most two were in similar habitat along this river about 5 km southwest of Brawley 6–17 June, and another was near the mouth of the New River 17 June. Three birds apparently summered in the valley in 2002 (*NAB* 56:487). The Ash-throated Flycatcher is a fairly common spring migrant through the Salton Sink from 24 March (1942, 2 at the south end; C. A. Harwell) to 26 May (1996, 2 at the south end; G. McCaskie). Fall migrants appear early, beginning 8 July (1972, south end; G. McCaskie); migration peaks from August through mid-September, with stragglers as late as 16 October (1987, 10 km west-northwest of Imperial; P. Unitt, SDNHM 44841). This species winters in the Imperial Valley once

every two to three years, with about 16 records from 13 November (1967, northwest of Niland; E. A. Cardiff, SBCM 36000) to 23 February (1985; *AB* 39:211). It winters regularly in desert scrub just outside the region, especially to the east (Garrett and Dunn 1981; Rosenberg et al. 1991; SDNHM specimens).

ECOLOGY. Breeding Ash-throated Flycatchers occupy dense stands of mature mesquite with scattered Fremont Cottonwood or ranch lands supporting cottonwood and other large trees. Considering the paucity of primary excavators, breeding Ash-throateds may be limited by a paucity of suitable nest cavities. Migrants can be found in any sort of vegetated habitat. Wintering birds are most often found in more xeric desert scrub, usually with mesquite, Ironwood, or Blue Palo Verde.

TAXONOMY. Two subspecies are frequently recognized (AOU 1957; Traylor 1979a), the grayer, larger nominate subspecies, widespread in the West, and the browner, yellower, smaller *M. t. pertinax* Baird, 1859, resident in southern Baja California. Like Lanyon (1963), we could not find consistent differences in plumage (SDNHM specimens). Unlike Lanyon (1961), we judge the average length of the wing, tail, and bill to be insufficiently different at the 75 percent level ($D_{ij} < -3.5$ for each character in both directions based on data in Lanyon 1961:table 2).

BROWN-CRESTED FLYCATCHER
Myiarchus tyrannulus magister Ridgway, 1884
Extremely rare breeder (May through August) in the Coachella Valley; 1 record for the Imperial Valley. The Brown-crested Flycatcher is a rare, local breeding visitor to southeastern California, where it has expanded its range in the past half-century (Johnson 1994b). Now there are probably 15–20 pairs breeding in California in any given year, although numbers along the lower Colorado River were once larger (Rosenberg et al. 1991). Regular breeding stations west of the Colorado River valley are Morongo Valley, Fort Piute, Mojave Narrows, Tecopa, and the South Fork of the Kern River. At each of these locales this species occupies riparian forest with numerous large Fremont

Cottonwoods and a dense understory of willows or mesquite.

Given these strict habitat requirements and its local distribution, it is surprising that this species reached the southern Coachella Valley as a breeder in 1978 and again from 1996 to 1998. A pair was observed near Mecca in May and June 1978 (*AB* 32:1209). A lone bird 28 April 1980 (*AB* 34:816), perhaps one of this pair, provided one of the earliest return dates for California. The species returned to the same location near Mecca 21 May–3 August 1996 (*FN* 50:997); young were observed on the final date. The adults returned in summer 1997, when they were last seen 24 August (*FN* 51:1054), and in May 1998 (*FN* 52:391). A pair also returned 15 May 1999 (S. Koonce), but their habitat was destroyed soon thereafter to make way for an expressway. The lone regional record away from breeding sites involved a calling bird in a large Fremont Cottonwood alongside the New River near Seeley 8 July 1995 (G. McCaskie).

ECOLOGY. Brown-crested Flycatcher nests near Mecca were in large Fremont Cottonwoods lining freshwater ponds. This habitat is the closest match in the region to their preferred gallery riparian forest elsewhere in southeastern California.

TAXONOMY. California records are of the large, pale form *M. t. magister*, which breeds from southern Arizona south through the length of western Mexico.

TROPICAL KINGBIRD
Tyrannus melancholicus [satrapa]
Cabanis and Heine, 1859
Three fall records (late September through October). The Tropical Kingbird is a rare but regular fall vagrant to California from Arizona or Mexico, chiefly to the coast (Mlodinow 1998b). It has been recorded inland in California only casually, typically in fall (Garrett and Dunn 1981). The three records for the Salton Sink, from near Niland 23 October 1976 (*AB* 31:224), at Finney Lake 1 October 1978 (*AB* 33:216), and near Fig Lagoon 22 September 2001 (G. Hazard), exemplify this pattern.

TAXONOMY. Nearly all California specimens of this kingbird are of the gray-chinned, pale-crowned *T. m. satrapa,* breeding throughout Mexico and north to southeastern Arizona (Traylor 1979b; Monson and Phillips 1981; Phillips 1994c). We infer that birds in the Imperial Valley were the same. However, a Tropical Kingbird taken from Southeast Farallon Island, California, 18–25 August 1973 (PRBO 713) was identified as one of the white-chinned, dark-crowned subspecies of South America, either *T. m. melancholicus* Vieillot, 1819, or *T. m. obscurus* Zimmer, 1937 (Mlodinow 1998b).

CASSIN'S KINGBIRD
Tyrannus vociferans Swainson, 1826
Casual winter vagrant (November–February); 4 fall (August to mid-October) records; 1 spring record. Despite its being fairly common to the northwest, Cassin's Kingbird is a casual fall and winter vagrant to the California desert, with a few spring records. There are 11 records for the Salton Sea region, 8 of them from winter. Winter dates extend from 5 November (2000, Wister; *NAB* 55:104) to 28 February (1975, south end; *AB* 29: 743), although the first two records for the area, of a bird collected at Brawley and another observed at Mecca, were on unreported dates during the winter of 1910–11 (van Rossem 1911). Other winter records are from Niland 19 December 2000–2 February 2001 (*NAB* 55:228), near Calipatria 21 December 1982 (*AB* 37:772), at the north end 29 December 1970 (*AFN* 25:503), and of a presumably returning bird at Finney Lake 29 January 1989 (*AB* 43:367), 17 February 1990 (G. McCaskie), and 19 January–10 February 1991 (*AB* 45:321). Migrants were at Rock Hill 29 May 1971, Salton City 3 August 1968, 3 km southeast of El Centro 22 September 2002 (K. Z. Kurland), and Niland 14–21 August 1976 (Garrett and Dunn 1981) and 15 October 2002 (G. McCaskie).

ECOLOGY. Most regional Cassin's Kingbirds have been observed in tall gum or Athel adjacent to open agricultural fields.

TAXONOMY. This species is monotypic, as *T. v. xenopterus* Griscom, 1934, of Guerrero, Mexico, is insufficiently distinct (Binford 1989:188).

WESTERN KINGBIRD
Tyrannus verticalis Say, 1823
Abundant breeder (early March through September); 1 winter record. The Western Kingbird (Fig. 56) is one of the most common and conspicuous breeding landbirds in the Coachella, Imperial, and Mexicali Valleys. It is one of the earliest summer visitors to return to California, with the first in the Salton Sink by 5 March (1989, 11 km west of Imperial; R. Higson, SDNHM 45987) and exceptionally by 28 February (1945, Holtville; S. G. Jewett, PSM 1031). Numbers build rapidly, with thousands present throughout the region by April. Breeding commences soon after arrival, with some birds on eggs by mid-April (e.g., 18 April 1992 at Unit 1; M. A. Patten) and most on nests by early May. The first young of the year fledge in May. Most depart by early September, the last 12 October (1996, Unit 1; S. von Werlhof) and from 14 to 16 October (2002, Niland; G. McCaskie). In fall 2000 up to 8 at Niland straggled to the unparalleled date of 24 October (*NAB* 55:104). This species is frequently reported in winter in southern California, but there are in fact only a few valid records in that season for anywhere in the western United States. One such record is of a bird carefully studied by numerous observers near Calipatria 18 December 1986 (*AB* 41:1289).

ECOLOGY. This species forages throughout the region's extensive agricultural land, placing its nests on substrata ranging from forks in Fremont Cottonwoods and dense Athel to utility poles and awnings on buildings.

EASTERN KINGBIRD
Tyrannus tyrannus (Linnaeus, 1758)
Casual spring (late May to mid-June) and fall (August–September) vagrant; 1 summer record. The Eastern Kingbird is a rare annual spring and fall vagrant to California, with spring records from late May through June and autumn records from early August to mid-September with a few into October (Manolis 1973). It has been recorded 11 times in the Salton Sea region, 5 times each in spring and fall and once on an anomalous date. Spring records include those of birds

FIGURE 56. The Western Kingbird is abundant and conspicuous throughout the agricultural regions of the Salton Sink in summer. Photograph by Kenneth Z. Kurland, July 2000.

from Finney Lake 26 May 2001 (*NAB* 55:356), near Niland 3 June 1989 (*AB* 43:537), and at the south end 4 June 1973 (*AB* 27:821), one photographed at Wister 14 June 1999 (*NAB* 53:432), and one near Mecca 18 June 1992 (*AB* 46:1179). The earliest fall records are from the south end 4 August 1968 (*AFN* 23:109) and 4 August 1977 (Garrett and Dunn 1981). Subsequent fall records have been on more expected dates, with individuals at Wister 10 September 1994 (G. McCaskie), at the south end 13 September 1964 (*AFN* 19:80), and near the mouth of the New River 21 September 1990 (*AB* 45:152). An Eastern Kingbird at a small lagoon near Obsidian Butte 10 July 1993 (*AB* 47:1151) may have been attempting to summer locally, although it could not be found on later dates.

ECOLOGY. Eastern Kingbirds have occurred in a variety of habitats but mostly those dominated by Saltcedar scrub with larger Fremont Cottonwood, mesquite, or gum intermixed.

SCISSOR-TAILED FLYCATCHER
Tyrannus forficatus (Gmelin, 1789)
One fall record. The Scissor-tailed Flycatcher is a casual straggler to California, with more than

a hundred records from throughout the year. There is, however, only one documented record for the Salton Sink, from Ramer Lake 29 August 1987 (Pyle and McCaskie 1992). Reports from Indio (Tinkham 1949), the Imperial Valley, Calexico, and Dos Palmas Spring (Garrett and Dunn 1981) are not sufficiently documented (Roberson 1993), although the Calexico bird was supposedly "photographed from 30 feet and watched for 30 minutes" (Roberson 1993).

LANIIDAE • *Shrikes*

NORTHERN SHRIKE
Lanius excubitor borealis Vieillot, 1808
Two winter records (late January). The Northern Shrike is a rare bird in California. As its name suggests, it has a distinctly northerly distribution, and in the West it rarely makes it south of the Great Basin during the winter. The two southernmost records for California are of immatures from the Imperial Valley, the first near the mouth of the New River 23–26 January 1971 (*AB* 25: 629), the second photographed near Westmorland 22–29 January 1989 (*AB* 43:367).

TAXONOMY. The only subspecies in North America is *L. e. borealis*, because the westerly *L. e.*

invictus Grinnell, 1900, is a synonym (Phillips 1986; *contra* Lefranc 1997). Relative to all Old World subspecies, *L. e. borealis* has a paler, more contrasting rump, more prominent supercilia, and heavier barring on the underparts.

LOGGERHEAD SHRIKE

Lanius ludovicianus Linnaeus, 1766 subspp.

Fairly common breeding resident; more numerous in winter (mid-July through March). The story of the Loggerhead Shrike in North America is mixed. The species has declined throughout much of its range, especially in the East (Peterjohn and Sauer 1995; Cade and Woods 1997), but it persists in large numbers over much of the West. The Salton Sink and the Central Valley support the highest densities in California; the Loggerhead Shrike is common to fairly common in both regions. Larger numbers occur in the irrigated valleys in winter, when the breeding population is augmented by an influx from the Mojave Desert and the Great Basin (or from the surrounding Sonoran Desert; see Garrett and Dunn 1981), mainly from mid-July through March. Breeding takes place from March (e.g., nest at Mecca 13 March 1908; MVZ 64) through August (e.g., 2 adults with 2 begging fledglings at Salton City 24 August 1991; M. A. Patten), with young hatching as early as mid-March (van Rossem 1911).

ECOLOGY. During the breeding season this species is often scarce in agricultural areas yet fairly common in desert scrub, nesting in thorny shrubs such as mesquite, Quail Brush, and Ironwood. At the Salton Sea this small predator feeds chiefly on Coleoptera (e.g., stomach contents of FMNH 151806, 162495, 162496), but some take Orthoptera and "worms" (e.g., FMNH 151807). It forages in open habitats ranging from desert scrub to agricultural fields with brushy borders. It generally uses short perches and occurs in fields with grassy cover.

TAXONOMY. Two intergrading subspecies occur in the Salton Sink. The medium gray *L. l. gambeli* Ridgway, 1887, with a short supercilium and dusky rump, occurs fairly commonly from the Coachella Valley south to at least the Imperial

Valley. *L. l. excubitorides* Swainson, 1832, of which *L. l. nevadensis* Miller, 1930, and *L. l. sonoriensis* Miller, 1930, are synonyms (Phillips 1986), is uncommon in the Imperial and Mexicali Valleys. It is pale gray above with a long supercilium and a white rump. This picture is complicated by an apparent turnover in subspecies occupying the region. Almost all older specimens (AMNH, LACM, MCZ, MVZ, SDNHM) are *L. l. excubitorides*, save for a few *L. l. gambeli* from the Coachella Valley, where there was presumably a hybrid zone. Loggerhead Shrikes now occupying the Coachella Valley match *L. l. gambeli*, including breeders (e.g., fresh juvenile from Coachella 5 June 1992; LACM 107329). Birds closer to *L. l. gambeli* now predominate in the Imperial Valley. We suspect that irrigation and development of the Coachella, Imperial, and Mexicali Valleys proved favorable to the more mesic-adapted *L. l. gambeli*, which subsequently spread throughout the Salton Sink at the expense of the more xeric-adapted *L. l. excubitorides*. The latter persists in desert scrub east of the sink, with dispersal yielding extensive intergradation. Phillips (1986:77) synonymized *L. l. gambeli* with *L. l. mexicanus* Brehm, 1854.

VIREONIDAE • Vireos

The Warbling and Cassin's Vireos are the only members of the Vireonidae with a significant presence in the Salton Sink, and the former far outnumbers the latter (Fig. 57). Cassin's is mainly a spring migrant through the Salton Sink, but the Warbling occurs in both seasons. The Plumbeous Vireo is a rare annual winter visitor; the other two species are vagrants (and two others have been reported erroneously).

BELL'S VIREO

Vireo bellii pusillus Cooper, 1861

Eight records, 5 in fall, 2 in winter, and 1 in spring. As a result of extensive destruction of riparian habitat and heavy brood parasitism by the Brown-headed Cowbird, Bell's Vireo has undergone serious declines throughout California (Goldwasser et al. 1980). With its designation as an endangered species, restriction of habitat destruction,

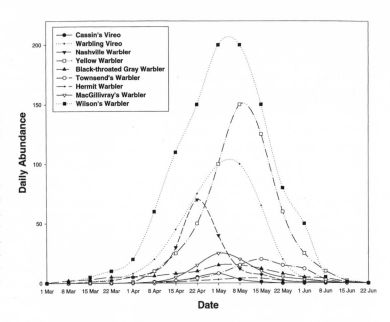

FIGURE 57. General peak abundance and timing of spring transient vireos (Vireonidae) and wood-warblers (Parulidae) through the Salton Sink. Wilson's Warbler is by far the most common landbird migrant through the region, followed by the Yellow Warbler, the Warbling Vireo, and the Western Wood-Pewee (see Fig. 55). Data are adapted from Michael A. Patten's field notes.

and extensive cowbird trapping, populations have recovered in some areas of cismontane southern California since the late 1980s (Brown 1993). Despite the formerly suitable breeding habitat around the Salton Sea and the formerly large population on the Colorado River (Rosenberg et al. 1991), a singing male collected at Mecca 26 March 1911 (UCLA 10697) provides the sole historical record for the region. This specimen was originally published as a Gray Vireo (van Rossem 1911) and was mistakenly treated as such for nearly 90 years (Grinnell and Miller 1944; Garrett and Dunn 1981). Bell's Vireo formerly bred along the Río Hardy southeast of the Mexicali Valley (Grinnell 1928), where specimens were taken 11 km east of Cerro Prieto 2 and 12 June 1918 (MVZ 53002, 53003). There are only seven recent records for the Salton Sink: individuals wintered at Ramer Lake 28 December 1963–8 February 1964 (McCaskie 1968) and at Brawley 10 November–20 December 1986 (AB 41:146, 331), and fall migrants were observed at Salton Sea NWR 22 August 1997 (S. B. Terrill), 19 August 2001 (B. G. Peterjohn), and 26 October 2002 (D. S. Cooper) and at Brawley 21 November 2002+ (G. McCaskie) and photographed at Sheldon Reservoir 24 November 2002 (K. Z. Kurland). One just outside the region

at Willow Hole, North Palm Springs, 24 March 1983 (AB 37:911) suggests the possibility of occasional spring migrants.

TAXONOMY. Two subspecies of Bell's Vireo breed in California: the Least Bell's Vireo, V. b. pusillus Cooper, 1861, designated endangered by the United States Fish and Wildlife Service, and the Arizona Bell's Vireo, V. b. arizonae Ridgway, 1903, designated endangered by the California Department of Fish and Game. V. b. arizonae is smaller in the bill, wing, and tail and more brightly colored, with a more olive (less gray) mantle and more yellow on the flanks (Ridgway 1904; Phillips 1991). The two subspecies may differ in song and habitat (L. R. Hays pers. comm.). The Mecca specimen is of V. b. pusillus. Without specimens the subspecies responsible for other records cannot be determined.

The ranges of V. b. pusillus and V. b. arizonae in California are based more on supposition than on direct evidence. Specimens from Death Valley have been identified as V. b. pusillus (Ridgway 1904; Grinnell 1923), but were taken in spring, when plumage can be considerably worn (Unitt 1985), thus confounding identification. Examination of specimens at LACM and SDNHM (M. A. Patten) and elsewhere (Allan R. Phillips in litt., Ned K. Johnson in litt.) indicates that available specimen evidence for defining the eastern extent of the range of *V. b. pusillus* is weak. It is generally believed that *V. b. arizonae* is confined to the lower Colorado River valley in Nevada and California, whereas *V. b. pusillus* occurs in cismontane southern California and on the western edge of the deserts, extending north into the Owens Valley and east into Death Valley, the Amargosa River valley, Fort Piute in the eastern Mojave Desert, and into the northern fringe of the Coachella Valley (Goldwasser 1978; Goldwasser et al. 1980; Franzreb 1987, 1989; Brown 1993; Small 1994). A specimen from Ash Meadows in western Nevada (USNM 137709) is even claimed to be *V. b. pusillus* (Linsdale 1936). Considering the biogeography of similarly distributed cismontane and transmontane species pairs (Grinnell and Miller 1944; Garrett and Dunn 1981), such as the *Callipepla* quails, Nuttall's and Ladder-backed Woodpeckers, and California (*Toxostoma redivivum*) and Crissal Thrashers, it is probable that *V. b. arizonae* occurs in the eastern Mojave Desert north through Death Valley. Spring birds in Death Valley and at Fort Piute are more brightly colored, with a greener back and yellower flanks. They are thus more like *V. b. arizonae* than are birds along the Mojave River or at Morongo Valley, which are grayer and thus more like *V. b. pusillus* (M. A. Patten). A 24 January specimen of *V. b. arizonae* from the Anza-Borrego Desert (Unitt 1985; Phillips 1991) shows that this subspecies can occur well west of its described range. The wintering birds, therefore, may have been either *V. b. arizonae* or *V. b. pusillus*. Of note in this regard, specimens cited above from southeast of the Mexicali Valley but well west of the Colorado River are of *V. b. arizonae* (Miller et al. 1957).

PLUMBEOUS VIREO

Vireo plumbeus plumbeus Coues, 1866

Rare winter visitor (October to early April); 3 fall records; 2 potential spring records. An individual of the *Vireo solitarius* complex wintering at the Salton Sea will generally be a Plumbeous Vireo, a widespread breeder in the Great Basin and the intermountain West (Johnson 1995a; Heindel

1996) that has steadily expanded its range southward in the past several decades. This species is annual in winter, with records from 23 September (2001, 3 km southeast of El Centro; K. Z. Kurland) to 10 April (2001, same bird 3 km southeast of El Centro; K. Z. Kurland), including a specimen taken 7 km northwest of Imperial 28 December 1995 (P. Unitt; SDNHM 49419). The bird wintering 3 km southeast of El Centro returned for four consecutive winters. By comparison, there is only one documented winter record of Cassin's Vireo, although the species winters annually in cismontane southern California. There are only three records of migrant Plumbeous Vireos, all from mid-September: 12 km northwest of Imperial 15 September 1991 (P. Unitt), 3 km southeast of El Centro 20 September 2000 (K. Z. Kurland), and at Westmorland 21 September 1975 (G. McCaskie). Considering the species' status elsewhere in the California deserts, small numbers may pass through the Salton Sink each fall, as is apparently the case along the Colorado River (Rosenberg et al. 1991; cf. Monson and Phillips 1981).

Although the Plumbeous Vireo is frequently reported as a spring migrant in southeastern California, there are virtually no physically documented records at that season; almost all reports pertain to dull Cassin's Vireos. Individuals at Salton Sea NWR 27 April 1996 and 8 km west of Westmorland 7 May 1993 (both G. McCaskie) were believed to be Plumbeous Vireos. These sight records are perhaps best considered tentative because there are no physically documented spring records of this species for the Salton Sink or for the lower Colorado River despite the huge numbers of migrant passerines that move through these regions in that season.

ECOLOGY. Most Plumbeous Vireos have been found in stands of planted pines, but some have occurred in mesquite thickets, Fremont Cottonwood, and various nonnative deciduous trees.

TAXONOMY. Because populations of the *Vireo solitarius* complex in southern Mexico and Central America were not studied by Johnson (1995a), their systematic position is unclear. The AOU (1998) treated *V. p. gravis* Phillips, 1991, *V. p.*

montanus van Rossem, 1933, and *V. p. notius* Van Tyne, 1933, as subspecies of the Plumbeous Vireo, even though plumages of the last two more closely resemble Cassin's Vireo (Phillips 1991). The nominate, *V. p. plumbeus,* is larger and grayer than all three taxa. Regardless of taxonomic treatment of southerly populations, *V. p. jacksoni* (Oberholser, 1974), of the Rocky Mountains, is a synonym of *V. p. plumbeus* (Browning 1978), as are *V. p. pinicolus* van Rossem, 1934, and *V. p. repetens* van Rossem, 1939, of western Mexico (Phillips 1991:192).

CASSIN'S VIREO

Vireo cassinii cassinii Xántus, 1858

Uncommon spring transient (mid-March to mid-May); rare fall transient (mid-September to mid-December); 1 winter record. With the split of the Solitary Vireo complex into three species, more attention is being focused on the status of Cassin's, Plumbeous, and Blue-headed Vireos. Cassin's Vireo, a widespread breeder in the Pacific states, is the expected spring transient. Even so, it is uncommon in that season from 20 March (1971, "Salton Sea"; *AB* 25:799) to 19 May (1999, female at Wister; M. A. Patten). Two seen 3 km southeast of El Centro 28 February 2002 and one there 12–16 March 2002 (both K. Z. Kurland) were exceptionally early (or wintered). Seldom are more than one or two seen in a day, but there is a slight peak in late April. Like some other passerines (e.g., the Hermit Warbler), most Cassin's Vireos migrate through the mountains in fall, so the species is rare in the desert in that season. It is especially rare in the Salton Sea region, presumably because most fall migrants moving through the desert are "guided" eastward to the lower Colorado River valley by the Transverse Ranges (see Fig. 57). Records extend from 15 September (1991, Finney Lake; M. A. Patten) to 19 December (2000, Calipatria; A. Howe, V. Howe), with most from late September to mid-October. The species' posited late migration notwithstanding (*AB* 26:656), a Cassin's Vireo at Westmorland 31 December 1971 was likely wintering (Garrett and Dunn 1981).

ECOLOGY. Migrant Cassin's Vireos are mostly noted in large trees, ranging from Fremont Cottonwood to various nonnative species such as Athel, gum, and pine.

TAXONOMY. The nominate subspecies, widespread on the Pacific Coast, is larger than *V. c. lucasanus* Brewster, 1891, of the Cape region of Baja California Sur, and its flanks are more olive (less yellow).

WARBLING VIREO

Vireo gilvus swainsonii Baird, 1858

Common spring (mid-March to early June) and fall (late July through October) transient; 2 winter records. The Warbling Vireo is a common, ubiquitous migrant in the lowlands of the West and is by far the most common vireo in the Salton Sink (see Fig. 57). It still breeds widely in California's foothills. With the Western Kingbird and Wilson's Warbler, this vireo is one of the first migrant passerines (other than swallows) to return to California in spring, with the first returning by 11 March (1999, 3 km southeast of El Centro; K. Z. Kurland) and exceptionally by 26 February (1972, south end; *AB* 26:656), 27 February (2001, near El Centro; *NAB* 55:229), and 5 March (2000, near Fig Lagoon; G. McCaskie). Migration peaks in late April and early May, with sometimes dozens seen in a single day (e.g., 30 at the mouth of the Whitewater River 1 May 1999; M. A. Patten), and continues until early June (e.g., 8 June 1989, 7 km northwest of Imperial; R. Higson, SDNHM 47529). An exceptionally late individual was at Wister 10–18 June 1999 (*NAB* 53:432). This species is likewise an early fall migrant, the earliest being observed 3 km southeast of El Centro 13 July 2002 (*NAB* 56:487) and at Wister 23 July 1988 (M. A. Patten). Fall migration peaks in September with the passage of immature birds. A few occur into mid-October, the latest 1 November (1988, 11 km southeast of Brawley; R. Higson, SDNHM 49379). The species occurs casually in winter anywhere in the United States, with two records in that season for the Salton Sink, from Niland 20 December 1976 (*AB* 31:881) and near Holtville 23 January 1999 (*NAB* 53:210).

ECOLOGY. Migrant Warbling Vireos are encountered anywhere that there is vegetation, but they are particularly common in more mesic settings such as the river courses (where they are especially common in Saltcedar), parks, and suburban neighborhoods.

TAXONOMY. *V. g. swainsonii* Baird, 1858, which is widespread throughout western North America, is the common subspecies recorded in California. This subspecies differs from the nominate, of eastern North America, in being slightly smaller overall, smaller-billed, and duller in coloration. The larger, darker-capped *V. g. brewsteri* Ridgway, 1903, of the Southwest, has been taken in coastal California on two occasions (Phillips 1991:215) but is apparently only a vagrant to the state. Subspecies in the *V. g. swainsonii* group (i.e., those in western North America) generally molt during migration, whereas the nominate subspecies generally molts on its breeding grounds before migration (Voelker and Rohwer 1998). There are slight differences in song (Sibley and Monroe 1990:455), there is about a 3 percent sequence divergence in mitochondrial DNA (Murray et al. 1994), and western birds accept eggs from the Brown-headed Cowbird, whereas eastern birds reject them (Sealy et al. 2000). These differences have led some to suggest that *V. g. gilvus* and the *V. g. swainsonii* groups are distinct species (Sibley and Monroe 1990; Phillips 1991; Voelker and Rohwer 1998).

RED-EYED VIREO
Vireo olivaceus (Linnaeus, 1766)
One fall record. The Red-eyed Vireo is a rare vagrant to California, but there are scores of records in fall from the coast and the northern desert. It is thus surprising that there is only one record for the Salton Sink, but it fits the species' pattern of fall occurrence elsewhere in the interior of California. Details of a sighting at Finney Lake 4 September 1972 (*AB* 27:123) were submitted to the California Bird Records Committee as pertaining to a Yellow-green Vireo, but the written description indicated that the bird involved was a first-fall Red-eyed Vireo (Winter 1973).

CORVIDAE • *Crows and Jays*

Corvids are generally poorly represented in the Salton Sink. Even the ubiquitous Common Raven, which is common and conspicuous throughout the desert Southwest, is uncommon in the Salton Sink. All other species are essentially vagrants, although small numbers of American Crows occur nearly annually in winter in the Coachella Valley.

WESTERN SCRUB-JAY
Aphelocoma californica (Baird, 1858) subspp.
Casual winter visitor (December to mid-March); 1 fall record. The Western Scrub-Jay is common in chaparral and live oaks in cismontane California and somewhat less common in the Great Basin and the Rocky Mountains, generally in pinyon-juniper. Birds from the Great Basin regularly disperse across the Mojave Desert in fall and winter, whereas coastal birds are more sedentary (Peterson 1991). Some of this dispersal brings individuals to the Salton Sea region, where this jay is a casual winter visitor from 20 December (1987, north end; *AB* 42:1132) to 12 April (1951, 2 at Salton Sea NWR from 8 January; *AFN* 5:227, 276), mostly to the Coachella Valley. Most records involve lone birds, but as many as five were at Niland 13 January–17 March 1973 (*AB* 27: 664). The exception to this pattern of winter dispersal is an immature at El Centro 10 August 1989 (R. Higson; SDNHM 45999).

ECOLOGY. On their rare visits to the Salton Sink scrub-jays frequent large trees, favoring Date Palm groves, especially those with an understory of dense scrub.

TAXONOMY. As expected on the basis of dispersal patterns (Peterson 1991), all records but one for the Salton Sink are of Woodhouse's Scrub-Jays from the northeast, easily distinguishable in the field from the California Scrub-Jays from the west by their muted patterning. A specimen from Brawley 27 January–18 February 1990 (M. Fenner; SDNHM 46372) is of the Great Basin subspecies, *A. c. woodhouseii* Baird, 1858, with gray underparts, a pale gray back, and the blue a pale powder blue. *A. c. woodhouseii* was called *A. c. nevadae* Pitelka, 1945, by the AOU (1957) and Behle (1985); they reserved *A. c. woodhouseii* for the eastern population. However, the eastern form was renamed *A. c. suttoni* by Phillips (1964), a move followed by Browning (1978, 1990). The disagreement arose over identification of the type specimen of *A. c. woodhouseii*, assigned by Pitelka (1945) to the darker Rocky Mountain subspecies, though not examined by him (*fide* Browning 1978), but asserted by Phillips (1964) to be the pale Great Basin form (Dickerman and Parkes 1997). *A. c. suttoni* (Phillips, 1964), which breeds as far west as eastern Utah (Behle 1985), might reach the Salton Sea on occasion. It has a darker gray back and the blue a deeper, purer blue than in *A. c. woodhouseii*, and it accounts for most lowland records of dispersing scrub-jays in Arizona (Phillips et al. 1964). It has been reported as far west as the Colorado River, at the Bill Williams Delta, Arizona (Phillips et al. 1964). May Canfield collected one on the California side, 3 km north of Bard, 3 January 1927 (SDNHM 32174), constituting the state's only record, previously unpublished, of this subspecies.

The single exception noted above is the August specimen from El Centro. It is *A. c. obscura* Anthony, 1899, a dark bright blue coastal subspecies with a complete breast band, brownish gray mantle, and white throat contrasting with a grayish belly. *A. c. obscura*, resident in the Peninsular Ranges, reaches desert riparian areas at Morongo Valley, Whitewater Canyon, and elsewhere near the northern edge of the region. It occurs throughout the Peninsular and Transverse Ranges, probably eastward to Eagle Mountain, where the drabber endemic *A. c. cana* Pitelka, 1951, has been described, although that subspecies is likely invalid (Peterson 1990).

PINYON JAY
Gymnorhinus cyanocephalus Wied, 1841 subsp.?
Two winter records, 1 of a flock that remained into summer. A gregarious montane species, the Pinyon Jay seldom wanders from its usual haunts of pinyon-juniper woodland. It has been recorded only twice in the Salton Sink, both times coinci-

dent with major incursions of other montane species. The first record was of a flock of ten photographed near Westmorland 31 December 1987–20 January 1988 (*AB* 42:322), six of which were seen irregularly 14 April–13 July 1988 (*AB* 42:1171). The second record was of a lone individual at El Centro 15 November 1996–16 March 1997 (*FN* 51:119, 804).

TAXONOMY. The Pinyon Jay is generally regarded as monotypic, though Phillips (1986:41) listed three subspecies differing in bill shape. Relative to the nominate subspecies, *G. c. rostratus* Brodkorb, 1936, is a wider-billed subspecies of mountains of southern California and northern Baja California and *G. c. cassini* (McCall, 1852) is a longer-billed subspecies of the Great Basin and the western Rocky Mountains. Without a specimen we cannot determine whether birds at the Salton Sea were from local mountains or the Great Basin.

CLARK'S NUTCRACKER
Nucifraga columbiana (Wilson, 1811)
Two fall records. Clark's Nutcracker is a high-mountain species that wanders irregularly to lowlands (Davis and Williams 1957). The only two records for the Salton Sink are from the southern Coachella Valley during major corvid invasions. A flock of 12 was observed west of Indio 17–18 October 1919 (Esterly 1920), and one was seen at Coachella 24 September 1935 (Clary and Clary 1936a).

AMERICAN CROW
Corvus brachyrhynchos hesperis Ridgway, 1887
Rare winter visitor (October through mid-May); formerly more numerous, now casual away from the Coachella Valley; 1 summer record. Few birds are more abundant across the United States than the American Crow, but this species is absent from much of the desert Southwest. It is a sporadic winter visitor to the Salton Sea, with flocks recorded in some winters (Garrett and Dunn 1981). As many as 40 were present in the Imperial Valley 5 November 1972–31 March 1973 (*AB* 27:122, 664), and van Rossem (1911) deemed the crow "common" around Brawley in the winter

of 1910–11. The species is more frequent in the Coachella Valley, where it was recorded three out of four years on average in the 1990s; in the Imperial Valley the average has been about one record (often of just a single bird) every five years. Records extend from 2 October (1999, 3 at the mouth of the Whitewater River; M. A. Patten) to 13 May (2000, Finney Lake; S. Koonce), with a single summer occurrence of six near the north end 5 July 1984 (*AB* 38:1062).

ECOLOGY. In the region, crows are most often found around ranches and feedlots, with a few occurring in towns.

TAXONOMY. There are no specimens for the Salton Sea. The small, thin-billed *C. b. hesperis,* of the Pacific slope and the western Great Basin, is the only subspecies known from California and has been collected in the lower Colorado River valley (Phillips 1986:69). As with the scrub-jays, occasional dispersants may be from the eastern Great Basin and the Rocky Mountains. The most likely candidate is *C. b. hargravei* Phillips, 1942, similar to *C. b. hesperis* but with longer wings and a longer tail; it breeds as close as north-central Arizona. We acknowledge that *C. b. hesperis* is likely a junior synonym of *C. caurinus* Baird, 1858, because type specimens of the latter are from Puget Sound, a contact zone between the taxa where intermediate birds predominate (Johnston 1961). The types of *C. caurinus* are either intermediate or closer to our current concept of *C. b. hesperis* (Johnston 1961; Rea in Phillips 1986). Whether or not specific status is accorded to the Northwestern Crow, it lacks a scientific name and requires designation of a new type, preferably from southeastern Alaska. Because that work has not been accomplished, we use conventional nomenclature.

COMMON RAVEN

Corvus corax clarionensis Rothschild and Hartert, 1902
Fairly common to uncommon breeding resident. Despite being a common bird in most of the desert Southwest, the Common Raven is noticeably less common in the Salton Sink. It is most numerous in the southern Coachella Valley (e.g., 25 around Mecca and Thermal 4 May 1999;

M. A. Patten). It is closer to uncommon in the Imperial and Mexicali Valleys, where maxima seldom reach double digits. Exceptional was a flock of about 400 just west of Seeley 5 November 1994 (R. Higson), the source of our two recent specimens. This species breeds in parts of the Coachella Valley and locally at the fringes of the Imperial and Mexicali Valleys. Breeding takes place from March through June.

ECOLOGY. As noted by Garrett and Dunn (1981), the raven largely avoids the heavily cultivated agricultural areas of the Coachella, Imperial, and Mexicali Valleys. Most birds in the region occupy the desert scrub bordering the western and eastern sides of the Salton Sea. They are most often noted in Creosote scrub on the eastern side of the sea, from Thermal to Bombay Beach, frequently perching (and occasionally nesting) atop utility poles. Nests have also been placed on water tanks, grain silos, and similar structures with flat areas on which sticks may be piled.

TAXONOMY. Rea (1986a) recognized three subspecies in North America: the large, heavy-billed *C. c. principalis* Ridgway, 1887, of Alaska, the Pacific Northwest, Canada, the Northeast, and the Appalachians; the small, thin-billed *C. c. clarionensis,* of California and Baja California; and *C. c. sinuatus,* of the Great Basin, the interior Southwest, and mainland Mexico. *C. c. sinuatus* is nearly as large as *C. c. principalis* but has a thinner, shorter bill. All five specimens from the Salton Sink are small and thus belong to *C. c. clarionensis* (Rea 1983:200, 1986a:214): from Mecca, an adult female 31 March 1908 (MVZ 770; wing chord 385 mm); from Calexico, a male 5 November 1909 (MCZ 58168; wing 386 mm) and a female 15 November 1909 (MCZ 58169; wing 382 mm); and from 4 km west of Seeley 5 November 1994, an immature male (SDNHM 49018; wing 390 mm) and an immature female (SDNHM 49019; wing 360 mm).

ALAUDIDAE • Larks

HORNED LARK

Eremophila alpestris (Linnaeus, 1758) subspp.
Common breeding resident. The Horned Lark is a common breeding resident throughout the

region, mainly in fallow or lightly used agricultural areas, open desert, and alkaline flats with scattered low vegetation. Breeding numbers are difficult to gauge as pairs are widely dispersed. Breeding activity begins early, with birds nesting by late January or February and fledging young as early as March. In winter, when flocks often number in the hundreds or even thousands, the Horned Lark seems to be more abundant. At least in the past there was an influx of winter visitors from the north or northeast; now, flocking of the local population may be more responsible for the apparent increase (see below).

ECOLOGY. As noted by Behle (1942:277), this lark flourishes, principally in wetter years, in sand hills and alkaline flats but also "occurs abundantly in the cultivated areas of the Imperial Valley as well as in the truly desert sections." It can be found just about anyplace in the region where there is open soil, from roadsides to plowed fields to the open desert. Breeders shun heavy agriculture for open desert and fallow fields.

TAXONOMY. Horned Larks of the Salton Sink have long been included in *E. a. leucansiptila* (Oberholser, 1902), a treatment amply supported by early specimens. Indeed, Behle (1942) wrote that "the specimens that show the characters [of *leucansiptila*] best developed are from the Salton Sea region and Imperial Valley of California." As one would expect of an open-country bird in this intensely insolated area, this Colorado Desert endemic is the palest Horned Lark of North America. The back feathers are broadly edged with pinkish buff, and in males the nape is tinged vinaceous. Recent specimens, however, disrupt this simple picture. A few are as pale as the historical population, but most are somewhat darker, some shockingly so. These birds clearly represent the local population, as they include specimens still in body molt collected 4 km south-southeast of the mouth of San Felipe Wash 24 July 1994 (P. Unitt; R. Higson; SDNHM 48908–10). SDNHM 48909 is so dark above and so heavily streaked on the breast as to suggest *E. a. insularis*, of the Channel Islands, the dark extreme of the species. In

many of these specimens the darkness is caused largely by an extension of dark centers of the upperpart feathers at the expense of pale edges, though pale edges themselves are often darker than in early specimens of *E. a. leucansiptila*. We can only speculate that conversion of the Imperial Valley from open desert scrub to irrigated farmland abruptly darkened the valley's background color, causing natural selection to shift gears, enhancing the fitness of dark variants or immigrants. The situation deserves intensive study as a possible example of rapid evolution in progress.

E. a. actia (Oberholser, 1902), of cismontane California, is a rare breeding resident in the Coachella Valley, forming a narrow hybrid zone with *E. a. leucansiptila* (Behle 1942:266), and so paralleling the Song Sparrow (Patten 2001; see also the Song Sparrow account). As with the sparrow, the coastal subspecies is the darker one; *E. a. actia* differs mainly in its darker cinnamon nape and darker, browner upperparts. Local specimens include two from Thermal 27 January 1918 (UCLA 11967, 11968) and one from Mecca 17 April 1908 (MVZ 959). We know of no recent Coachella Valley specimens, so we cannot assess changes there.

At least formerly, numerous Horned Larks moved south into the Salton Sink during the winter. According to Behle (1942), two subspecies predominate, *E. a. leucolaema* Coues, 1874, of the Rocky Mountains, and *E. a. ammophila* (Oberholser, 1902), of the Mojave Desert. The former, similar to *E. a. leucansiptila* but with a medium gray-brown back and first reported by Dickey and van Rossem (1922; UCLA 14695, 12512), has been recorded from 21 December (1921, Kane Spring; FMNH 173817) to 27 January (1918, Thermal; UCLA 11969). It has also been taken in northeastern Baja California (Grinnell 1928; Behle 1942), suggesting that it is a widespread winter visitor to the Colorado Desert. *E. a. ammophila*, with a browner back and a cinnamon nape, has been recorded from 1 December (1891, Carrizo Creek; F. Stephens, SDNHM 668, 669) to 1 February (1913, south end; Behle 1942, UCLA 10028). Additional subspecies are rare winter visitors to the region. There are four records of *E. a. enthymia* (Oberholser, 1902), of the northern Great Plains (Dickey and van Rossem 1924; Behle 1942), from Kane Spring 26 December 1924 (UCLA 14696) and 13 January 1923 (UCLA 12509, 12510) and the south end 1 February 1913 (UCLA 10037). This subspecies resembles *E. a. leucansiptila* but has a paler yellow throat and a white supercilia, and the pale gray back lacks pinkish tones. Last, specimens from the Imperial Valley 27 December 1924 (Behle 1942; MVZ 81327) and Kane Spring 21 December 1927 (G. P. Ashcroft; FMNH 173870) are *E. a. lamprochroma* (Oberholser, 1932), of the western Great Basin, characterized by its vinaceous nape (male only) and rather dark gray-brown back.

Curiously, none of the recent Imperial Valley specimens at SDNHM (*n* = ca. 30) suggests migration from the north or northeast, though many were collected when such migrants might be expected. The heterogeneity of the specimens may mask such variation to some extent, but the variation is in the direction of coastal subspecies, not interior ones. Might migration of Horned Larks into the Salton Sink now be reduced in comparison with that in the early twentieth century? Might warmer winters obviate their need to migrate this far south?

HIRUNDINIDAE • Swallows

In no group is the impressive landbird migration through the Salton Sink more visible than in the swallows. Tree Swallows in particular can be incredibly abundant, with concentrations in the many thousands. Cliff, Northern Rough-winged, and Barn Swallows also move through the area in large numbers, with the first two having a sizable breeding presence. The Barn Swallow has also bred. The largest numbers of migrating Bank Swallows in southern California are noted in the Salton Sink, and modest numbers of the Violet-green Swallow also pass through in spring (with fewer in fall). In addition to small numbers of the Purple Martin, a species that does not occur annually in the region, the Imperial Valley has hosted the only Cave Swallows recorded in California.

PURPLE MARTIN
Progne subis subis (Linnaeus, 1758)
Casual spring transient (mid-April to mid-May); 4 fall (late August to mid-September) and 2 "summer" records. The hefty Purple Martin has declined precipitously as a breeder in California and elsewhere in the Southwest. When this species was more common as a breeder in California's mountains, it was probably more frequent at the Salton Sea. However, all regional records of this species are from since the decline. It is now a casual spring transient, averaging one record every two to three years from 13 April (1975, north end; *AB* 29:910) to 23 May (1999, first-spring male at Mecca; M. A. Patten). Generally only single individuals are encountered, but five were seen at Salton City 23 April 1988 (B. E. Daniels; M. A. Patten). A female at the north end 16 June 1984 (*AB* 38:1062) was probably a late spring transient, but the date is anomalous. Likewise, one photographed at Red Hill 24 July 1979 (*AB* 33:898) defies categorization, although it may have been an extremely early fall migrant as it was associated with a large push of migrant Tree Swallows (M. A. Patten). Even so, four (2 males and 2 females) at the Salton Sea State Recreation Area 23 August 1976 (M. S. Zumsteg), a female at Hot Mineral Spa 7 September 2002 (H. King), an adult male near Wister 9 September 2002 (J. L. Dunn), and two at the north end 14 September 1982 (*AB* 37:225) provided the only definite fall records for the region.

TAXONOMY. All specimens from California belong to the widespread nominate subspecies. The larger, paler-throated (in females) *P. s. arboricola* Behle, 1968, of the Great Basin, was mistakenly attributed to California by Phillips (1986:8).

TREE SWALLOW
Tachycineta bicolor (Vieillot, 1808)
Abundant transient (mid-February to mid-May, early July to mid-November); fairly common winter visitor; rare to casual in summer. The Tree Swallow is the most numerous swallow wintering in the southwestern United States. It is fairly common

in the Salton Sink through the winter (particularly in the Imperial Valley), and it is by far the most likely swallow to be encountered in that season, often by the dozens. Were it not for the even larger numbers of migrants, its movements would be difficult to detect. However, this species is also the most numerous transient of the family. Spring migrants occur from mid-February (e.g., 8 at Mecca 21 February 1999; M. A. Patten) to 23 May (1999, 2 at Mecca; M. A. Patten), with a peak from early March to mid-April, when thousands move through the region (e.g., 5,000 at the south end 13 April 1991; M. A. Patten). Most have passed through by the end of April. Fall migration is equally impressive. Transients are noted from 5 July (1991, 150 at the south end; M. A. Patten) to mid-November. Fall movement peaks in August; for example, 100,000 swallows, mostly Tree and Cliff, were at the south end in late August 1953 (AFN 8:43) and "thousands" were near Westmorland 8–9 August 1987 (J. O'Brien). Many are still moving through in September and October (e.g., 300 in the Imperial Valley 25 October 1998; M. A. Patten). Nonbreeders occasionally remain through the summer (e.g., 2 at the south end 16 June 1979; G. McCaskie), but the species is unexpected in that season.

ECOLOGY. The staggeringly large flocks gather most often near water, sometimes on utility wires near Alfalfa fields. These flocks invariably contain lesser numbers of other swallows, most often Cliff and Northern Rough-winged, although all of the swallow species recorded in the region have been found in Tree Swallow flocks. Both swallows and swifts are most often observed en masse during periods of heavy cloud cover or strong winds.

VIOLET-GREEN SWALLOW

Tachycineta thalassina thalassina (Swainson, 1827)
Uncommon spring transient (mid-February through May); rare fall transient (mid-August through October); 2 winter records. In California the Violet-green Swallow breeds largely in montane coniferous forest, although it uses deciduous trees in valleys throughout the Great Basin (Ryser 1985)

and cliff faces locally along the Colorado River (Rosenberg et al. 1991) and in northeastern Baja California (Patten et al. 2001). Large numbers pass through the lowlands of the West during spring migration, particularly toward the coast. Surprisingly, the Violet-green Swallow is much less common as a transient through the desert. In the Salton Sink it is generally outnumbered by all other swallows except the Purple Martin (and sometimes the Bank Swallow). Spring migrants appear in mid-February. A male near Bombay Beach 14 February 1961 (SBCM 36315) is the earliest, except for two at the south end 22 January 1983 (AB 37:339), presumably exceptionally early migrants rather than wintering birds (see below). Small numbers trickle through in spring until the end of April, with a shallow peak in March, roughly a hundred at the south end 5 March 1978 (G. McCaskie) being the highest single-day count for the region. Spring migrants (or birds forced out of higher altitudes by bad weather?) are rarely noted in May, with one near Oasis 25 May 1996 (G. McCaskie) being the latest. The Violet-green Swallow is much rarer in fall, with records from 8 August (1981 and 1987, 1 at Finney Lake and 2 near Westmorland; G. McCaskie, M. A. Patten) to 29 October (1987, 6 birds 6 km west-southwest of Thermal; P. Unitt). Lone birds at the north end 29–31 December 1971 (AB 26:251; Garrett and Dunn 1981) and 17 December 1977 (AB 32:884) represent the only valid winter records for the region. Other claims, such as those by van Rossem (1911) and that of three at the south end 31 December 1965 (AFN 20:377) likely pertain to the Tree Swallow. Similarly, the Violet-green Swallows "still at Ramer Lake in mid-June" 1959 (AFN 13: 402) were undoubtedly misidentified as there are no summer records for the Salton Sink.

ECOLOGY. This species is most often noted with large flocks of Tree Swallows, especially those congregated near standing water.

TAXONOMY. Populations in the United States are of the large nominate subspecies, of which the allegedly smaller, greener *T. t. lepida* Mearns, 1902, is a synonym (Phillips 1986). Birds breeding along the lower Colorado River (Rosenberg et al. 1991) may be intermediates between the nominate subspecies and the small *T. t. brachyptera* Brewster, 1902, of the southern Gulf of California (Phillips 1986:15), but no specimens are available. Given the range of the latter subspecies, possibly north to the southern Colorado Desert of northeastern Baja California (Patten et al. 2001), it may occasionally occur in the region, but there are no records for the United States.

NORTHERN ROUGH-WINGED SWALLOW
Stelgidopteryx serripennis (Audubon, 1838) subspp.
Common transient (especially) and breeder (early February through October); uncommon winter visitor. The Northern Rough-winged Swallow and the Cliff Swallow are the only regularly breeding swallows at the Salton Sea. Like the Tree and Barn Swallows, the Rough-winged is also a common transient through the region, just as it is through much of the West. Records of breeders and transients extend from early February (e.g., 5 at the south end 5 February 2000; G. McCaskie) to late October (e.g., Obsidian Butte 24 October 1999; G. S. LeBaron). Numbers peak during migration periods, particularly from March to mid-April and from mid-July through August (e.g., 1,500 at the south end 13 April 1991: M. A. Patten), but thousands stay to breed. The Northern Rough-winged Swallow is outnumbered as a breeder by the Cliff Swallow, although the Northern Rough-winged is more widespread and not colonial. Nesting begins early, many pairs occupying suitable holes (or pipes) by March, and continues into May. Small numbers winter each year, with most records from December and January. The Northern Rough-winged is outnumbered during winter only by the Tree Swallow. These two species are the only expected swallows in that season, although the Barn Swallow is becoming increasingly frequent.

ECOLOGY. Migrant Northern Rough-wingeds may occur anywhere in the region but are particularly common over water. Breeders nest in earthen embankments, mainly the sides of irrigation ditches with vertical or near vertical walls (Hurlbert 1997). They dig their own holes or use small drainage pipes.

TAXONOMY. Two subspecies are generally recognized in the western United States (Brodkorb 1942): the darker, more northern nominate, *S. s. serripennis,* and the paler, more southern *S. s. psammochrous* Griscom, 1929. In adults, at least, the difference is slight, less than that arising from plumage wear and fading. Most Salton Sea specimens are of *S. s. psammochrous,* including two coming into breeding condition, a male 30 April 1989 (SDNHM 47424) and a female 10 April 1989 (SDNHM 47189). The single winter specimen, taken 30 December 1988 (SDNHM 45513), is also of *P. s. psammochrous.* The two darkest specimens, as dark or darker than six specimens from Oregon and northwestern California, were collected 26 February (SDNHM 46884) and 23 March 1989 (SDNHM 47279). Rea (1983) reported that the subspecies are best differentiated in juvenile plumage; the single regional juvenal specimen, taken 24 September 1989 (SDNHM 47603), is clearly *S. s. psammochrous.*

BANK SWALLOW
Riparia riparia riparia (Linnaeus, 1758)
Uncommon transient (late March to early May, early July through September); casual in winter. Among the swallows regularly occurring in California, the Bank Swallow is one of the latest to arrive in spring (along with the Purple Martin). Unlike the swallows that are common in the region, which arrive in late February, this species does not arrive until late March, with 21 March (1992, Unit 1; G. McCaskie) being the earliest date. Rarely are more than a few encountered, although double-digit counts can be achieved during peak passage, from mid-April through mid-May. The maximum counts, of 50 at the mouth of the Whitewater River 2 May 1987 (M. A. Patten) and near Red Hill 8 May 1971 (J. L. Dunn), were exceptional. The latest spring date is 23 May (1999, 10 at Mecca; M. A. Patten). Fall passage is more in time with that of other swallows. The earliest Banks typically arrive in early July, with the first Tree Swallows. Five at Fig Lagoon 20 June 1981 (G. McCaskie) provided the earliest fall record. Fall migration peaks in early

August (e.g., 50 near Westmorland 8–9 August 1987; M. A. Patten) but tapers off quickly, with few moving through in September and the latest record being 3 October (1986, 2 at the mouth of the Whitewater River; B. E. Daniels). In winter this species is casual in the region, as it is anywhere in the United States. Winter birds are recorded once every three to four years from 9 November (1963, south end; Garrett and Dunn 1981) to 14 February (1999, Wister; G. McCaskie), with the majority in December, suggesting that they could be extremely late migrants. One remained at the now defunct Imperial Warm Water Fish Hatchery 10–21 December 1989 (*AB* 44:330), and one or two were observed at the south end during December 1971 and January 1972 (*AB* 26:656).

ECOLOGY. Like other swallows migrating through the region, this species may be encountered just about anywhere. Even so, it is most numerous near fresh water.

TAXONOMY. The widespread nominate subspecies is the only one recorded in North America and in most of Europe, as both *R. r. maximiliani* (Stejneger, 1885) and *R. r. ijimae* (Lönnberg, 1908) are synonyms (Peters 1960; Phillips 1986; Gibson and Kessel 1997). It is medium-brown above with a well-defined wide breast band.

CLIFF SWALLOW

Petrochelidon pyrrhonota (Vieillot, 1817) subspp.
Abundant transient and breeder (February through early October). Among the earliest migrant landbirds to return to California in spring is the Cliff Swallow. Many appear in the Salton Sink by mid-February, the earliest arrival date being 29 January (1972, south end; *AB* 26:656). Thousands are present by March, when nesting begins. The first juveniles fledge in May (e.g., 19 May 1999, 2 near Obsidian Butte; M. A. Patten). As a breeding species the Cliff Swallow is a recent colonist at the Salton Sea. Twenty-five nests under a bridge at the south end 1 May 1977 provided the "first evidence of breeding in the Imperial Valley" (*AB* 31:1191). An adult female 11 km east of Cerro Prieto 13 June 1928 (MVZ

53000) hints at the possibility of nesting near the region well before the 1970s. Apparent breeders have been taken in the Imperial Valley from 10 April (1989; R. Higson, SDNHM 45821) to 24 September (1989; R. Higson, SDNHM 47602), the latest apparent migrant noted 9 October (1994, 5 at the south end; M. A. Patten). In late summer and early fall, augmented by transients from the north, numbers can reach the tens of thousands (e.g., 10,000 at the south end 8 July 1972 and 20,000 there 4 August 1973; G. McCaskie, see also the Tree Swallow account). This species is virtually unrecorded after October in California, even during mild winters. One collected at Brawley 18 December 1910 (van Rossem 1911) and one observed at Rock Hill 4 December 1971 (*AB* 26:656) provide two of few winter records for California. Thus, a claim that the Cliff Swallow was "common about reservoirs and flooded fields" 1 December 1910–14 January 1911 (van Rossem 1911) is unfounded.

ECOLOGY. In the Salton Sink, Cliff Swallows nest only under bridges and in culverts crossing open water. Various bridges across the Alamo and New Rivers in the Imperial Valley are especially productive nest sites, but even some narrow drainage canals with concrete culverts passing under roads are used (e.g., under Davis Road at Wister). Migrants and foraging birds are most often noted over water.

TAXONOMY. There appear to be only two subspecies of the Cliff Swallow in the western United States, the larger, more northern nominate subspecies and the smaller, more southern *P. p. tachina* Oberholser, 1903; *P. p. hypopolia* Oberholser, 1920, is a synonym of *P. p. pyrrhonota* (Browning 1992). Breeders belong to *P. p. tachina*, differing from the nominate subspecies in its shorter wings and perhaps more cinnamon forehead (Behle 1976). Other alleged color differences (e.g., paler cheeks and throat) are not consistent, as many specimens of *P. p. tachina* have dark cheeks and a dark throat and some of *P. p. pyrrhonota* have pale cheeks. Most specimens from the Salton Sea are small (wing chord 100–107 mm). However, the nominate subspecies, breeding in the West along the

Pacific Coast and in much of California west of the Sierra Nevada, is a common migrant throughout the state. It passes through the region in large numbers during spring and fall; for example, migrants were collected 12 km northwest of Calipatria 1 May 1937 (MVZ 71521, 71522, 71524, 71525; wing chord 109–111.5 mm).

CAVE SWALLOW

Petrochelidon fulva pallida Nelson, 1902

Three records, 2 in spring (May) and 1 in fall (August). The only records of the Cave Swallow for California are from the Imperial Valley. The first was a bird photographed among a large flock of migrant swallows near the mouth of the New River 8 August 1987 (Patten and Erickson 1994). There are two additional sight records from spring 1995, of adults at Sheldon Reservoir 6 May and at Wister 21 May (McCaskie and San Miguel 1999). Within the past two decades the Cave Swallow has expanded its range in Texas and New Mexico and has bred as far west as Tucson, Arizona (Huels 1984).

TAXONOMY. There are no specimens of the Cave Swallow for California or Arizona, but the 1987 bird and the second individual of 1995 showed the characters of the expected Mexican subspecies *P. f. pallida* (= *pelodoma* Brooke, 1974, if the species is placed in *Hirundo* because *pallida* is preoccupied in that genus). This pale, buff-rumped, gray-flanked subspecies has occurred widely as a vagrant throughout the United States, even reaching southeastern Canada (McNair and Post 2001).

BARN SWALLOW

Hirundo rustica erythrogaster Boddaert, 1783

Common transient (mid-February to mid-May, early July to mid-November); rare winter visitor; formerly bred. Few passerines are more familiar than the cosmopolitan Barn Swallow, which is common across North America. It is principally a transient in the Salton Sink. Spring migrants pass through from 16 February (1985, 3 at Wister; B. E. Daniels) to 25 May (1995, 5 at the south end; M. A. Patten), with a peak in late April and May (e.g., 75 at Mecca 23 May 1999; M. A. Patten).

A spectacular concentration of 25,000 was noted at Red Hill 8 May 1971 (J. L. Dunn), and a late concentration of 500 was noted 26 May 1996 (G. McCaskie). Fall migrants have been detected as early as 9 July (1988, 2 at Salton City; M. A. Patten). In contrast to other swallows in North America, this species has a distinctly protracted fall migration, running commonly to the end of October, the latest migrant detected 20 November (1988, Finney Lake; B. E. Daniels). In the Salton Sink the Barn Swallow's abundance is roughly equal in spring and fall, whereas along the lower Colorado River fall migrants greatly outnumber spring ones (Rosenberg et al. 1991). This species is rare in winter anywhere in the United States, but in southern California and southern Arizona records have been increasing since the mid-1990s (*FN* 52:236, 259). It is nonetheless a rare (but approaching uncommon) winter visitor to the region, with records from 1 December (2001, 2 near Obsidian Butte through 17 December; G. McCaskie) to 5 February (2000, Fig Lagoon; *NAB* 54:222), including one dating from 18 December 1910 (at Brawley; van Rossem 1911). Flocks of 42 at Mecca 31 January 1998 (*FN* 52:259) and 40 at the south end 8 December 2002 (G. McCaskie) were exceptional. Most individuals in the flock at Mecca were in active molt (M. A. Patten), indicating that they were not migrating.

The Barn Swallow breeds in cismontane southern California (Garrett and Dunn 1981; Lee 1995) but is a scarce, local breeder in transmontane southern California. It formerly nested in the Imperial Valley, at least in the 1970s. Nesting was first noted near Westmorland May–July 1973, when two pairs each raised two broods (*AB* 27:919). By summer 1976 it nested at three localities in the Imperial Valley (*AB* 30:1002), but within a few years nesting had ceased. Breeding took place mainly in spring and early summer, although one pair still had young in the nest on the late date of 13 August (1976, south end; *AB* 28:950).

ECOLOGY. The cup-shaped mud nest of the Barn Swallow is remarkably similar to that of the Black Phoebe in both appearance and placement.

Thus, breeders require shaded vertical walls over water for nest sites. Nests in the Imperial Valley were placed below bridges over drainage canals, locations now occupied by the burgeoning Cliff Swallow population. Migrant Barn Swallows are most frequently noted over water but may be encountered almost anywhere in the region.

TAXONOMY. The New World subspecies is *H. r. erythrogaster*, distinguished from the various Old World forms by its deep chestnut underparts and the reduction of the breast band to mere patches of dark blue at the sides of the breast. It is the only subspecies collected in California, despite sight reports of white-bellied birds that suggest *H. e. gutturalis* Scopoli, 1786, of eastern Asia (e.g., Huntington Beach 10–11 June 1989; *AB* 43:1369).

PARIDAE • *Chickadees and Titmice*

Aside from the Verdin, the related families of the various titmice, nuthatches, and creepers are essentially vagrants or rare, irregular winter visitants to the Salton Sink. The Verdin, by contrast, is one of the few members of these families that occurs in the desert. It is common in riparian thickets and dense desert scrub throughout the region.

MOUNTAIN CHICKADEE
Poecile gambeli baileyae (Grinnell, 1908)
Casual winter visitor (October through January). As its name indicates, the Mountain Chickadee is generally a denizen of higher elevations. This common species of the West stages irregular incursions into lowlands in fall and winter. Even so, it is scarce in the desert Southwest, although it breeds in lowland riparian forest in the Mojave Desert at Morongo Valley (M. A. Patten) and along the Mojave River at Victorville (*AB* 38:1062; Myers 1993). There are six records for the Salton Sink from 5 October (1955, near Niland; *AFN* 10:59) to 29 January (1984, north end; *AB* 38:358). Most individuals have been recorded in the Coachella Valley, with two records involving flocks of at most 12 at Coachella 8–15 December 1935 (Clary and Clary 1936b) and 10 at the north end 4 January 1997

(*FN* 51;638). One at Mecca 6–7 October 1955 (*AFN* 10:59) occurred the same year as the one near Niland, suggesting an incursion into the desert that fall and winter. This species has reached the Imperial Valley only twice, near Niland (see above) and 7 km northwest of Imperial 16 December 1990 (R. Higson; SDNHM 47587). Of interest is a female just east of the Salton Sink at Regina 26 February–5 March 1978 (*AB* 32:400; SBCM 30285). Garrett and Dunn (1981:268) stated that this species had been recorded "at least three times in the vicinity of Mecca," but we are aware of only two records for that area prior to 1980.

TAXONOMY. Wanderers into the Coachella Valley are *P. g. baileyae*, breeding in mountains of coastal central and southern California, including most of the Transverse and Peninsular Ranges, and in the lowland desert locales mentioned above. *P. g. abbreviatus* (Grinnell, 1918) and *P. g. grinnelli* (van Rossem, 1928) were merged into *P. g. baileyae* by Phillips (1986). *P. g. baileyae sensu lato* has a darker gray back, grayer flanks, and a larger bill than *P. g. inyoensis* Grinnell, 1918, breeding in the desert ranges of eastern California but not known to wander south to the Sonoran Desert. *P. g. atratus* (Grinnell and Swarth, 1926), of the Sierra Juárez and the Sierra San Pedro Mártir, has a darker gray mantle and restricted white in the supercilia that is often clouded with black flecking. The specimen from Regina is of *P. g. baileyae*.

To classify the Imperial Valley specimen, we measured the mantle paleness of the two southern subspecies using a Minolta CR-300 Chroma Meter, with the L metric yielding lower values for darker coloration. Foxing, the tendency for specimens to become redder and paler with age, was corrected for each skin (SDNHM; $n = 32$) using the slope from a linear regression of mantle color of specimens from the Laguna Mountains ($n = 8$) onto year. The year of the newest specimen (1996) was used as the reference. All values reported below are the corrected L, which is measured $L - 0.0231 \times (1996 - \text{year of collection})$. Superciliary extent was judged qualitatively. Although previously undescribed as such, the

TABLE 13
Mantle Coloration and Supercilia Pattern of the Mountain Chickadee
(Poecile gambeli) in Southern California and Northern Baja California

POPULATION	MINIMUM *L*	MAXIMUM *L*	MEAN *L*	STANDARD DEVIATION *L*	SUPERCILIA
P. g. baileyae	34.92	39.86	36.41	1.39	Broad, clear white
Laguna Mountains	33.91	37.12	35.30	1.16	Moderate, clouded
P. g. atratus	32.27	34.49	33.40	0.70	Narrow, restricted

Note: Reported values of *L*, a measure of mantle darkness as assessed with a Minolta CR-300 Chroma Meter, were corrected for foxing. Reference specimens of *P. g. baileyae* (*n* = 13) were from Lebec, Kern County, south to the Cuyamaca Mountains, San Diego County. Reference specimens of *P. g. atratus* (*n* = 10) were from the Sierra Juárez and the Sierra San Pedro Mártir.

population in the Laguna Mountains of southern San Diego County is intermediate between *P. g. baileyae* and *P. g. atratus* in mantle color and superciliary extent (Table 13). Interestingly, the specimen from the Imperial Valley matches this intermediate population in having the mantle fairly dark (*L* = 34.78) and the supercilia somewhat restricted and clouded (white flecked with black).

OAK TITMOUSE
Baeolophus inornatus affabilis
Grinnell and Swarth, 1926
Five records for late fall and winter (late November to mid-February). The Oak Titmouse inhabits oak and riparian woodlands of much of coastal California and the Little San Bernardino Range just north of the Salton Sink. It has been recorded five times in the Salton Sea region, four times in the southern Coachella Valley and once in the Imperial Valley. This last record was of a bird taken at the Vail Ranch, about 9 km north of Westmorland, 22 February 1956 (CSULB 1652). The others were of lone birds collected at the mouth of the Whitewater River 26 November 1964 (SBCM 36667), photographed and later collected at Mecca 29–31 December 1990 (SBCM 52614), and observed 12 km west of Mecca 26 November 1994 (G. McCaskie) and near the north end 4 January 1997 (*FN* 51:638).

TAXONOMY. The bird from Mecca was published as *B. i. transpositus* Grinnell, 1928 (*AB* 45: 322); an artifact of comparison of foxed and un-foxed specimens, it is a synonym of *B. i. affabilis*, of the southern coast (Phillips 1986; Cicero 1996). The Whitewater River specimen is badly foxed but matches *B. i. affabilis* of like specimen age. The Imperial Valley specimen is also *B. i. affabilis* (C. Cicero pers. comm.). The paler *B. i. mohavensis* (Miller, 1946), endemic to the Little San Bernardino Mountains and occurring as close as Morongo Valley in southern San Bernardino County, should be borne in mind for future records. Phillips (1986:93) considered it a synonym of *B. i. affabilis*, but Cicero (1996:156, 2000) considered it valid. The similar Juniper Titmouse, *B. ridgwayi* (Richmond, 1902), has been recorded no closer than the central Mojave Desert (M. T. Heindel pers. comm.). See Cicero (2000) for notes on the nomenclature of this species.

REMIZIDAE • Penduline Tits and Verdins

VERDIN
Auriparus flaviceps acaciarum Grinnell, 1931
Common breeding resident. The furtive, noisy Verdin (Fig. 58), a characteristic species of the desert Southwest, is a common breeding resident throughout the Salton Sea region, reaching peak abundance in native habitats and along well-vegetated rivers and channels. Breeding dates are difficult to determine because of this species' penchant for constructing nests purely for roosting. However, most breeding takes place in spring, mainly from March (e.g., see van Rossem 1911; and MVZ 6412) to early June, although

FIGURE 58. Characteristic of thorny desert shrubs now largely eliminated from the Salton Sink, the Verdin has maintained its numbers by adapting to Saltcedars and garden trees. This bird is a juvenile. Photograph by Kenneth Z. Kurland, August 2001.

breeding has been confirmed from January to October elsewhere on the Colorado Desert (R. L. McKernan). Fledglings are first noted by mid-June and are common by July.

ECOLOGY. Breeding Verdins have probably undergone a habitat shift in the region. Through much of its desert range the Verdin favors mesquite, Catclaw, Blue Palo Verde, and Smoke Tree as nesting trees, but these trees are now scarce at the Salton Sea. Nevertheless, this species remains common because it now frequently uses Saltcedar as a nest tree. As a result it is now most numerous along the rivers and on wetlands that border desert scrub, although as noted by Grinnell and Miller (1944), its dependence on water is illusory. Some occur in suburban neighborhoods with adequate tree and shrub cover.

TAXONOMY. Verdins in the southwestern United States belong to *A. f. acaciarum,* charac-terized by its longer tail, sandy-tan (less gray) flanks, and paler dorsal coloration in comparison with those of the various subspecies of Baja California, New Mexico, Texas, and mainland Mexico (Phillips 1986:98).

AEGITHALIDAE • *Long-tailed*
 Tits and Bushtits

BUSHTIT
Psaltriparus minimus melanurus
 Grinnell and Swarth, 1926
Casual fall and early winter visitor (mid-August through mid-March) to the Coachella Valley; 1 spring record. The highly gregarious Bushtit is a common bird of chaparral, coastal sage scrub, and oak and riparian woodlands in cismontane California. It is less common but widespread in the Great Basin and the interior Southwest. This species is a casual fall and winter visitor to the

California deserts, including the Coachella Valley, where there are seven records from fall and winter from 16 August (1980, 4–5 near the north end through 22 October; *AB* 35:227) to 18 March (2000, 2 along the Whitewater River near Mecca; M. A. Patten). Other records are of 4 at the north end 19 December 1982 (*AB* 37:771), 1 there 22 December 1985 (*AB* 40:1008), "several" near Mecca 29 December 1970 (Garrett and Dunn 1981), a flock of 25 at Oasis 2–14 January 1993 (*AB* 47:968; SBCM 54242–43), and 2 at Indio 4 March 2000 (M. A. Patten); there is an anomalous spring record of 4 at the mouth of the Whitewater River 1 May 1999 (M. A. Patten). One at Salton City 30 September 1987 (*AB* 42:137) furnished the only record for the Salton Sink away from the Coachella Valley.

ECOLOGY. Bushtits have occurred in Saltcedar, Quail Brush scrub, and citrus orchards.

TAXONOMY. Although it has been said that Salton Sea "records probably pertain to the interior subspecies" (Garrett and Dunn 1981), there are no specimens supporting this claim. Furthermore, the January 1993 (see below) and May 1999 birds were distinctly brownish and thus clearly not *P. m. plumbeus* (Baird, 1854), the Lead-colored Bushtit, of the Great Basin. The specimens from Oasis, a male (SBCM 54243) and a female (SBCM 54242) taken 14 January 1993, are of *P. m. melanurus*, of the Peninsular Range and the coastal slope of California from Los Angeles southward (Rea in Phillips 1986). Most Bushtits occurring in the Salton Sink likely belong to this subspecies, characterized by its dark brown crown, gray cheeks, and brownish-gray flanks (males only). Occasional birds might belong to *P. m. sociabilis* Miller, 1946, of the Little San Bernardino and Eagle Mountains west to Morongo Valley, which has a sooty crown, gray cheeks, and paler tan flanks (again, males only). It may be that the latter is not a valid subspecies but instead should be merged with *P. m. californicus* Ridgway, 1884, of the Central Valley and probably south to the Mojave River valley. Phillips (1986) merged *P. m. sociabilis* with *P. m. plumbeus* despite the browner plumage of the former.

RED-BREASTED NUTHATCH
Sitta canadensis Linnaeus, 1766

Rare fall (especially) and winter visitor (mainly November through March). A denizen of boreal forests of high latitudes and elevations, the Red-breasted Nuthatch occasionally stages impressive irruptions into the lowlands throughout the continent during the fall, with many staying through the winter. Such irruptions account for most of the records for the Salton Sea region, although a few have occurred during non-irruption years, for example, at Red Hill 16 October 1966 (SBCM 36693) and Westmorland 30 September 1972 (SBCM 36697). Incursions were noted during the falls of 1935 (Clary and Clary 1936b) and 1973 (*AB* 28:109) and during the winters of 1984–85 (*AB* 39:211) and 1996–97 (*FN* 51:119). Even during a major irruption to lowland southern California the numbers seldom reach high levels in the Salton Sink, maxima being 20 at Brawley 9 November 1996 (G. McCaskie) and at most 10 there during the winter of 1984–85 (*AB* 39:211). The latter group also supplied the first winter record for the region, as all previous records involved fall vagrants. Birds have occurred as early as 16 September (1973, south end; *AB* 28:109) and stayed as late as 24 April (1988, 2 at Brawley; M. A. Patten). An exceptionally early individual was observed at Brawley 20 August 2000 (C. McGaugh).

ECOLOGY. This nuthatch is invariably found in ornamental pines and other nonnative conifers, making Brawley and El Centro particularly good locales.

TAXONOMY. The Red-breasted Nuthatch is monotypic because *S. c. clariterga* Burleigh, 1960, is not diagnosable (Banks 1970; Phillips 1986).

WHITE-BREASTED NUTHATCH
Sitta carolinensis Latham, 1790

Eight fall records (late July to mid-December). The White-breasted Nuthatch is a casual postbreeding visitor to California deserts well away from breeding areas. There are only eight records for the Salton Sink, two from the Coachella Valley

FIGURE 59. The White-breasted Nuthatch seldom wanders into the California deserts, but the Salton Sink has had eight records. Photograph by Michael A. Patten, 1 October 2000, Wister.

and six from the Imperial. Both Coachella Valley records are from 1935, one from Coachella 10 September and one northwest of Mecca 15 December (Clary and Clary 1936b). In the Imperial Valley, individuals were seen 8 km west of Westmorland 7 November 1970 (*AB* 25:110) and 6 August 1976 (*AB* 31:224), calling birds were heard at Brawley 28 November 1996 (*FN* 51:119) and Salton Sea NWR 31 July 1999 (*NAB* 53:432), a silent bird was photographed at Wister 30 September–1 October 2000 (Fig. 59; *NAB* 55:104), and a calling bird was heard along the New River 5 km west of Calipatria 19 December 2000 (B. E. Daniels).

ECOLOGY. Unlike those of the previous species, records of the White-breasted Nuthatch are not confined to conifers. Instead, this species has occurred in Fremont Cottonwood, Athel, large ornamental broadleaf trees in suburban neighborhoods or at ranch yards, and even in stands of mesquite.

TAXONOMY. Some California desert records of this species are of the long-billed, white-breasted Great Basin subspecies, *S. c. tenuissima* Grinnell, 1918. This subspecies and the similar *S. c. nelsoni* Mearns, 1902, of Arizona, utter rapid,

staccato calls that are readily separable from the slower, longer calls of the short-billed, brown-breasted *S. c. aculeata* Cassin, 1856, of the Pacific Coast. Calling birds at Brawley and Salton Sea NWR were *S. c. aculeata* (M. A. Patten; G. McCaskie), but the one near Calipatria was apparently either *S. c. tenuissima* or *S. c. nelsoni* (B. E. Daniels). Other records cannot be assigned to subspecies without a specimen or notes on vocalizations.

CERTHIIDAE • *Creepers*

BROWN CREEPER

Certhia americana Bonaparte, 1838 subspp.

Casual fall and winter visitor (mid-November to mid-March). A fairly common bird of coniferous forests through much of California (and indeed across the temperate portion of the continent), the Brown Creeper disperses in small numbers out of its breeding habitat during the winter. It is only a casual late fall and winter visitor to the Salton Sea region. Records extend from 14 November (1986, 6 at Brawley through 19 November; *AB* 41:145) to 17 March (1973, Niland since 23 December 1972; *AB* 27:664), with most in November.

ECOLOGY. The creeper requires large trees for foraging, so area records are from wooded portions of cities (e.g., southwest Brawley) or large Athel and other trees along rivers, ditches, or at ranches.

TAXONOMY. The single specimen from the region, collected 11.5 km northwest of Imperial 21 November 1986 (P. Unitt; SDNHM 44520), is one of four California specimens of *C. a. americana* Bonaparte, 1838, which breeds east of the Rocky Mountains in Canada and the northeastern United States. The specimen was identified by its pale tawny rump, buff-tinted crown streaks, and short bill (Unitt and Rea 1997). It is likely, however, that two additional subspecies, both known from several specimens from the Mojave Desert (Grinnell and Miller 1944; Unitt and Rea 1997), also reach the area, probably more frequently than *C. a. americana*. These subspecies are *C. a. montana* Ridgway, 1882, breeding in the Rocky Mountains and the Great Basin, and *C. a. zelotes* Osgood, 1901, breeding in the Cascades, the inner Coast Ranges, the Sierra Nevada, and montane southern California. The former differs from *C. a. americana* in its whiter streaks, less rufous back, and longer bill. The latter differs from both *C. a. montana* and *C. a. americana* in its darker cinnamon rump.

TROGLODYTIDAE • Wrens

Dense thickets and riparian scrub in the Salton Sink harbor the Cactus, Bewick's, and House Wrens, the last two principally in winter. Marshes and river courses support large numbers of the Marsh Wren, a common breeding resident. With the small numbers of Rock Wrens that winter annually, observers can readily record five species of wrens in a small portion of the Salton Sink in winter, a feat seldom attainable elsewhere in California.

CACTUS WREN
Campylorhynchus brunneicapillus anthonyi
(Mearns, 1902)
Fairly common breeding resident. Although a characteristic species of the deserts of the Southwest, Baja California, and western Mexico, the Cactus Wren is at most a fairly common, somewhat local breeding resident in the Salton Sea region. The local nature of its distribution is related to its habitat requirements (see below); thus, this species is most numerous along rivers, at the edges of larger towns, and at areas supporting significant expanses of vegetation (e.g., Wister, Ramer and Finney Lakes, Fig Lagoon). Breeding takes place principally in spring and summer from late March (e.g., a nest with 5 eggs at Indio 22 March 1931; SBCM/WCH 4027) through July, with peak activity in April and May. Nonetheless, breeding has been documented elsewhere on the Colorado Desert from January to October (R. L. McKernan). Young fledge by late May and early June.

ECOLOGY. There is little cactus in the Salton Sink, so this wren occupies mesquite thickets, washes with stands of Blue Palo Verde, and dense Saltcedar scrub along rivers and irrigation canals. Some occur in dense Quail Brush scrub, especially if it borders stands of tamarisk, and others occur in suburban neighborhoods and parks. Like the Verdin, this species frequently constructs nests for roosting.

TAXONOMY. We follow Rea and Weaver (1990) in recognizing *C. b. anthonyi* and thus restricting *C. b. couesi* Sharpe, 1881, no farther west than central Arizona. With this treatment, all Cactus Wrens in California away from the south coast are *C. b. anthonyi*, characterized by its narrow black markings on rich buff flanks, round breast spots, constricted white markings on the back, and black inner webs to the rectrices.

ROCK WREN
Salpinctes obsoletus obsoletus (Say, 1823)
Rare winter visitor (mid-September to mid-March); has summered. The lack of suitable habitat at the Salton Sea limits the occurrence of the Rock Wren in the region. It is a fairly common species through much of the desert Southwest, but the ocean of agriculture surrounded by desert scrub of little topographical relief characterizing the Salton Sink largely excludes this species. Not surprisingly, then, it is a rare winter visitor from 12 September (2000, 3 km southeast of El Centro;

K. Z. Kurland) to 12 March (1988, Whitewater River near Mecca; M. A. Patten), including specimens from north of Niland 30 December 1962 (SBCM 36768) and Coachella 22 January and 22 February 1934 (LACM 18465, 18466). It is a casual nonbreeding summer visitor, mainly at Rock Hill and Red Hill, with summer records averaging a mere one or two per decade. R. P. Henderson noted the species at San Sebastian Marsh 25–31 May 1977, suggesting breeding in the rugged terrain surrounding that area; farther west it commonly breeds in the Anza-Borrego Desert, at least in wetter years (Unitt 1984).

ECOLOGY. The moniker Rock Wren is no misnomer. This species prefers areas with rock outcrops and is thus most often noted at locales such as Red Hill, Rock Hill, and Obsidian Butte. Earthen embankments, steep-sided ditches, and expanses of concrete (e.g., lining large reservoirs) also suffice, allowing this species to be encountered throughout the region.

TAXONOMY. Rock Wrens throughout mainland North America north of southern Mexico are of the nominate subspecies, diagnosed by its paler, grayer plumage (see Phillips 1986).

CANYON WREN

Catherpes mexicanus (Swainson, 1829) subsp.?
One fall record. A Canyon Wren seen in Niland 25 November 1989 (*AB* 44:163), far from its normal habitat, constitutes the sole record for the Salton Sink. This species is recorded sporadically on the fringes of the Coachella Valley near the north end of the Salton Sink, but all are above sea level, even though some are reported on the Salton Sea (north) Christmas Bird Count (e.g., 22 December 1985; *AB* 40:1008).

TAXONOMY. Birds from the edge of the Coachella Valley are *C. m. conspersus* Ridgway, 1873, a somewhat pale, fairly rufous subspecies. Without a specimen the subspecific identity of the individual in Niland is unknown. It may have been *C. m. pallidior* Phillips, 1986, a palecrowned, less rufous subspecies from the intermountain West that has strayed to eastern California (Phillips 1986:169).

BEWICK'S WREN

Thryomanes bewickii (Audubon, 1827) subspp.
Fairly common to uncommon winter visitor (early September to mid-April); uncommon to rare breeder in the Coachella Valley and at San Sebastian Marsh. Despite perilous declines in the East, Bewick's Wren remains a common bird through much of the West and is a characteristic bird of chaparral and lowland riparian habitats in southern California. As a breeder in the Salton Sink, however, it is restricted to the southern Coachella Valley, occurring at least at Coachella, Thermal, Mecca, and Indio. Probably no more than 10 pairs breed in the area (e.g., only 7 singing males were located throughout the southern Coachella Valley 4 May–13 July 1999; M. A. Patten). Nesting has been documented from February through June (R. L. McKernan). There is no evidence of breeding in the Imperial or Mexicali Valleys, but a molting adult collected 7 km northeast of Imperial 29 August 1992 (R. Higson; SDNHM 48186) indicates occasional summering in those valleys. The species is apparently resident at San Sebastian Marsh, where singing males were noted 25–31 May 1977 (R. P. Henderson) and 24 March 1989 (P. Unitt). Wintering Bewick's Wrens occur throughout the Salton Sink, being fairly common to uncommon from 4 September (2001, 3 km southeast of El Centro; K. Z. Kurland) to 17 April (1988, 2 at the mouth of the Whitewater River; M. A. Patten).

ECOLOGY. Breeding of Bewick's Wren in the region may be limited because of a restriction of suitable nesting sites. In the Coachella Valley this species occupies both ranches with mature Fremont Cottonwood and dense mature stands of mesquite; a singing male along the Whitewater River channel 31 May–13 July 1999 occupied dense thickets of Saltcedar with a few Goodding's Black Willows (M. A. Patten). Wintering birds are more catholic, occupying virtually any scrubby habitat, native or not.

TAXONOMY. Breeders in the region, the summering bird from the Imperial Valley, and the vast majority of wintering birds are *T. b. charienturus* Oberholser, 1898, of cismontane southern

California. It is the darkest and brownest of the Bewick's Wrens, with dark central rectrices partly masking the black barring. A few wintering birds are the transmontane *T. b. eremophilus* Oberholser, 1898, paler grayish brown above with paler central rectrices showing distinct black barring. Only two Salton Sink specimens, from 7 km northwest of Imperial 26 October 1996 (P. Unitt; SDNHM 48929) and 11 km southeast of Brawley 19 January 1989 (R. Higson, SDNHM 45884), are *T. b. eremophilus;* all 23 others are *T. b. charienturus. T. b. eremophilus* is the predominant wintering form on the Colorado River (Swarth 1916; Linsdale 1936; Phillips et al. 1964), so the Salton Sink marks the eastern limit of the normal winter range of *T. b. charienturus.*

HOUSE WREN

Troglodytes aedon parkmanii Audubon, 1839
Uncommon winter visitor (September through April); has bred. An exuberant songster, the House Wren is common throughout North America. In the Southwest it is a bird of riparian forest and oak woodland. It breeds fairly commonly at a number of locales fringing the northern edge of the Coachella Valley (e.g., Whitewater Canyon, Morongo Valley) but is essentially only an uncommon migrant through, and winter visitor to, the Salton Sink. Migrants and wintering birds occur from 28 August (1994, Poe Road; G. McCaskie) to 1 May (1909, female at the south end; MVZ 8188). A juvenile along the New River 3 km north-northwest of Seeley 4 August 1985 (*AB* 39:963; SDNHM 43916) and an apparent pair collected there 18 August 1989 (R. Higson; SDNHM 47555, 47556) constitute the only evidence of breeding in the region. Another adult was noted at Fig Lagoon 1 August 2002 (G. McCaskie).
ECOLOGY. Wintering House Wrens occupy dense scrub, especially near water (though water is not essential), such as stands of Saltcedar, Goodding's Black Willow, and Arrowweed. They also occur in suburban, ranch land, and orchard habitats throughout the region. Breeding birds occupied a remnant patch of riparian woodland with Fremont Cottonwood, now largely elimi-

nated. The lack of suitable nest holes in most of the remaining riparian habitat undoubtedly discourages nesting efforts.
TAXONOMY. The widespread western subspecies is *T. a. parkmanii.* It differs from the nominate subspecies, of the East, in being grayer below and drabber (less rufescent) above, with dusky dorsal barring.

WINTER WREN

Troglodytes troglodytes (Linnaeus, 1758) subspp.
Casual late fall and early winter vagrant (mid-November through December). The diminutive Winter Wren is a rare bird in the desert Southwest, especially so in the Sonoran Desert, where most records are from late fall and winter. It is casual in the Salton Sink, with only about 12 records in the narrow window from 20 November (1988, Finney Lake; B. E. Daniels) to 24 December (1986, Brawley from 22 December; *AB* 41:331).
ECOLOGY. Winter Wren records are from dense, scrubby habitats (e.g., Saltcedar, Common Reed, mesquite), usually with available water.
TAXONOMY. There are no specimens for the Salton Sea. Elsewhere in southern California and in Arizona, specimens are either the dark rufous *T. t. pacificus* Baird, 1864, of the Pacific Northwest, or the medium dark, browner *T. t. salebrosus* Burleigh, 1959, of the northwestern Great Basin ranges (Rea 1986b). Save for one, winter wrens in the Salton Sink have looked and sounded like birds from western populations. One at Wister 5–22 December 1998 (*NAB* 53:210) was *T. t. hiemalis* Vieillot, 1819, of the East, on the basis of its call (K. L. Garrett), a sweet Song Sparrow–like "tcheep" rather than the harsh Wilson's Warbler–like "chimp" uttered by all of the subspecies in the West.

MARSH WREN

Cistothorus palustris (Wilson, 1810) subspp.
Common breeding resident; numbers augmented in winter (mid-September to early April). With the Common Yellowthroat and the Song Sparrow, the Marsh Wren is one of a trio of common riparian and marshland passerines in the Salton

Sink. Each is absent from much of the desert Southwest, being confined to substantial river courses and lacustrine habitats. This wren has a spotty distribution in the Colorado Desert of California; it is common only at and around the Salton Sea and otherwise occurs at scattered oases (e.g., Big Morongo Canyon Preserve). Breeding takes place mainly from late March through June, with nesting activity peaking in May (e.g., a nest 12 km northwest of Calipatria 13 May 1937; MVZ 2597). Birds nest well into summer; for example, there was a nest with six eggs near Westmorland 24 June 1970 (WFVZ 149312) and an adult carrying food along the Alamo River 12 July 1999 (M. A. Patten). Young fledge between May and August. Males sing throughout the year, regardless of breeding condition. Winter brings an influx of birds from the Great Basin and the interior Southwest, making numbers higher from mid-September to early April (Unitt et al. 1996).

ECOLOGY. Breeding Marsh Wrens are associated with marsh habitats along the edges of rivers, ponds, and lakes. Nests are placed mainly in cattail, bulrush, and Common Reed. Birds forage in these substrates and in adjacent thickets of dense Saltcedar. The species seldom strays far from water. Almost all wintering birds and migrants occupy similar marshland habitats, though occasional migrants appear in dry, weedy brush, in isolated patches of trees, and in parks, ranch yards, and well-wooded neighborhoods.

TAXONOMY. The breeding subspecies throughout central California and the California desert, including the Salton Sea and the lower Colorado River valley, is *C. p. aestuarinus* (Swarth, 1917), distinguished from migrants from farther north by its darker rump and scapulars, buffier underparts, shorter wings, and blacker crown. *C. p. deserticola* Rea, 1986, type locality 3.5 km north-northwest of Seeley, is not differentiated sufficiently from *C. p. aestuarinus* over most of the latter's range, with specimens matching *C. p. deserticola* originating even from southern Oregon (Unitt et al. 1996). In winter the paler, larger subspecies of the Great Basin and the Rocky Mountain region invade. Of 13 recent specimens

of migrants, most are *C. p. plesius* Oberholser, 1897. The earliest in fall is from 11 km east of Calexico 5 October 1986 (P. Unitt; SDNHM 44592); records from elsewhere in southern California imply that the fall arrival of *C. p. plesius* at the Salton Sea should be expected in the third week of September. We have not studied specimens from the breeding range of *C. p. plesius* to assess whether *C. p. pulverius* (Aldrich, 1946), breeding along the east side of the Cascade Range and the Sierra Nevada, is adequately differentiated from it, as recognized by Phillips (1986). Three or 4 of 13 winter specimens from the Imperial Valley (SDNHM) have extensively white underparts and pale tawny rump and scapulars, approaching SDNHM 43469 (Owens Lake, Inyo County, 20 September 1984) and so implying *C. p. pulverius*. We wish to emphasize that *C. p. paludicola* Baird, 1858, of the coastal Pacific Northwest, is unknown from the region. Although it is attributed to the Sonoran Desert by the AOU (1957), Miller et al. (1957), and others, there is in fact no evidence that this sedentary brown-backed subspecies with a reduced black-and-white back patch has even reached California (Unitt et al. 1996).

REGULIDAE • *Kinglets*

The Old World warblers and their kin, including the kinglets and the gnatcatchers, form a conspicuous, albeit not diverse, component of the landbird fauna in the Salton Sink. Both the Ruby-crowned Kinglet and the Blue-gray Gnatcatcher are fairly common winter visitors, furtively nabbing arthropods in thickets, riparian belts, and neighborhoods throughout the region. The Black-tailed Gnatcatcher is also fairly common, but it is a breeding resident (and thus the only species present during summer).

GOLDEN-CROWNED KINGLET
Regulus satrapa Lichtenstein, 1823 subsp.?
Casual late fall and winter visitor (late October through February). The tiny Golden-crowned Kinglet is a denizen of boreal coniferous forests in the far north and of high mountains in the West. It stages irregular irruptions into the

lowlands of the Southwest, including the deserts of California. It is casual in the Salton Sea region, with records every two to three years on average from 22 October (1977, near Westmorland; AB 32:259) to 6 March (1982, 2 at Brawley; G. McCaskie), with most in November and December. Generally only one to four birds are encountered, but up to eight were at Finney Lake 26 December 1986–25 January 1987 (AB 41:331), a winter that brought numerous other montane species to the California lowlands.

ECOLOGY. These kinglets are most frequently encountered near the apexes of tall trees, principally ornamental pines and Athel.

TAXONOMY. There are no specimens for the Salton Sea (contra AB 28:109). Most fall and winter birds taken in the deserts of southeastern California are R. s. apache Jenks, 1935, of which R. s. amoenus van Rossem, 1945, is a synonym (Phillips 1991). This large, pale, gray subspecies breeds in the northern Coast Range, the Sierra Nevada, and high mountains of southern California north to southeastern Alaska. Salton Sink records presumably pertain to this subspecies. The similar nominate subspecies, from the East, with a less distinct supercilium, bolder wing covert and secondary fringes, and a stubbier bill, has been taken in Death Valley, southern Nevada, and southern Arizona (Phillips 1991); it may account for some regional records.

RUBY-CROWNED KINGLET
Regulus calendula calendula (Linnaeus, 1766)
Fairly common winter visitor (late September to mid-May). The Ruby-crowned Kinglet is one of the more common wintering passerines in coastal California and Baja California. It is somewhat less common in the Salton Sea region, though it is nonetheless a conspicuous, prevalent component of the wintering avifauna where there are trees. Records extend from 24 September (1994, 3 km west of Westmorland; R. Higson, SDNHM 48947) to 14 May (1999, Salton Sea NWR; G. McCaskie), with most present from October through April.

ECOLOGY. Any tree or shrub seems to suffice as a foraging or roosting site for this species,

but it seems partial to mesquite, tamarisk, and pines.

TAXONOMY. Wintering birds throughout southern California are of the nominate subspecies, which breeds across the continent, including in the Cascades, the Great Basin Ranges, the Sierra Nevada, and higher mountains of southern California. R. c. cineraceus Grinnell, 1904, is a synonym. The nominate subspecies is paler, grayer-backed, and whiter-bellied than R. c. grinnelli Palmer, 1897, breeding in southern Alaska and coastal British Columbia and wintering south to the north coast of California, rarely as far south as San Diego.

SYLVIIDAE • Old World Warblers and Gnatcatchers

BLUE-GRAY GNATCATCHER
Polioptila caerulea obscura Ridgway, 1883
Fairly common winter visitor (mid-August to mid-May). The Blue-gray Gnatcatcher breeds fairly commonly over much of temperate North America, including the chaparral belt of California. It is a short-distance migrant, with most wintering birds reaching only the southern United States and northern Mexico. In the Salton Sea region birds have returned as early as 17 August (2000, Salton Sea NWR; G. McCaskie), but most arrive in mid-September. By winter the Blue-gray Gnatcatcher is more common than the Black-tailed. Most depart by mid-April, with one 10 May 1997 (G. McCaskie) being the latest.

ECOLOGY. As noted by van Rossem (1911), this species is "as a rule found in trees while [P. melanura is] more often found in the low brush." Wintering birds reach their peak abundance in thickets dominated by Saltcedar and Common Reed that pervade the river courses, multitudinous drainage ditches, and various other mesic habitats. It occurs with the Black-tailed in this habitat during winter but is less numerous than that species in saltbush or Iodine Bush scrub. Both species occur in mesquite thickets.

TAXONOMY. The subspecies occurring throughout the West is P. c. obscura, which includes P. c. amoenissima Grinnell, 1926, as a

FIGURE 60. The Black-tailed Gnatcatcher remains fairly common wherever stands of native desert shrubs persist, and like the Verdin, it has begun to frequent stands of Saltcedar. Photograph by Kenneth Z. Kurland, May 2001.

synonym (Monson and Phillips 1981; Phillips 1991). All California specimens are of this drabber, browner subspecies. Even so, the bluer (within age and sex classes) nominate subspecies, of the East, is migratory and thus may reach California as a vagrant (Phillips 1975b).

BLACK-TAILED GNATCATCHER
Polioptila melanura lucida van Rossem, 1931
Fairly common breeding resident. Although it tends to be uncommon over much its range in the southwestern United States and northern Mexico, the Black-tailed Gnatcatcher (Fig. 60) is a fairly common breeding resident in the Salton Sea region, just as it is in the neighboring Anza-Borrego Desert to the west. It reaches peak abundance along the rivers and in areas with expansive desert scrub. During the winter the Blue-gray Gnatcatcher is often equally or more abundant along the rivers and larger irrigation canals (i.e., more mesic habitats), but the Black-tailed is always more abundant in desert scrub and in dry washes. Breeding begins in early March and continues into July, with most activity from late March (van Rossem 1911; PSM 15842) through May.

ECOLOGY. Although this species occurs extensively in thickets of Saltcedar and natural willow riparian woodland along rivers and drainage ditches, it is generally outnumbered in this habitat during winter by the Blue-gray Gnatcatcher. However, the Black-tailed Gnatcatcher prevails in stands of mesquite, in desert scrub dominated by Quail Brush, and along washes supporting Blue Palo Verde. Nests are typically placed in dense shrubs, especially saltbush, mesquite, and Saltcedar, but one pair nested 3 m above the ground on a bough of a mature Fremont Cottonwood at Mecca in mid-June 1999 (M. A. Patten).

TAXONOMY. Gnatcatchers in the Sonoran Desert are *P. m. lucida*, which differs from the disjunct nominate subspecies, of the Chihuahuan Desert, in having a pale base to the mandible, more white at the edge of the outer rectrices, and a paler, browner mantle.

TURDIDAE • *Thrushes*

Presumably as a result of the lack of extensive wooded habitats, thrushes are only modestly represented in the Salton Sink, with but few species and generally low abundances. The Hermit Thrush and the American Robin are both regular in winter, but aside from occasional incursions of the latter, both tend to be rather uncommon. Likewise, Swainson's Thrush is an uncommon spring transient. The bluebirds, Townsend's Solitaire, and the Varied Thrush are rarer still, none of them occurring annually (although the Western Bluebird's occurrence is nearly annual).

WESTERN BLUEBIRD
Sialia mexicana Swainson, 1832 subspp.

Rare to casual winter visitor (early November through March). The Western Bluebird is a common bird in cismontane California up through the oak belt. It is a rare, nearly annual winter visitor to the southern Coachella Valley, but it is a casual visitor to the Imperial Valley in that season. Records extend from 7 November (1918, male and female at Indio; LACM 2848, 2849) to 31 March (1942, near Mecca; C. A. Harwell). Most occurrences are of small flocks (2–5 birds), with a maximum of 20 at the golf course in Mexicali 14 December 1994 (T. E. Wurster) and 27 at three locations on the Salton Sea (south) Christmas Bird Count 19 December 2000 (*fide* G. McCaskie). In addition, a flock of 12 was observed at Brawley 7 February 1982 (G. McCaskie), 10–12 were at Ramer Lake 30 January–24 February 1990 (G. McCaskie et al.), and 10 were near Salton Sea NWR 22 December 1998–19 March 1999 (M. A. Patten et al.).

ECOLOGY. This bluebird is most often recorded in mesquite (especially those festooned with Desert Mistletoe) or tamarisk thickets but has also been found in orchards in the vicinity of Coachella.

TAXONOMY. Two females found at Ramer Lake 28 December 1993 (P. Unitt; SDNHM 48753, 48754) are the rustier-backed *S. m. bairdi* Ridgway, 1894, which breeds in mountains of Arizona and elsewhere in the interior Southwest. It has been recorded throughout the Colorado and Mojave Deserts in winter and is the expected subspecies in the Imperial and Mexicali Valleys. Eight specimens (LACM) from the Coachella Valley are of the less rusty *S. m. occidentalis* Townsend, 1837, of cismontane California; occasional individuals in the Imperial Valley may belong to this subspecies.

MOUNTAIN BLUEBIRD
Sialia currucoides (Bechstein, 1798)

Rare, irregular winter visitor (mid-November to early March); absent most years. The Mountain Bluebird is a species of the Great Basin that occasionally stages winter incursions into lowlands throughout the Southwest. It is a highly unpredictable bird in the Coachella and Imperial Valleys, absent more years than it is present. When it occurs, however, it can be fairly numerous, with flocks numbering in the many dozens in some years, in the hundreds in others (e.g., more than 750 on 28 and 29 December 1971 and 1,174 on 19 December 2000; *AB* 26:521, *NAB* 55:229), and reaching an unparalleled 3,000–4,000 in the Imperial Valley 20 January 1968 (G. McCaskie). Records extend from 9 November (1986, male and 2 females 14 km north-north-east of Plaster City; P. Unitt, SDNHM 44825, 44826, 44845) to 5 March (1989, female 12 km west-northwest of Imperial; R. Higson, SDNHM 47298).

ECOLOGY. This species feeds in barren dirt fields and in open agricultural fields with sparse vegetation as long as stakes, low shrubs, dried Russian Thistle, or utility lines are available for perches.

TOWNSEND'S SOLITAIRE
Myadestes townsendi townsendi (Audubon, 1838)

Casual fall and winter visitor (early October through February); 1 spring record. The elegant Townsend's Solitaire frequents montane coniferous forests throughout the West. It is a rare to irregularly uncommon fall migrant through the Mojave Desert but is less frequent in the Sonoran Desert, being casual in the Salton Sea region. Most birds encountered in the region are fall transients from 2 October (1997, 3 km southeast of El Centro; K. Z. Kurland) to mid-November (e.g., Brawley 10 November 1996; M. A. Patten). Three together at Hot Mineral Spa 21 November 2000 (G. McCaskie) were late migrants. Wintering birds are less frequent, averaging one every two to three years, with dates from late November (e.g., El Centro 28 November 1996; M. A. Patten) to 22 February (1992, Ramer Lake; S. Goldwasser). A male taken northeast of Calipatria 7 April 1962 (E. A. Cardiff; SBCM 36964) represents the only spring record for the region.

ECOLOGY. The solitaire is most often encountered in tall deciduous trees, especially Fremont Cottonwood with Desert Mistletoe or other sources of berries. A female near Indio 10 January 1890 (SDNHM 1681) was taken in a Date Palm orchard (F. Stephens).

TAXONOMY. Solitaires in the United States are the nominate subspecies, a paler form with buffier markings on the wings.

SWAINSON'S THRUSH
Catharus ustulatus (Nuttall, 1840) subspp.
Uncommon (to occasionally fairly common) spring transient (mid-April to mid-June); 2 fall records.
Swainson's Thrush is a common spring migrant in the western Mojave Desert, but it is fairly common to uncommon elsewhere in the desert Southwest in that season. The Salton Sea region is one of those areas where it is less common. The window of occurrence is wide, from 7 April (1992, 10 km west of Westmorland; R. Higson, SDNHM 48185) to 10 June (1989, 11 km southeast of Brawley; R. Higson, SDNHM 46092), with a peak during the first three weeks of May. However, it is unusual to tally more than a few in a day, with double digits seldom achieved, although on some spring days it can be fairly common (e.g., 20 in the Imperial Valley 16 May 1999; M. A. Patten). There are only two fall records (*contra* Garrett and Dunn 1981:59), both recent: individuals 3 km southeast of El Centro 29 September 2000 and at Wister 30 September 2000 (both *NAB* 55:104).

ECOLOGY. Migrant Swainson's Thrushes seek shaded areas with heavy tree cover. Thus, most are encountered in the older portions of large towns (e.g., suburban neighborhoods, parks, cemeteries) and in patches of vegetation in natural settings (e.g., the headquarters of Wister and Salton Sea NWR, Ramer and Finney Lakes).

TAXONOMY. All specimens from the Salton Sea region are of the richly rufescent nominate subspecies, of the Pacific Northwest. *C. u. oedicus* (Oberholser, 1899), the paler, less rufescent subspecies breeding from the Sierra Nevada south through southern California, presumably moves through the region in small numbers as it winters largely in western Mexico (Phillips 1991).

HERMIT THRUSH
Catharus guttatus (Pallas, 1814) subspp.
Uncommon transient and winter visitor (mid-September to mid-May). The Hermit Thrush is a common winter visitor throughout the coastal slope of California, particularly in the southern portion. Moreover, this species is a fairly common migrant through much of the desert Southwest. However, it is an uncommon bird in the Salton Sea region, where it occurs as both a migrant and a winter visitor. Records extend from 21 September (1975, Westmorland; G. McCaskie) to 18 May (1991, 8 km southwest of Westmorland; S. von Werlhof), with most being from mid-October through mid-April.

ECOLOGY. Shaded lawns and gardens on ranches and in suburbia, cemeteries, and parks host most Hermit Thrushes. Some occur in habitats less modified by humans, particularly those with dense, mesic thickets of Saltcedar or mesquite.

TAXONOMY. The pattern of occurrence of the Hermit Thrush in the Salton Sink is a faint echo of that in coastal southern California, to the west, where the species is common. Mirroring its status nearer the coast, the nominate subspecis, *C. g. guttatus*, breeding in southwestern and south-central Alaska, is the most frequent subspecies, accounting for five of seven recent specimens. Two of these specimens are somewhat intermediate toward *C. g. vaccinius* (Cumming, 1933), SDNHM 47597 in its darker back, SDNHM 49830 in its heavier breast spots and darker gray flanks. Neither, however, combines all the distinctions of *C. g. vaccinius,* which breeds in coastal mainland British Columbia and is regular but less common than *C. g. guttatus* and *C. g. nanus* (Audubon, 1838) in winter in cismontane southern California. The earliest of these five specimens, from 11 km northwest of Imperial 25 September 1994 (SDNHM 48949), nearly matches the earliest sight record. A fall migrant from the Crucifixion Thorn grove 12 km southeast of Ocotillo 6 October 1986 (SDNHM 44617) is *C. g. nanus,* resembling *C. g. guttatus* in its underpart pattern but more rufous above. This subspecies, breeding in southeastern Alaska, was renamed *C. g. osgoodi* by Phillips (1991), but Dickerman and Parkes (1997) argued for retention of the traditional name *C. g. nanus.* A spring migrant 12 km south of Seeley 30 March 1989 (SDNHM 45915) is the small, pale, grayish, lightly spotted *C. g. slevini* (Grinnell, 1901). *C. g. slevini* breeds in the northwestern United States south to coastal central California and winters in western Mexico, so it occurs in southern California primarily as a migrant. Interestingly, the relatively large, gray, heavily spotted *C. g. sequoiensis* (Belding, 1899), which breeds from the Sierra Nevada south in high mountains to the Sierra San Pedro Mártir (Erickson and Wurster 1998) and migrates through the eastern Mojave Desert, has not been recorded in the Salton Sink or in northeastern Baja California (Patten et al. 2001). The lone winter specimen of *C. g. sequoiensis* (Belding, 1899) reported by Grinnell and Miller (1944) from the lower Colorado River 2 January 1931 (MVZ 81417) is actually *C. g. slevini* (M. A. Patten). Given the Colorado River record and a specimen from San Diego 18 February 1971 (SDNHM 37879) of *C. g. slevini,* or *C. g. jewetti* (Phillips 1962), if it is not a synonym of *C. g. slevini,* it is possible that *C. g. slevini* winters rarely as far north as the Salton Sink. A *C. g. slevini* from San Felipe in northeastern Baja California 12 April 1926 (Grinnell 1928; MVZ 48388) was clearly a migrant.

AMERICAN ROBIN
Turdus migratorius propinquus Ridgway, 1877
Winter visitor (mid-October through April), irregularly uncommon to common; has bred at Brawley and Indio. Compared with the remainder of California, the Sonoran Desert generally experiences a dearth of American Robins. This species is abundant across most of Canada and the United States but is generally an uncommon winter visitor at the Salton Sea. Records extend from 18 October (2000, 3 at Wister; G. McCaskie) to 1 May (1894 and 1995, New River 4 km northwest of Calexico and the Highline Canal east of Calipatria 1 May 1995; USNM 133411, LSUMZ 166710), with most present from mid-November through March. One at the south end 7 May 1988 (G. McCaskie) was exceptionally late. In most winters numbers are small (15–20 birds), but in a few they are much larger (e.g., 150 in the Imperial Valley 26 January 1992; M. A. Patten).

This species has bred occasionally in the well-wooded suburban neighborhood of southwestern Brawley since the mid-1990s. The presence of a pair including a singing male 2 August 1992 (M. A. Patten) was the first evidence that birds were summering in the area. A pair was building a nest 16 April 1994 (FN 48:989); they tended the nest until 25 June 1994 (G. McCaskie), but the eventual outcome of the effort is unknown. Breeding was not recorded in 1995 or 1996, but a pair was observed gathering nest material 5 April 1997 (R. Higson). Nesting was also documented in a residential part of Indio in April 1983 (R. L. McKernan). The robin occasionally breeds elsewhere in the California desert (Garrett and Dunn 1981), but breeding is otherwise unknown in the Salton Sink.

ECOLOGY. The nesting site in Brawley features abundant well-watered lawns and leafy trees, including Fremont Cottonwood, sycamore, maple, gum, fig, and other nonnative species. Wintering birds occupy similar habitats. Sizable congregations are frequently seen tending fruiting California Fan Palm (especially) and clumps of Desert Mistletoe.

TAXONOMY. Robins breeding in California and wintering in the Southwest are the large, pale *T. m. propinquus,* whose rectrices have small white corners.

VARIED THRUSH
Ixoreus naevius [meruloides] (Swainson, 1832)
Casual fall (mainly) and winter vagrant; 1 spring record. A striking species of old-growth forests of the Pacific Northwest, the Varied Thrush is rare in the desert Southwest, with most records being from the Mojave Desert and the Great Basin. It is only casual at the Salton Sea, with most records for fall from 3 October (1975, north end; *AB* 30:128) to 29 November (1957, Mecca; *AFN* 12:60). There are only three winter records, of one found dead (but not preserved) at Niland 28 December 1972 (*AB* 27:664), the first for the Imperial Valley, followed by one at the north end 3 January 1998 (*fide* C. McGaugh) and a female at Finney Lake 11 March 1978 (*AB* 32:400). Records are fairly evenly split between the Coachella and Imperial Valleys. The single spring record is of a bird near Holtville 12–13 April 1984 (*AB* 38:961).

ECOLOGY. Wooded parklands, residential areas, and Date Palm groves have supported all Varied Thrushes in the region.

TAXONOMY. There are no extant specimens for the Salton Sea, but we presume that records pertain to *I. n. meruloides,* a widespread winter visitor to California that breeds in the northern portion of the species' range. Only this subspecies has been collected in western Nevada (Linsdale 1936) and southern Arizona (Phillips et al. 1964), and it is the more frequent in coastal southern California. Males are similar in all subspecies, but females of *I. n. meruloides* differ from those of the nominate form, which breeds south to northwestern California and is partially migratory, in being paler and grayer.

MIMIDAE • *Mockingbirds and Thrashers*

Thrashers have a checkered history in the Salton Sink. Probably the most numerous mimid in the region at the turn of the twentieth century was Le Conte's Thrasher. The spread of agriculture and suburbia eliminated most of the saltbush scrub favored by this species, essentially extirpating it. The Crissal Thrasher suffered a similar demise, although small numbers persist in remaining mesquite thickets (sometimes with mixed Saltcedar) at scattered locales in the Coachella, Imperial, and Mexicali Valleys. However, the novel human-created habitats resulting from extensive irrigation brought the Northern Mockingbird to the region, where it is now the most common mimid. The Sage Thrasher, an uncommon spring transient and rare winter visitor, is the only regularly occurring mimid whose status has remained essentially unchanged over the past century.

GRAY CATBIRD
Dumetella carolinensis (Linnaeus, 1766) subsp.?
One fall record. The Gray Catbird is a scarce vagrant to the far West, mainly during migration. There are nearly a hundred records for California, but one at the Highline Canal 10 km northeast of Calipatria 26 October 1999 (Rogers and Jaramillo 2002) provided the only record for the Salton Sink. One at Ocotillo 13–17 June 1995 (Garrett and Singer 1998) was just west of the region.

TAXONOMY. Although the Gray Catbird is generally regarded as monotypic, Phillips (1986) recognized *D. c. ruficrissa* Aldrich, 1946, from west of the Great Plains. This subspecies allegedly differs from the nominate in its paler, more rufescent crissum. If recognized, most California records are probably assignable to this westerly subspecies.

NORTHERN MOCKINGBIRD
Mimus polyglottos polyglottos (Linnaeus, 1758)
Common breeding resident. Cultivation of the Coachella, Imperial, and Mexicali Valleys favored expansion of the Northern Mockingbird into the Salton Sink (Arnold 1980). This species was unknown in the region shortly after the turn of the twentieth century (Grinnell 1911b). It was first recorded along the New River 25 November 1906

(F. Stephens; SDNHM 1486), it was breeding by March 1911 (van Rossem 1911), and it was established and widespread by the 1930s (Arnold 1935). This species is now a common breeding resident throughout the agricultural and otherwise settled portions of the Salton Sink. Breeding begins in late March (van Rossem 1911) and peaks from April through June, when males defend their nesting territory vigorously, even against Loggerhead Shrikes (Arnold 1935) and other tenacious species.

ECOLOGY. As noted by Arnold (1980), with the advent of suburbia this species expanded its range significantly. It was formerly reported to breed in native desert scrub in the Coachella Valley (Arnold 1935), but the species is now seldom encountered away from habitats modified by humans (Rosenberg et al. 1987). Thus, it occurs mainly in suburban vegetation, ranch yards, parks, and orchards. It is one of the most commonly seen species in gardens and yards in the region. Some birds occur in riparian woodland along the rivers, especially the Whitewater River, and in mesquite thickets laden with fruiting Desert Mistletoe (Ohmart 1994).

TAXONOMY. *M. p. leucopterus* (Vigors, 1839) is a synonym of the nominate subspecies (Phillips 1961). Thus, birds throughout continental North America north of the Isthmus of Tehuantepec belong to *M. p. polyglottos*, characterized by its largely white primary coverts and grayish breast.

SAGE THRASHER

Oreoscoptes montanus (Townsend, 1837)

Uncommon spring transient (late January to mid-April); rare winter visitor (late to mid-November through February); casual fall transient (September–October). The Sage Thrasher is probably the earliest spring migrant passerine in the West. By February, and sometimes by mid-January (Gilman 1907), spring migrants are already heading north to their breeding grounds in the vast ocean of sagebrush in the Great Basin. Migrants have been detected from 17 January (1999, Imperial; B. Miller) to 12 April (1987, Finney Lake; G. McCaskie), with a peak in late February and early

March. One at the south end 30 April 1967 (G. McCaskie) was exceptionally late. Generally only one to two birds are encountered in spring, but six were counted near Oasis 6 February 1999 (M. D. Better), as well as 3 km southwest of Westmorland 23 March 2000 (G. McCaskie). Schneider (1928) considered it "common" in the Imperial Valley on 29 February 1928 and saw it "in flocks" not far north of the region at Palm Springs during mid-February that year. This species winters in northern Mexico and the southern portions of the Southwest, including around the Salton Sea. However, it is rare, but annual, in the region during fall and winter, with records from 12 November (1996, 5 km west of Westmorland; P. Unitt; SDNHM 49940) to 24 February (1990, Ramer Lake from 29 January; K. A. Radamaker et al.). Most records are for December and January. Oddly, despite being a regular migrant elsewhere in the California deserts, the Sage Thrasher is only a casual fall transient through the Salton Sea region from 1 September (1969, Rock Hill; G. McCaskie) to 24 October (1976, 6 km northwest of Niland; P. F. Springer).

ECOLOGY. As if mirroring its preference on the breeding grounds, this species favors scrub habitats around the Salton Sea. It is most often encountered in saltbush scrub but also occurs frequently in mesquite thickets.

BROWN THRASHER

Toxostoma rufum longicauda (Baird, 1858)

Three winter records (mid-December to mid-March). The Brown Thrasher is a scarce vagrant to California from the eastern United States, with fewer than ten records annually. It has been recorded three times around the Salton Sea, at Heise Springs 15 January–18 March 1972 (Winter 1973), at Finney Lake 26 February 1978 (*AB* 32:400), and near Calexico 13 December 1990 (*AB* 45:322).

TAXONOMY. All California specimens are the paler western subspecies *T. r. longicauda*, so it presumably accounts for the Salton Sea records. Although it is generally larger, its size overlaps greatly with that of the nominate subspecies, of eastern North America (Rand and Traylor 1950).

BENDIRE'S THRASHER

Toxostoma bendirei (Coues, 1873)

Four winter records (mid-September to early March). Bendire's Thrasher is a casual migrant and winter visitor to the Colorado Desert, with four records for the Salton Sea (England and Laudenslayer 1989). All regional records are best treated as pertaining to wintering birds. Reports of one near Niland 12 November 1967 (*AFN* 22:90) and 2 March 1968 (*AFN* 22:479) were from the same location, so we infer that they involved the same wintering bird. Other records are of birds photographed at Salton Sea NWR 1 November 1964–25 January 1965 (McCaskie and Prather 1965) and Salton City 14 September 1993–16 February 1994 (*AB* 48:153; *FN* 48:249) and of one carefully studied at the north end 30 December 1995 (*FN* 50:848). The bird in 1964–65 was originally published as a Curve-billed Thrasher, but photographs revealed a pale base to the mandible and a relatively short, straight bill (Roberson 1993). Of note just outside the area covered here is a specimen collected from Palm Springs 8 April 1885 (Stephens 1919; SDNHM 1507); this presumed spring migrant likely passed through the Salton Sink. The identity of one reported near Seeley (*FN* 51:928) is questionable.

ECOLOGY. These thrashers have been observed in tamarisk and mesquite scrub or in residential yards with similar habitat elements.

TAXONOMY. This species is monotypic, as both *T. b. candidum* van Rossem, 1942, and *T. b. rubricatum* van Rossem, 1942, are invalid (Phillips 1962).

CURVE-BILLED THRASHER

Toxostoma curvirostre palmeri (Coues, 1872)

Four winter records (mid-December to mid-April). The Curve-billed Thrasher is a casual vagrant to California, but 4 of the 14 records are from the Salton Sink. There are two records from Finney Lake, of one photographed 25 January–13 April 1976 and a second individual seen 31 January–22 March 1976 (Luther et al. 1979; *AB* 30:768). One was seen near the mouth of the New River 20 December 1979–4 February 1980 (Binford

1985), and one was photographed at Brawley 21 January–3 March 1990 (Patten and Erickson 1994). Additional sight reports from Brawley (*AB* 29:123) and the south end (published as a Curve-billed or Bendire's Thrasher; *AB* 28:535) lack documentation and are thus unacceptable (Roberson 1993). Furthermore, the Curve-billed Thrasher record published by McCaskie and Prather (1965) pertained to a Bendire's Thrasher (see the Bendire's Thrasher account).

ECOLOGY. Each Curve-billed Thrasher occupied Saltcedar or Quail Brush scrub near water, although the bird in Brawley occasionally frequented Pomegranate thickets near a defunct Date Palm orchard.

TAXONOMY. California specimens (five at SDNHM from Bard, on the lower Colorado River, from 1916 to 1925) are of the relatively unspotted *T. c. palmeri* of Arizona and Sonora, the expected subspecies given its westerly distribution. *T. c. palmeri* is resident west to within a mere 30 km of the Colorado River (Rosenberg et al. 1991).

CRISSAL THRASHER

Toxostoma crissale coloradense van Rossem, 1946

Uncommon breeding resident. The Crissal Thrasher is nowhere common, although it approaches that status in the lower Colorado River valley, not far to the east (Rosenberg et al. 1991). By contrast, it is an uncommon breeding resident in the Salton Sea region. Unfortunately, because the extensive conversion of native habitats to agricultural lands and the rapid spread of Saltcedar, this species is now local and declining around the Salton Sea. Mecca, Coachella, Finney and Ramer Lakes, Fig Lagoon, and portions of the Whitewater, New, and Alamo Rivers support almost all of the regional population. As with Le Conte's Thrasher, nesting activity begins early in the year. As an example, three nests in the Coachella Valley already contained either eggs or nestlings by 12 February 1933 (Hanna 1933b), so young would fledge by early March. However, nesting continues later in the spring, with some birds still on eggs through April (e.g., Mecca 29 April 1908; MVZ 60).

ECOLOGY. Desert riparian and wash habitats are favored by this species, but typically only if there is sufficient mesquite and sufficient ground cover. The species also occurs in Ironwood and sometimes in thickets of Saltcedar (Hunter et al. 1988), but at the Salton Sea a general truth is that where there is no mesquite there will be no Crissal Thrashers.

TAXONOMY. *T. c. coloradense* is endemic to the western half of the Sonoran Desert (i.e., the Colorado Desert) and thus is the only subspecies in California and northeastern Baja California. It is the palest dorsally of three subspecies (Phillips 1986).

LE CONTE'S THRASHER
Toxostoma lecontei lecontei Lawrence, 1851
Casual postbreeding visitor (mid-March to mid-November); formerly an uncommon breeding resident in the Coachella Valley and probably also in the Imperial Valley. Whereas the Northern Mockingbird reaped the benefits of cultivation of the Coachella, Imperial, and Mexicali Valleys, Le Conte's Thrasher suffered the costs. This species once occurred throughout the open desert of the Salton Sink and definitely bred in the southern Coachella Valley around Thermal, Mecca, and probably Indio, where a pair were taken 8 November 1918 (LACM 92812, 92813). Breeding began early, with a nest in a cholla containing three eggs by 31 January 1932 (Coachella Valley; Hanna 1933a) and others in late January of other years (Hanna 1937). Given the early nesting schedule, it is not surprising that juveniles had appeared by April, as at Mecca 13 April (1908; MVZ 794) and 28 April (1908; MVZ 1079) and Thermal 26 April (1914; UCLA 256). The population at Mecca seemed to be thriving around the turn of the twentieth century, with numerous specimens collected between 1889 (MVZ 10916) and 1913 (MVZ 103568), but it has long since been extirpated.

The historical status of Le Conte's Thrasher's in the Imperial Valley is less clear. All specimens were collected from 14 September (1952, 16 km northwest of Westmorland; SBCM 36876) to 16 April (1909, male along the New River 8 km northwest of Westmorland; MVZ 8082). Given its late winter nesting in the Coachella Valley, there is no reason to assume that records from the Imperial Valley were of wintering birds, but there is no breeding evidence either. A widespread population in the Imperial Valley is suggested by records of six at the south end 25–27 November 1933 (LACM 18194–97, 22764–65), two at Kane Spring 8 and 11 December 1933 (LACM 18214–215), and single birds at a nonspecific valley locale 12 December 1910 (SDNHM 24766), at Holtville 25 December 1934 (UCLA 41050), 20 km east of Calexico 29 December 1938 (MVZ 74948), and along the New River 8 km northwest of Westmorland 14 April 1909 (MVZ 8081). Presumably the species was formerly a nesting resident but was driven out by urbanization and cultivation on a large scale (see Laudenslayer et al. 1992).

In the past three to four decades the species has been only an apparent postbreeding dispersant to the Salton Sink, with all recent records from 11 March (1974, El Centro; LACM 85414) to 21 November (1967, 2 near Niland; *AFN* 22: 91) and most in August and September at Salton City. It is now recorded but once every four to five years on average. Presumably birds seen at Salton City dispersed from undeveloped desert scrub just west of the region; Le Conte's Thrasher remains an uncommon resident in the Anza-Borrego Desert and along the eastern flank of the Santa Rosa Mountains, west of Oasis (R. L. McKernan). A specimen taken 14 km north-north-east of Plaster City 9 November 1986 (P. Unitt; SDNHM 44827) was in suitable breeding habitat, and R. P. Henderson noted the species at San Sebastian Marsh 25–31 May 1977.

ECOLOGY. That this species is colored like sand is no coincidence. In the Mojave and Sonoran Deserts it occurs almost exclusively in open Creosote and Mojave Yucca scrub, particularly around sand dunes or sandy washes. Much of this sort of habitat was destroyed around the Salton Sea through extensive irrigation, although some persists along the western edge of the Imperial Valley, toward the Anza-Borrego Desert. The most recent bird noted at Salton City

(7 August 1993; M. A. Patten) was in open salt-bush scrub.

TAXONOMY. Mensural analyses by Sheppard (1996) notwithstanding, there are two subspecies in California, the nominate, in the southeastern desert, and *T. l. macmillanorum* Phillips, 1965, in the San Joaquin Valley (Browning 1990). The latter is diagnosed by its darker dorsal coloration, not by its size, which explains why Sheppard (1996) found no differences. Even so, the subspecies may yet prove to be invalid given that preliminary colorimetric analyses found little difference among populations in the Southwest (Zink et al. 1997).

MOTACILLIDAE • *Wagtails and Pipits*

AMERICAN PIPIT

Anthus rubescens pacificus Todd, 1935

Common winter visitor (mid-September to mid-May). The American Pipit is a widespread winter visitor in the Southwest both on the coast and inland. It is common in the Salton Sea region in that season, with records from 20 September (1981, 2 at the south end; G. McCaskie) to 12 May (1968, Whitewater River delta; *AFN* 22:577, SBCM 37280) and the majority present from mid-October to mid-April. It reaches its peak abundance in the Coachella, Imperial, and Mexicali Valleys but also occurs along the shore of the Salton Sea.

ECOLOGY. The agricultural areas of the region provide an abundance of suitable habitat for this species. Fallow fields or fields with short grass (e.g., Perennial Rye, Wheat), Alfalfa, cabbage, or other low-growing crops are ideal, particularly when irrigated. Smaller numbers occur on barnacle beaches and mudflats on the shore of the sea, and some are found foraging along the edge of freshwater impoundments.

TAXONOMY. Specimens from the Salton Sea at SDNHM match the somewhat heterogeneous subspecies *A. r. pacificus,* of central and southern Alaska and the northern Rocky Mountains, of which *A. r. geophilus* Lea and Edwards, 1950, is a synonym (AOU 1957; Parkes 1982; Gibson and Kessel 1997; cf. Phillips 1991:146). They are not

gray enough dorsally nor lightly streaked enough ventrally to be *A. r. alticola* Todd, 1935, of the southern Rocky Mountains, the Great Basin, and the Sierra Nevada. Likewise, their dorsal coloration is not the dark rich brown of the nominate subspecies, of boreal Canada and the northeastern United States.

SPRAGUE'S PIPIT

Anthus spragueii (Audubon, 1844)

Three winter records. Sprague's Pipit is a casual fall and winter vagrant to California, with about 30 records from the northern deserts and the coastal slope. It regularly winters in grasslands of southeastern Arizona and northern Sonora (Monson and Phillips 1981; Russell and Monson 1998). Because suitable habitat, in the form of fallow agricultural fields, is extensive in the Salton Sink, it is somewhat surprising that there are but three Salton Sea records. A bird flying over Brawley 29 November 1986 (Langham 1991) provided the first. A flock of up to 12, by far the largest recorded in California, in fields of dormant Perennial Rye 10 km north of Calipatria 10 January–8 March 1998 (Erickson and Hamilton 2001) could perhaps have been predicted given the species' range and habitat requirements. At least eight were photographed in fields east of Calipatria 17 December 2002–25 January 2003+ (CBRC 2002-217). A report from near Westmorland (*AB* 41:331) lacks documentation.

BOMBYCILLIDAE • *Waxwings*

CEDAR WAXWING

Bombycilla cedrorum Vieillot, 1808

Generally uncommon but erratic winter visitor (late August through May). A showy bird of more northern climes, the Cedar Waxwing (Fig. 61) is uncommon in much of the desert Southwest. Its status is similar around the Salton Sea, where it is an uncommon winter visitor from 31 August (1986, 3 km north-northwest of Seeley; P. Unitt) to 27 May (2000, 2 near Kakoo Singh Reservoir; M. A. Patten), with most present from October to mid-April. Fifty at Salton Sea NWR 24 May 1962 (*fide* D. V. Tiller) constituted a notable late

FIGURE 61. A winter visitor, the Cedar Waxwing is usually uncommon in the Salton Sink, feeding most heavily on the fruits of fan palms. Photograph by Kenneth Z. Kurland, November 1997.

PHAINOPEPLA

Phainopepla nitens lepida van Tyne, 1925

Uncommon winter visitor and breeder (late September to mid-May); rare in late spring and early summer (mid-May to early June); generally absent in late summer and early fall (mid-June through mid-September). Although widespread and common through much of the desert Southwest, the Phainopepla is uncommon around the Salton Sea. Like the Crissal Thrasher, this species is a rather local breeding resident in the Salton Sea region. It is concentrated in the southern Coachella Valley, particularly around Mecca, Coachella, and Thermal. It was formerly more numerous; for example, 100 or more were in the southern Coachella Valley 22 March 1942, including a number of nesting pairs (C. A. Harwell). It is now unusual for counts to reach 20 birds. This species breeds around San Sebastian Marsh and in scattered locales in the Imperial and Mexicali Valleys with native patches of mesquite. As suggested above, breeding activity peaks in late winter and early spring, with fledglings appearing by the end of March (Gilman 1903; van Rossem 1911). A young female already in postjuvenal molt taken 5 km northeast of Imperial 25 April 1987 (P. Unitt; SDNHM 44683) shows how early the young disperse.

Breeders disperse away from the Salton Sink shortly after nesting. Most of them have departed by mid-May, with records as late as 27–28 May (2000, 3 in the Imperial Valley; M. A. Patten) and 9 June (1990, Wister; M. A. Patten). The Phainopepla is generally absent from the Salton Sink after May and before a hypothesized fall influx of wintering birds from breeding locales in the Mojave Desert and in foothills of the Peninsular and Transverse Ranges (Crouch 1943). Wintering birds arrive by the end of September (e.g., at least 3 in mesquite thickets at Mecca 30 September 2000; M. A. Patten). There are a few records for earlier in September, such as a molting female at Niland 3 September 2000 (G. McCaskie) and adult males at Mecca 10 Sep-

concentration, although 11 remained at Brawley through the summer of 1987 (*AB* 41:1488). This species periodically stages massive invasions (e.g., 1,910 at the north end 29 December 1970; *AB* 25:504), but winter numbers are generally low, with no more than a few dozen birds present throughout the region.

ECOLOGY. Orchards and suburban neighborhoods with an abundance of fruiting trees hold most waxwings. This species normally avoids native habitats except mesquite thickets with Desert Mistletoe in fruit (Ohmart 1994).

TAXONOMY. This species is monotypic, as neither *B. c. larifuga* Burleigh, 1963, nor *B. c. aquilonia* Burleigh, 1963, is consistently diagnosable (Mayr and Short 1970; Browning 1990; Phillips 1991; cf. Behle 1985).

tember 2000 (M. A. Patten) and 5 km southwest of Brawley 11 September 2000 (G. McCaskie). There are but three midsummer records, of two at Wister 15 June–1 July 2001 (G. McCaskie), two in the Imperial Valley 19–27 June 1999 (M. A. Patten), and one at Brawley 23 July 1989 (M. A. Patten). One at Ramer Lake 24 August 1996 (S. von Werlhof) furnished the sole Salton Sink record for that month. The scattering of records from June through mid-September suggests that the Phainopepla occasionally summers in the region, though some September individuals may have been migrants.

ECOLOGY. The Phainopepla has a symbiotic relationship with the Desert Mistletoe (see Ohmart 1994), the mistletoe parasitizing mainly Honey and Screwbean Mesquites, Ironwood, and Catclaw. Not surprisingly, then, this species breeds only in mesquite thickets, and most wintering birds are found in similar habitat. Indeed, the decline of mesquite throughout the Salton Sink has no doubt led to the regional decline of this species. Some migrants and dispersants occur in suburban neighborhoods, ranch yards, and orchards, where they forage on nonnative fruits.

TAXONOMY. Birds in the West, including those in California, Baja California, the Great Basin, Arizona, and Sonora, belong to the small subspecies *P. n. lepida*.

PARULIDAE • Wood-Warblers

The Salton Sink is an important migratory corridor for wood-warblers and other passerines during spring. Thousands of warblers move through the region, with the Orange-crowned, Yellow, and Wilson's Warblers being particularly common (see Fig. 57). The sink is much less heavily used in fall, although large numbers of these three species (and MacGillivray's Warbler) also occur in that season. By contrast, the Nashville, Townsend's, Hermit, and Black-throated Gray Warblers, all fairly common to uncommon in spring, are seldom recorded in fall. Winter brings an abundance of Yellow-rumped Warblers. That species, the Orange-crowned Warbler, and the Common Yellowthroat are the only species

that winter in numbers in the region, although the Imperial Valley hosts a tiny wintering population of American Redstarts. With the extirpation of Lucy's Warbler, only the yellowthroat and the Yellow-breasted Chat breed in the Salton Sink, the latter uncommonly. As in much of the West, a variety of wood-warblers generally associated with eastern North America have been recorded in the region, principally in migration, with a few records for winter. Yet vagrant warblers are notably scarcer in the Salton Sink than elsewhere in the southern California desert.

BLUE-WINGED WARBLER
Vermivora pinus (Linnaeus, 1766)
One fall record. It is not surprising that the Blue-winged Warbler has been found at the Salton Sea only once, for it is a casual vagrant anywhere in California, with fewer than 30 records statewide. The sole record is of a male at Finney Lake 13 September 1988 (Pyle and McCaskie 1992).

TENNESSEE WARBLER
Vermivora peregrina (Wilson, 1811)
Six records, 3 in spring (May) and 3 in fall (October to mid-December). The Tennessee Warbler is a rare annual vagrant to California and the West. Like almost all of the "eastern" wood-warblers that have a more northerly distribution, this species is much more frequent on the coastal slope than it is in the interior. There are but six records for the Salton Sea, three of spring vagrants: one at Finney Lake 4 May 1969 (*AFN* 23: 627), a female at the Vail Ranch, 9 km north of Westmorland, 11 May 1974 (*AB* 28:853), and a singing male at Salton Sea NWR 28 May 1994 (K. L. Garrett). Birds at the north end 17 December 1977 (*AB* 32:884) and Niland 18–21 December 1984 (*AB* 39:211) were probably late fall vagrants. Additionally, one was photographed 3 km southeast of El Centro on an unknown date in October 2000 (K. Z. Kurland).

ORANGE-CROWNED WARBLER
Vermivora celata (Say, 1823) subspp.
Common transient and winter visitor (mid-July to mid-May). The Orange-crowned Warbler is a

common bird in the West, in contrast to its status in the East. It is one of the most numerous warblers around the Salton Sea, being a common migrant and winter visitor. Fall migrants arrive early, with one at the mouth of the Whitewater River 12 July 1986 (M. A. Patten) being the earliest. The majority of birds arrive from mid-August through September. The Orange-crowned Warbler is common through the winter, but numbers climb during spring, with an influx of migrants (e.g., 45 at Brawley 24 April 1988; M. A. Patten) from March to mid-May. A male 7 km west-southwest of Westmorland 21 May 1995 (P. Unitt; SDNHM 49239) is the latest. A female at Salton Sea NWR 12 June 1994 (K. C. Molina; LACM 108414) and a bird at the Salton Sea State Recreation Area 23 June 1998 (*FN* 52:504) were anomalous. The former was likely an exceptionally late spring migrant (see below); the latter may have been an early fall migrant, a late spring migrant, or a dispersant from one of the breeding populations in local mountains.

ECOLOGY. Wintering Orange-crowneds are most numerous in thickets of Saltcedar and Common Reed along the rivers, irrigation canals, and ditches. They also frequent similar mesic habitats with Athel, Goodding's Black Willow, and Honey Mesquite. Migrants are also common in these habitats, but they are slightly more common in suburban neighborhoods, parklands, and ranch yards.

TAXONOMY. Three of the four described subspecies have occurred at the Salton Sea, two of them regularly in large numbers. *V. c. lutescens* (Ridgway, 1872), of the Pacific Coast, characterized by bright yellow, lightly streaked underparts and yellow-green upperparts, is the predominant form, accounting for about 60 percent of the population. It arrives earlier in fall and departs earlier in spring than *V. c. orestera*, with records from 12 July (1986, cited above) to 16 May (1999, Ramer Lake; M. A. Patten) and peak numbers present from mid-August through April. *V. c. orestera* Oberholser, 1905, of the Great Basin, characterized by moderately yellow, lightly streaked underparts, drab yellow-green upper-

parts, and a grayish head in the female, accounts for about 40 percent of the population. Records extend from 10 September (2000, mouth of the Whitewater River; M. A. Patten) to 21 May (1995, cited above). The earliest date for a fall specimen is 13 September (1992, female 12 km south of Seeley; R. Higson, SDNHM 48202). An exceptionally early fall transient was observed at Mexicali 1 September 2000 (R. A. Erickson). The June specimen cited above is of *V. c. orestera* and thus a late migrant. Spring migrant Orange-crowned Warblers after the first week of May are virtually always *V. c. orestera*. The nominate subspecies, of Alaska through eastern North America, is characterized by pale, often blotchy yellow, lightly streaked underparts, pale gray-green upperparts, and a contrasting gray head in both sexes. It has been recorded three times in late fall but is probably regular in small numbers. A male was taken 11 km northwest of Imperial 4 October 1987 (P. Unitt; SDNHM 45096), and females were taken 7 km northwest of Imperial 19 October 1989 (R. Higson; SDNHM 47677) and 3 km north-northwest of Seeley 25 October 1986 (P. Unitt; SDNHM 44814).

NASHVILLE WARBLER
Vermivora ruficapilla ridgwayi van Rossem, 1929
Fairly common spring transient (late March to mid-May); uncommon to rare fall transient (August to early October); casual in winter. Most wood-warblers (and other passerines) move through the Salton Sea region in much higher numbers in spring than in fall, but few more strikingly so than the Nashville Warbler. This species is a fairly common transient in spring, when records extend from 24 March (1991, male at Brawley; M. A. Patten, G. McCaskie) to 20 May (1995, 6 km west-southwest of Westmorland; R. Higson, SDNHM 49198). Peak passage is in late April, when the species is occasionally common (e.g., 85 at Brawley 24 April 1988; M. A. Patten). In stark contrast, fall records are few, all involving only one or two birds. Records extend from 30 July (2000, 3 km southeast of El Centro; G. McCaskie) to 5 October (2000, Wister; K. Z. Kurland), with a slight peak in early September and a specimen

of a male taken north of Westmorland 2 October 1948 (E. A. Cardiff; SBCM 681). As anywhere in the West, this species is a casual winter visitor to the region, with records averaging about one every two to three years from 26 November (1999, Wister; B. Miller) to 3 March (1999, Mexicali; Patten et al. 2001). One at Wister 18 October 2000 (G. McCaskie) either was attempting to winter or was an exceptionally late fall migrant.

ECOLOGY. Well vegetated residential areas, ranch yards, orchards, and dense riparian scrub are used by Nashvilles, but the first attract most of the migrants, whereas the last harbors most wintering birds.

TAXONOMY. Birds west of the Rocky Mountains are *V. r. ridgwayi*, differing from the nominate subspecies, of the East, in its grayer back, brighter yellow rump, more rapid tail wagging, and hard call note (Dunn and Garrett 1997). There is as yet no specimen evidence that the nominate subspecies has occurred in California (see Phillips 1975b), but it undoubtedly occurs, probably annually, as a vagrant given its breeding range (west through central Saskatchewan) and highly migratory habits.

VIRGINIA'S WARBLER

Vermivora virginiae (Baird, 1860)

Seven records, 4 in fall (mid-August to mid-September), 2 in winter, 1 in spring. Virginia's Warbler is an uncommon and local breeder in the desert mountains of eastern California. It is a rare migrant anywhere in southern California, with even fewer records for winter. Contrary to some popular accounts (e.g., Small 1994), this species is virtually unrecorded as a spring migrant in most of the interior of California. The only credible spring record for the Salton Sea was furnished by an individual in the company of numerous migrant landbirds at Finney Lake 7 May 1972 (*AB* 26:810). By contrast, Virginia's Warbler is found regularly in small numbers in fall, along the coast and in the interior, from mid-August through September. Even so, there are only four records in fall for the Salton Sea region, from Finney Lake 17 August 1968 (*AFN* 23:110), Brawley in the

"fall" of 1972 (*AB* 27:123), the south end 14 September 1973 (*AB* 28:109), and Red Hill 3 September 1987 (*AB* 42:138). There are two winter records, both involving birds that returned for multiple winters. One was seen at Finney Lake 17 February 1989 (*AB* 43:368) and 2 December 1989–17 March 1990 (*AB* 44:331), and another was noted at Brawley 17–24 February 1990 (*AB* 44:331) and 10 November 1990–31 January 1991 (*AB* 45:322).

LUCY'S WARBLER

Vermivora luciae (Cooper, 1861)

Casual vagrant; formerly an uncommon breeder (mid-March to mid-August). A dynamic "wee gray mite" (Phillips 1968), Lucy's Warbler has had a plastic distribution in the Southwest over the past century. It has spread in some areas, especially to the north and east (Phillips 1968), but has withdrawn from others. This species formerly bred at the north end of the Salton Sea, in the vicinity of Mecca, where it was first detected 29 March 1911 (van Rossem 1911; UCLA 10702), and Coachella. A small colony persisted in stands of mesquite in southern Coachella until 1985 (R. L. McKernan). The enlarged testes of a singing male collected at Silsbee 8 April 1909 (Grinnell and Miller 1944; MVZ 8118) suggest former breeding in the Imperial Valley as well. The species bred just east of the Imperial Valley, northeast of Holtville, as recently as 1994 and 1995 (R. L. McKernan). Lucy's Warbler is now only a casual straggler to the Salton Sink, with only five records away from Mecca since the mid-1960s. Single birds were at the south end 17 May 1969 (*AFN* 23:627) and 28 September 1974 (*AB* 29:123), at Salton Sea NWR 27 July–22 August 1997 (*FN* 51:1054; S. B. Terrill), and 5 km southwest of Brawley 14 July 2001 (P. A. Ginsburg) and 2 June–1 August 2002 (*NAB* 56:487).

ECOLOGY. Mesquite is a favored nesting tree for this warbler. The spread of Lucy's Warbler in parts of the Southwest is probably directly related to the spread of mesquite in overgrazed grasslands (Phillips 1968). In contrast, mesquite, especially trees large enough to provide suitable nest cavities, has declined drastically in much of southeastern California (Grinnell and Miller 1944:397), either by direct removal or as a result of competition with Saltcedar.

NORTHERN PARULA
Parula americana (Linnaeus, 1758)

Casual vagrant in winter (mid-December to early April), spring (late April to early June), and fall (mid-September through November). In the Salton Sink, as elsewhere in California, the Northern Parula is a rare spring and fall migrant and winter visitant. Records have increased in recent decades, and it now breeds in most years along the central coast (Patten and Marantz 1996). It is one of few "eastern" wood-warblers that occurs more frequently in spring than in fall (Roberson 1980), and migrants greatly outnumber wintering or summering birds. It is therefore surprising that half of the Salton Sea records involve wintering birds. There are nine winter records extending from 14 November (2000, immature female at Wister through 21 March 2001; *NAB* 55:104) to 9 April (1997, male at Brawley from 2 February; *FN* 51:804). A female at Wister 1 January–12 February 2002 (C. McFadden et al.) possibly had returned from the previous winter. The other six winter records are of birds at Niland 13–28 December 1978 (*AB* 33:316) and 18 December 1979–16 January 1980 (*AB* 34:308, 662), a female near Red Hill 21 December 1989 (*AB* 44:331), birds near the mouth of the Whitewater River 2 January–6 March 1993 (*AB* 47:302) and at Mecca 21 January–23 February 2002 (C.-T. Lee et al.), and a male near the mouth of the Alamo River 22 March 1969 (E. A. Cardiff; SBCM 37446). More typical of records statewide were individuals at Ramer Lake 26 April 2002 (D. W. Aguillard) and 3 km southeast of El Centro 16–17 May 2000 (K. Z. Kurland), a female 6.5 km west-southwest of Westmorland 21 May 1995 (P. Unitt; SDNHM 49185), a male 12 km northwest of Imperial 26 May 1991 (R. Higson; SDNHM 47745), and a singing male at Wister 9 June 2000 (*NAB* 54:424). In fall, a female was photographed 3 km southeast of El Centro 12 September 2000 (K. Z. Kurland), and late migrants were an immature female near the mouth of the New River 2 November 1993 (*AB* 48:153) and a bird at Wister 29 November 1998 (*NAB* 53:106). A report of a female at Ramer Lake (*AB* 45:153), though perhaps correct, is supported by details that do not conclusively eliminate other warbler species.

ECOLOGY. Parulas have been observed in either tall Athel or Fremont Cottonwood or in tall ornamental trees in residential neighborhoods.

YELLOW WARBLER
Dendroica petechia (Linnaeus, 1766) subspp.

Common spring (April to mid-June) and fall (mid-July through October) transient; rare winter visitor (late October through March). The Yellow Warbler rounds out the trio of common migrant warblers in the Southwest that are colored bright yellow, the others being the Orange-crowned Warbler and Wilson's Warbler. The Yellow is a common spring and fall migrant through the Salton Sea region, with a few birds wintering annually. Spring records extend from 3 April (1994, adult male at Salton Sea NWR; G. McCaskie) to 9 June (1990, south end; G. McCaskie), with a peak from late April through the third week of May (e.g., 75 on 20 May 1995; G. McCaskie). An adult male at Mecca 31 March 1908 (C. H. Richardson Jr.; MVZ 649) was exceptionally early and perhaps wintered locally. This species is one of southern California's earliest migrant landbirds in fall. The first transients appear in mid-July, with one 11 July 1998 being the earliest (G. McCaskie). Fall movement peaks from mid-August to mid-September but tails off quickly thereafter. The latest record of a migrant is of one found 7 km northwest of Imperial 19 October 1989 (R. Higson; SDNHM 47678). This species was first recorded in winter, when one was found at the south end 13 February 1966 (*AFN* 20:460), although note the March specimen above. It has since proved regular in small numbers, with generally two to four found each year (cf. *AB* 44:331), mainly in the Imperial Valley. Winter records extend from 24 October (1999, Alamo River near Red Hill through at least 21 December; M. A. Patten et al.) to 27 March (1994, mouth of the New River from 16 December 1993;

G. McCaskie). Surprisingly, it is unknown as a breeder in the region, even historically. A few pairs formerly bred on the Río Hardy southeast of the Mexicali Valley (Grinnell 1928).

ECOLOGY. Migrant Yellow Warblers occur just about anywhere that there is vegetation. They tend to occupy trees and taller shrubs more often than either the Orange-crowned or Wilson's Warbler, and they have an especial penchant for Fremont Cottonwood, willows, and other deciduous trees.

TAXONOMY. The common migrant and wintering subspecies in the West is *D. p. morcomi* Coale, 1887, to which the above account applies. This subspecies is by far the most common migrant through the Salton Sea region, accounting for 27 of 34 recent specimens from the Imperial Valley (SDNHM). No specimens have been collected in winter, but we suspect that wintering birds—bright yellow during that season—are *D. p. morcomi* as well. We cannot consistently distinguish *D. p. brewsteri* Grinnell, 1903, so we consider it a synonym of *D. p. morcomi* (Behle 1948; AOU 1957; *contra* Browning 1994). *D. p. morcomi* is similar to *D. p. aestiva*, of the East, with moderately yellow coloration, a bright (but relatively pale) olive dorsum, and a yellow forehead. The distinctive *D. p. sonorana* Brewster, 1888, was formerly common on the lower Colorado River (where it is now rare; R. L. McKernan) and along the Río Hardy east of Cerro Prieto (Grinnell 1928), about 10 km southeast of the Mexicali Valley (where it is now extirpated). It is mostly yellow dorsally with fine chestnut streaking ventrally and a wholly yellow crown. There is no evidence that it occurred in the Salton Sink.

Small numbers of *D. p. rubiginosa* (Pallas, 1811), breeding along the Pacific coast from south-central Alaska to southern British Columbia, pass through the region in spring and fall, in both seasons later, on average, than *D. p. morcomi*. The sink's five spring males of *D. p. rubiginosa*, identified by their darker greenish backs and foreheads, extend from 29 April (1989, 7 km northwest of Imperial; R. Higson, SDNHM 47403) to 8 June (1989, 7 km northwest of Imperial; R. Higson, SDNHM 47530). Yet *D. p.* *morcomi* evidently migrates late as well: SDNHM 47531, collected at the same time as 47530, is a female *D. p. morcomi*. Identifying the subspecies confidently in fall, when all age and sex classes must be considered separately, requires better comparison specimens than we have seen. Nevertheless, SDNHM 48216 and 48217, an adult female and an adult male, respectively, from 3 km west of Westmorland 18 October 1992 (P. Unitt), have plumages that are notably darker and less yellow than corresponding plumages of *D. p. morcomi* and so apparently are *D. p. rubiginosa*. Specimens in the San Diego Natural History Museum of apparent *D. p. rubiginosa* from elsewhere in southern California and northern Baja California were taken from 30 September to 21 November.

CHESTNUT-SIDED WARBLER

Dendroica pensylvanica (Linnaeus, 1766)

Casual winter vagrant (mid-November to mid-April); 4 fall records (early October to early November). Chestnut-sided Warbler records for the Salton Sea region defy explanation. Elsewhere in California and the West this species is a rare fall vagrant and a casual spring and winter vagrant (Garrett and Dunn 1981). In the case of the Salton Sea, on the other hand, a mounted specimen from near Niland 5 October 1952 (Cardiff and Cardiff 1953b; SBCM M1919), one banded at Salton Sea NWR 9–16 October 1993 (*AB* 48: 153), and immatures at Brawley 2 November 1993 (*AB* 48:153) and Hot Mineral Spa 18 November 2000 (E. A. Cardiff) furnished the only four fall records. The remaining 12 records are for winter, with dates ranging from 9 November (1986, Brawley into February 1987; *AB* 41:146, 331) to 21 April (1990, Finney Lake from 11 February and molting into alternate plumage before departing; *AB* 44:331, 498). The bird at Brawley had been present the previous winter, 22 February–30 March 1986 (*AB* 40:335); it thus provided one of few records for anywhere in the Southwest of a returning "eastern" warbler. Apart from immatures near Mecca 24–29 December 1970 (*AB* 25:630) and 10 March 2001 (D. Guthrie) and at the north end 27 December 1974 (*AB* 29:580) and 23 December 1976 (*AB* 31:880), all area records of the Chestnut-sided Warbler are from the Imperial Valley.

ECOLOGY. Many Chestnut-sideds have been seen in Saltcedar or Common Reed scrub along rivers or canals, but some have been noted in residential neighborhoods, mesquite scrub, or Fremont Cottonwood.

MAGNOLIA WARBLER
Dendroica magnolia (Wilson, 1811)
Four fall records (late September through November). The Magnolia Warbler is a rare vagrant to California, but it is regularly recorded in the interior, especially in the Mojave Desert, north of the Transverse Ranges. Thus, it is surprising that only four have been found at the Salton Sea, all in fall. An immature of unknown sex was collected 7 km northwest of Imperial 29 September 1989 (R. Higson; SDNHM 47390), one was seen at Seeley 25 October 1986 (*AB* 41:146), an immature male was photographed 3 km southeast of El Centro 28–29 October 2000 (K. Z. Kurland), and one was observed near Westmorland 28 November 1982 (*AB* 37:225).

CAPE MAY WARBLER
Dendroica tigrina (Gmelin, 1789)
Three records, 1 in late fall and 2 in winter. The Cape May Warbler is a rare to casual vagrant to California, with a few winter records. Its occurrence in California is strongly tied to outbreaks of the Spruce Budworm, *Choristoneura fumiferana* (Patten and Burger 1998), and records have been decreasing statewide for two decades. This warbler has been recorded thrice at the Salton Sea, twice in winter and once in fall. The first two regional records were during the species' peak, in the 1970s, when one was observed at Finney Lake 5–20 March 1978 (*AB* 32:1211) and another was photographed there 1 November 1979 (*AB* 34:203). The only other area record was of an immature male at Wister 28 December 1993–8 January 1994 (*FN* 48:249). The identity of one reported along the Alamo River near Red Hill 24 February 1996 (*FN* 50:225) is questionable.

BLACK-THROATED BLUE WARBLER
Dendroica caerulescens caerulescens Gmelin, 1789
Four fall records (mid-October to early November). The Black-throated Blue Warbler is a rare fall vagrant to California, although it is one of the few "eastern" wood-warblers that are recorded almost as frequently in the interior of southern California as they are on the coast. There are four records for the Salton Sink, all from mid-October to early November, fitting the pattern elsewhere in the state. A male was at the north end 10 October 1975 (*AB* 30:129), individuals were 8 km west of Westmorland 13 October 1968 (*AFN* 23:110) and 7 November 1970 (*AFN* 25:111), and a male was 3 km southeast of El Centro 16 October 1993 (K. Z. Kurland).

TAXONOMY. All California records pertain to the widespread nominate subspecies. Indeed, the blacker *D. c. cairnsi* Coues, 1897, of the southern Appalachian Mountains, is extremely unlikely to occur anywhere in the West.

YELLOW-RUMPED WARBLER
Dendroica coronata (Linnaeus, 1766) subspp.
Common winter visitor (early September through mid-May). In winter the Yellow-rumped Warbler is the most numerous and widespread warbler in the United States. Around the Salton Sea its winter numbers exceed those of both the Orange-crowned Warbler and the Common Yellowthroat. It is common from 6 September (1992, 11 km southeast of Brawley; P. Unitt, SDNHM 48153) to 21 May (1995, Wister; M. A. Patten), with most occurring from early October through April. Two distinct groups occur in the region (see below), with Audubon's Warbler constituting the majority.

ECOLOGY. This species occupies just about any wooded or scrub habitat in the region, native or not. Small numbers even forage in agricultural fields adjacent to ditches lined with scrubby Saltcedar. Open saltbush or Creosote scrub is the only habitat in the region that the Yellow-rumped Warbler avoids, although it may occur in isolated mesquite thickets even in these areas. Still, it reaches peak abundance in stands of gum in suburban neighborhoods, parks, and riparian areas with large trees.

TAXONOMY. The common (sub)species wintering in the Salton Sink is Audubon's Warbler, *D. [c.] auduboni* (Townsend, 1837), in which we include *D. [c.] memorabilis* Oberholser, 1921, as a synonym (*fide* Paynter 1968 and Hubbard 1970), although we lack the necessary specimen material to judge its validity ourselves. Some 2–3 percent of wintering birds are the Myrtle Warbler, *D. [c.] coronata,* in which we similarly include *D. c. hooveri* McGregor, 1899, as a synonym (*fide* Monroe 1968; Paynter 1968; Rea 1983). The Myrtle Warbler has been recorded from 5 November (1955, 15 km north of Westmorland; E. A. Cardiff, SBCM 37524) to 16 May (1999, first-spring male near Obsidian Butte; M. A. Patten). It is distinguished from Audubon's Warbler by its white throat, fewer rectrices with white patches, a conspicuous supercilium, its different throat pattern, its call and song, and various other features (Dunn and Garrett 1997; Pyle 1997).

The Yellow-rumped Warbler as currently recognized (AOU 1998) comprises three distinct groups: (1) the Myrtle Warbler, breeding across northern North America from Alaska to Newfoundland and the Northeast; (2) Audubon's Warbler, *D. [c.] auduboni* and *D. [c.] nigrifrons* Brewster, 1889, breeding in western North America from British Columbia to northwestern Mexico; and (3) Goldman's Warbler, *D. [c.] goldmani* Nelson, 1897, resident in southern Mexico and northern Guatemala. Subspecies of Audubon's Warbler differ in the extent of black on the underparts and mantle, especially in males in breeding plumage, with *D. [c.] auduboni,* of the Pacific Coast, grayest, grading to blacker in the Great Basin and the Rocky Mountains; *D. [c.] nigrifrons,* of northwestern Mexico, is blacker still. In Goldman's Warbler the gray is largely replaced by black and the yellow throat border is replaced by a white one. Songs and calls differ among the three groups, and each group is diagnosable by plumage, even in the field, in all age and sex classes (Howell and Webb 1995; Dunn and Garrett 1997; Pyle 1997).

No one has conducted research on species limits of the disjunct Goldman's Warbler, but ex-

tensive research has been conducted on species limits of the Myrtle and Audubon's Warblers. The two groups were universally recognized as separate species until Hubbard (1969) published an analysis of hybridization in the Rocky Mountains of British Columbia (see Campbell et al. 2001). On the basis of apparent introgression about 75 km either side of a narrow contact zone, Hubbard concluded that the two forms showed no tendency toward assortative mating. Barrowclough (1980) later provided genetic evidence supporting Hubbard's conclusion. Subsequent genetic work paints a somewhat different picture, with divergence of mitochondrial DNA restriction fragments and sequence between Myrtle and Audubon's Warblers comparable to that between Townsend's and Hermit Warblers and Eastern and Spotted Towhees (Bermingham et al. 1992; Klicka and Zink 1997). More or less genetic divergence is meaningful only if we accept the notion that all speciation occurs in allopatry, and no one knows how much genetic divergence defines a species. Nonetheless, these additional data, in concert with the concordant and relatively sharp differences in phenotype and voice, imply that the Myrtle and Audubon's Warblers, at least, are biological species (Zink and McKitrick 1995).

BLACK-THROATED GRAY WARBLER
Dendroica nigrescens (Townsend, 1837)
Fairly common spring transient (late March through May); uncommon fall transient (late August to mid-November); casual in winter; 2 summer records. A widespread breeder in the more xeric montane coniferous forests of the West, the Black-throated Gray Warbler is a fairly common migrant through the Southwest. It is a somewhat early spring migrant, with dates ranging from 21 March (1911, 8 at Mecca; van Rossem 1911) to 31 May (1999, singing male at the mouth of the Whitewater River; M. A. Patten) and a peak during the second half of April and the first half of May. A bird at the Salton Sea 11 March 1976 (*AB* 30:886) probably had wintered locally. This species is less common in fall, with records from 24 August (1991, south end; M. A. Patten) to 13 November

(1994, Brawley; G. McCaskie) and peak movement from mid-September to mid-October. It is casual in winter in the region, with about 17 records from 2 November (1993, Brawley through 15 January 1994; *FN* 48:249, G. McCaskie) to 31 March (1985, 2 in Brawley from 1 December 1984; *AB* 39:211, G. McCaskie), with a high count of 3–4 at Ramer (not Finney) Lake 8–9 February 1964 (*AFN* 18:388). Surprisingly, this pinewoods species has summered once in the Salton Sink, when an apparently injured female remained in a yard 3 km southeast of El Centro from 3 July to 19 August 2001 (*NAB* 55:483; K. Z. Kurland). Remarkably, another (or the same?) bird appeared in the same yard 29 July 2002 (*NAB* 56:487).

ECOLOGY. Like other migrant warblers, this species may be encountered anyplace in the region where there is sufficient tree growth. It seems partial to Athel, mesquite, and stands of broadleaf trees, especially native Fremont Cottonwood and Goodding's Black Willow.

BLACK-THROATED GREEN WARBLER
Dendroica virens virens (Gmelin, 1789)
One late fall record. The Black-throated Green Warbler is a rare fall and casual spring and winter vagrant to California from the eastern United States. Only one has been recorded at the Salton Sea, a probable late fall vagrant at the north end 23 December 1976 (*AB* 31:375).

TAXONOMY. California records are of the widespread nominate subspecies, which breeds widely in northeastern North America. The weakly differentiated and probably invalid *D. v. waynei* Bangs, 1918, of lowland pine woodlands of Virginia and the Carolinas, is an unlikely candidate for vagrancy to the West.

TOWNSEND'S WARBLER
Dendroica townsendi (Townsend, 1837)
Fairly common spring transient (mid-April to early June); rare fall transient (late August to mid-November); casual in winter (late November through March). Unlike the Black-throated Gray, Townsend's Warbler breeds in mesic coniferous forests, particularly old-growth forests of the Pacific Northwest. It is a fairly common spring migrant and a rare fall migrant through the Salton Sea region, a status typical of the California deserts. Spring dates extend from 17 April (1982, south end; G. McCaskie) to 6 June (1999, 2 at Finney Lake; *NAB* 53:432), with a peak during the first half of May. A male 3 km southeast of El Centro 6 April 2002 (*NAB* 56:358) was exceptionally early. Given that most of its somewhat late fall migration in the West is along the immediate coast and through the mountains, it is not surprising that there are few fall records for the Salton Sink. Indeed, the species is scarce in fall anywhere in the California desert. Regional records extend from 26 August (1999, 3 km southeast of El Centro; K. Z. Kurland) to 12 November (1989, male 7 km northwest of Imperial; R. Higson, SDNHM 47668), with most from mid-September to mid-October. Exceptionally early was a male photographed 3 km southeast of El Centro 13 August 2001 (K. Z. Kurland). This species has been recorded on nine occasions in winter, from 28 November (1970, near Westmorland; G. McCaskie) to 25 March (1978, immature at Brawley; G. McCaskie). Other records are of individuals at the north end 17 December 1977 (*AB* 32:884), Calipatria 17 December 2002 (*fide* G. McCaskie), the south end 26 December 1981 (*AB* 36:758), Salton Sea NWR 7 January 1995 (*FN* 49:200), and Brawley 30 January 1983 (*AB* 37:339), a female along the Alamo River near Red Hill 22 December 1987 (*AB* 42:1136), and three at Ramer Lake 15 January 1966 (G. McCaskie).

A Townsend's × Hermit Warbler, a hybrid produced on a broad front of range contact in the Pacific Northwest (Rohwer and Wood 1998) and regularly noted in California during migration (Dunn and Garrett 1997:321), was at El Centro 2 May 2001 (P. A. Ginsburg; G. McCaskie).

ECOLOGY. Most migrant Townsend's Warblers are encountered in Athel and Saltcedar. In contrast to the Black-throated Gray, but like the Hermit, in suburban areas Townsend's Warbler is more likely to be encountered in planted pines than in broadleaf deciduous trees.

HERMIT WARBLER
Dendroica occidentalis (Townsend, 1837)

Uncommon spring transient (mid-April to mid-June); casual to rare fall transient (mid-August to early October). In general the Hermit Warbler is rare in the desert Southwest. The majority migrate via mountain corridors (Miller and Stebbins 1964:215; Weathers 1983:197), particularly during fall. The species is an uncommon spring migrant through the Salton Sea region, with small numbers passing through each year (probably averaging fewer than 20 individuals per spring, with higher numbers in some years). Most apparently overfly the region, many making landfall along the eastern escarpment of the southern Peninsular Ranges (e.g., in spring 1999 and 2000; P. Unitt). Records for the Salton Sink extend from 17 April (1982, south end; G. McCaskie) to 11 June (1989, female 7 km northwest of Imperial; R. Higson, SDNHM 47381), with a peak during the last week of April and the first half of May. Generally only one or two birds are noted per day even during peak movement, with a maximum of ten in the Imperial Valley 12 May 1963 (G. McCaskie). The few fall records for the Salton Sink (only ca. 15 are documented) extend from 16 August (2002, photographed 3 km southeast of El Centro; K. Z. Kurland) to 7 October (1971, south end; G. McCaskie) and include a specimen of a male north of Westmorland 27 September 1959 (E. A. Cardiff; SBCM 37716).

ECOLOGY. Like most migrant passerines, this species can be found in virtually any trees in the region. Even so, the Hermit Warbler seems to have a propensity for planted Athel and pines: more than 90 percent of records are of individuals in these trees.

BLACKBURNIAN WARBLER
Dendroica fusca Müller, 1776

One fall record. An immature male Blackburnian Warbler collected along the Alamo River 11 km southeast of Brawley 18 October 1992 (*AB* 47:151; SDNHM 48188) provides the only record for the Salton Sink. This species is a rare but regular fall vagrant to coastal California, but it is far scarcer in the interior West.

YELLOW-THROATED WARBLER
Dendroica dominica [dominica] (Linnaeus, 1766)

One winter record. The Yellow-throated Warbler is a rare vagrant to California, with most records from migration (Dunn and Garrett 1997). However, the single record for the Salton Sea is from winter, when one was photographed at Ramer Lake 17 December 1995–16 March 1996 (Garrett and Singer 1998).

TAXONOMY. Most records of the Yellow-throated Warbler for California (more than 90% of ca. 90) are of the short-billed, white-lored, white-chinned westerly subspecies *D. d. albilora* Ridgway, 1873 (Garrett and Dunn 1981; Dunn and Garrett 1997), the Sycamore Warbler. However, although there are no specimens, most winter and some late fall records (Craig 1972) for California are of the large-billed, yellow-lored nominate subspecies. The bird at Ramer Lake was no exception; careful study revealed that it indeed had yellow lores and a mostly yellow chin. Sight records do not exclude the weakly differentiated *D. d. stoddardi* Sutton, 1951, of coastal Alabama and the panhandle of Florida. This subspecies differs from the nominate form only in that its bill is usually slightly longer and thinner. Given its limited range, it is extremely unlikely to occur in California.

PINE WARBLER
Dendroica pinus pinus (Wilson, 1811)

One fall record. An immature female Pine Warbler was collected 3 km west of Westmorland 13 October 1991 (R. Higson; SDNHM 47864). This species is a casual migrant through California, with all but a handful of the some 60 records being from the coastal slope in fall and winter. The presence of a male at Regina, just east of the Salton Sea region, 4–26 February 1978 (Luther et al. 1983) suggests that the species could winter in the stands of pines or tamarisk in the residential areas of the Imperial and Coachella Valleys.

TAXONOMY. All California records pertain to the widespread nominate subspecies. The longer-billed, duller subspecies *D. p. florida* (Maynard, 1906) is unrecorded away from peninsular

Florida. Two other subspecies are endemic to the West Indies.

PRAIRIE WARBLER
Dendroica discolor discolor (Vieillot, 1808)

Two fall records. The Prairie Warbler is a rare but regular fall vagrant to California, mostly on the coastal slope. There are only about a dozen records for the southern California desert, two of them from the Salton Sink: at the south end 24 October–2 November 1977 (*AB* 32:263) and at Hot Mineral Spa 25 (not 22) November 2000 (*NAB* 55:105).

TAXONOMY. The nominate subspecies is the only one recorded as a vagrant. The slightly grayer *D. d. paludicola* Howell, 1930, endemic to the southeastern Atlantic coast and mangrove swamps of southern Florida, is extremely unlikely to occur in the West.

PALM WARBLER
Dendroica palmarum palmarum (Gmelin, 1789)

Casual winter (mid-December to mid-April) and fall (mid-October to early November) vagrant; 2–3 spring records. Along with the American Redstart, the Blackpoll Warbler, the Black-and-white Warbler, and the Northern Waterthrush, the Palm Warbler is among the most frequently encountered "eastern" wood-warblers on the West Coast (Roberson 1980; Patten and Marantz 1996); for example, more than a thousand individuals were recorded in California during the fall of 1993 alone (Dunn and Garrett 1997). This species is much rarer in the interior West than along the coast, but it is still encountered annually in the deserts. The majority of records are for fall, from late September through mid-November, but many individuals linger into early January, and the species is found regularly in winter. Winter records predominate in the Salton Sink, with dates extending from 29 October (2000, Niland until 22 December; P. A. Ginsburg) to 10 April (1963, mouth of the New River from 22 February; *AFN* 17:359, 435). On average, the Palm Warbler is recorded in one of three winters. In contrast to its status in the rest of California, this warbler is much rarer at the Salton Sea during fall mi-

gration. Records number but five, from the south end 11 October 1986 (*AB* 41:146), Brawley 22 October 1986 (*AB* 41:146), Niland 29 October 2000 (*NAB* 55:105), the mouth of the Whitewater River 3 November 2001 (C. McGaugh), and Finney Lake 7 November 1979 (*AB* 34:203). Likewise, there are three spring records, of birds at the north end 28 April 1980 (*AB* 34:816), near the mouth of the New River 10 (not 11) April 1999 (*NAB* 53:331; S. Guers), and banded at Wister 16 May 1999 (S. Guers); the mid-April bird may have wintered locally. There are few records statewide for that season. Interestingly, many records are from mid-April to early May, coinciding with the bird's normal migration schedule in the East and suggesting that a few winter in western Mexico.

ECOLOGY. In California this species is invariably found in low, weedy vegetation, particularly in mesic settings. It shows a particular affinity for scrubby Saltcedar, Common Reed, and Arrowweed in the Salton Sea region.

TAXONOMY. There are no specimens for the Salton Sea region, but all local birds showed characters of the widespread Western Palm Warbler, *D. p. palmarum*. The Yellow Palm Warbler, *D. p. hypochrysea* Ridgway, 1876, with yellow underparts and more olive upperparts, has been taken in California in spring and fall, but records of it are outnumbered by those of the nominate subspecies by a ratio of hundreds to one.

BAY-BREASTED WARBLER
Dendroica castanea (Wilson, 1810)

One spring record. There is a single Salton Sea record for the Bay-breasted Warbler, a scarce vagrant to California occurring mainly when populations are elevated in response to outbreaks of the Spruce Budworm, *Choristoneura fumiferana* (Patten and Burger 1998). A singing male was observed at the south end of the Salton Sea 11 June 1982 (*AB* 36:895).

BLACKPOLL WARBLER
Dendroica striata (Forster, 1772)

Two records. Even though the Blackpoll Warbler is extremely rare inland in California, it is nonetheless one of the most frequently encountered "eastern" warblers in fall along the California coast (McCaskie 1970b), with more than 120 recorded annually (Dunn and Garrett 1997). It is therefore surprising that there is only one fall record for the Salton Sea, of a bird seen near Mecca 16 September 1976 (Garrett and Dunn 1981). The Blackpoll is casual in the Southwest in spring, but a male was seen 8 km southwest of Westmorland 18–20 May 1991 (R. Higson; S. von Werlhof).

CERULEAN WARBLER
Dendroica cerulea (Wilson, 1810)
One fall record. California's first Cerulean Warbler was an immature male collected near the mouth of the Alamo River 1 October 1947 (Hanna and Cardiff 1947; SBCM 37584). This species remains one of the scarcest vagrant parulids to reach California, with only about 15 records to date.

BLACK-AND-WHITE WARBLER
Mniotilta varia (Linnaeus, 1766)
Casual spring (early April to mid-June) and fall (mid-August to late November) transient and winter visitor (mid-December to early March). Grinnell and Miller (1944) recognized that the Black-and-white Warbler was not truly a vagrant to California but rather a scarce migrant, and in the past half-century this view has been solidified. This species is among the most frequently encountered "eastern" wood-warblers throughout the western United States (Dunn and Garrett 1997), with many spring and fall records during times of normal migration in the East (rather than the typically later dates of occurrence shown by most vagrants). The status of this species at the Salton Sea mirrors its status elsewhere in the interior West. It is a scarce spring and fall migrant, with a few recorded in winter. There are about 17 spring records extending from 6 April (1922, male at Thermal; Wyman 1922, LACM 4717) to 14 June (1999, female photographed at Wister; *NAB* 53:432), with many in late April and early May. Autumn records number about the same and extend from 15 August (2000, female photographed 3 km southeast of El Centro through 18 August;

K. Z. Kurland) to 28 November (1974, 8 km west of Westmorland; G. McCaskie). There are nine winter records, with one at Salton Sea NWR 15 December 2000–4 March 2001 (*NAB* 55:229; B. Miller) providing the date span. Other winter records are from El Centro 17 December 2001–7 January 2002 (M. San Miguel); Brawley 15–29 January 1994 (*FN* 48:249), 20 January 1983 (*AB* 37:339), and 1–22 February 1986 (*AB* 40:335); the south end 21 December 1974 (*AB* 29:581) and 27 January–24 February 1973 (*AB* 27:664); 8 km west of Westmorland 25 January 1976 (*AB* 30:768); and at Ramer Lake 31 January 1965 (*AFN* 19:417). All records have involved lone birds.

ECOLOGY. This species is associated with large trees, on which it forages in a nuthatch-like fashion. The tree species seems unimportant, though many have been observed in Fremont Cottonwood and Athel.

AMERICAN REDSTART
Setophaga ruticilla (Linnaeus, 1766)
Rare winter visitor (October to mid-April) and fall transient (September and October); casual spring transient (mid-May to mid-June); 1 summer record. The American Redstart is a rare migrant and winter visitor to the West (Dunn and Garrett 1997). Hundreds are found throughout California annually, and the species has bred in the northwestern part of the state on several occasions (McCaskie et al. 1988). It winters regularly in small numbers at the south end of the Salton Sea (McCaskie 1970a), particularly along the Alamo River near Red Hill and around the mouth of the New River. Usually no more than five to six are present (e.g., winter of 1972–73; *AB* 27:664), two to four being a more usual number; the maximum is ten individuals at the south end 21 December 1982 (*AB* 37:772). Known wintering birds have arrived as early as 25 September (1994, adult male at Wister through at least January 1995; M. A. Patten et al.) and have departed as late as 27 April (2001, Wister from November 2000; W. Wehtje). Curiously, this redstart is unrecorded in the Mexicali Valley, and there are but five winter records for the southern Coachella Valley, where individuals were at the north end

19 December 1973 (*AB* 28:534), 30 December 1965 (*AFN* 20:376), and 3 January 1998 (*fide* J. Green), two immature males were near Mecca 17–23 December 1989 (*AB* 44:331), and an immature male was at Mecca 31 January 1998 (M. A. Patten).

This species is otherwise a rare spring and fall migrant through the Salton Sink, the area lying close to a hypothesized migratory route for the species (Pulich and Phillips 1953). Spring migrants have occurred from 15 May (1952, female near Niland; Cardiff and Cardiff 1953a, SBCM M1571) to 17 June (1969, Finney Lake; *AFN* 23:696). A first-spring male near the mouth of the New River 18 April 1992 (M. A. Patten) and females at the south end 19 April 1969 (*AFN* 23:627) and at Wister 28 April 2000 (G. Mc-Caskie) either were early spring migrants or had wintered locally. Records of fall migrants extend from 2 September (1962, 8 km west of Westmorland; *AFN* 17:71) to 31 October (1971, female near Westmorland; *AB* 26:123), with a female collected 20 km north of Westmorland 3 October 1948 providing the first record for the region (Cardiff and Cardiff 1949; SBCM 37667). One near Mecca 13 September 1958 (*AFN* 13:67) and a female at Oasis 28 October 1998 (R. L. Mc-Kernan; SBCM 55611) are the only migrants recorded away from the Imperial Valley. The sole summer record is of a female at Salton Sea NWR 13 July–24 August 1996 (*FN* 50:998; S. von Werlhof).

ECOLOGY. The vast majority of wintering redstarts occupy Saltcedar thickets along river courses, but some occur in Honey Mesquite and Fremont Cottonwood, likewise in mesic settings.

TAXONOMY. This species is monotypic, as *S. r. tricolor* Müller, 1776, is a synonym (Mayr and Short 1970).

PROTHONOTARY WARBLER
Protonotaria citrea (Boddaert, 1783)
Two fall records (early September). The Prothonotary Warbler is a rare vagrant to California, with about two-thirds of the more than 120 records from fall (Dunn and Garrett 1997). The only two Salton Sea records are of birds seen within

two days of each other, at the mouth of the White-water River 3 September 1987 and at Red Hill 5 September 1987 (Pyle and McCaskie 1992).

OVENBIRD
Seiurus aurocapilla aurocapilla (Linnaeus, 1766)
Five records, 3 in spring and 1 each in fall and winter. Even though the Ovenbird occurs as a regular vagrant elsewhere in California, particularly in the interior, there are only five records for the Salton Sink. A female collected near the southeastern edge of the Salton Sea 3 October 1948 was just the third recorded in California (Cardiff and Cardiff 1949; SBCM 37593). Individuals seen at Finney Lake 3 May 1975 (*AB* 29:910) and photographed at Wister 18 May 1999 (*NAB* 53:331), at El Centro 2 June 2000 (B. Krause), and 3 km southeast of El Centro 16 November 1994–17 January 1995 (*FN* 49:200) provided the other records. The last is particularly noteworthy because the vast majority of Ovenbirds in the West are encountered during migration; the species is casual in winter (Roberson 1980; Garrett and Dunn 1981).

TAXONOMY. Virtually all specimens from California and elsewhere in the West, including the bird from the Salton Sink, are of the nominate subspecies (Miller 1942; Unitt 1984; Monson and Phillips 1981), of which *S. a. furvior* Batchelder, 1918, and *S. a. canivirens* Burleigh and Duvall, 1952, are synonyms. However, the paler *S. a. cinereus* Miller, 1942, the breeder along the eastern edge of the Rocky Mountains, has been taken twice in Arizona (Monson and Phillips 1981) and therefore might be expected to reach the Salton Sink.

NORTHERN WATERTHRUSH
Seiurus noveboracensis (Gmelin, 1789)
Casual visitor in winter (mid-December to early March) and fall (mid-September to mid-October). The Northern Waterthrush is one of the most frequently encountered "eastern" wood-warblers in western North America, particularly in the desert Southwest. Despite what appears to be a great deal of prime habitat for this species, however, it was unrecorded in the Salton Sink until

30 December 1970, when one was discovered at the mouth of the New River, where it remained until 6 February 1971 (*AB* 25:630; Binford 1971). Increased coverage in recent years has shown it to be a casual winter visitor to the region. It is recorded once every two to three winters from 18 December (2001, near the mouth of the Alamo River; G. C. Hazard) to 9 March (1996, Alamo River near Red Hill; *FN* 50:225). The only fall records are of individuals near Westmorland 20 September 1981 (G. McCaskie), 3 km southeast of El Centro 28 September 1999 and 3 October 1996 (both K. Z. Kurland), at the mouth of the New River 7 October 1971 (*AB* 26:123), and at Hot Mineral Spa 18 October 2000 (G. McCaskie).

TAXONOMY. This species is monotypic, with both *S. n. notabilis* Ridgway, 1880, and *S. n. limnaeus* McCabe and Miller, 1933, being synonyms of the nominate form (Eaton 1957; Molina et al. 2000). Monson and Phillips (1981) recognized the slightly grayer, darker *S. n. notabilis*, which is ascribed to the westerly breeding populations, despite acknowledging its weak differentiation from the nominate subspecies.

LOUISIANA WATERTHRUSH
Seiurus motacilla (Vieillot, 1808)
One fall record. The first Louisiana Waterthrush for California (and the only one for 77 years) was collected at Mecca 17 August 1908 (Miller 1908; MVZ 1105). This species is an early migrant, with several more recent records of vagrants in the Southwest in late July and August (Dunn 1988; Rosenberg et al. 1991).

MACGILLIVRAY'S WARBLER
Oporornis tolmiei tolmiei (Townsend, 1839)
Fairly common spring (late March through May) and fall (mid-August to early October) transient. The skulking MacGillivray's Warbler is a fairly common migrant throughout the lowlands of the Southwest, but it can be difficult to observe because of its retiring habits. Like most migrant landbirds in the region, it is slightly more numerous in spring than in fall, although in this case the difference is not striking. Spring migrants pass through from 23 March (1982, "Salton

Sea"; *AB* 36:893) to 26 May (1996, 3 at the south end; G. McCaskie), with a peak during late April and early May. Fall records extend from 11 August (1989, 7 km northwest of Imperial; R. Higson, SDNHM 47544) to 12 October (1975, south end; G. McCaskie), with a peak during September (e.g., 10 at the south end 25 September 1994; M. A. Patten). There are no winter records for the Salton Sink.

ECOLOGY. Because of its habits, this near terrestrial species requires dense scrub, native or not, that provides extensive ground cover. Furthermore, it prefers areas with somewhat moist soil or with extensive leaf litter.

TAXONOMY. All California specimens except one are of the nominate subspecies, breeding from the Pacific Northwest south to central and, sparingly, southern California. *O. t. intermedia* Phillips, 1947, is a synonym (Monson and Phillips 1981). The subspecies *O. t. monticola* Phillips, 1947, and *O. t. austinsmithi* Phillips, 1947, are grayer above and paler, more greenish yellow below. They are weakly differentiated from each other, differing only in the latter's shorter tail; they are probably best synonymized (e.g., AOU 1957). Both breed and migrate to the east of California, with *O. t. monticola* being the southerly of the two. Given that *O. t. austinsmithi* breeds west to eastern British Columbia and southeastern Oregon, occasional strays may occur in eastern California. A male MacGillivray's Warbler taken in Tubb Canyon, in the Anza-Borrego Desert, 14 April 1938 (F. Wood; LACM 73980) is dull grayish olive on the mantle and rump and is small (tail 51 mm, wing chord 57 mm), fitting *O. t. austinsmithi*. Specimens from the Salton Sink are brightly colored and larger, and even in central Arizona the majority of specimens are of the nominate subspecies (Rea 1983).

COMMON YELLOWTHROAT
Geothlypis trichas (Linnaeus, 1766) subspp.
Common breeding resident; numbers augmented in winter (September–April). With the Marsh Wren and the Song Sparrow the Common Yellowthroat forms a trio of small, common denizens of marsh-

land and riparian woodland in the arid Southwest. The yellowthroat is a common breeding resident in mesic habitats throughout the Salton Sink. Numbers are augmented in winter, mainly from September through April, by breeders from the Mojave Desert, the Great Basin, and other, more northerly climes. Like many marshland species, the yellowthroat nests a bit later than desert scrub species. Breeding begins in late February or early March (van Rossem 1911), with the first young fledging by late April. Many are still nesting into July, perhaps mostly second broods by then.

ECOLOGY. Marshes, river courses, and irrigation drains with stands of Southern and Broadleaved Cattail and bulrush provide a haven for this species. It occurs also in Common Reed, but much less frequently. Many occupied sites support dense thickets of Saltcedar and Arrowweed and scattered Goodding's Black Willow.

TAXONOMY. Geographic variation in the Common Yellowthroat is complex and is clouded by extensive individual variation. Several subspecies have been described for southern California, but we follow Ridgway (1902), Phillips et al. (1964), and Marshall and Dedrick (1994) in treating *G. t. scirpicola* Grinnell, 1901, of southern California, and *G. t. arizela* Oberholser, 1899, of the Pacific Northwest, as synonyms of *G. t. occidentalis* Brewster, 1883. We do not see that *G. t. scirpicola* or *G. t. arizela* is well differentiated, if at all, from *G. t. occidentalis* (contra Behle 1950), although specimens in SDNHM from the ranges of *G. t. occidentalis sensu stricto* and *G. t. arizela* are few. With this taxonomic treatment, both breeders in the region and the great majority of migrants are *G. t. occidentalis*.

One specimen from the Salton Sink, a male taken 7 km northeast of Imperial 2 January 1989 (R. Higson; SDNHM 45725), clearly is not *G. t. occidentalis*. The yellow on its throat is paler (more lemon yellow, less orangish) than in *G. t. occidentalis* and does not extend as far down the breast. The back and crown are conspicuously grayer than in all other specimens from the region, lacking the contrast between the greenish back and the brownish crown prevailing in *G. t.*

occidentalis. The specimen is a good match for a male collected 1.5 km north of Potholes, on the lower Colorado River, 3 April 1930 (SDNHM 12802), reported by Monson and Phillips (1981) as *G. t. yukonicola* Godfrey, 1950. Neither the AOU (1957) nor Paynter (1968) recognized *G. t. yukonicola* as distinct from *G. t. campicola* Behle and Aldrich, 1947, but these two specimens from Imperial County are considerably grayer above and less yellow below than a male from Lawrence County, South Dakota, not far from the type locality of *G. t. campicola*, in Rosebud County, Montana. Whether or not *G. t. campicola* and *G. t. yukonicola* are distinguishable and whether or not Imperial County specimens represent the latter, we infer that the Salton Sea region marks the northwestern corner of the winter range of the population breeding in north-central North America east of the Rocky Mountains. Behle (1985) reported only migrants of *G. t. campicola* from Utah, as did Phillips et al. (1964) from Arizona, except for one winter specimen from Yuma. Both subspecies are as migratory as other warblers from central Canada, as illustrated by an immature male, tentatively identified by Allan R. Phillips as *G. t. yukonicola*, that came aboard a boat 160 km south of Socorro Island 20 November 1974 (SDNHM 39586) and agrees closely with these two Imperial County specimens. A couple of females collected in the spring from the Salton Basin have rather little yellow on the underparts and could be *G. t. campicola* or *G. t. yukonicola* as well, though the color of their upperparts does not differ from that of *G. t. occidentalis*.

Some male Common Yellowthroats observed in the Salton Sink (e.g., mouth of the Whitewater River 8 July 2000; M. A. Patten) have the underparts nearly completely yellow, thus showing the characters of *G. t. chryseola* van Rossem, 1930. This partly migratory subspecies, breeding in the American Southwest and in northwestern Mexico, has not been collected in California but has strayed to northwestern Arizona and coastal Texas (Phillips et al. 1964; Oberholser 1974). Specimens are needed to determine whether the yellow-bellied birds are *G. t. chryseola* or variants of *G. t. occidentalis*.

WILSON'S WARBLER

Wilsonia pusilla (Wilson, 1811) subspp.

Common spring (early March through May) and fall (early August through October) transient; casual in winter (early December to mid-February). Through most of the West, Wilson's Warbler vies for the title of most common migrant passerine. Even if it does not capture it, it is certainly one of the four most common migrant warblers, along with the Orange-crowned, the Yellow, and the Yellow-rumped (see Fig. 57). It is a conspicuous, common migrant through the Salton Sea region (Guers and Flannery 2000). Spring records extend from 5 March (2000, male at Fig Lagoon; G. McCaskie) to 31 May (1997, Brawley; G. McCaskie), with a peak from late April through mid-May; for example, there were 250 at the south end 5 May 1990 (M. A. Patten), 200 at the south end 4 May 1996 (G. McCaskie), and more than 150 in the Imperial Valley 16 May 1999 (M. A. Patten). In fall, Wilson's Warbler is less numerous, but it is still common from 4 August (2000, male 3 km southeast of El Centro; G. McCaskie) to 31 October (1971, south end; G. McCaskie), with a peak in mid-September. Given its abundance as a migrant and its relatively close wintering grounds in western Mexico, this species is surprisingly rare in the United States in winter. There are nine winter records for the Salton Sea. A male that settled 3 km southeast of El Centro 6 December 2001–18 February 2002 (K. Z. Kurland et al.) provided the date span for local occurrences. Other records are from the south end 16–26 December 1972 (*AB* 27:664), 18 December 1973 (*AB* 28:535), 21 December 1974 (*AB* 29:581), and 28 December 1971 (*AB* 26:522); near Mecca 29 December 1971 (*AB* 26:521) and 2 January 1994 (*FN* 48:249); at Oasis 30 December 2000 (R. L. McKernan); and at Ramer Lake 15 January 1966 (*AFN* 20:461). Some of these birds may have been late fall migrants.

ECOLOGY. During peak migration hardly a bush or tree lacks a Wilson's Warbler, with birds frequenting everything from the shortest Iodine Bush scrub to the tallest Red River Gum.

TAXONOMY. The vast majority (probably more than 90%) of Wilson's Warblers migrating through the desert Southwest are of the bright golden Pacific Coast subspecies *W. p. chryseola* Ridgway, 1902, a pattern mirrored by species such as the Willow Flycatcher, the Western Flycatcher, and MacGillivray's Warbler. In contrast to those species, however, an appreciable minority of migrant Wilson's Warblers are of the interior and Alaska subspecies *W. p. pileolata* (Pallas, 1811), which differs from *W. p. chryseola* in its duller coloration, with a more olive (less golden) back and less intensely yellow underparts, and the rounder fore edge of its cap. Like the interior subspecies of the Orange-crowned Warbler, the Yellow Warbler, and the White-crowned Sparrow, *W. p. pileolata* tends to migrate later than its Pacific Coast counterpart. Of 36 recent spring specimens (25 at SDNHM, 11 at SBCM), 28 are of *W. p. chryseola* collected from 12 March (1989, 7 km northwest of Imperial; SDNHM 47281) to 14 May (1989, 7 km northwest of Imperial; SDNHM 47501). Seven are of *W. p. pileolata*, extending from 4 May (1996, 3 km west of Westmorland; SDNHM 49521) to 27 May (1989, 7 km northwest of Imperial; SDNHM 47482). All seven fall specimens at SDNHM and the two at SBCM are of *W. p. chryseola*, extending from 26 August (1989; SDNHM 47657) to 8 October (1989; SDNHM 47590, 47591). There are fall specimens of *W. p. pileolata* at SDNHM from the nearby lower Colorado River valley, however, collected from 17 September (1924, 3 km north of Bard; SDNHM 33438) to 14 October (1925, 3 km north of Bard; SDNHM 33441), again implying a later migration. None of the winter records is supported by a specimen, so the subspecies occurring then is unknown; only *W. p. pileolata* has been recorded in that season in Arizona (Phillips et al. 1964).

PAINTED REDSTART

Myioborus pictus pictus (Swainson, 1829)

Four records, 2 in winter and 2 in fall (September). There are only four Salton Sea records of the handsome Painted Redstart, a rare vagrant to California from Arizona or Mexico. The first two records, supported by photographs, are of a bird at Salton Sea NWR 26 September 1991 (*AB*

FIGURE 62. The Painted Redstart is a casual vagrant to the Imperial Valley, with only three records. Photograph by Kenneth Z. Kurland, January 1999, Brawley.

46:152) and one at Brawley 29 November 1998–19 March 1999 (Fig. 62; *NAB* 53:106, 53:210). The third record is from El Centro 31 October 1999–18 March 2000 (*NAB* 54:106, 222), and the fourth is from near Fig Lagoon 14 September 2001 (J. Kuhn). Records for the Salton Sink are anomalous: most for elsewhere in the California deserts are from early spring, especially April (Garrett and Dunn 1981). This species was found at the Brock Research Center, just east of the region, 16–19 April 1975 (*AB* 29:911) and 13 September 1981 (*AB* 36:220).

TAXONOMY. California specimens are of the expected nominate subspecies (with extensive white in the tail and white tertial fringes), from southeastern Arizona south through Mexico, so we presume that Salton Sea records are also of the nominate subspecies.

YELLOW-BREASTED CHAT
Icteria virens auricollis (Deppe, 1830)
Rare spring (mid-April through May) and fall (late August through September) transient; rare breeder (formerly common). Devastation wrought upon riparian woodlands across California has seriously reduced the populations of many species,

including the Yellow-breasted Chat (Remsen 1978). This species formerly bred commonly at the Salton Sea (Grinnell and Miller 1944; Garrett and Dunn 1981). We know of breeding or probable breeding in the late 1990s at only four sites, home to at most six pairs, one along the Whitewater River near the north end, one or two in the vicinity of Wister, one along the New River 7 km southwest of Brawley, and one or two along the New River 4 km east of Fig Lagoon. Three territorial males at the New River locales were the only ones present in the Salton Sink during the summer of 2000, posing the worrisome specter that the decline continues and that the species may soon be extirpated as a breeder. Territorial birds return by the end of April (e.g., a singing male 4 km east of Fig Lagoon 30 April 2000; G. McCaskie). Breeding begins in May (e.g., a male defending territory accompanied by a female at Wister 19–23 May 1999, M. A. Patten; nest at Mecca 24 May 1916, H. H. Heath, MVZ 3059), with fledging by July (e.g., a juvenile from the Wister pair noted 18 July 1999; C. A. Marantz). Migrants are scarce but now account for most regional records. Breeders or transients have occurred as early as 22 April (1908, female at Mecca; J. Grinnell, MVZ 946). Spring migrants have been noted as late as 21 May (1995, Finney Lake; M. A. Patten), with most in early May. Chats are even less numerous in fall, with the few records extending from 29 August (1989, male 7 km northwest of Imperial; R. Higson, SDNHM 47666) to 5 October (1952, 5 km west of Niland; E. A. Cardiff, SBCM 37638).

ECOLOGY. This species generally occupies only native riparian habitats, even some that are narrow and meagerly vegetated. Thickets of Fremont Cottonwood, willow, and mesquite are ideal but have become a scarce commodity at the Salton Sea and elsewhere in the desert Southwest with the rapid colonization of nonnative Saltcedar (Ellis 1995; Cleverly et al. 1997).

TAXONOMY. Only the Long-tailed Chat, *I. v. auricollis*, of which *I. v. longicauda* Lawrence, 1853, is a synonym, has been recorded with certainty in California. This subspecies, widely distributed in the West, differs from the nominate, of the East, in its tail being longer than the wing and its upperparts being grayer and less olive. Eastern birds likely occur as vagrants in California (Phillips 1975b), but as yet there are only a handful of inconclusive sight reports.

THRAUPIDAE • *Tanagers*

Peak diversity in the tanager family is centered in the American tropics, with few species reaching the northern temperate region. Only two species occur regularly in the lowlands of southern California: the Summer Tanager, a former breeder but current rare vagrant or scarce migrant in the Salton Sink, and the Western Tanager, a common migrant.

SUMMER TANAGER

Piranga rubra (Linnaeus, 1758) subspp.
Casual vagrant in late fall and early winter (mid-October through January) and spring (early June); 1 fall record; formerly bred in the Coachella Valley.
In spite of expanding its range westward into southern California since the late 1950s (Johnson 1994b), the Summer Tanager has been extirpated as a breeder in the Salton Sink. It formerly

bred in the Coachella Valley but has not done so since the late 1960s; for example, a female at the north end 10 August 1968 (G. McCaskie) and a pair nesting near Mecca from May to July 1969 (*AFN* 23:627). Similarly, it formerly bred southeast of the Mexicali Valley, where a series was taken 11 km east of Cerro Prieto 29 May–14 June 1928 (Grinnell 1928; MVZ 52992–99). Its current status at the Salton Sea is as a casual late fall and winter vagrant, records averaging one every two years from 4 October (2002, female 3 km southeast of El Centro; K. Z. Kurland) to 21 January (1985, Brawley; *AB* 39:211). A first-spring male at Red Hill 2 June 1990 (R. E. Webster) and a female 13 km east-northeast of Calipatria 2 June 1996 (E. A. Cardiff; SBCM 55399) represent the only recent spring records for the region. The sole record of a definite fall vagrant is of a female 8 km west of Westmorland 21 September 1975 (G. McCaskie).

ECOLOGY. Summer Tanagers breeding in the West require riparian forest dominated by Fremont Cottonwood. The destruction of ranch lands in the Coachella Valley that supported this type of habitat has undoubtedly been responsible for the disappearance of this species as a breeder. Migrants, like most other passerines, may use virtually any trees, native or nonnative.

TAXONOMY. Tanagers formerly breeding in the Salton Sink, including the series from near Cerro Prieto, were *P. r. cooperi* Ridgway, 1869, larger, paler, and in the female yellower below and grayer above than eastern *P. r. rubra*. *P. r. ochracea* Phillips, 1966, of central Arizona, is similar to *P. r. cooperi* and is sometimes synonymized with it (see, e.g., Paynter and Storer 1970), but it differs in the juvenile's being browner and grayer. Like some other migratory breeders in the Southwest (e.g., the *E. t. extimus* subspecies of the Willow Flycatcher), *P. r. cooperi* is virtually unknown as a migrant. Instead, nearly all Summer Tanagers recorded away from breeding areas are of the highly migratory nominate subspecies, which is smaller, darker, and in the female more orange below and more olive above. We thus presume that the late fall and winter records for the Salton Sea pertain to the nomi-

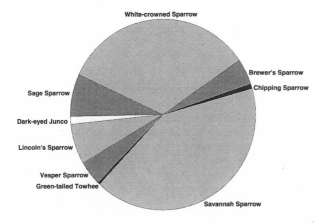

White-crowned Sparrow

Brewer's Sparrow

Chipping Sparrow

Sage Sparrow

Dark-eyed Junco

Lincoln's Sparrow

Vesper Sparrow
Green-tailed Towhee

Savannah Sparrow

FIGURE 63. Relative abundance of migratory emberizid sparrows wintering in the Salton Sink. Not included are the Lark Sparrow, which now also breeds extensively in the region, and Abert's Towhee and the Song Sparrow, both common residents, and various species that are rare or vagrants. Data are adapted from Christmas Bird Counts and field notes of Michael A. Patten.

nate subspecies, although *P. r. cooperi* has been collected on at least two occasions in coastal southern California (Willett 1933; Grinnell and Miller 1944), indicating some vagrancy in that subspecies. Nonetheless, specimens of an immature female from near Westmorland 31 October 1970 (*AB* 25:111; SBCM 37987) and an adult female from 4 km south of Calipatria 28 December 1993 (R. W. Dickerman; SDNHM 48725) are of the expected *P. r. rubra*.

WESTERN TANAGER

Piranga ludoviciana (Wilson, 1811)

Common spring (early April to mid-June) and fall (mid-July to mid-November) transient; 3 winter records. The Western Tanager is a widespread, common breeder throughout the more montane and forested areas of the West. It is a common transient virtually everywhere in California, including through the Salton Sea region. It is slightly more numerous in spring, on the basis of birds counted per day. Spring migrants occur from 4 April (2001, adult male 3 km southeast of El Centro; K. Z. Kurland) to 12 June (1999, adult male at the mouth of the Whitewater River; M. A. Patten), with a peak in late April and early May (e.g., 50 at Brawley 24 April 1988; M. A. Patten). Fall migration extends from 14 July (2001, male at Wister; *NAB* 55:484) to 16 November (1990, female at Brawley from 10 November; G. McCaskie), with a peak during early and mid-September (e.g., 20 at the south end 3 September 1987; B. E. Daniels). Adults move through

in July and August, followed by birds of the year from late August on. A male in the Imperial Valley on an unknown date in March 1937 (B. L. Clary; LACM 87593), two birds together at Brawley 20 January 1983 (*AB* 37:340), and a calling bird there 8 December 1996 (*FN* 51:804) provided the only winter records for the region, three of few for California in that season away from the immediate coast.

ECOLOGY. Migrant tanagers may be encountered just about anywhere, but they seem to concentrate in areas with large trees. In suburban neighborhoods they seem to be especially partial to flowering gum trees. They are also prevalent in orchards.

EMBERIZIDAE • *Emberizids*

Sparrows are widespread and abundant in the Salton Sink in winter, with numerous species and subspecies (Fig. 63). The Savannah Sparrow and the White-crowned Sparrow are particularly common, but there are also large numbers of Brewer's, Vesper, Sage, Lincoln's, and others. By contrast, only Abert's Towhee and the Song Sparrow are widespread, common breeders. Small numbers of the Lark Sparrow recently colonized the Salton Sink; the species now breeds locally in both the Coachella and the Imperial Valleys.

MCCOWN'S LONGSPUR

Calcarius mccownii (Lawrence, 1851)

Rare winter visitor (mid-November to mid-February); 1 fall record. In general McCown's Longspur is by far the rarest of the three longspurs regularly occurring in California. It is most frequent in the southern half of the state, particularly in the desert, but still rare in the Imperial Valley. It was more numerous in the 1960s and 1970s (Winter 1973), when small flocks were found regularly, but it remains an annual winter visitor. Two birds 8 km north of Westmorland 30–31 January 1965 were the first recorded in the Imperial Valley (McCaskie 1966). Usually only one to three birds are found, but as many as 20 were seen near Westmorland 24 November 1966–11 February 1967 (*AFN* 21:80, 460; SBCM 39051), up to 6 were seen at Salton Sea NWR 6–24 January 1981 (*AB* 35:337), and 4 were observed near the south end 9 December 1990 (*AB* 45:322). Aside from an exceptionally early female at Salton City 13 October 1991 (*AB* 46:152), dates of occurrence range from 18 November (1973, near Westmorland; *AB* 29:111) to 20 February (1972, 1–3 in the vicinity of Calipatria from 8 January; *AB* 26:657). There are no records for the Coachella and Mexicali Valleys.

ECOLOGY. McCown's Longspur is typically found among large flocks of Horned Larks (sometimes with other longspurs) in plowed or barren fields, particularly in slightly xeric, sparsely vegetated settings.

LAPLAND LONGSPUR

Calcarius lapponicus alascensis Ridgway, 1898

Rare winter visitor (mid-November to mid-February). The Lapland Longspur is a scarce, nearly annual winter visitor to the Imperial Valley. It is the most common wintering longspur in the northern half of California but is generally outnumbered by the Chestnut-collared in the southern half, particularly in the Mojave Desert. Around the Salton Sea, however, the Lapland is much the more frequent. The first area record was of a female collected 7 km east of Calipatria 11 February 1939 (McLean 1969), but the species was not recorded again until 23 January 1965 (McCaskie 1966). Since then it has been found in roughly two of every three winters and could well be annual in small numbers; it was more numerous in the 1960s and 1970s. Usually fewer

than five birds are encountered, with dates extending from 20 November (1986; *AB* 41:147) to 22 February (1965; male collected from up to 20 birds northwest of Westmorland from 23 January; McCaskie 1966, SBCM M3678). Other large (for southern California) flocks recorded were as many as 25 at Salton Sea NWR 10 December 1972–27 January 1973 (*AB* 27:665), up to 20 birds 8 km north of Westmorland 28 December 1968–15 February 1969 (*AFN* 23:523; SBCM 38296, 38297), and up to 20 at Salton Sea NWR in January and February 1972 (*AB* 26: 657). There are no records for the Coachella and Mexicali Valleys.

ECOLOGY. Lapland Longspurs are generally found with flocks of Horned Larks in plowed fields, especially those with a little moisture or even puddles of standing water.

TAXONOMY. The western North American subspecies *C. l. alascensis* is the only one recorded in California. It differs marginally from the nominate, *C. l. lapponicus* (Linnaeus, 1758), in being slightly paler, although some variation is clinal (Rising 1996).

SMITH'S LONGSPUR

Calcarius pictus (Swainson, 1832)

One winter record. Smith's Longspur is a bird of the central United States and Canada. It is casual in the Southwest, where there are eight records, six from California and one each from Arizona and Nevada, chiefly from fall (September and October). The only winter record for the Southwest is from the Salton Sink, of a bird 4 km northeast of Calipatria 31 December 2001–12 February 2002 (CBRC 2002-002).

CHESTNUT-COLLARED LONGSPUR

Calcarius ornatus (Townsend, 1837)

Casual winter visitor (early November to early March). In general the Chestnut-collared is the most common and frequently encountered longspur in southern California. In the Salton Sea region, however, both McCown's and the Lapland have been recorded more frequently. Like other longspurs, this species is unrecorded in the Coachella and Mexicali Valleys, but unlike those species, it most often occurs in large flocks. Since the first Salton Sink record, of a bird 8 km north of Westmorland 23–31 January 1965 (McCaskie 1966), this species has been recorded on only 11 occasions in the Imperial Valley, with dates ranging from 20 November (1966, 8 km north of Westmorland; *AFN* 21:80) to 9 March (1996, 2–3 at Unit 1 from 24 February; *FN* 50: 225). Other record are of 1 near Niland 28 November 1965 (*AFN* 20:93), 4 near Westmorland 4 December 1971 (*AB* 26:657), 1 at the south end 21 December 1974 (*AB* 29:581), as many as 45 about 10 km south-southeast of Brawley 28 January–11 February 1995 (*FN* 49:200), at least 35 at Wister 29 January–17 February 1994 (*FN* 48:250), 18–20 at the south end 1–2 February 1989 (*AB* 43:368), 30 there 1 March 1987 (*AB* 41:332), and amazing flocks of 150 individuals 4 km northeast of Calipatria 31 December 2001–26 January 2002 (T. Easterla et al.) and again 17 December 2002 (*fide* G. McCaskie).

ECOLOGY. Like California's other regularly occurring longspurs, the Chestnut-collared is typically found in fields, but it is less likely to associate with Horned Larks because it prefers denser vegetation. Thus, unlike its congeners in the region, this species tends to avoid barren fields. Instead, it occupies fields with grass or weed stubble, often quite dry. Fields where all three species converged had a mixture of grass or weed cover and open ground.

GREEN-TAILED TOWHEE
Pipilo chlorurus (Audubon, 1839)
Rare to uncommon winter visitor (September to mid-May), with a slightly higher number during spring migration (April–May). Unlike the other towhees, which are somewhat skulking but generally readily seen, the Green-tailed is secretive and inconspicuous. Even so, this flashy species is a fairly common breeder in montane chaparral and a migrant through the deserts, including the Salton Sea region. Small numbers regularly winter in the region, concentrating where vegetation is dense. Regional records extend from 2 September (1996, Imperial Valley; P. Unitt) to 19 May (1999, Salton Sea NWR; M. A. Patten).

Generally only one or two birds are encountered on a typical winter day, but five were seen at the south end 18 December 1979 (*AB* 34:662). There is a slight peak during spring migration, mainly from mid-April through early May.

ECOLOGY. In contrast to the Spotted Towhee (see below), this species is somewhat less likely to be found in suburbia. However, it is more likely to be encountered in mesquite and Saltcedar scrub along rivers or ditches or in more natural oasis settings (those dominated by Fremont Cottonwood and willows rather than by nonnative trees). Many are encountered in flocks of White-crowned Sparrows.

SPOTTED TOWHEE
Pipilo maculatus Swainson, 1827 subspp.
Rare winter visitor (October through mid-April). The Spotted Towhee, the western representative of the transcontinental Rufous-sided Towhee (*P. erythrophthalmus*) complex, is a common bird in cismontane chaparral and semimontane scrub and woodland throughout the West. This species is quite rare in the deserts, where it is generally encountered as a migrant. Following the first, an adult male along the New River near Westmorland 8 October 1949 (Cardiff 1956; SBCM 36560), we learned that small numbers winter in the region annually. In contrast to the slightly more numerous Green-tailed Towhee, the Spotted Towhee does not exhibit a perceptible peak during migration. Records extend from 5 October (2000, 2 at Wister; K. Z. Kurland) to 22 April (1985, Unit 1; B. E. Daniels).

ECOLOGY. Most Spotted Towhees are located in mesic "oasis" settings, be they artificial, such as suburban neighborhoods, parks, and orchards, or natural, vegetated with stands of Fremont Cottonwood or thickets of mesquite.

TAXONOMY. *P. m. curtatus* Grinnell, 1911, with pale rufous sides, heavy white spotting, short wings, and in females a dark slate head, breeds in the northwestern Great Basin south to Mono Lake. It is the principal subspecies wintering along the lower Colorado River (Grinnell and Miller 1944; Phillips et al. 1964). A male taken 11 km southeast of Brawley 18 October 1987

(P. Unitt; SDNHM 44856) was identified as *P. m. curtatus* by Allan R. Phillips; three specimens taken by E. A. Cardiff 14 km north of Niland, males 12 November 1961 (SBCM 39060) and 25 March 1962 (SBCM 38948) and a female 24 March 1962 (SBCM 38947), also are of this subspecies. *P. m. montanus* Swarth, 1905, with even paler rufous sides, heavy white spotting, long wings, and the head of females blackish, breeds from the southern Great Basin west to eastern California from the White Mountains through the Providence Mountains. It too winters south of its breeding range and is the predominant form in Arizona, although mainly east of the lower Colorado River valley (Phillips et al. 1964), and is thus far unrecorded in the Salton Sink. *P. m. megalonyx* Baird, 1858, of cismontane southern California, with moderately dark rufous sides, moderate white spotting, a dark rump in adult males, and large feet, is essentially sedentary, so it is unexpected in the region. Yet the 1949 specimen is of *P. m. megalonyx* (Cardiff 1956), a subspecies that has reached south-central Arizona on several occasions (Phillips et al. 1964).

CALIFORNIA TOWHEE
Pipilo crissalis senicula Anthony, 1895

One winter record. The California Towhee is a resident of the cismontane chaparral belt of California and is nearly unknown as a vagrant away from there. Limited wanderings have been documented in southern Oregon (Gilligan et al. 1994) and eastern California (Grinnell and Miller 1944), and the species occurs regularly on the northern fringe of the Coachella Valley, well above sea level, at Deep Canyon (Weathers 1983), Palm Springs (Patten and Rotenberry 1998), and Morongo Valley (Patten 1995a). Thus the male that P. I. Osburn collected at Mecca 29 December 1908 (FMNH 178185) is one of few vagrant California Towhees recorded anywhere. By contrast, we believe that the three "seen well and heard" at the north end (*AFN* 24:453) were misidentified, and there is no basis for claims from Finney or Ramer Lake (California Department of Fish and Game 1979).

TAXONOMY. Too many subspecies of the California Towhee have been described north of the Mexican border. We recognize only three. *P. c. petulans* Grinnell and Swarth, 1926, is the rich brown, chestnut-crowned, buff-breasted coastal form north of Monterey Bay. *P. c. crissalis* (Vigors, 1839) is the pale gray-brown form with a dull rusty crown and a grayish breast from Shasta south through the western slope of the Sierra Nevada, the Central Valley, and the Pacific coast from Monterey Bay south to Oxnard. Thus the nominate form includes *P. c. bullatus* Grinnell and Swarth, 1926, *P. c. carolae* McGregor, 1899, *P. c. kernensis* Grinnell and Behle, 1937, and *P. c. eremophilus* van Rossem, 1935, as synonyms. Finally, *P. c. senicula* is the darker gray-brown, more unicolored form on the south coast and inland to the desert edge. The Mecca specimen is not as dark as some specimens of *P. c. senicula* from coastal San Diego County, but it is darker than nearly all of *P. c. crissalis* and is matched by SDNHM specimens of *P. c. senicula* from the eastern slope of the Peninsular Ranges above the Anza-Borrego Desert.

ABERT'S TOWHEE
Pipilo aberti aberti Baird, 1852

Common breeding resident. A Sonoran Desert endemic with a somewhat local distribution, Abert's Towhee (Fig. 64) is a common breeding resident in the Salton Sink. It is easily found in any dense shrubbery and is particularly common along the rivers and in thickets around marshes (e.g., at Wister) and lakes in the Imperial Valley. Most nesting takes place in spring, with egg dates from 2 March (1941, Coachella; WFVZ 33146) to 22 May (1916, Mecca; MVZ 3275). Breeding is largely finished by July but may take place as late as 16 September (1949, nest with 4 eggs north of Westmorland; Cardiff and Cardiff 1950, SBCM E1136).

ECOLOGY. This towhee nests in Saltcedar and mesquite thickets along the rivers and ditches, around marshes, and in dense saltbush scrub in mesic sites. It is somewhat less numerous, though still common, in towns, ranch yards, and orchards. Foraging sites always have adequate ground cover and are generally well shaded. Albert's Towhee avoids open desert scrub.

FIGURE 64. Abert's Towhee is common in dense shrubbery, both native and nonnative, throughout the Salton Sink, an area that represents a substantial fraction of the species' limited range. Photograph by Kenneth Z. Kurland, January 2001.

TAXONOMY. Paralleling geographic variation in the Crissal Thrasher, the palest subspecies of Abert's Towhee occurs in the western Sonoran Desert. California birds are the pale cinnamon nominate subspecies. This subspecies was also named *P. a. dumeticolus* van Rossem, 1946, but Phillips (1962) demonstrated that the holotype of *P. aberti* was from the pale western population rather than the dark eastern population, which he named *P. a. vorhiesi*. This change in nomenclature has not gained universal acceptance (see Hubbard 1972; Browning 1990; Pyle 1997), but it was accepted by Paynter and Storer (1970: 179) and Dickerman and Parkes (1997). Furthermore, the American Ornithologists' Union (1998) followed Phillips (1962) in restricting the type locality of *P. aberti* to Gila Bend, Arizona, meaning that the name *P. a. aberti* applies to the pale western population.

CASSIN'S SPARROW
Aimophila cassinii (Woodhouse, 1852)
One spring record. The only Salton Sea record of Cassin's Sparrow, a casual spring vagrant to southern California, is of a silent bird observed

near the mouth of the Whitewater River 2 May 1978 (Binford 1983). That wet year saw a modest incursion of this species into southeastern California, including as many as 15 singing males in the Lanfair Valley, in the eastern Mojave Desert, 21 May–7 June (Binford 1985).

AMERICAN TREE SPARROW
Spizella arborea ochracea Brewster, 1882
Five late fall and winter records (late November through mid-February). The American Tree Sparrow is a rare late fall and winter visitor to California. It occurs only casually south of Death Valley, so the five records for the Salton Sea are some of the southernmost in the state. All five fit the established pattern of occurrence in California (McCaskie 1973), with a female collected near the mouth of the New River 28 November 1968 (SDNHM 36898) and individuals seen near Westmorland 26–29 December 1972 (*AB* 27:665) and 28 January 1973 (*AB* 27:665), at Niland 22–29 December 1987 (*AB* 42:323), and near the mouth of the New River 17 February 2001 (*NAB* 55:229).

TAXONOMY. The specimen is of the larger, paler western subspecies, *S. a. ochracea*, and indeed all California specimens are of this geographically expected subspecies.

CHIPPING SPARROW
Spizella passerina Bechstein, 1798 subspp.

Uncommon winter visitor (September to mid-May). A familiar bird throughout North America, the Chipping Sparrow is nowhere common in the desert Southwest, even though it winters across the whole of this vast region. Records in the Salton Sink extend from 31 August (1986, San Sebastian Marsh; P. Unitt, SDNHM 44421) to 14 May (1996, 3 birds 12 km east of Calipatria; SBCM 55045–47), with most from late September to mid-April. Flocks are generally small (at most 5 birds), but some of as many as 20 birds have been encountered (e.g., Riverview Cemetery in Brawley 19 January 1999; G. McCaskie).

ECOLOGY. This species tends to form pure flocks or flocks with other *Spizella* in parkland settings with extensive lawns or fields hosting weedy grasses. It is especially common in suburban areas and in ranch yards but also occupies field borders containing dried thistle or native desert plants, mainly during migration. A few flock with *Zonotrichia*, the Chipping Sparrow being more likely to do so than is Brewer's Sparrow.

TAXONOMY. Because we have not seen specimens from the boreal zone of Alaska and Canada, identified by Kenneth C. Parkes (pers. comm.) as the breeding range of *S. p. boreophila* Oberholser, 1955, we cannot comment on the validity of this subspecies. Monson and Phillips (1981) synonymized it with *S. p. arizonae* Coues, 1872, which breeds widely in western North America. All but 1 of 12 SDNHM specimens in basic plumage from the Imperial Valley have paler sandy brown backs with narrow blackish streaks, matching *S. p. arizonae sensu stricto*. The exception is from 3 km north-northwest of Seeley 31 October 1994 (P. Unitt; SDNHM 49065), which has a deeper, more rufous brown back with broader dark streaks. It closely resembles two specimens from coastal San Diego County identified by Parkes as *S. p. boreophila*. Its back is not nearly dark enough or rusty enough for it to belong to the eastern nominate subspecies, reported by Phillips et al. (1964) as far west as Yuma, on the lower Colorado River, but still unknown from California.

CLAY-COLORED SPARROW
Spizella pallida Swainson, 1832

Four records, 2 in spring and 1 each in fall and winter. Oddly, although the Clay-colored Sparrow is a rare to uncommon fall vagrant or scarce migrant in California, breeding in northeastern British Columbia and wintering in southern Baja California, there is only one fall record for the Salton Sink, of an immature at the Salton Sea State Recreation Area 11 September 1996 (C. A. Marantz; P. Keller). It occurs casually in spring and winter in the state, yet these seasons account for the other three records for the Salton Sink. Birds in alternate plumage were seen at Ramer Lake 9 May 1987 (*AB* 41:490) and at Wister 11 May 2000 (*NAB* 54:328). In winter, an apparent immature was with a large flock of Chipping and Brewer's Sparrows at Sheldon Reservoir 23–30 January 1988 (*AB* 42:323; P. Unitt).

BREWER'S SPARROW
Spizella breweri breweri Cassin, 1856

Common winter visitor (September to mid-May). Called drab or colorless by some, the subtle Brewer's Sparrow is a beautiful songster that breeds in the Great Basin and winters throughout the desert Southwest. It is a common winter visitor to the Salton Sea region from 31 August (2000, near Fig Lagoon; G. McCaskie) to 11 May (1909 and 2000, one 10 km west of Imperial and 6 at Wister; MVZ 8225, G. McCaskie). One 10 km southwest of Niland 13 June 1953 (E. A. Cardiff; SBCM 36557) was anomalous. The species is generally encountered in small flocks, but its preferred habitat (see below) is seldom surveyed, so its true abundance in the region tends to be underestimated. Nevertheless, it is possible to tally hundreds on Christmas Bird Counts (e.g., 256 at the north end 22 December 1985; *AB* 40:1008).

ECOLOGY. In the developed portions of the Coachella, Imperial, and Mexicali Valleys,

Brewer's Sparrows congregate in weedy areas and open fallow fields. Thus, their distribution gives the appearance of being local. However, they are also widespread in open desert scrub, whether dominated by saltbush or by Creosote. Small numbers occupy parks in towns, typically associating with flocks of Chipping or White-crowned Sparrows.

TAXONOMY. With one exception all California specimens are of the nominate subspecies, the common breeder in sagebrush throughout the Great Basin locally south to the Transverse Ranges of southern California. The exception is an *S. b. taverneri* Swarth and Brooks, 1925, taken at San Luis Rey, San Diego County, 14 February 1962 (Rea 1967; DEL 27230). The wintering range of *S. b. taverneri,* commonly known as the Timberline Sparrow, is largely unknown but is presumed to lie well to the east of California (Rotenberry et al. 1999). The Timberline Sparrow breeds in the Canadian Rockies north to east-central Alaska and south to northern Montana (B. Walker pers. comm.). If it reaches the Salton Sea region, it is only as a vagrant in migration. This subspecies differs from the nominate in its darker, grayer coloration with greater contrast between the gray breast and the white belly; heavier dorsal streaking; bold head pattern, including bolder supercilia; heavier nape streaking; larger size; and smaller, more slender, and darker bill. Some consider these distinctions, combined with differences in song and breeding habitat (Doyle 1997), enough to warrant full species status for the Timberline Sparrow (e.g., Sibley and Monroe 1990), a move arguably supported by recent genetic work (Klicka et al. 1999). But the birds breeding in northern California that show "tendencies toward the characters of *taverneri*" (Grinnell et al. 1930) need study before such a step is taken.

BLACK-CHINNED SPARROW
Spizella atrogularis cana Coues, 1866

Four records, 2 in spring and 2 in fall. Although the Black-chinned Sparrow is a common breeder in chaparral in cismontane California, it is extremely rare as a migrant outside its breeding

habitat anywhere in California (Patten, Unitt, et al. 1995). The only records for the Salton Sea are of such migrants, one near the south end 13 August 1974 (*AB* 29:124), a singing male at Finney Lake 13 May 1972 (*AB* 26:814), a female at the Highline Canal east of Calipatria 30 April 1996 (E. A. Cardiff; SBCM 55195), and a bird at Fig Lagoon 12 August 2002 (K. Z. Kurland).

TAXONOMY. Two subspecies breed in California. *S. a. cana* Coues, 1866, characterized by a dark gray head, dark rufous mantle, and short tail (less than 68 mm in the male, less than 64 mm in the female), occupies the Coast Range of central California south through cismontane southern California. *S. a. caurina* Miller, 1929, described as darker still, is a synonym of *S. a. cana* (Phillips et al. 1964) or at best weakly differentiated (Paynter and Storer 1970). Monson and Phillips (1981) later recognized *S. a. caurina* but synonymized *S. a. cana* with *S. a. evura* Coues, 1866 (cf. Ridgway 1901:323), of the eastern desert ranges east through Arizona and characterized by a pale gray head, a pale rufous mantle, and a long tail. No reasons were provided for this synonymy, but we assume that it stemmed from Phillips's discovery that both *S. a. cana* and *S. a. evura* winter in the Cape District of Baja California Sur, the type locality of the former. SDNHM specimens confirm that both subspecies winter in Baja California Sur, but we do not feel this finding negates the use of the name *S. a. cana* for the small, dark coastal subspecies. Furthermore, given that *S. a. caurina* is tenuous, we feel it best to recognize only a small, dark coastal subspecies (*S. a. cana*) and a larger, pale interior subspecies (*S. a. evura*). The Imperial Valley specimen has a dark rufous mantle, a dark gray head, and a short tail (61.8 mm), matching specimens from the San Bernardino Mountains (SBCM). It is thus *S. a. cana*. This subspecies is an early migrant in both spring and fall, so the August bird was almost certainly *S. a. cana*. By contrast, the late date of the Finney Lake bird hints that it may have been *S. a. evura*. Use of Coues as the authority for *S. a. cana* follows convention (e.g., AOU 1957; Paynter and Storer 1970), despite *S. a. cana*'s being essentially a *nomen nudum* when the name was first applied (Deignan 1961:656).

VESPER SPARROW

Pooecetes gramineus confinis Baird, 1858

Fairly common winter visitor (mid-September to mid-April). A denizen of grasslands and sagebrush, the Vesper Sparrow is a widespread winter visitor to the lowlands in the desert Southwest. It reaches peak abundance in the Colorado River valley and in the dry grasslands of southeastern Arizona but is nevertheless fairly common around the Salton Sea. Records extend from 18 September (2002, near El Centro; G. McCaskie) to 8 April (1908, 2 at Mecca; MVZ 801, 802). A few likely remain later in the spring, although the species is an early migrant.

ECOLOGY. The Vesper Sparrow is often noted with flocks of Savannah Sparrows, although it seemingly does not associate with that species directly. Most birds occur at the edges of fallow and overgrown agricultural fields or in vacant lots covered with weeds. Generally it requires some open ground, even if just a dirt road. This species also occurs in Creosote scrub, especially in low-lying areas that accumulate moisture and so more grasses and weeds. Some flock with Sage Sparrows in saltbush scrub.

TAXONOMY. Large *P. g. confinis*, gray above and white below, of the Great Basin (and locally south to the San Bernardino Mountains), accounts for all regional records. *P. g. definitus* Oberholser, 1932, is a synonym, and the browner *P. g. altus* Phillips, 1964, of the Four Corners region (Browning 1990), is not known from California. Although small *P. g. affinis* Miller, 1888, brown above and buff below, of the Pacific Northwest, reaches (reached?) coastal southern California in winter (Unitt 1984), it is unknown in the interior in that season, and with the decline of populations and the loss of winter habitat, its winter range may have retracted northward.

LARK SPARROW

Chondestes grammacus strigatus Swainson, 1827

Fairly common breeder; numbers augmented in winter (late September to April). The Lark Sparrow (Fig. 65) has a spotty distribution in the West, especially in the desert, where breeding birds tend to be restricted to mesic habitats, usually modified by humans. Indeed, this species has been drawn to a number of desert locales that have recently been cultivated and irrigated. For example, it began to breed in orchards near Blythe, in the lower Colorado River valley, in the mid-1970s (Rosenberg et al. 1991). We do not know when the Lark Sparrow began to nest in orchards in the Coachella Valley, but Garrett and Dunn (1981) suggested that it did in the late 1970s. The species is now a well-established, fairly common breeder there, particularly at Oasis, Coachella, Mecca, and southern Indio. This sparrow began nesting around ranch yards in the Imperial Valley in the mid-1980s; it is now an uncommon breeder in that valley, and it may also breed in the Mexicali area. Nesting begins in March and April, with pairs still mating in early May (e.g., a copulating pair near Mecca 1 May 1999; M. A. Patten). Juveniles fledge by late May (e.g., an adult with 2 fledglings near Westmorland 26 May 1996; G. McCaskie), with some fledglings appearing well into summer (e.g., 5 juveniles at Salton Sea NWR 14 August 1994; M. A. Patten). Numbers increase in winter, so that the species is fairly common throughout the Salton Sink from late September to April (the presence of breeders hinders determination of arrival and departure dates).

ECOLOGY. This sparrow's nesting appears to be restricted to orchards (particularly citrus in the Coachella Valley) and ranch yards with large trees. Wintering birds occupy a broader range of habitats, from parks to weedy fields to orchards.

TAXONOMY. Birds west of the Rocky Mountains are *C. g. strigatus*, characterized by its pale coloration relative to the nominate subspecies.

BLACK-THROATED SPARROW

Amphispiza bilineata deserticola Ridgway, 1898

Rare fall and winter visitor (mid-August through March), mainly to the Coachella Valley (where it is irregularly uncommon). The handsome Black-throated Sparrow is a characteristic species of the desert Southwest, so it may seem surprising that

FIGURE 65. The Lark Sparrow exemplifies the dynamism of many bird distributions in the Salton Sink. Since colonizing the Imperial Valley only about 1985, the species has become fairly common there. Photograph by Kenneth Z. Kurland, November 2000.

it is so rare around the Salton Sea. Habitat is the key, with the region supporting none of the vegetation type this species prefers (see below). Thus, this sparrow is but a casual visitor to most of the region from 4 August (1973, immature at the mouth of the New River; *AB* 28:111) to 2 April (1908, female at Mecca; MVZ 678), with most records from August through October and a late bird near the mouth of the New River 2 May 1999 (S. Guers). In some years small flocks are found wintering on the fringe of the Coachella Valley (e.g., 36 near the north end 2 January 1993; *AB* 47:968), at or near sea level.

ECOLOGY. In southeastern California this species reaches its peak abundance in desert scrub dominated by Joshua Tree, Mojave Yucca, cholla, and other spiny plants on well-drained rocky or gravelly soils, a habitat type not found around the Salton Sea. This habitat preference undoubtedly explains the species' scarcity in the region. Records from the Imperial Valley are from habitat islands of nonnative Saltcedar and suburban areas (e.g., Calipatria) with planted Fremont Cottonwood, Red River Gum, or other trees. Occasional flocks in the Coachella Valley occur in desert scrub dominated by Quail Brush or in dense weed fields, often dominated by Russian Thistle.

TAXONOMY. California birds are of the large, pale subspecies *A. b. deserticola,* also distinguished by the reduced white in the outermost rectrix and the browner dorsal coloration compared with the nominate subspecies, of Texas and the southern Great Plains, or the small, dark *A. b. grisea* Nelson, 1898, of central Mexico.

SAGE SPARROW
Amphispiza belli nevadensis Ridgway, 1873
Fairly common winter visitor (late August through April). Despite the lack of sage and sagebrush in the Salton Sink, the Sage Sparrow is a fairly common winter visitor. Indeed, given the rarity of the Black-throated Sparrow in the region, the Sage is the only *Amphispiza* sparrow one is likely

to encounter. Records extend from 27 August (2000, Salton City; G. McCaskie) to 27 April (1999, banded at Wister; S. Guers), with the majority present from mid-September to early April. It is most common in undeveloped habitats supporting extensive halophytic scrub around the Whitewater River delta, Niland, Salton City, and the San Felipe Creek drainage.

ECOLOGY. This species forms large flocks in scrub dominated by saltbush and Iodine Bush and in weedy fields dominated by Russian Thistle, especially in areas with open patches of silty soils. It commonly associates with both Brewer's and White-crowned Sparrows but is often found in pure flocks too. It avoids irrigation, generally entering only citrus orchards and vineyards bordering desert scrub.

TAXONOMY. Two pale interior subspecies of the Sage Sparrow are generally recognized: the large *A. b. nevadensis* (Ridgway, 1873), of the Great Basin, with *A. b. campicola* Oberholser, 1946, a synonym; and the small *A. b. canescens* Grinnell, 1905, of California's Central Valley and Mojave Desert. Both subspecies reportedly winter in the Colorado Desert (Grinnell and Miller 1944; Phillips et al. 1964; Rea 1983), the latter more frequently in the Salton Sink (van Rossem 1911). Sex for sex, these subspecies differ significantly in mean mensural characters (Johnson and Marten 1992), particularly wing chord. Wing chord is broadly clinal, however, being shortest in the western San Joaquin Valley and longest in the northern Great Basin (Patten and Unitt 2002). Despite genetic differentiation (Johnson and Marten 1992), the populations cannot be diagnosed morphologically at the 75 percent level (Patten and Unitt 2002).

LARK BUNTING
Calamospiza melanocorys Stejneger, 1885
Rare winter visitor (mid-November to mid-February); casual in spring (late March to early May). California is on the fringe of the normal range of the Lark Bunting. This species appears in the southeastern part of the state surprisingly frequently, with records throughout the year, though concentrated in winter (Wilbur et al. 1971). Winter

records are especially prevalent around the Salton Sea, which is not surprising given that this species is regular in that season in the Sonoran Desert, not far east (or south) of the region. Winter records extend from 17 November (2000, 8 km east of Imperial; *NAB* 55:105) to 26 February (1984, at most 3 at Niland from 20 December 1983; *AB* 38:359), with some apparently staying into April (see below). Most records are from the eastern side of the Imperial Valley. The only winter records away from the Imperial Valley are of a male at Mecca 17 December 1989–20 January 1990 (*AB* 44:331; SBCM 52616), a flock of about 20 at Thermal 3 January 1922 (Hoffmann 1922; note that Hoffmann 1923 listed the date as 1 January 1922), and a male collected along the western shore of the Salton Sea 15 February 1940 (SDNHM 27934). Most records involve 1–3 birds, but flocks of about 20 have been noted twice, once at Thermal (see above) and again near Westmorland 3 January 1923 (Hoffmann 1923). There are four to five spring records, of a male at Salton City 27–29 March 1983 (*AB* 37:914), one (age and sex unknown) at Finney Lake 21 April 1990 (*AB* 44:498), and 2 adult females 5 km north of Calexico 6 May 1921 (UCLA 5112, 5113). Although not published as such (cf. Hoffmann 1922), the flock of about 20 at Thermal in 1922 seemingly stayed into mid-April, as a series of 12 males and 3 females were collected there 11–14 April 1922 (LACM 4699–4713).

ECOLOGY. Most Lark Buntings have been found in Quail Brush or Russian Thistle scrub bordering agricultural areas. Incursions into California are strongly related to rainfall, with large numbers invading in wet years.

SAVANNAH SPARROW
Passerculus sandwichensis (Gmelin, 1789) subspp.
Abundant winter visitor (mid-July to mid-May); breeds at Cerro Prieto. The Savannah Sparrow is a widespread, familiar species occurring in grasslands and fields throughout North America. Its occurrence at the Salton Sea is complex because two distinct populations, one from the Great Basin (especially) and one from the Gulf of Cal-

FIGURE 66. A nonbreeding visitor to the Salton Sink from the Gulf of California, the Large-billed Sparrow engages in one of the most unusual migrations of any North American landbird. After declining nearly to zero from 1975 to 1985, its numbers have increased abruptly, and it recently began breeding in the region. Photograph by Kenneth Z. Kurland, February 2001.

ifornia, have a significant presence in the region. Savannah Sparrows from the Great Basin are common to abundant winter visitors to the Salton Sink, the first arriving by 27 August (1983, 5 at the south end; G. McCaskie), the last departing by 26 May (1996, south end; G. McCaskie). A bird at the south end 4 August 1990 (G. McCaskie) was exceptionally early, and most wintering birds depart by the end of April. Many thousands occur throughout the Coachella, Imperial, and Mexicali Valleys during peak abundance, from October through March.

Another population from the Gulf of California, the Large-billed Sparrow (Fig. 66; see below), occurs as a fairly common postbreeding dispersant from mid-July though February. Willet (1930) reported a female collected at the edge of the Salton Sea, near Mecca, 23 February 1930 (LACM 16913) as the first winter record for the area, but one was taken at Mecca 21 December 1908 (SDNHM 11742), seven were taken there in the winter of 1910–11 (van Rossem 1911), two others were collected near there 31 January 1913 (MVZ 105864, 105865), and two were taken near

Westmorland 14 January 1924 (MCZ 258019, 258020). Numbers fluctuate greatly from year to year, and from the mid-1970s to the mid-1980s they dropped to almost zero (AB 39:211). A bird at the north end 3 September 1987 drew comment because the subspecies had become "virtually accidental" in California (AB 42:139). Since the mid-1980s, however, the Large-billed Sparrow has enjoyed a reversal of fortune, leading to an unprecedented count of more than 100 around the south end 22 November 1989 (AB 44:165). Dozens are now found annually around the south end in winter, principally at Obsidian Butte and in the vicinity of the mouth of the New River. Molina and Garrett (2001, pers. comm.) found the Large-billed Sparrow breeding at Cerro Prieto 7 June 1997 and in summer 2000. A singing male at Salton Sea NWR 4 June 1998 (K. L. Garrett) and 2 singing males in a flock of 16 at Obsidian Butte 21 December 1999 (M. A. Patten) were perhaps harbingers of nesting at the Salton Sea.

ECOLOGY. Agricultural lands of the region constitute ideal wintering habitat for wintering birds from the Great Basin, which occur in just about any field with some weed or grass cover. They are particularly common where there is some water nearby, whether in a flooded portion of a field or in weedy ditches (Hurlbert 1997). Most Large-billed Sparrows occur along the shore of the Salton Sea in low, wet scrub of Iodine Bush, saltbush, and Bassia, a few in similar halophytic scrub around ponds elsewhere (as at duck clubs), and some in weedy ditches near the sea. Some forage on barnacle beaches and rock outcrops along the shore.

TAXONOMY. The vast majority of birds wintering in the region are *P. s. nevadensis* Grinnell, 1910, a medium-sized subspecies with pale brownish gray upperparts distinctly streaked with fuscous and the breast finely streaked blackish. The subspecies breeds in the Great Basin and winters widely in southern California and western Arizona. A few wintering birds (probably fewer than 10%) are of the subspecies *P. s. anthinus* Bonaparte, 1853, from Alaska and northwestern Canada, which differs in fresh plumage from *P. s. nevadensis* primarily in its rusty brown upperparts with an olive cast and the rusty borders of the dark brown breast streaks. *P. s. anthinus* has been recorded from 21 September (1994, immature female 3 km north-northwest of Seeley; P. Unitt, SDNHM 49107) to 28 April (1908, male at Mecca; MVZ 1058). A female taken 3 km north-northwest of Seeley 29 September 1985 (P. Unitt, SDNHM 44280) is, surprisingly, *P. s. brooksi* Bishop, 1915, the small subspecies with finely streaked underparts and a buffy nape that breeds in the coastal Pacific Northwest and normally winters near the coast. Grinnell and Miller's (1944) attribution of the dark *P. s. alaudinus* Bonaparte, 1853, resident on the coast of central California, to Mecca 27 April 1908 is erroneous. Supporting specimens (MVZ 913, 914) are highly worn and so cannot be distinguished safely from the much more likely *P. s. nevadensis*.

P. s. rostratus Cassin, 1852, the Large-billed Sparrow, is a fairly common postbreeding fall and winter visitant to the region. It typically arrives in late July, exceptionally 1 July 2000 (C. A. Marantz), 7 July 1990 (*AB* 44:1188), and the isolated June record cited above. The latest record is the 23 February 1930 record cited above. Peak numbers are encountered from November through early January. This subspecies breeds in a limited area around the Río Colorado delta, at the northern edge of the Gulf of California (Oberholser 1919). It is markedly different from *P. s. nevadensis*, being easily distinguished in the field by its heavy bill, sandy gray mantle faintly streaked with rufous brown, and light brown breast streaking. Described as a full species, it was lumped with the Savannah Sparrow following discovery of its breeding range and various morphologically intermediate subspecies in Baja California. Recent proposals to restore it to species rank (Zink et al. 1991; Sibley and Monroe 1993:77) have yet to address the spectrum of Savannah Sparrows south of the border between the United States and Mexico.

GRASSHOPPER SPARROW
Ammodramus savannarum perpallidus (Coues, 1872)
Casual spring (late April to mid-June) and late fall (mid-November to mid-December) transient; 2 winter records. The Grasshopper Sparrow is a locally common breeder in grasslands and open scrub throughout cismontane California, and it is probably a resident in the southerly portion of its range. It is extremely rare in the transmontane portion of California, mostly as a fall migrant. There are three records of the Grasshopper Sparrow from late fall or early winter for the Imperial Valley. One was collected 7 km north of Niland 9 November 1963 (McCaskie et al. 1967, SDNHM 30787), one was near Westmorland 12 December 1964 (McCaskie et al. 1967), and one was near Niland 11 November 1970 (*AFN* 25: 112). Interestingly, there are also three records for spring and summer, seasons when the species is far less frequently recorded in the interior. One was photographed near the south end 3 May 1988 (*AB* 42:483). The other two records involved singing males in Alfalfa fields, the first a remarkable flock of up to six together near the south end 17–23 June 1984 (*AB* 38:1063), the second of a lone bird at Fig Lagoon 27 April

1996 (*FN* 50:998). Unexpected because of species' paucity in California in winter, a Grasshopper Sparrow at the Whitewater River delta 5 January 2002 (E. A. Cardiff et al.) and four at Niland 17 December 2002+ (P. A. Ginsburg) provided the only winter records for the California desert.

ECOLOGY. Most Grasshopper Sparrows in the region have been associated with agricultural fields, generally either fallow, weedy fields in fall or Alfalfa fields in spring.

TAXONOMY. The specimen, an immature of unknown sex, is of the widespread western subspecies *A. s. perpallidus*. This subspecies is the only one recorded in California and throughout much of the West. It differs from *A. s. ammolegus* Oberholser, 1942, of southeastern Arizona and northern Sonora, in its browner (less rufescent) wings and from *A. s. pratensis* (Vieillot, 1818), of the East, in its slightly grayer mantle fringes and duller buff breast.

LE CONTE'S SPARROW

Ammodramus leconteii (Audubon, 1843)

One winter record. Le Conte's Sparrow has reached California as a vagrant 30 times, with records fairly evenly split between the coast and the desert. Seventy percent of the records are from late fall, though the species has wintered in the Mojave Desert. The sole record for the Sonoran Desert is of a bird wintering at Niland 17 December 2002–24 January 2003+ (P. A. Ginsburg et al.; CBRC 2003-002).

FOX SPARROW

Passerella iliaca (Merrem, 1786) subspp.

Casual late fall and early winter visitor (October through mid-March). Despite being fairly common as a breeder in most of the montane West, as a migrant through the Mojave Desert and the southern Great Basin, and as a winter visitor to the coastal slope, the Fox Sparrow is scarce in the Sonoran Desert. It is a casual winter visitor to the Salton Sink, with only one record every three years from 5 October (1952, west of Niland; Cardiff 1956, SBCM 38381) to 22 March (1959, east of Calipatria; SBCM 38383). In most winters

none are found, but four were seen at the north end 4 January 1997 (*FN* 51:639), and five reached the Imperial Valley 25 October–17 December 2002+ (G. C. Hazard et al.).

ECOLOGY. Most Fox Sparrows are encountered in dense vegetation, whether *Atriplex* thickets along the rivers or well-shaded yards in suburban neighborhoods. In the Salton Sink they associate with flocks of White-crowned Sparrows.

TAXONOMY. The Fox Sparrow is highly variable geographically and perhaps comprises several incipient species (Zink 1986, 1994; Rising 1996; Zink and Kessen 1999). Unfortunately, most of the few records for the Salton Sea region are sight records, prohibiting subspecific identification. Worse, some were never even identified to one of the four (sub)species groups (*sensu* Zink 1994; Zink and Kessen 1999), the *P. i. iliaca* group (Red Fox Sparrow), east of the Rocky Mountains; the *P. i. unalaschcensis* group (Sooty Fox Sparrow), of southern Alaska and the coastal Pacific Northwest; the *P. i. schistacea* group (Slate-colored Fox Sparrow), of the Great Basin and the Rocky Mountains; and the *P. i. megarhynchus* group (Thick-billed Fox Sparrow), of the Cascades, the Sierra Nevada, and the Transverse and Peninsular Ranges. Fox Sparrows of each group have been recorded in the Salton Sink. Individuals at Finney Lake 30 January 1961 (G. McCaskie) and 6 km east of Imperial 13 January 1990 (R. Higson; S. van Werlhof) furnished the only two records for the Salton Sink of the Red Fox Sparrow, the lone Fox Sparrow having the dorsum streaked boldly with rust and the wings and underparts extensively reddish. Only the westerly *P. i. zaboria* Oberholser, 1946, has been collected in California and the Southwest. It differs from the nominate, *P. i. iliaca*, in being grayer with a browner malar, but Rising (1996:180) questioned these distinctions (cf. Phillips et al. 1964).

Four Sooty Fox Sparrows of unknown subspecies have been observed at the mouth of the New River, one 23 January 1971 and two 7 October 1971 (G. McCaskie). Another was at Salton Sea NWR 16–17 December 2002 (B. E. Daniels). *P. i. unalaschcensis* (Gmelin, 1789), of southwestern Alaska; *P. i. insularis* Ridgway, 1900, of Kodiak and adjacent islands; and *P. i. sinuosa* Grinnell, 1910, of the Gulf of Alaska region, are the three prevalent subspecies in cismontane southern California (Willett 1933; Unitt 1984). Each represents a segment of a cline from grayer, lightly spotted, and large-billed at the western end (*P. i. unalaschcensis*) to browner, more heavily spotted, and small-billed at the eastern (*P. i. sinuosa*). The New River birds could also have been *P. i. altivagans* Riley, 1911, a poorly understood and highly variable form variously classified in the Sooty, Red, or Slate-colored groups. It breeds in the northern Rocky Mountains, with its range basically representing a broad contact zone between the Sooty Fox Sparrow (*P. i. fuliginosa* Ridgway, 1899), the Red Fox Sparrow (*P. i. zaboria*), and the Slate-colored Fox Sparrow (*P. i. olivacea* Aldrich, 1943). Its plumage is much like that of a Sooty Fox Sparrow, but it generally shows some obscure reddish streaking on the mantle and some rusty streaks ventrally and has rusty uppertail coverts.

Slate-colored Fox Sparrows are the best-represented at the Salton Sea and in the desert Southwest in general. The 5 October–22 March date range applies to birds in this group. The 1952 and 1959 specimens cited above and others from near the mouth of the New River 11 January 1969 (SBCM 38363) and Mexicali 1 February 1929 (UMMZ 167282) are *P. i. schistacea* Baird, 1858, with which *P. i. canescens* Swarth, 1918, and *P. i. swarthi* Behle and Selander, 1951, are synonymous (Phillips et al. 1964). Another *P. i. schistacea* was collected not far east of the sink, along the Río Alamo 32 km southwest of Pilot Knob (Grinnell 1928; MVZ 52470). This subspecies, which is widespread in the Great Basin and the Rocky Mountains, is characterized by its small bill, unstreaked gray head and back, and black ventral spotting. It is only "weakly differentiated" from *P. i. olivacea* (Phillips et al. 1964), which, if recognized, may occur in the region given its scattered records for Arizona.

Although it breeds closest to the Salton Sink, in the Transverse and Peninsular Ranges south to the Sierra San Pedro Mártir (Erickson and Wurster 1998), the Thick-billed Fox Sparrow has been recorded just once. As a group it is similar in plumage to the Slate-colored Fox Sparrow, differing in its thicker bill. A bird taken north of Westmorland 24 October 1947 (Cardiff 1956; SBCM 38394) is *P. i. megarhynchus* Baird, 1858, of the Sierra Nevada and southern Cascades, with which *P. i. fulva* Swarth, 1918, and *P. i. monoensis* Grinnell and Storer, 1917, are synonymous on the basis of intergradation in bill size with *P. i. schistacea*. Despite its name, *P. i. megarhynchus* has but a moderately thick bill; it is otherwise like *P. i. schistacea*. *P. i. stephensi* Anthony, 1895, the subspecies with the heaviest bill, breeds in the nearby Peninsular Ranges yet is unrecorded in the Salton Sink.

SONG SPARROW
Melospiza melodia (Wilson, 1810) subspp.
Common breeding resident, numbers slightly augmented in winter. Much of North America is home to one or several forms of the highly diverse Song Sparrow (Fig. 67). Boasting 25 diagnosable subspecies (Patten 2001), this species is the champion of geographic variation among North American birds. However, virtually all Song Sparrows in the Salton Sea region are of a single common resident subspecies. Nesting begins in late February or early March, with the first juveniles appearing by late April. Most pairs produce at least two broods, so nesting continues into early July.

ECOLOGY. Virtually any wet area in the region that harbors dense stands of Saltcedar will also harbor Song Sparrows. Indeed, more than 90 percent of these sparrows occur in stands of Saltcedar, although in the Coachella Valley there is often at least a minor mixture of Goodding's Black Willow. The species is particularly common in this habitat along the major river channels and around the large lakes (e.g., Finney and Ramer Lakes, Fig Lagoon). Some birds occur in mixed stands of Arrowweed and Southern and Broad-leaved Cattails, generally containing some Saltcedar. Likewise, a few occur in Iodine Bush scrub with lesser amounts of Saltcedar intruding.

FIGURE 67. The pale, rufous-streaked subspecies of the Song Sparrow resident in the Salton Sink, *Melospiza melodia fallax*, pictured here, is being infiltrated by the dark, black-streaked coastal subspecies, *M. m. heermanni*. This spread of coastal subspecies into the range of desert ones, evident also in the Loggerhead Shrike and the Horned Lark, may have been induced by the large-scale conversion of desert to irrigated agriculture. Photograph by Kenneth Z. Kurland, May 2000.

TAXONOMY. Of the 52 named subspecies only 25 are valid (Patten 2001). The common breeder in the Salton Sink is *M. m. fallax* (Baird, 1854), a small, pale clay-gray, rusty-streaked subspecies endemic to the Sonoran Desert. Both *M. m. virginis* Marshall and Behle, 1942, and *M. m. bendirei* Phillips, 1943, are synonyms of *M. m. fallax* (Phillips 1943; Dickerman and Parkes 1997). So too is *M. m. saltonis* Grinnell, 1909, whose type locality is Mecca. This subspecies is not distinguishable from *M. m. fallax,* many Salton Sink specimens being indistinguishable from southeastern Arizona specimens (Patten 2001). Phillips et al. (1964) recognized *M. m. saltonis,* considered it weakly differentiated, and assigned specimens taken as far east as the New Mexico–Arizona border to this sedentary subspecies, whose breeding range is ascribed to the lower Colorado River valley and the Salton Sink. Similarly, many specimens from the lower Colorado River (MVZ; SDNHM) match some definitions of *M. m. fallax sensu stricto* in their slightly darker and browner mantle and crown, yet this locale is supposed to harbor only *M. m. saltonis.* Even though *M. m. fallax sensu stricto*

averages a slightly longer wing and tail (Twomey 1947), measurements overlap broadly. During summer months *M. m. fallax* is the only Song Sparrow in most of the Salton Sink.

A few summer birds in the southern Coachella Valley show characters of *M. m. heermanni* Baird, 1858, of which *M. m. cooperi* Ridgway, 1899, is a synonym (Patten 2001). *M. m. heermanni* is grossly darker than *M. m. fallax,* streaked olive gray, brown, and blackish above and blackish below, with a slightly shorter tail. Its range is from central California to cismontane southern California, with breeding populations in the northernmost Coachella Valley (Whitewater Canyon, Mecca Creek), at Morongo Valley, and at Palm Springs (e.g., adult male 24 April 1889; MCZ 241213). Riparian habitat along the Whitewater River at Indio and Thermal supports a small hybrid zone between *M. m. fallax* and *M. m. heermanni* (Patten 2001), with a few specimens, such as one taken from Mecca 22 April 1908 (J. Grinnell; MVZ 907) and three from 2 km southeast of Thermal 5 April 1994 (P. Unitt; SDNHM 48708, 48869, 48870), being *M. m. fallax* × *M. m. heermanni* hybrids. *M. m. heermanni* is also a rare winter visitor to the region, with as many records in that season from the south end as from the Whitewater River and vicinity (e.g., 3 km north-northwest of Seeley 26 November 1989; R. Higson, SDNHM 47595, 47596). After *M. m. fallax,* it is by far the next most likely subspecies encountered in the region.

At least one other subspecies of the Song Sparrow occurs during fall and winter. *M. m. montana* Henshaw, 1884, which includes *M. m. fisherella* Oberholser, 1911, as a synonym (Patten 2001), is a gray form with thin black breast streaks with pale rufous fringes and a medium gray dorsal color. It is widespread in the Great Basin and the Rocky Mountains and occurs throughout the Mojave Desert as a fall migrant and winter visitor. This form is the most common and most widespread subspecies wintering in Arizona (Phillips et al. 1964), and it has been recorded regularly along the lower Colorado River (Grinnell and Miller 1944; SDNHM specimens). It may be regular in small numbers around the Salton Sea, but there are only 11 records extending from 23 October (1947, 19 km north of Westmorland; SBCM 39077) to 29 April (1937, 12 km northwest of Calipatria; Grinnell and Miller 1944, MVZ 71774). Other records include six specimens from the Mecca vicinity (see Grinnell and Miller 1944) from 27 December (1908; SDNHM 11743) to 14 March (1908; MVZ 602) and individuals at the south end 28 October 1940 (UMMZ 166935), 3 km west of Niland 29 October 1940 (UMMZ 166948), and 5 km west of Niland 4 April 1953 (SBCM 39079). The lack of *M. m. montana* among recent specimens hints at the possibility that its winter range may have retracted northward. *M. m. merrilli* Brewster, 1896, of the interior west of the Cascades, is a dark brown and gray form with blackish breast streaks haloed in rufous. It is scarce in southern California and has not been recorded at the Salton Sea, but specimens have been taken on either side, at Yaqui Well, in the Anza-Borrego Desert, 13 October 1936 (Unitt 1984; SDNHM 17255) and 1.5 km north of Potholes, on the lower Colorado River, 23 October 1924 (SDNHM 34964).

LINCOLN'S SPARROW

Melospiza lincolnii (Audubon, 1834) subspp.

Fairly common winter visitor (mid-September through mid-April). A common bird of the West, Lincoln's Sparrow is a familiar species in California, albeit a somewhat secretive one. It is a fairly common winter visitor in the desert Southwest, with Salton Sea area records from 18 September (1933, Coachella; LACM 18177) to 13 April (1991, 2 at Brawley; M. A. Patten). A few may remain later in spring, as migrants are often detected well into May in the Mojave Desert. Generally its numbers are small, but Lincoln's Sparrow is widespread enough to yield impressive tallies (e.g., 64 at the south end 18 December 1988; *AB* 43:1176), provided an effort is made to find it.

ECOLOGY. Nary a weedy ditch with standing water lacks this species. It also frequents dense undergrowth in thickets of Saltcedar and hedges in suburban neighborhoods and ranch yards. It rarely ventures more than a meter from the ground, remaining in dense grass or under cover and revealing itself only by call. It does not flock per se, but some are seen with *Zonotrichia* flocks.

TAXONOMY. Most birds wintering in the Salton Sink belong to the large nominate subspecies (wing chord more than 60.5 mm), of which *M. l. alticola* (Miller and McCabe, 1935) is a synonym (Wetmore 1943; Phillips 1959). A small percentage (fewer than 5%) of wintering birds are the small *M. l. gracilis* (Kittlitz, 1858), with wide black streaks on the crown, mantle, and uppertail coverts. It breeds in southeastern Alaska and coastal British Columbia and winters in central California. A male taken 3 km north-northwest of Seeley 1 January 1995 (P. Unitt; SDNHM 49157) is unequivocally *M. m. gracilis,* being both extensively black dorsally and small (wing chord 56.5 mm). Another male taken from the same locale 29 September 1985 (T. Ijichi; SDNHM 44266) is as black above but is too large (wing chord 64.3 mm). *M. m. gracilis* has also reached the lower Colorado River (Grinnell and Miller 1944; Phillips et al. 1964) and the

Valle de Trinidad in north-central Baja California (Grinnell 1928).

SWAMP SPARROW

Melospiza georgiana (Latham, 1790) subspp.

Rare winter visitor (late September to early April); 3 spring records. Although primarily a bird of eastern North America, the Swamp Sparrow is regular in fall and winter throughout the West, with hundreds recorded in California annually. This species is frequent in small numbers throughout the Mojave Desert in fall and winter but is less numerous in the Sonoran Desert. Even so, it is a rare annual winter visitor to the Salton Sea region. An adult male (SBMNH 6383) and an immature male (SBMNH 6414) collected by D. Chappel at Westmorland 21 September 1992 provided the earliest fall arrival dates and were exceptionally early for anywhere in southern California. Fall migrants generally appear after about 25 September even in the Mojave Desert. Presumed wintering birds have been recorded as late as 2 April (1953, male west of Niland; Cardiff 1961, SBCM 2162). In most winters one to three birds are located, but as many as seven were at the south end, mainly around the Alamo River, during the winter of 1990–91 (*AB* 45:322). There are three records of spring vagrants, one banded at Wister 23 April 1999 (S. Guers), a female collected northwest of Westmorland 25 April 1953 (Cardiff 1961; SBCM 38521), and a bird observed at the north end 9 May 1964 (McCaskie et al. 1967).

ECOLOGY. This species focuses on wet habitats, occurring in marshes, in ditches with dense weedy cover, and in Saltcedar scrub, particularly with Common Reed intermixed, along the rivers.

TAXONOMY. All California specimens but one are the westerly subspecies *M. g. ericrypta* Oberholser, 1938, including the two spring specimens cited above (Cardiff and Cardiff 1954). This subspecies differs from the nominate in being paler and grayer, with bolder white edgings to the mantle feathers. The sole exception is a specimen of the darker, browner, less contrasty *M. g. georgiana* taken west of Niland 1 February 1953

(Cardiff 1961; SBCM 38522). The nominate subspecies has also reached the lower Colorado River valley (Phillips et al. 1964).

WHITE-THROATED SPARROW

Zonotrichia albicollis (Gmelin, 1789)

Casual winter visitor (early December to early April). A familiar, common bird of woodlands and gardens in eastern North America, the White-throated Sparrow is a rare fall and winter visitor to the West Coast, though hundreds occur in California annually. Although regular in the Mojave Desert, it is only a casual winter visitor to the Salton Sea region, with records averaging one every three to four years since the first near Calipatria 24 January 1954 (*AFN* 8:272) and at the Vail Ranch, near the mouth of the New River, 10 December 1966 (*AFN* 21:460; SBCM 38606). Records extend from 7 December (1969, south end; *AFN* 24:541) to 11 March (1978, near Niland; G. McCaskie), with two at Coachella 17 March–7 April 1991 (K. A. Radamaker; M. A. Patten) being exceptionally late.

ECOLOGY. This species occurs in weed patches, dense saltbush scrub, yards, and neighborhoods, virtually always in the company of large flocks of the White-crowned Sparrow.

HARRIS'S SPARROW

Zonotrichia querula (Nuttall, 1840)

Casual winter visitor (December through mid-March); 1 spring record. A bird of the Great Plains, Harris's Sparrow is rare on either coast of North America. It is regular in California, with about 20 each year in fall and winter. In the desert, however, records decrease abruptly south of the northern Mojave Desert. There are only about 15 winter records for the Salton Sea region from 2 December (2000, Wister until 24 January 2001; *NAB* 55:229) and 17 March (1973, 2 birds 2 km west of Calipatria from 19 February; *AB* 26:665, G. McCaskie). Additional records are of individuals near the mouth of the New River 11 January 1969 (*AFN* 23:523), at the south end late December 1971 (*AB* 26:522) and 20 December 1976 (*AB* 31:881), at Niland 21

December 1982–20 January 1983 (*AB* 37:340), and near Brawley 6 December 1986 (*AB* 41:331) and at most two at Brawley 21 January–24 February 1990 (P. Pryde et al.). As many as 5 were in the Imperial Valley during the winter of 1972–73 (*AB* 27:526, 665; SBCM 38530, 39088), a season that produced a remarkable 40 individuals throughout southern California. An adult at Cattle Call Park, in Brawley, 21–28 April 1990 (*AB* 44:499) provided the sole spring record.

ECOLOGY. Most Harris's Sparrows occupy Quail Brush and Saltcedar scrub bordering lawns or open ground. Like the other rare species of *Zonotrichia* in the region, Harris's is generally found with flocks of wintering White-crowned Sparrows.

WHITE-CROWNED SPARROW
Zonotrichia leucophrys (Forster, 1772) subspp.
Common winter visitor (mid-September through mid-May). The most common sparrow wintering in most of the Southwest is the White-crowned, yet in the Salton Sink the wintering population of the Savannah Sparrow is larger. Even so, the White-crowned Sparrow is common, occurring from 20 September (1999, Imperial Valley; B. Mulrooney) to 20 May (1995, Salton Sea NWR; G. McCaskie). Some birds pass through the region as migrants only. The White-crowned Sparrow reaches its maximal abundance in partly developed areas, but it is outnumbered by the Savannah Sparrow in open agricultural habitats and generally by Brewer's and Sage Sparrows in desert scrub.

ECOLOGY. This species is virtually always noted in large flocks. It is common in suburban neighborhoods, parks, ranch yards, and any dense, shrubby vegetation bordering rivers, irrigation canals, or the edges of marshes. In towns it frequently flocks with House and Chipping Sparrows, though the latter is uncommon. In weedy and fallow fields and in dense scrub bordering towns it often forms mixed flocks with Brewer's and Sage Sparrows.

TAXONOMY. Two subspecies occur in the Salton Sink. By far the commonest, accounting for more than 98 percent of the wintering pop-

ulation, is *Z. l. gambelii* (Nuttall, 1840), of Alaska and northwestern Canada. It is distinguished by its white (adult) or tan (first winter) lores, pink bill, pale gray breast, and gray mantle feathers narrowly streaked with dark purplish rufous. Records extend from 21 September (1975, 2 at the south end; G. McCaskie) to 20 May (cited above), although most arrive in early October and depart before the end of April. *Z. l. oriantha* Oberholser, 1932, of the mountain West, is similar in overall plumage and bare-part coloration but has black (adult) or dark brown (first winter) lores that break the supercilium. Records extend from 20 September (cited above) to 19 May (2000, Wister; G. McCaskie). It arrives on average slightly earlier than *Z. l. gambelii,* so many White-crowned Sparrows encountered during late September are *Z. l. oriantha.* A few remain through the winter (e.g., 8 km north of Holtville 21 February 1933; LACM 76464). This subspecies is most frequently noted during spring migration. By late April most *Z. l. gambelii* have departed, and during the first half of May about 80 percent of the White-crowned Sparrows in the Salton Sink are migrant *Z. l. oriantha.* Thus, the vast majority of regional records of *Z. l. oriantha* extend from 29 April (1989, 7 km northwest of Imperial; R. Higson, SDNHM 47401) to 19 May (cited above).

One specimen of *Z. l. pugetensis* Grinnell, 1928, has been collected in the Salton Sink, an adult female 7 km northwest of Imperial 30 September 1989 (R. Higson; SDNHM 47635). This subspecies, originating in the Pacific Northwest, is distinguished as *Z. l. gambelii* by its yellowish bill and brown mantle feathers streaked with black. An adult female *Z. l. pugetensis* from even farther east, collected 3 km north of Bard along the lower Colorado River 22 October 1924 (L. M. Huey; SDNHM 9617), represents the only other record for southeastern California.

GOLDEN-CROWNED SPARROW
Zonotrichia atricapilla (Gmelin, 1789)
Rare winter visitor (mid-November through April). In winter the Golden-crowned Sparrow is confined mainly to the chaparral and scrub of coastal

California and northwestern Baja California, but a few reach the deserts (e.g., Morongo Valley) regularly. The first regional record was of an immature male collected at Brawley 18 December 1910 (van Rossem 1911). It was followed by a female northeast of Calipatria 7 April 1962 (E. A. Cardiff; SBCM 38616). Since that time the species has been recorded nearly annually from 16 November (1991, Mecca Beach; T. E. Wurster) to 3 May (1975, Salton Sea NWR from 26 April; G. McCaskie). Generally only single individuals are encountered, but five were observed at the north end 29 December 1990 (AB 45:995). Most records are of immatures.

ECOLOGY. Golden-crowned Sparrows are invariably found among the large wintering flocks of White-crowned Sparrows that occur throughout the area in thickets and weeds.

DARK-EYED JUNCO

Junco hyemalis (Linnaeus, 1758) subspp.

Uncommon fall transient and winter visitor (late September through March). The highly diverse Dark-eyed Junco is an uncommon fall transient and winter visitor to the Salton Sink, where it is substantially less common than elsewhere in California. Records extend from 26 September (1970, Red Hill; AB 25:112) to 3 April (1954, 2 north of Westmorland; E. A. Cardiff, SBCM 38673, 38674). There is a slight peak from mid-October through the end of December, which suggests that many juncos occurring in the region are fall transients or perhaps do not survive in the marginal habitat the sink has to offer this species.

ECOLOGY. Although it sometimes mixes with flocks of White-crowned Sparrows or even *Spizella*, this species tends to form small, pure flocks. It may occur in stands of any kind of shade tree, feeding on the ground below and taking refuge in the canopy above. It favors areas with short grass. In the Coachella Valley juncos frequently use citrus orchards.

TAXONOMY. As expected in the West, the Oregon Junco group of subspecies accounts for more than 90 percent of juncos occurring in the Salton Sink. As a group, the Oregon Junco has a blackish (male) or gray (female) hood with a convex lower border that contrasts with a browner back and richly colored flanks. Oregon Juncos have been recorded from 26 September (1970, south end; G. McCaskie) to 3 April (1954, cited above). The majority (perhaps 60–70%) reaching the region are *J. h. montanus* Ridgway, 1898 (Cardiff 1956), of the northern Great Basin, characterized by cinnamon-brown flanks, a slaty (male) or gray (female) hood, and a dark grayish brown back. This subspecies was called *J. h. shufeldti* by Phillips (1962), Phillips et al. (1964), and Monson and Phillips (1981), a move supported by Browning (1990). Most of the remaining Oregon Juncos are *J. h. thurberi* Anthony, 1890, of the southern Cascades and montane California, especially in the Coachella Valley, such as the male at Mecca cited above. It is similar to *J. h. montanus*, but males have a blacker head, and both sexes have a pinkish tan back. A small number (fewer than 5%), such as a male from Sheldon Reservoir 3 February 1989 (R. Higson, SDNHM 45937), are *J. h. shufeldti* Coale, 1887, of the coastal ranges of Washington and Oregon, distinguished by its rich coffee-brown back. This subspecies was renamed *J. h. simillimus* by Phillips (1962) because he believed that the type of *J. h. montanus* actually applied to the population named *J. h. shufeldti*.

The Slate-colored Junco group, breeding in boreal forests from Alaska to New England and in the Appalachians, is a distant second to the Oregon Junco in abundance in the Salton Sink. This group differs in its more uniform plumage and the concave lower border to the hood. Small numbers are sometimes noted, such as five at the south end 30 December 1970 (AB 25:504) and four there 26 December 1981 (AB 36:758), but generally only one or two are noted among small flocks of Oregon Juncos. Records extend from 25 October (1969, Niland; SBMNH 1639) to 30 March (1975, Niland; G. McCaskie). On the basis of specimens from throughout the Sonoran Desert (SDNHM; SBCM; Phillips et al. 1964), probably more than 90 percent are *J. h. cismontanus* Dwight, 1918, including six of seven specimens at SBCM (Cardiff 1956) and all three at SDNHM. It breeds in the northern Rocky Mountains and has a back and flanks of gray or brown (mainly immature females) that contrast with the slate-gray hood. The name *J. h. connectens* Coues, 1884, cannot apply to this population because the type is a female *J. h. hyemalis*. Phillips (1962) renamed the population *J. h. henshawi* even though it is clear that *J. h. cismontanus* already applies to it (Browning 1990; Dickerman and Parkes 1997). The remainder of Slate-colored Juncos in the region are *J. h. hyemalis*, with one definite record from northwest of Westmorland 5 February 1955 (Cardiff 1961; SBCM 38632). The nominate subspecies breeds east of the Rocky Mountain divide from northern Alaska to the Northeast. Males are uniform slate, and females are uniform gray or brownish gray.

The Gray-headed Junco, *J. h. caniceps* (Woodhouse, 1853), is a rare visitor in late fall and winter. It breeds in the southern Great Basin and the Rocky Mountains west to eastern California and south to central Arizona. Unlike the preceding forms, it is not sexually dimorphic; both sexes have medium gray flanks and hood with a bright rusty back, contrasty black lores, and a pink bill. Records extend from 26 September (1970, Red Hill; *AB* 25:112) to 17 March (1956, Westmorland; E. A. Cardiff, SBCM 38715) and include several specimens, such as three published by Cardiff (1956) and an immature female 3 km west of Westmorland 19 October 1991 (R. Higson; SDNHM 47865). Interestingly, the majority of records are from November. The Pink-sided Junco, *J. h. mearnsi* Ridgway, 1897, has a similar status at the Salton Sea, being a rare fall transient from 31 October (1954, male northwest of Westmorland; Cardiff 1956, SBCM 2426) to 22 December (1987, near Calipatria; *AB* 42:1137), with most records for December. Of note is a specimen just to the east of the Salton Sink, at Regina 20 December 1983 (P. Unitt; SDNHM 42846). The Pink-sided Junco breeds in the Rocky Mountains north of *J. h. caniceps*. Males and females are similar. It is a large junco with a pale gray or blue-gray hood contrasting with broad, pinkish buff flanks, blackish lores (mainly in males), and a brown back.

CARDINALIDAE • *Cardinals, Saltators, and Allies*

In contrast to the Emberizidae, the cardinalids are only modestly represented in the Salton Sink. Only the Blue Grosbeak breeds, and it, the Black-headed Grosbeak, and the Lazuli Bunting are the only species regularly encountered (the last 2 as migrants, especially in spring).

PYRRHULOXIA
Cardinalis sinuatus [fulvescens] (van Rossem, 1934)
Four records, 2 in winter and 1 each in spring and summer. The Pyrrhuloxia is a casual visitor to California from Arizona or Mexico, with four acceptable records for the Salton Sea area. A bird wintering at Heise Springs (11 km west-northwest of Westmorland) in three consecutive years furnished the first verified record for California (McCaskie 1971a). It was present 24 February–8 March 1971, 31 December 1971–27 March 1972, and 22 January–23 March 1973 (Winter 1973; Dunn 1988). In addition, a male was photographed at Calipatria 17 December 1972–19 February 1973 (Dunn 1988), and singing males were observed near Westmorland 18 July 1974 (Luther et al. 1979) and photographed at El Centro 28 May–5 June 1996 (McCaskie and San Miguel 1999). Additional reports from Mecca (*AFN* 7:236), the south end (*AB* 28:854), and Westmorland (*FN* 50:988) lack documentation and are thus unacceptable.

TAXONOMY. There are no specimens of the Pyrrhuloxia for California, but we assume birds occurring there to be of the Arizona and western Mexico subspecies *C. s. fulvescens*, which is larger and browner-backed than the nominate subspecies, from New Mexico, Texas, and eastern Mexico.

ROSE-BREASTED GROSBEAK
Pheucticus ludovicianus (Linnaeus, 1766)
Casual spring (May) and fall (mid-August to mid-November) vagrant; 2 winter records. The Rose-breasted Grosbeak is a rare to uncommon vagrant or scarce transient and winter visitor throughout California. It occurs in the same habitats as the Black-headed Grosbeak and has bred on rare oc-

casions, often in mixed pairs with its congener (Garrett and Dunn 1981). The majority of records for the Salton Sea are of migrants. There are four spring records, of a male at Brawley 12 May 2001 (W. Widdowson), a first-spring male at Salton Sea NWR 20 May 1995 (G. McCaskie), a male at the south end 25 May 1970 (*AFN* 24:646), and a bird there 26 May 1978 (*AB* 32:1057). There are nine fall records for the Imperial Valley from mid-August to mid-November. Single birds were seen at Finney Lake 9 August 1997 (adult male; D. Powell), 12 August 1972 (*AB* 27:124), 13 September 1988 (immature; J. L. Dunn), and 13 October 1991 (K. A. Radamaker). Unitt collected immature females 11 km southeast of Brawley 25 September 1988 (SDNHM 45341) and 18 October 1987 (SDNHM 44855). An immature male at Wister 20–22 November 1999 (E. A. Cardiff) and a female there 21 November 1999 (G. McCaskie) were late vagrants. Last, a female was photographed 3 km southeast of El Centro 18–26 October 2001 (K. Z. Kurland). A male at Brawley 11 February 1990 (*AB* 44:331) and a female at Niland 22 December 1992 (*AB* 47:302) furnished the only winter records for the California deserts.

ECOLOGY. Rose-breasted Grosbeaks have been observed in ornamental plantings, gum, tamarisk, Fremont Cottonwood, Pomegranate, and other introduced trees.

BLACK-HEADED GROSBEAK
Pheucticus melanocephalus maculatus
    (Audubon, 1837)
Common spring transient (late March through May); uncommon fall transient (late July through mid-September); 4 "summer" records. One of the most common migrant landbirds in the lowlands of the West is the Black-headed Grosbeak. Large numbers move through the Salton Sea in spring en route to a breeding range that fringes the northern edge of the Coachella Valley, not far from the Salton Sink. Spring passage is from 27 March (1972, "Salton Sea"; *AB* 26:808) to 31 May (1999, male at Mecca; M. A. Patten), with a peak during late April and early May (e.g., 25 at Brawley 24 April 1988; M. A. Patten).

This species is casual in summer. A male collected along the Alamo River 11 km southeast of Brawley 10 June 1989 (R. Higson, SDNHM 45997) was in breeding condition (testes 6 × 5 mm each). Two at the south end 30 June 1990 (G. McCaskie; M. A. Patten) and males 3 km southeast of El Centro 1 July 1999 and 3 July 2002 (K. Z. Kurland; *NAB* 56:488) were also likely summering but may have been extremely early fall migrants. The Black-headed Grosbeak is far less common in fall than in spring, with records from 13 July (1971, 12 at the south end; G. McCaskie) to 18 September (1994, 3 km south of Seeley; P. Unitt) and peaking in the first half of August. A few probably occur later in the fall given that the species moves through the California desert into mid-October (Garrett and Dunn 1981; M. T. Heindel unpubl. data; M. A. Patten pers. obs.). It is generally uncommon in fall, but 30 were seen in Mexicali 1 September 2000 (R. A. Erickson).

ECOLOGY. Migrant Black-headed Grosbeaks occur in well-planted areas, especially those supporting nonnative fruit trees such as Pomegranate, citrus, and various drupes. They are particularly common in partly flooded citrus orchards bordering Date Palm orchards, and they frequent stands of gum and mulberry. During peak movement large numbers also occur in riparian woodlands dominated by Saltcedar, Goodding's Black Willow, Fremont Cottonwood, and Athel.

TAXONOMY. Two subspecies, *P. m. maculatus* (Audubon, 1837), of the Pacific Coast, and *P. m. melanocephalus* (Swainson, 1827), of the Great Basin and the Rocky Mountains, appear to be well differentiated in bill size and shape, the nominate subspecies having a thicker bill with the base of the mandible expanded. Grinnell (1914) identified specimens of spring migrants from the lower Colorado River as *P. m. melanocephalus*, but specimens from the Salton Sink are clearly smaller-billed *P. m. maculatus*, including the June male, a female at the same locale 6 May 1989 (R. Higson; SDNHM 47431), and an immature female 3 km north-northwest of Seeley 4 August 1985 (P. Unitt; SDNHM

43907). Phillips (1994b) questioned the validity of *P. m. maculatus,* asserting that "recognition of subspecies rests entirely on bill size [and] most of this variation is mosaic, not clinal." We follow the norm (e.g., Grinnell and Miller 1944; AOU 1957; Rea 1983) and the evidence in the SDNHM collection in recognizing *P. m. maculatus,* yet we acknowledge that a thorough quantitative study is lacking.

BLUE GROSBEAK

Passerina caerulea salicaria (Grinnell, 1911)

Fairly common breeder (mid-April through mid-October); 2 late fall or winter records. The jumbled song of the Blue Grosbeak (Fig. 68) pervades lowland riparian woodlands and deciduous forests over most of the southern half of the United States and northern Mexico. This species is a common breeder in riparian habitats in the desert Southwest, but it is only a fairly common breeder in the Salton Sea region. Records extend from 13 April (1983, "Salton Sea"; *AB* 37:911) to 12 October (1921, immature male 5 km north of Calexico; UCLA 5928), with most birds present from May to early September. Nesting occurs mostly in May and June, with juveniles appearing by mid-June and July. There are few records after October for anywhere in California, but one is from Calipatria 26 December 1981 (*AB* 36: 758), and another (a male) is from Wister 29 November 2002 (P. E. Gordon).

ECOLOGY. This species reaches its peak abundance in dense stands of Saltcedar at the edge of water, either running or standing, particularly if there is a healthy mixture of Goodding's Black Willow. It avoids stands of Common Reed and cattail.

TAXONOMY. Specimens from the Salton Sink are of the small-billed, westernmost subspecies, *P. c. salicaria.* Although the larger-billed *P. c. interfusa* Dwight and Griscom, 1927, has been attributed to southeastern California (AOU 1957), it is in fact not known to reach the state; even populations along the Colorado River are typical of *P. c. salicaria* (van Rossem 1945; Phillips et al. 1964; *n* = 28 SDNHM specimens).

FIGURE 68. By adapting to Saltcedar, the Blue Grosbeak, a species found in riparian scrub, has been able to remain common in the Salton Sink while other riparian species have declined. Photograph by Kenneth Z. Kurland, June 1999.

LAZULI BUNTING

Passerina amoena (Say, 1823)

Fairly common to uncommon spring (early April through mid-June) and fall (mid-July through September) transient. Whether as a breeder or as a migrant, the Lazuli Bunting is a familiar bird through much of the West, although it is only a migrant through the Salton Sea region. It occurs in spring from 1 April (1998, Coachella Valley; *AB* 36:893) to 10 June (1999, female 3 km southeast of El Centro; K. Z. Kurland) and in fall from 19 July (1997, male at the south end; G. McCaskie) to 6 October (2000, Calipatria; G. McCaskie). It is fairly common during spring migration, which peaks during the last week of April and the first half of May, although even then double-digit counts are unusual. It is slightly less numerous in fall, but migration is more protracted in that season, with a peak from mid-August into early September. Some records involve singing males accompanied by females, but there is no evidence of breeding in the region or, indeed, of any oversummering birds.

ECOLOGY. This species may be encountered wherever migrant passerines are attracted, from suburban neighborhoods and parks to orchards and mesic scrub. It is particularly numerous in Saltcedar scrub along ditches or river courses and in extensive mesquite thickets near water.

INDIGO BUNTING
Passerina cyanea (Linnaeus, 1766)
Casual spring vagrant (mid-April to June); 2 winter records; 1 summer record. In some years the Indigo Bunting occurs frequently enough in the northern deserts and on the coast of California to be considered almost uncommon in spring and fall. Despite its status elsewhere, however, this species is quite rare in the southern deserts. There are only seven records for the Salton Sea region, four of them for spring. A male taken at Mecca 11 April 1908 (MVZ 811) was the first for California, although the specimen was not correctly identified until more than 50 years later (Thompson 1964). It remains one of the earliest spring records for the state. More seasonable males were at Fig Lagoon 4 May 1998 (E. R. Caldwell), near Niland 5 May 1984 (AB 38:965), east of Calipatria 14 May 1994 (SBCM 55152), and at Wister 22 May 1994 (M. A. Patten). A singing first-spring male was at Salton Sea State Recreation Area 1 June 1976 (M. S. Zumsteg). Remaining regional records defy the general pattern of spring and fall vagrancy observed elsewhere in the West. A male at Brawley 19–20 January 1991 (AB 45:322) provided California with one of its few winter records; a male at Brawley 18 March 1978 (Garrett and Dunn 1981) also was likely wintering. The presence of an adult male at the south end 13 July 1995 (FN 49:983) was most unusual, although this species has summered in riparian habitats throughout southern California and has bred along the nearby lower Colorado River (Rosenberg et al. 1991).

ECOLOGY. Indigo Buntings have occurred in thickets of Saltcedar or mesquite or in parkland and other suburban settings.

PAINTED BUNTING
Passerina ciris (Linnaeus, 1758)
One fall record. The Painted Bunting is a casual fall vagrant to California, with about 50 acceptable records, mainly of immatures. Despite numerous fall records for the Mojave Desert, an immature at Calipatria 6 October 2000 (NAB 55:105) provided the only Salton Sink record of apparently wild Painted Buntings. Two adults at Niland, a male 22–24 December 1998 (NAB 53: 210; M. A. Patten) and a female 22 December 1998–18 January 1999 (NAB 53:210), were judged to be escapees from captivity given their age and the winter date (Erickson and Hamilton 2001). This gaudy species, females as well as males, is the most numerous cagebird offered for sale in *mercados* of northern Baja California (Hamilton 2001).

TAXONOMY. Two subspecies of the Painted Bunting are generally recognized: the nominate subspecies, of the southeastern United States, and *P. c. pallidior* Mearns, 1911, of western Mexico and parts of the southwestern United States. However, mensural and color differences between the populations are not consistent, so that the recognition of subspecies is not warranted (Thompson 1991).

DICKCISSEL
Spiza americana (Gmelin, 1789)
One fall record. The Dickcissel is a casual spring and rare fall vagrant to California. The region's only record is of a bird at the south end of the Salton Sea 28 November 1964 (AFN 19:82), the latest record of a fall vagrant for southern California (Garrett and Dunn 1981).

ICTERIDAE • Blackbirds

Among all families of passerines, it is the blackbirds whose economics have been studied most intensively. Since they are largely granivorous, massive flocks of Red-winged Blackbirds,

Brewer's Blackbirds, and Brown-headed Cowbirds can do extensive damage to grain crops once kernels have ripened. Indeed, blackbirds have been identified as a "serious problem" in the agricultural areas of the Imperial Valley (Garlough 1922). That the cowbirds, Brewer's Blackbird, and the Great-tailed Grackle have increased substantially in the Salton Sink over the past half-century has probably compounded this perception. More than balancing the scales, however, has been the marked decline of the Red-winged and Yellow-headed Blackbirds as breeders in the region with the loss of many marshes.

BOBOLINK

Dolichonyx oryzivorus (Linnaeus, 1758)

Two fall records; 1 spring record. Like the Dickcissel's, the status of the Bobolink at the Salton Sea belies its status elsewhere in California. This species is a rare spring and uncommon fall (with flocks of as many as 50) vagrant to California (Garrett and Dunn 1981), but there are only two records for the Salton Sea, both of which fall outside the established pattern of occurrence elsewhere in the state. One observed near the north end 28 October 1977 (*AB* 32:263) was later than most in California, although records do extend to mid-November (Garrett and Dunn 1981). A molting adult male at Niland Marina County Park 31 July 1965 (McCaskie and DeBenedictis 1966) provided the earliest fall record for California, although given its state of molt it may have been summering locally. Additionally, a male was near Seeley on 4 and 5 June 2002 (*NAB* 56:359).

RED-WINGED BLACKBIRD

Agelaius phoeniceus (Linnaeus, 1758) subspp.

Abundant breeding resident. An abundant species throughout its North American range, the Red-winged Blackbird has found the expansive agriculture of the Coachella, Imperial, and Mexicali Valleys to its liking (Howell 1922a). It likely occurred in small numbers in the Salton Sink prior to flooding and irrigation but doubtless became much more common subsequently. Its tendency

to form large flocks during winter gives the impression of increased abundance in that season, but there is little broad-scale movement of birds within the Salton Sink and almost no evidence of winter influxes of birds breeding farther north (*contra* Garlough 1922). Breeding begins in February and March, with nest building from mid-March through April. Egg dates range from 3 April (1931, near Calipatria; WFVZ 5513) to 3 June (1933, 16 km north of Westmorland; WFVZ 18084). Juveniles fledge as early as May, more typically in June or July.

ECOLOGY. Most breeding Red-winged Blackbirds occupy sloughs, ditches, rivers, lake edges, and marshes dominated by cattail or sometimes Common Reed. Lesser numbers nest in cultivated fields with sufficient plant cover (e.g., Asparagus, corn). Some colonies are mixed with Yellow-headed Blackbirds and Great-tailed Grackles, but generally the Red-winged Blackbird greatly outnumbers both, often forming pure colonies. It occasionally nests in other settings, even in tall Fremont Cottonwood (Howell 1922a). Marshes dominated by cattail and Saltcedar serve as roosting sites. The Red-winged Blackbird's penchant for foraging on crop grains, ranging from corn to millet, has earned it enemies among farmers (Garlough 1922; Howell 1922a). It forms massive flocks during winter, often in association with Brewer's Blackbirds, Great-tailed Grackles, and Brown-headed Cowbirds. Partly flooded fields and cattle feedlots support especially high numbers.

TAXONOMY. The abundant breeding resident is *A. p. sonoriensis* Ridgway, 1887, of the Sonoran Desert, characterized by its long, thin bill and the female's being pale and finely streaked. *A. p. thermophilus* van Rossem, 1942, is a synonym. A female collected at Calipatria 8 November 1921 (USNM 287678) is *A. p. californicus* Nelson, 1897 (Phillips et al. 1964), of California's Central Valley, characterized by its thick bill and black belly. An adult male taken from Calipatria 21 October 1921 (USNM 287672) is also *A. p. californicus*, characterized by its thick bill and minimal yellow bordering the red wing coverts. There are two Salton Sink records of *A. p. neutralis* Ridgway,

FIGURE 69. A male *Agelaius phoeniceus neutralis* Red-winged Blackbird from the Salton Sink (*center*), collected 10 km west of Imperial on 6 May 1909 (MVZ 8205). Note the short bill with a thick base, forming an almost equilateral triangle, as on the breeding male *A. p. neutralis* from San Timoteo Canyon, Riverside County, California (*left*). By contrast, the resident breeding subspecies of the Salton Sink, *A. p. sonoriensis*, such as this male from the Imperial Valley (*right*), has a long, slender bill.

1901, of the southern Pacific Coast, one an adult male 10 km west of Imperial 6 May 1909 (Fig. 69; van Rossem 1926; MVZ 8205). The subspecies differs in its short, thick bill and in the female's being heavily streaked and having a grayer mantle. It has also reached the lower Colorado River (SDNHM 33507) and Arizona (Phillips et al. 1964; Rea 1983; USNM 239868). A female taken 7 km east of Brawley about 29 April 1966 (SDSU 1632) has an extensive dark belly and heavy breast streaks. It is either *A. p. neutralis* or *A. p. californicus;* although worn, it better matches the former. The collection date is not on the tag, but other specimens from the same locale and with close catalog numbers were taken 29 April 1966. Although it is a likely candidate for wintering in the Salton Sink, there are no records of the western Great Basin subspecies, *A. p. nevadensis* Grinnell, 1914, characterized by its thin bill and the female's being heavily streaked and having a rustier mantle. Claims that Red-winged Blackbird populations in the Imperial Valley are augmented by winter visitors from the Great Basin (Garlough 1922) are thus baseless. *A. s. nevadensis* winters in Arizona west to the Colorado River, with records from Yuma (Phillips et al. 1964; UA 659) and near Bard (SDNHM 10119, 10149), so it may occasionally reach the Salton Sink.

TRICOLORED BLACKBIRD
Agelaius tricolor (Audubon, 1837)
Two spring records (March). There are but two records for the Salton Sea of the highly gregarious Tricolored Blackbird. Adult males near Finney Lake 12 March 1978 (*AB* 32:401) and near Calexico 3 March 1980 (*AB* 34:308) could have been spring wanderers. A claim of a flock of 100 on the Salton Sea (south) Christmas Bird Count (*AFN* 20:377) is erroneous.

WESTERN MEADOWLARK
Sturnella neglecta neglecta Audubon, 1844
Common winter visitor (mid-September to mid-April); fairly common breeder. The cheerful song of the Western Meadowlark brightens most grasslands in the West. This species is generally a fairly common resident in southern California where suitable habitat remains, with higher numbers in winter as birds from the colder, northerly parts of its range move south (Garrett and Dunn 1981; Rosenberg et al. 1991). This summary reflects its status in the Salton Sea region. Although it is a fairly common breeder, with most nesting activity taking place from February through June, it is a common winter visitor from roughly mid-September to mid-April. As with many species largely resident in the region,

determination of exact arrival and departure dates for the winter influx is nearly impossible. In addition, this species' tendency to form large flocks in winter compounds the impression of increased abundance in that season.

ECOLOGY. Native grasslands are absent from the Salton Sink, but the meadowlark does not require them. Instead, it readily uses fields, whether abandoned, undeveloped, lying fallow, or actively farmed (but prior to harvest). It requires dense grasses, weeds, or crops, favoring Asparagus. Nonbreeders gather in flocks that forage in virtually any sort of field with some plant cover, from agricultural to extensive lawns. Some congregate along roadsides.

TAXONOMY. In California the nominate subspecies is perhaps confined to the southeastern deserts (Monson and Phillips 1981; Rea 1983). It differs from *S. n. confluenta* Rathbun, 1917, of the Pacific Northwest and much of California, in having narrower, more distinct barring on the rectrices and inner secondaries, paler upperparts with less black, paler yellow underparts, and small flank spots. Some winter birds may be *S. n. confluenta*, but there are no definite records. Thus, even the winter influx appears to be of birds from the Mojave Desert and the Great Basin.

YELLOW-HEADED BLACKBIRD

Xanthocephalus xanthocephalus (Bonaparte, 1826)
Fairly common breeder (mid-March to early October); uncommon to fairly common winter visitor.
Populations of the Yellow-headed Blackbird have been rather cyclic at the Salton Sea. In the Sonoran Desert this species is confined to marshes of the Salton Sink and the lower Colorado River valley. Regional records date back to the beginning of ornithological exploration of the area (e.g., van Rossem 1911), but it was apparently only a transient and winter visitor until the 1950s. A small colony discovered near Calipatria 15 May 1952 contained at least 15 nests, providing the first breeding evidence for the Salton Sink (Cardiff 1956; SBCM E1577, E1578). Within a year the Yellow-headed Blackbird was discovered nesting in the Coachella Valley, with a small colony at Mecca in early April 1953 (*AFN* 7:292). The

breeding population grew rapidly and probably peaked during the 1970s. In spring 1971 two colonies at the Salton Sea hosted more than 800 breeders, 300 at a marsh 8 km south of Coachella and more than 500 at Ramer Lake (Crase and DeHaven 1972). At the time the population at the Salton Sea was nearly half that in all of California (Crase and DeHaven 1972: table 1). Thousands probably bred in the Salton Sink in the mid- to late 1970s.

The Yellow-headed Blackbird is scarcer now, being but a fairly common breeder in most years and uncommon in others. The current breeding population probably numbers under 200 pairs, although about 500 were recorded at Finney Lake 24 March 1991 (M. A. Patten). Presumed breeders are present from 17 March (1991, 5 km southwest of Coachella; M. A. Patten) to 2 October (1999, 5 at the mouth of the Whitewater River; M. A. Patten). Males begin to defend territory in late March and early April. Nesting begins in mid-April, with eggs laid by late April and hatching by early May. Active nests have been found as late as 3 June (1962, near Mecca; WFVZ 148895). Most breeders head southward, usually leaving the species uncommon in the Salton Sink through the winter. Winter counts often reach no more than a few dozen, but large numbers remain in milder years, such as 1,000 near Finney Lake 5 February 2000 (G. McCaskie), 2,402 on the Salton Sea (south) Christmas Bird Count 21 December 1974 (*AB* 29:581), and a remarkable 5,000 at the south end 6 January 1973 (G. McCaskie).

ECOLOGY. This species seems to be more particular than the Red-winged Blackbird in its choice of nesting habitat, requiring freshwater marshes or ponds with dense cattail. It generally avoids ditches or plants other than cattail. Nesting colonies are often pure, but some breed adjacent to Red-winged Blackbird colonies. The Yellow-headed forages in agricultural fields, often partly flooded ones. Current nesting colonies are at Fig Lagoon, Finney and Ramer Lakes, and Wister. Most wintering birds are mixed with large flocks of Red-winged Blackbirds, particularly at cattle feedlots.

BREWER'S BLACKBIRD

Euphagus cyanocephalus minusculus Grinnell, 1920

Common winter visitor (mainly October through March); locally fairly common (and increasing) breeder. Brewer's Blackbird is a widespread bird of suburbia and farmlands throughout the West. It is principally a winter visitor to the desert Southwest, including the Salton Sea region. In that season it is common (and increasing) in agricultural lands of the Coachella, Imperial, and Mexicali Valleys, often congregating in large flocks, sometimes of thousands. Flocks of wintering birds occur from 31 August (2000, ca. 150 in the Imperial Valley; G. McCaskie) to 19 March (2000, ca. 100 at Oasis; M. A. Patten), with the majority present from late September through February. Arrival and departure dates are now difficult to determine because of the substantial presence of breeders. Brewer's Blackbird breeds locally in the Coachella Valley (Garrett and Dunn 1981), mostly in the vicinity of Oasis, Coachella, Thermal, and Mecca. Breeding in the Imperial Valley was first suspected in 1987 (G. McCaskie) and was confirmed in 1988, when an adult male was observed feeding three fledglings at Brawley 7 May (*AB* 42:1341). It now breeds annually in small colonies scattered throughout the Imperial Valley, as in southwestern Brawley (mostly around Cattle Call Park), El Centro, Calipatria, Westmorland (3 ovulating females 2 April 1994; P. Unitt, SDNHM 48834–36), and Fig Lagoon (male with enlarged testes 20 May 1995; P. Unitt, SDNHM 49260). Breeding was suspected at Salton City by 1995 (M. A. Patten) and was confirmed there by the late 1990s (G. McCaskie; M. A. Patten). Breeding in the Mexicali Valley has not yet been documented, although birds have been noted there in early April (e.g., 2 at Mexicali 3 April 1993; T. E. Wurster). Nesting takes place early, with fledglings appearing by 2 May (1992, at Brawley; M. A. Patten).

ECOLOGY. In winter this species is found throughout the agricultural areas and parklands of the region. It is frequently encountered as a human commensal in cities and towns, often feeding in parking lots and planters of ornamental shrubs and flowers. In the Coachella Val-

ley it breeds in citrus orchards with some regular flooding from irrigation or in dense shrubs (such as Pomegranate) along ditches. In the Imperial Valley it breeds in pines, hedges of Oleander, and other ornamental plants with dense foliage near well-watered lawns, which the species requires for foraging.

TAXONOMY. The species is generally regarded as monotypic (AOU 1957; Miller et al. 1957). However, small, thin-billed birds with the female pale from California were named *E. c. minusculus,* and large, thick-billed birds with the female dark from the Pacific Northwest to the Great Lakes were named *E. c. brewsteri* (Audubon, 1843), with large, thick-billed birds with the female pale from the Great Basin and Rocky Mountains being the nominate subspecies (Oberholser 1974:1006; Rea 1983). The allegedly darker *E. c. aliastus* Oberholser, 1932, of the Great Basin, is a synonym of the nominate subspecies (Rea 1983). Specimens from the Salton Sink agree with *E. c. minusculus.* Nevertheless, there is substantial variation in size, with birds as small as *E. c. minusculus* collected in eastern Arizona (Phillips et al. 1964) and even in western Texas (Oberholser 1974:832). Coloration depends heavily on season because wear greatly affects the darkness of the plumage, such that birds from one population can match birds from any other. We await a thorough revision, but in the meantime we follow Rea (1983) in recognizing three subspecies.

GREAT-TAILED GRACKLE

Quiscalus mexicanus (Gmelin, 1788) subspp.

Common breeding resident; a recent colonist. Even at the species level the rapid expansion of the range of the Great-tailed Grackle (Fig. 70) has been dramatic and fascinating (Phillips et al. 1964; Phillips 1968), and at the subspecies level it has been yet more complex. The species had poured north into central Arizona by the mid-1930s (Phillips 1950) and marched west and north from there. It was first recorded at the Salton Sea 18 July 1964 (*contra* Garrett and Dunn 1981, who listed the date as 18 June), less than two months after the first California record, at West Pond, on

FIGURE 70. Following the invasion of two subspecies and a rapid increase in the 1980s, the Great-tailed Grackle has become one of the Salton Sink's most common landbirds. Photograph by Richard E. Webster, December 2001.

the lower Colorado River (McCaskie and DeBenedictis 1966). Five years passed before the next regional record, of a pair at Ramer Lake 17 May 1969 (*AFN* 23:627). Three adults carrying food at Ramer Lake 14 July 1973 (*AB* 27:920) provided the first nesting evidence. Double-digit counts were commonplace in the Imperial Valley by summer 1976 (*AB* 30:1005). Data are lacking, but the status of the Great-tailed Grackle in the adjacent Mexicali Valley was presumably the same, although the species has not extended south beyond the limits of agriculture (Patten et al. 2001). The species spread slowly into the Coachella Valley, where it remained rare into the mid-1980s (Garrett and Dunn 1981; Glenn 1983). By the late 1980s it was a common breeding resident throughout the developed portions of the Salton Sink. For example, a total of about 600 were tallied on the Salton Sea (south) and Salton Sea (north) Christmas Bird Counts, 22 December 1996 and 4 January 1997, respectively (*FN* 51:639, 643), even though the count circles miss most of the occupied habitat. Breeding activity is concentrated in spring and early summer, mainly from April through July.

ECOLOGY. This grackle has been aided immensely by human modification of the environment. Shunning natural desert scrub or riparian woodland, it is found almost exclusively in agricultural fields, orchards, and suburban settings (Rosenberg et al. 1987). It also forages along the shores of the sea, lakes, and marshes and in cattle feedlots. Large mixed flocks of grackles, Red-winged and Brewer's Blackbirds, and Brown-headed Cowbirds are a common sight around feedlots and at roosts. Nesting birds occupy cattail in marshes and dense foliage of any kind near their foraging habitat, especially Saltcedar and Common Reed.

TAXONOMY. The region's Great-tailed Grackles have been of two subspecies. Forerunners of the invasion were the small *Q. m. nelsoni* (Ridgway, 1901), of western Sonora (Phillips 1950), further diagnosed by the paler, tawny females. All early California records are of this subspecies, as are some recent ones, such as those of females near Mecca 17 June 1990 (WFVZ 49094) and 14 July 1991 (WFVZ 49512) and a male there 15 July 1991 (WFVZ 49513). By the late-1980s, concomitant with the species' spread into the Coachella Valley, the large *Q. m. monsoni* (Phillips, 1950), of Chihuahua and southeastern Arizona, had begun to invade. Females of this subspecies are darker and browner. *Q. m. monsoni* displaced *Q. m. nelsoni* in some areas around the Salton Sea and interbred extensively with the already established smaller subspecies, mirroring the pattern documented in Arizona, where the two subspecies likewise formed an amalgam (Phillips et al. 1964; Rea 1969). Thus, although some individuals currently in the Salton Sink can be assigned to either *Q. m. nelsoni* or *Q. m. monsoni*, many are intermediate between the two subspecies.

BRONZED COWBIRD

Molothrus aeneus loyei Parkes and Blake, 1965

Rare and local breeder (mid-April to early September); 2 winter records. Beginning at the turn of the twentieth century the brood-parasitic Bronzed Cowbird (Fig. 71) rapidly colonized the Southwest (Phillips 1968; Hubbard 1978), apparently

FIGURE 71. The Bronzed Cowbird's spread into the Salton Sink, beginning in 1956, has been slow. But since 1984 it has accelerated with the colonization of an additional host, Brewer's Blackbird, coming from the opposite geographic direction. The bird shown here is a recently fledged juvenile. Photograph by Kenneth Z. Kurland, August 1998, near El Centro.

as a result of expanding agriculture and suburbia. It is thus a relatively recent colonist of southeastern California and the Salton Sea region. The first record for California was of one observed at Lake Havasu 29 May 1951 (Monson 1954, 1958), and the species has been a regular, albeit rather rare breeder in the lower Colorado River valley ever since (Rosenberg et al. 1991). Expansion into the Salton Sea region took place at a more modest pace. A female collected 10 km north of Westmorland 22 April 1956 was the first found in California away from the Colorado River (Cardiff 1961; SBCM 37971). The next regional record came 12 years later, with a female seen near Mecca 13 July 1968 (*AFN* 22:650). It was not until the mid-1980s that the Bronzed Cowbird finally established itself in the Salton Sink. Up to three individuals, including an adult male, were discovered at Cattle Call Park, in Brawley, 12–18 July 1984 (*AB* 38:1063), and a female was seen at Niland 1–2 June 1985 (*AB* 39:963). As many as 15 individuals (8 August 1998; G. McCaskie) have since been recorded at and around Cattle Call Park. Breeding was confirmed there when two fresh juveniles were found in 1989 (*AB* 43:1369), and a juvenile was being fed by a Brewer's Blackbird 21–22 June 1991 (*AB* 45:1163). The Cowbird is now a regular breeder at Brawley. By 1996 small numbers were summering and presumably breeding at Calipatria and Niland (*FN* 50:998). At Salton Sea Beach a fresh juvenile was observed 16 July 1998 (M. A. Patten) where a lone male had been observed 6 July 1996 (*FN* 50:998). At El Centro three molting juveniles were noted 2 September 1999 (D. Sebesta) and young were being fed by orioles during the summers of 1999 and 2000 (K. Z. Kurland). Furthermore, five adults were together at Oasis 4 July 1992 (*AB* 46:1180), and displaying males have been noted at Wister, beginning with one 1 May 1993 (G. McCaskie). Until 2002, area records extended from 12 April (1998, male at Brawley; *FN* 52:392) to 10 September (2000, molting male at Brawley; J. L. Dunn). Four at Brawley 2 March 2002 (*NAB* 56:359) were remarkably early for anywhere in California, whereas up to three males at Calipatria 18–22 September 2002 (G.

McCaskie) were quite late. Winter records are of two adult males and an adult female at Calipatria 21 November 2000 (*NAB* 55:105) and nine at Mexicali 13 January 2003 (R. A. Erickson).

ECOLOGY. Bronzed Cowbirds breeding in the Salton Sink are associated with human-modified habitats, particularly parks, ranch yards, and residential areas with lawns. Open areas with short grass seem to be a factor limiting their spread in the region.

TAXONOMY. California and Arizona birds are *M. a. loyei*, of western Mexico, a subspecies larger than the nominate form, of southern Texas and eastern Mexico. In addition to the differences in size, females of *M. a. loyei* are much paler and browner.

BROWN-HEADED COWBIRD
Molothrus ater (Boddaert, 1783) subspp.
Common breeding resident; numbers augmented in winter (September–March). The common thread running through the blackbird tapestry in the desert Southwest is the substantial invasion of many species, most notably the Brown-headed Cowbird. Although the incursion of Brewer's Blackbirds and Great-tailed Grackles has been impressive, the establishment of breeding colonies of the Yellow-headed Blackbird significant, and the increase in the numbers of Hooded and Bullock's Orioles noteworthy, these stories pale in comparison with this cowbird's dramatic march west. Outside the Great Basin it was basically unknown in California away from the lower Colorado River valley until the 1890s. From 1889 to 1896 scattered individuals were recorded in southern California (Laymon 1987). By the 1900s it was breeding and established in many locales, including the Salton Sink (Rothstein 1994). By the 1920s it had colonized central California, and by the 1930s it had laid claim to much of northern California. Thus, in only five decades this species went from being unknown as a breeder west of the Colorado River valley and the Sierra Nevada to being a common breeder in most of the state (Rothstein 1994).

The Brown-headed Cowbird is now a common breeding resident in the Salton Sink, where agriculture has created an enormous amount of suitable foraging habitat. As with the Red-winged Blackbird, flocks congregate during the months when birds are not breeding. Flocks of adults begin to form in early August (e.g., 250 at the south end 8–9 August 1987; M. A. Patten) and persist into March, when birds disband and occupy riparian and marshland habitats, where they breed. Egg laying occurs mainly from mid-April to mid-June; young are fed by foster parents into September. There is a minor influx of birds from the Great Basin in winter, from September through March.

ECOLOGY. Much has been written about the devastation wrought by the brood-parasitic Brown-headed Cowbird upon breeding birds in riparian habitats in the West (see Morrison et al. 1999). The most heavily affected species in California and Arizona, the Willow Flycatcher, Bell's Vireo, the Yellow Warbler, and the Summer Tanager (Garrett and Dunn 1981; Rosenberg et al. 1991), were never part of the breeding avifauna of the Salton Sink, so the cowbird's impact has been less severe in this region. There are few data on host use in the Salton Sink, but favored hosts elsewhere include the Common Yellowthroat, the Song Sparrow, the Blue Grosbeak, and the Red-winged Blackbird (Friedmann 1963). A juvenile was begging from an Abert's Towhee at Mexicali 3 September 1993 (T. E. Wurster). Breeders are tied to mesic habitats and are most numerous along rivers, irrigation canals, and edges of marshes and lakes. Nonbreeders frequent cattle feedlots and any agricultural habitat, where they form large flocks with other blackbirds. Smaller numbers occur in suburban settings, such as neighborhoods and parks.

TAXONOMY. *M. a. obscurus* (Gmelin, 1789), the Dwarf Cowbird, which enjoyed the enormous population explosion and range expansion in the twentieth century, is the common subspecies in the Salton Basin in all seasons, as it is elsewhere in southern California and southern Arizona. Eight of 11 recent specimens are of this subspecies, of which *M. a. californicus* Dickey and van Rossem, 1922, is a synonym. *M. m. obscurus* is easily the smallest of the three subspecies, showing little overlap in wing chord with the nominate subspecies, of the East, and no overlap with the larger, somewhat duskier *M. a. artemisiae* Grinnell, 1909, of the Great Basin and Rocky Mountain region. It further differs from *M. a. ater* in its bill being slender and evenly tapered rather than thick and swollen with a flattened culmen. In the Salton Sink, *M. a. artemisiae* occurs less commonly as a migrant or winter visitor. Two specimens, both females, were collected 11 km north-northwest of Holtville 16 October 1988 (R. Higson, SDNHM 45780; wing 100.8 mm) and 7 km northwest of Imperial 12 March 1989 (R. Higson, SDNHM 46006; wing 101.6 mm). A male of the nominate subspecies, *M. a. ater,* taken 4 km northwest of Imperial 15 January 1989 (R. Higson, SDNHM 45836) is apparently only the second for California, following one from San Diego (Unitt 1984). It is too large to be of *M. a. obscurus* (wing 107.0 mm) and has the swollen maxilla distinguishing *M. a. ater* from *M. a. artemisiae.* The nominate subspecies has also been reported in Arizona west to Tucson (Phillips et al. 1964) and at the Gila River Indian Reservation (Rea 1983).

ORCHARD ORIOLE
Icterus spurius spurius (Linnaeus, 1766)
Four winter records (mid-November through March); 2 fall records. The Orchard Oriole is a rare but regular fall and winter vagrant to California, with 6 records in these seasons for the Salton Sea. Two fall records are of females at Salton Sea NWR 9 September 1968 (*AFN* 23:112) and near Westmorland 31 October–4 December 1971 (*AB* 26:124). The other four records are from winter, with single birds seen near Mecca 13 January 1954 (McCaskie and DeBenedictis 1966) and near Niland 18 November 1973–30 January 1974 (*AB* 28:110) and a female seen at Niland 22 December 1998–23 January 1999 (*NAB* 53:211). A singing male photographed at El Centro 25 March 1964 (Hubbard 1965) probably had wintered locally.

ECOLOGY. These orioles have occurred in planted figs, citrus trees, and other fruit-bearing ornamentals, as well as in flowering gums.

TAXONOMY. There are no specimens for the region, but all California specimens have been attributable to the widespread, richly colored nominate subspecies, and other subspecies are unlikely to reach California.

HOODED ORIOLE
Icterus cucullatus nelsoni Ridgway, 1885
Uncommon migrant and local breeder (mid-March to mid-October); rare in winter (mid-November to early March). The Hooded Oriole is generally outnumbered by Bullock's Oriole in much of their shared range in the desert Southwest. This general rule prevails in the Salton Sink, where the Hooded Oriole is uncommon as a migrant and somewhat local breeder. Birds have been detected as early as 13 March (2002, 3 km southeast of El Centro; K. Z. Kurland), and most depart by mid-September, with one 7 km northwest of Imperial 14 October 1989 (R. Higson; SDNHM 47606) being the latest. Nesting begins in early April and continues through July. Migrants are detected principally in April and August, so numbers are slightly higher during those months. This species is rare but nearly annual in winter from 22 November (1955, Mecca; *AFN* 10:59) to 7 March (1984, adult male at Brawley from 28 January; *AB* 38:359).

ECOLOGY. For nesting this species heavily favors palms (Garrett and Dunn 1981), particularly fan palms. Extensive Date Palm groves in the southern Coachella Valley are thus particularly attractive and support much of the regional population. The lack of commercial Date Palm orchards in the Imperial and Mexicali Valleys tends to restrict this species to planted fan palms in the large towns, along roadways, and in other developed areas.

TAXONOMY. Birds in the Southwest are of the pale, short-billed *I. c. nelsoni,* the females of which are further distinguished from those of the nominate subspecies by their flanks being concolorous with the belly rather than contrasting gray (Phillips et al. 1964).

BULLOCK'S ORIOLE
Icterus bullockii (Swainson, 1827)
Common breeder and migrant (mid-March through September); casual in winter. Woodlands of the

West, from lowland riparian to foothill oak–sycamore, are home to Bullock's Oriole. It is generally a numerous bird throughout its range, including at the Salton Sea, where it is a common nesting species. Presumed breeders are present from 11 March (1909, female 10 km west of Imperial; MVZ 8214) to 27 September (2002, Ramer Lake; G. McCaskie), with an early adult male at Wister 4 March 2000 (J. Brandt) and another just outside the region, 3 km east of Plaster City, 9 March 1989 (R. Higson; SDNHM 45663). Most birds have arrived by April, when many migrants are passing through the area (e.g., 50 at Brawley 24 April 1988; M. A. Patten), yet nesting begins with the first arrivals, such as a mated pair in El Centro 27 March 1998 (J. Marshall). Most breeders have departed by early August, but a few may remain later, when birds that have bred farther north are passing through. This species is casual in winter at the Salton Sea, with the species recorded an average of once every two to three years from 13 November (1999, immature male at Mecca; M. A. Patten) to 5 March (1937, immature male 12 km northwest of Calipatria; MVZ 71720). Generally only one individual is found, but three were at the north end 27 December 1975 (*AB* 30:608).

ECOLOGY. Nesting sites for this oriole are in large trees, whether nonnative (e.g., Blue Gum) or native (e.g., Fremont Cottonwood), generally near water and with adjacent open patches (parks, fields, yards) for foraging. Breeding is thus concentrated around settled areas. Wintering birds are found almost exclusively in exotic trees, especially flowering gum and Date Palms, but some have occurred in stands of large Fremont Cottonwoods in ranch yards.

TAXONOMY. Bullock's Orioles breeding in the southern part of the species' range were described as *I. b. parvus* van Rossem, 1945, differing only in its smaller size. We reevaluated this distinction with specimens at SDNHM. Nineteen males from San Diego and Imperial Counties, annotated as breeding or collected in June or July, juveniles excluded, have an average wing chord of 95.2 mm (range = 90.9–98.7; s = 1.95).

Eight females meeting the same criteria have an average wing chord of 91.2 mm (range = 89.3–94.5; s = 1.67). Of six specimens (3 males, 3 females) of the nominate, *I. b. bullockii*, of Oregon or northeastern California, three are beyond the range of *I. b. parvus* and four lie outside the zone encompassing 90 percent of that subspecies. But two, though near the large end of variation in *I. b. parvus*, would not be confidently identified as *I. b. bullockii*, implying that the two samples are only 67 percent separable, a figure under the threshold desirable for recognition of subspecies (Amadon 1949). A definitive answer awaits analyses of larger samples. If two subspecies are recognized, the identification of two of the five recent regional specimens is equivocal. One early fall migrant, a second-year male from the Alamo River 11 km southeast of Brawley 5 August 1989 (SDNHM 46105) is small (wing chord 92.5 mm). Of the remaining two specimens, taken from the same location 20 May 1989, the adult male (SDNHM 47521), apparently nonbreeding, is large (wing chord 99.1 mm), whereas the female (SDNHM 47522), just coming into breeding condition, is small (wing chord 87.7 mm).

Allegedly yellower Bullock's Orioles from eastern Texas were named *I. b. eleutherus* Oberholser, 1974, but there is broad overlap between it and the nominate subspecies, such that *I. b. eleutherus* is not diagnosable (Browning 1978). With the Black-backed Oriole, *I. abeillei* (Lesson, 1839), accorded species status (AOU 1998), the synonymy of *I. b. parvus* and *I. b. eleutherus* renders Bullock's Oriole monotypic.

BALTIMORE ORIOLE
Icterus galbula (Linnaeus, 1758)
Casual fall (mid-September through November) and spring (April and May) vagrant. The widespread Baltimore Oriole, of eastern North America, is a rare but regular vagrant to California and the West, particularly in fall. It has been recorded eight times at the Salton Sea. The two spring records are of a female at Finney Lake 29 April 1972 (S. B. Terrill) and a singing male 6.5 km west-southwest of Westmorland 20 May 1995

(R. Higson; SDNHM 49184). A female taken 18 km north of Westmorland 2 April 1953 (E. A. Cardiff; SBCM 37771) likely had wintered locally but may have been an early spring vagrant. Three of five fall records are from September: an adult male at Finney Lake 9–11 September 1990 (*AB* 45:154), a female there 13 September 1988 (*AB* 43:171), and an immature male at Salton Sea NWR 25 September 1994 (*FN* 49:103). Later were an immature female at Wister 24–27 October 2002 (G. McCaskie) and an adult male at Salton Sea NWR 27–30 November 1964 (McCaskie and DeBenedictis 1966).

ECOLOGY. Date Palms, flowering gums, and large Athel have hosted the Baltimore Orioles appearing in the region.

SCOTT'S ORIOLE
Icterus parisorum Bonaparte, 1838
Casual visitor (mainly winter). Woodlands of Joshua Tree and open pinyon-juniper in the Mojave Desert and taller stands of Mojave Yucca in the Sonoran Desert host the showy Scott's Oriole. Because no such habitat exists in the Salton Sink it is but a casual visitor to the region. There are three records of presumed transients: a male at Salton Sea Beach 26 April 1997 (G. McCaskie); an immature male at Red Hill 27 July 1974 (G. McCaskie); and a female near the mouth of the New River 24 August 1991 (*AB* 46:153). Scott's Oriole is casual through most of California in winter, although small numbers are regular in that season along the eastern base of the southern Peninsular Ranges, as in the Anza-Borrego Desert (Massey 1998). There are five winter records for the Salton Sink, of six birds on the Salton Sea (north) Christmas Bird Count 17 December 1989 (*AB* 44:989), one on that count 22 December 1985 (*AB* 40:1008), an adult male at Niland 22 December 1975 (*AB* 30:609), one at the south end 11 February 1967 (*AFN* 21:459), and an adult male at Calipatria 21 December 1999–9 January 2000 (*NAB* 54:223).

ECOLOGY. Scott's Orioles have occurred in stands of large Fremont Cottonwood and other deciduous trees, in flowering gum, and in or-

chards. Birds wintering in the Anza-Borrego Desert favor Date Palms and California Fan Palms.

FRINGILLIDAE • *Fringilline and Cardueline Finches and Allies*

Finches give the appearance of being well represented in the Salton Sink, but only the House Finch is even locally common. The Lesser Goldfinch breeds irregularly, its numbers increasing in winter. All other finches are rare in the region, several of them irruptive vagrants.

PURPLE FINCH
Carpodacus purpureus californicus Baird, 1858
Rare, irregular winter visitor (September to mid-April). Although widespread and common in the oak belt of California, the Purple Finch seldom reaches the desert floor. Small numbers of this species wander irregularly to the Salton Sea region in fall and winter, with records once every two to three years on average. Records extend from 28 August (1986, up to 15 at Brawley through 20 March 1987; *AB* 41:147) to 13 April (1975, near Mecca; G. McCaskie), but most occur from November through mid-March. In a typical incursion year there are at most 10 birds, but as many as 35 were at Brawley 16 February–2 March 1985 (G. McCaskie) and an amazing concentration of up to 80 was at Brawley during the winter of 1984–85 (*AB* 39:211).

ECOLOGY. Purple Finches occurring in the region occupy planted pines and other ornamental trees in well-vegetated, settled areas.

TAXONOMY. The only regional specimen, a male found at Thermal 11 November 1914 (LACM 8044), is *C. p. californicus*, the olive-streaked buff subspecies of the Pacific Coast. Of the brown-streaked white nominate subspecies, *C. p. purpureus* (Gmelin, 1789) of the East, there is one California specimen (female at Santa Miguel Island 11 May 1976; SBMNH 3506) and there are several sight records (e.g., Furnace Creek Ranch, in Death Valley, 18–21 November 1987; *AB* 42:139), so it may occur in the Salton Sink as a casual vagrant.

CASSIN'S FINCH

Carpodacus cassinii Baird, 1854

Casual, highly irregular late fall and winter vagrant (early November to early March). A high-mountain species of coniferous forests of western North America, Cassin's Finch is an extremely rare, irregular wanderer to lowlands outside its normal range. In winter this species shows strong irruptive patterns, with years passing between invasions (Bock and Lepthien 1976). Reflecting this pattern, Cassin's Finch has been recorded in the Salton Sea region in only five years: in the winter of 1984–85, the fall of 1986, the fall of 1996 and the winter of 1996–97, the winter of 1997–98, and the fall of 2000. An influx of various fringillids brought up to 25 Cassin's Finches to Brawley 26 January–2 March 1985 (*AB* 39: 211), followed by 1 at Calexico 27 January 1985 (*AB* 39:211) and 2 at the north end 16 February 1985 (*AB* 39:211). Fall 1986 found single birds near Seeley 9 November (*AB* 41:147) and at Brawley 10–14 November (*AB* 41:147), whereas the fall of 1996 and the winter of 1996–97 yielded two to three at Brawley 9–10 November (*FN* 51:119) and another at Niland 26 December (*FN* 51:643). A female was collected at Oasis 27 December 1997 (R. L. McKernan; SBCM 55538). Finally, six to seven birds were at Niland 2–14 November 2000 (*NAB* 55:105).

ECOLOGY. Cassin's Finches found in the region are associated with planted conifers, mostly large pines.

HOUSE FINCH

Carpodacus mexicanus frontalis (Say, 1823)

Common breeding resident. Few birds are more common in the West than the House Finch. As expected, then, it is a common bird in the Salton Sink that may be encountered anywhere in the region except on the open water of the Salton Sea and along its immediate shore, and it occurs even along the shore on occasion. Peak abundance is reached in developed areas, where flocks of hundreds are frequently encountered in towns and agricultural areas with scattered trees. Breeding takes place mainly in spring and summer, with peak activity from March through June. However, pairs frequently attempt several broods and may nest well into September.

ECOLOGY. This species' adaptability is the key to its great abundance in the West and its rapid spread following introduction in the East. It readily occupies human settlements, nesting even in structures such as homes and utility poles. Large gums, planted palms, and citrus orchards also support breeders; it may nest just about anywhere in suburban neighborhoods and in parklands and ranch yards. It is common in agricultural areas with suitable perches (trees, wires, etc.), where it often gathers in large feeding flocks. Numbers thin substantially in desert scrub, although it is still recorded frequently even in this comparatively bleak habitat.

TAXONOMY. Finches of the mainland Southwest are the large *C. m. frontalis*, with diffuse red coloration. Populations are highly variable in the West because much of the birds' red coloration is dependent on acquiring carotenoid pigments from their diet (Hill 1993; Hill and Montgomerie 1994). This variability has led to the unfortunate naming of several additional subspecies, such as *C. m. grinnelli* Moore, 1939, and *C. m. solitudinis* Moore, 1939, both synonyms of *C. m. frontalis*.

RED CROSSBILL

Loxia curvirostra Linnaeus, 1758 subsp.?

Casual, highly irregular late fall and winter vagrant (mid-November to early May). The Red Crossbill, like many granivorous birds dependent on conifer seeds, stages occasional incursions into areas well away from its normal range (Bock and Lepthien 1976). It has appeared at the Salton Sea in only three years, two of which coincided with occurrences of Cassin's Finches. The area's first record was of two individuals in the southeastern Imperial Valley 28 November 1950 (*AFN* 5:228). Two subsequent records are from Brawley, with at most 10 from 25 November to 3 May 1985 (*AB* 39:105, 212, 351) and at least 20 from 2 November 1996 to 3 May 1997 (S. von Werlhof; *FN* 51:119, 927). Also part of the greatest crossbill irruption in decades were seven at Calipatria 26 December 1996 (*FN* 51:643).

ECOLOGY. Like Cassin's Finch, crossbills appearing in the Salton Sea region are invariably associated with planted pines.

TAXONOMY. There are no specimens for the Salton Sea region, nor are there taped vocalizations of the Red Crossbill's calls. Therefore it is impossible to determine to which subspecies or semispecies or sibling species the recorded birds belonged. Geographic variation in the Red Crossbill presents one of the greatest taxonomic and systematic challenges in North America (Griscom 1937; Monson and Phillips 1981; Payne 1987; Groth 1993). If proximity to breeding areas is any indication, then birds appearing at the Salton Sea should be of Groth's (1993) "type 2," a larger form breeding throughout the Sierra Nevada and presumably also the Transverse and Peninsular Ranges of southern California. This form most closely coincides with the named subspecies *L. c. pusilla* Gloger, 1834, of which both *L. c. benti* Griscom, 1937, and *L. c. grinnelli* Griscom, 1937, are probably synonyms and *L. c. bendirei* Ridgway, 1884, is perhaps a synonym (Groth 1993:97). The measurements of specimens from either side of the Salton Sink, a male taken from Borrego Springs 31 October 1996 (SDNHM 49665; mandible width 9.5 mm, wing chord 90.0 mm) and a female taken 22 km north of Blythe 20 August 1992 (SDNHM 48237; mandible width 9.5 mm, wing chord 89.5 mm), match the measurements of *L. c. pusilla*.

PINE SISKIN
Carduelis pinus pinus (Wilson, 1810)
Rare winter visitor (late October to mid-April); 4 spring records. The Pine Siskin is the rarest *Carduelis* in the Salton Sea region. Generally only a few birds are encountered, but the region is occasionally graced with large flocks, such as more than 100 at Salton City 20 February 1993 (S. von Werlhof), 50 at Brawley 30 January 1993 (G. McCaskie), and 28 on the Salton Sea (north) Christmas Bird Count 2 January 1993 (*AB* 47: 968). Records extend from 19 October (1963, female west of Niland; SBCM 38086) to 22 April (1985, Brawley; B. E. Daniels), including additional specimens from 3 km west of Westmor-

land 10 November 1992 (SDNHM 48278) and from Mecca 29–30 March 1911 (UCLA 10322, 10323). Twelve in the Imperial Valley 13 October 2000 (G. McCaskie) were exceptionally early. There are also four records from May: two at Brawley 5 May 1985 (G. McCaskie), single birds at El Centro 5 May 1990 (G. McCaskie) and Salton Sea NWR 19 May 1999 (C. A. Marantz), and three birds 8 km southwest of Westmorland 19 May 1999 (M. A. Patten). We presume that these late birds were spring migrants, as the species winters occasionally in Baja California (Unitt et al. 1992; Patten et al. 1993).

ECOLOGY. Although sometimes found with flocks of Lesser Goldfinches, the Pine Siskin is frequently noted alone or in pure flocks, most often in heavily planted suburban neighborhoods sporting large nonnative pines. Some occur in mature Athel, and a few have been noted in mature Fremont Cottonwood. The crops of birds collected at Mecca (cited above) were filled with cottonwood seeds (van Rossem 1911).

TAXONOMY. U.S. birds are of the smaller, less yellow, more streaked nominate subspecies (see Unitt et al. 1992), further distinguished from *C. p. macropterus* (Bonaparte, 1850), of Mexico, by its paler crown.

LESSER GOLDFINCH
Carduelis psaltria psaltria (Say, 1823)
Fairly common winter visitor (mid-September to mid-May); uncommon, apparently irregular breeder. The common, widespread goldfinch of the Southwest is the Lesser. In most of the Sonoran Desert, including the Salton Sink, it is principally a winter visitor but breeds locally, especially along the lower Colorado River (Rosenberg et al. 1991). In the Salton Sea region, breeding is generally confined to shade trees of neighborhoods and ranches. Breeding occurs from March (e.g., "nests well under way" at Mecca 30 March 1911; van Rossem 1911) through summer (e.g., 20 apparently nesting at Brawley 14 July 1984; G. McCaskie). The species may not breed in the Salton Sink every year, as it has not been observed in some summers. Records away from breeding locales provide a gauge of when wintering birds

arrive (e.g., 8 km southwest of Westmorland 10 September 1995; S. von Werlhof) and depart (e.g., 3 at Wister 22 May 1994; M. A. Patten); an exceptionally early fall migrant was noted 3 km southeast of El Centro 24 August 2001 (K. Z. Kurland). The majority occur from early October to late April. Peak numbers in winter seldom exceed a few dozen birds (e.g., 50 at the south end 23 January 1988; G. McCaskie), though more than a hundred were seen at Salton City 20 February 1993 (S. von Werlhof).

ECOLOGY. Breeding goldfinches, requiring water as well as food and nest sites, favor neighborhoods, ranches, cemeteries, and other locales with large trees, ranging from Fremont Cottonwood and Goodding's Black Willow to gum, Athel, mulberry, and other planted exotics. They nest in native mesquite thickets in the southern Coachella Valley (e.g., a nest with 4 eggs 11 km east of Mecca 31 March 1973; WFVZ 47956). Wintering birds are more widespread but still concentrate around trees.

TAXONOMY. The widespread olive-backed subspecies of the Pacific Coast is generally called *C. p. hesperophilus* (Oberholser, 1903). But Phillips et al. (1964) reported that the type of *C. psaltria,* from eastern Colorado and now apparently lost (Miller et al. 1957:321), had a green back, and Ridgway (1901:114) noted that males throughout the western United States have "olive-green" backs. *C. p. mexicanus* Swainson, 1827, of northern Mexico and southern Texas, has black-backed males. It hybridizes with the nominate subspecies in southern Arizona, creating an intergrade population formerly called *C. p. arizonae* (Coues, 1874). These subspecies also hybridize along a broad front from northern Mexico to Texas (Paynter and Storer 1970), apparently influencing the population north to Colorado. In the United States, therefore, variation in back color is clinal, with olive predominating on the Pacific Coast and mixed black and olive predominating in Colorado and northern Texas. Accordingly, the recognition of *C. p. hesperophilus* is problematic. Males with black backs have been taken in Colorado (Ridgway 1901), photo-graphed at far-flung locales from Missouri (Robbins and Easterla 1992) to Maine (*AB* 47:240), and observed in California, for example, at La Tuna Canyon 9 May 1992 (K. F. Campbell) and Arcadia 6 February 1994 (M. A. Patten). Whether such birds are *C. p. mexicanus,* intergrades, or represent a color morph (Phillips et al. 1964:188) is unknown. No black-backed males have been noted in the Salton Sea region.

LAWRENCE'S GOLDFINCH
Carduelis lawrencei Cassin, 1850

Rare spring transient (March and April); irregularly rare to uncommon in winter (early November to mid-February); 1 fall record. Lawrence's Goldfinch is a bird of the oak belt in cismontane California and northwestern Baja California. It stages irregular winter influxes into the eastern Sonoran Desert, particularly southern Arizona and northern Sonora (Monson and Phillips 1981; Russell and Monson 1998), reaching the former nearly annually. This species also winters irregularly in extremely small numbers at the Salton Sea, with dates of wintering birds from 4 November (1984, 21 km east of Calipatria; P. Unitt, SDNHM 43459) to 14 February (1998, Whitewater River near Mecca; M. A. Patten). Two near Westmorland 9 October 1977 (G. McCaskie) we presume to have been fall transients. A few wintering birds may stay into March, when the presence of spring migrants clouds departure dates. Even though Lawrence's Goldfinch is casual in that season, some winters see large numbers; for example, there were 25 males near Niland 12 December 1987 (*AB* 42:324) and 40 at the north end 2 January 1993 (*AB* 47:968). Also, van Rossem (1911) found it common in the Imperial Valley and around Mecca 1 December 1910–31 March 1911 (MVZ 105545; UCLA 10315; YPM 4590). There is an increase from mid-February through April (e.g., 70 at the south end 11 March 1954; *AFN* 8:272), apparently reflecting spring transients returning from wintering areas to the southeast, the latest being at Mecca 1–4 May 1999 (M. A. Patten). Five at a ranch 8 km southwest of Westmorland 12 May 1996

(G. McCaskie) and a female and a singing male at the same locale in late May 1999, the former 19 May, the latter 23 May (M. A. Patten), hinted at the possibility of nesting. Breeding is unconfirmed for the region, however, although this species has nested opportunistically in Arizona (Monson and Phillips 1981), at the Brock Research Station, just east of the region (Garrett and Dunn 1981), and in the lower Colorado River valley (Rosenberg et al. 1991).

ECOLOGY. Opportunistic and nomadic, wintering Lawrence's Goldfinches seek patches of seed-bearing forbs, so they are found primarily in weedy fields, if not heard only in flight high overhead. Some frequent extensive stands of tamarisk near water.

AMERICAN GOLDFINCH
Carduelis tristis salicamans (Grinnell, 1897)
Rare winter visitor (November through mid-May). Normally a denizen of cismontane California, the Great Basin, and points east, the American Goldfinch is a rather rare winter visitor in the desert Southwest, including around the Salton Sea. Although it occurs nearly annually, it is scarce. The first regional occurrences were relatively recent, with four at Calipatria 1 January 1949 (*AFN* 3:144) and one 4 km northwest of Westmorland 5 February 1955 (Cardiff 1956). Records extend from 2 November (2000, 3 at Niland; G. McCaskie) to 10 May (1986, 2 at Wister; G. McCaskie), with most from mid-December through February. Generally numbers range only from 1 to 3 birds, but occasional flocks are noted, such as 50 at Brawley 30 January 1993 (G. McCaskie), 40 at Wister 16 April 1987 (B. E. Daniels), 30 at the south end 20 November 1976 (G. McCaskie), and 25 at Mecca 6 February 1993 (M. A. Patten). Notably, approximately 100 were scattered throughout the Imperial Valley 23 January 1988 (G. McCaskie; M. A. Patten).

ECOLOGY. Preferring mature riparian woodland, this species must make do with poor facsimiles of this habitat on its rare excursions to the Salton Sink: remnant stands of Fremont Cottonwood and Goodding's Black Willow and large ornamental shade trees. It forages in weedy vegetation near such trees.

TAXONOMY. All California records are apparently of the small, dark Pacific Coast subspecies, *C. t. salicamans,* including the three local specimens, two from 7 km northwest of Imperial 29 December 1988 (M. A. Holmgren; SDNHM 45497, 45498) and one west of Niland 19 February 1966 (E. A. Cardiff; SBCM 38117). This subspecies has been recorded only once in Arizona, at Parker 17 May 1948 (Phillips et al. 1964). In that state the large, pale *C. t. pallidus* Mearns, 1890, the breeder in the Great Basin and the Rocky Mountains, is the regular subspecies (Rea 1983), so it should be anticipated in southeastern California. There are several records for central Arizona of the mid-sized, dark *C. t. tristis* (Linnaeus, 1758), of eastern North America (Monson and Phillips 1981; Rea 1983), also distinguished by reduced white in the wings. It is as yet unknown from California but could occur as a vagrant.

EVENING GROSBEAK
Coccothraustes vespertinus (Cooper, 1825) subsp.?
Three records. In California, Evening Grosbeaks are normally found in the high mountains of the northern and central parts of the state, but they sporadically wander into the lowlands of southern California in fall and winter, occasionally in spring (presumably on their return north after winter dispersal). Two of the Salton Sea area records are from late fall and winter, one of a bird photographed at Salton Sea NWR 17–18 October 1990 (*AB* 45:154) and another of a bird seen at Brawley 10 December 1990 (*AB* 45:323). A male near Calipatria 7 May 1996 (*FN* 50:334) provided the only spring record. A female at Plaster City 1 February 1987 (*AB* 41:332) was barely west of the edge of the Salton Sink.

TAXONOMY. California specimens are of the largest-billed subspecies, *C. v. montanus,* whose females have the brownest breasts. It breeds in the Sierra Nevada and other high ranges in California. We include the supposedly slightly darker

C. v. brooksi Grinnell, 1917, of the Rocky Mountains, as a synonym (see Phillips et al. 1964:182). Lacking specimens from the Salton Sink, however, we cannot assign records to *C. v. montanus, C. v. mexicanus* Chapman, 1897, a slender-billed subspecies breeding as close as southeastern Arizona, or *C. v. vespertinus,* a short-billed subspecies recorded as a winter vagrant in southern Arizona (Phillips et al. 1964:183).

NONNATIVE SPECIES

Included below are those species that are not native to the region but have become an established part of the avifauna either through expansion from nonnative populations or through successive introductions as game animals. The many species of birds that have been encountered only one to several times as escaped exotics are not included.

RING-NECKED PHEASANT
Phasianus colchicus Linnaeus, 1758

Whether the Ring-necked Pheasant is established in the region or merely persists through repeated game releases is unclear. The species was established in the Mexicali Valley by 1922 (Hart et al. 1956) and has always been more common on the Mexican side of the border, where "habitat is better" (Leopold 1959). It remains an uncommon resident in the Mexicali Valley (Patten et al. 1993, 2001). In the Imperial Valley the pheasant appears to be established in small numbers around Wister, in extensive Alfalfa and weedy fields, where groups of up to ten are occasionally noted. Stubble and grass fields in the vicinity of Mount Signal, Brawley, and El Centro are the only other locales for which there are repeated recent records, although individuals have been noted throughout much of the Imperial Valley, suggesting that the species is established in various places. The Ring-necked Pheasant was more common during the 1950s, with estimates of 19 per km^2 in the southern half of the valley (Leopold 1959). It is seldom noted in the Coachella Valley.

FERAL PIGEON
Columba "livia" Gmelin, 1789

The extent to which any populations of the Feral Pigeon are truly established in North America is unknown given that repeated escapes and releases greatly augment the pigeon's numbers every year yet populations are not increasing (Johnston and Garrett 1994). The situation at the Salton Sea is no different. These pigeons are common to abundant human commensals around all the major cities in the region, particularly Indio, Brawley, El Centro, and Calexico. They occur less frequently, but fairly commonly, in agriculture areas throughout the Coachella, Imperial, and Mexicali Valleys.

Although clearly derived from Rock Dove stock, this pigeon has been bred extensively for show, racing, and so on, and virtually no birds occurring in North America exhibit the phenotypic characters of true, wild Rock Doves. In that respect, referring to these birds as Rock Doves would be equivalent to referring to domestic dogs as Gray Wolves in a document on mammals.

SPOTTED DOVE
Streptopelia chinensis (Scopoli, 1786)

A species native to southeastern Asia, the Spotted Dove was introduced into coastal southern California in the early 1900s (Hardy 1973). It quickly became established on the coast from Santa Barbara south to Tijuana. It was first recorded in the Coachella Valley 21 March 1942 (C. A. Harwell et al.); many were there by the 1960s (Johnston and Garrett 1994). This valley now harbors a small population, but the species is only uncommonly recorded in the southern portion, exclusively in orchards and towns. The species has been recorded once in the Imperial Valley, near Calipatria 28 November 1982 (*AB* 37:224). Attribution of the Spotted Dove "south to the Imperial Valley" in the 1950s (*AFN* 8:270) is baseless.

EUROPEAN STARLING
Sturnus vulgaris Linnaeus, 1758

The European Starling, a species deliberately introduced from Europe into eastern North America in the 1890s, quickly spread across the continent. It was first recorded in the Imperial Valley 24 November 1956 (*AFN* 11:62) and 9 February 1957 (Cardiff 1961; SBCM 36361) and was first found breeding 8.5 km east of Holtville 4 May 1958 (Rainey et al. 1959; CSULB 2068). It is now an abundant resident breeder throughout the urban, suburban, and agricultural lands of the Coachella, Imperial, and Mexicali Valleys. Indeed, this species is seldom encountered far from human habitation anywhere in North America. This starling is a cavity nester that has severely reduced the populations of numerous native species (Zeleny 1969; Weitzel 1988), but the extent to which it has negatively affected species in the Salton Sea region is not known. Flocks of many thousands are frequently encountered in winter as they congregate to forage (often with large flocks of native blackbirds) in cattle feedlots and agricultural fields.

HOUSE SPARROW
Passer domesticus (Linnaeus, 1758)

Like the European Starling, the House Sparrow was deliberately introduced into North America during the mid-nineteenth century, but in the case of the House Sparrow birds were released not only on the East Coast but also in San Francisco and Salt Lake City (Robbins 1973). This species quickly spread throughout North America. It was established widely in California by 1915 (Grinnell and Miller 1944) and had definitely reached the Imperial Valley by December 1910 (van Rossem 1911). It is now a common breeding resident throughout the settled portions of the Salton Sea region.

HYPOTHETICAL LIST

For the species listed below, we found documentation wanting or insufficient to allow us to include them on the list of birds we consider to have occurred naturally at the Salton Sea. Species are included in this list for one of the following reasons: (1) identification was well established, but natural occurrence was questionable in that the species' ability to reach Cali-

fornia under its own power was judged to be highly unlikely (e.g., the Purplish-backed Jay); (2) documentation or further research or observation uncovered a definite misidentification (e.g., the Anhinga); (3) identification was not established to a sufficient degree, but the species is a good candidate for natural occurrence (e.g., the Rufous-backed Robin); (4) identification was not established, and the species is an unlikely candidate for natural occurrence (e.g., the Masked Duck); or (5) baseless regional claims of a species have crept into the literature (e.g., the Northern Goshawk). In several cases (e.g., the Red-necked Stint and the Slaty-backed Gull) documentation includes physical evidence that supports the claimed identification, but the records have not been endorsed by the California Bird Records Committee.

BULWER'S PETREL
Bulweria bulweri (Jardine and Selby, 1828)

A Bulwer's Petrel reported at the mouth of the Whitewater River 10 July 1993 (*AB* 47:1149) would have been the first record for North America. Although there was no doubt that a procellariiform was observed, the documentation was not sufficient to rule out the similar congener, Jouanin's Petrel, *B. fallax* Jouanin, 1955, or other small, dark procellariiforms (Patten and Minnich 1997; Garrett and Singer 1998). This enigmatic petrel was not documented in North American waters until 1998, when birds were photographed on Monterey Bay, California, 17 July (*FN* 52:498), and off Hatteras, North Carolina, 8 August (LeGrand et al. 1999).

ANHINGA
Anhinga anhinga (Linnaeus, 1766)

Written details of a female Anhinga reported at Finney Lake 17 November 1999 (CBRC 1999-210) suggest that *A. melanogaster* Pennant, 1769, of the Orient, or *A. [m.] rufa* (Daudin, 1802), of Africa, was involved. Indeed, a bird that was doubtless this same individual was rediscovered and photographed at nearby Ramer Lake 19 June–22 July 2000 (S. Olson et al.). The bird proved to be an adult female African Darter (*A. [m.] rufa*), an obvious escapee from captivity. It was still present in the area 19 August 2001, when it was observed at Salton Sea NWR (G. McCaskie).

Escaped *Anhinga* species have been taken in the wild elsewhere in North America, such as a September 1927 specimen from Aurora, Colorado (DMNH 12296), long treated as *Anhinga anhinga* (e.g., Bailey and Neidrach 1965:92) until it was reidentified by the late Allan R. Phillips as an escaped *A. [m.] novaehollandiae* (Gould, 1847), of Australia and New Guinea (Andrews and Righter 1992:14). More recently, an adult male *A. melanogaster* was observed near Ensenada, Baja California, 24 September 2001 (R. A. Erickson). Documentation and review of all claims of vagrant Anhingas in North America need to take escaped darters into consideration.

CHINESE POND HERON
Ardeola bacchus (Bonaparte, 1855)

A heron in a flooded field south of Brawley 21 January 1987 was reported as a Chinese Pond Heron, but the description did not eliminate a Cattle Egret in alternate plumage (Langham 1991). Although this Asian pond heron has reached Alaska once (Hoyer and Smith 1997), it is an unlikely candidate for vagrancy to California.

AGAMI HERON
Agamia agami (Gmelin, 1789)

An Agami Heron reported at Salton Sea Beach 24 August 1977 (Luther 1980) undoubtedly was misidentified. This sedentary, tropical species gets no closer to California than southern Mexico (Howell and Webb 1995).

CHILEAN FLAMINGO
Phoenicopterus chilensis Molina, 1782
LESSER FLAMINGO
Phoeniconaias minor (Geoffroy, 1798)

Both the Chilean Flamingo and the Lesser Flamingo were listed as occurring at Salton Sea NWR by the United States Fish and Wildlife Service (1993), with an "origin unknown" caveat. Both species have been recorded at the Salton Sea, as has the Greater Flamingo, *Phoenicopterus*

ruber Linnaeus, 1758, and they survive in the wild for many years. The origin of these birds is not unknown, however, as there is a zero probability that any flamingo would reach California as a naturally occurring vagrant. Given the high prevalence of captive flamingoes at zoological parks throughout the United States, it seems clear that all records in the West are of escapees (and that the vast majority of extralimital Greater Flamingoes recorded in North America are likely escapees as well).

RED-BREASTED GOOSE
Branta ruficollis (Pallas, 1769)

Two Red-breasted Geese in the Imperial Valley 28 December 1968–11 January 1969, one of which remained through 16 February 1969 (*AFN* 23:515, 519), were undoubtedly escapees from captivity. This handsome West Asian goose is an extremely unlikely candidate for vagrancy to North America, let alone California.

AMERICAN BLACK DUCK
Anas rubripes Brewster, 1902

An American Black Duck report from the Imperial Warm Water Fish Hatchery 11 November 1978 was not accepted by the California Bird Records Committee because of questionable natural occurrence (Binford 1985). A bird identified as an American Black Duck × Mallard hybrid was at the south end 15–20 May 1977 (*AB* 31:1047), and a similar hybrid (or possible pure American Black Duck) was at the mouth of the Whitewater River 4 January–22 February 1997 (*FN* 51:801). There remains only one valid record of the American Black Duck for California, of a bird collected at Willows, Glenn County, 1 February 1911 (Grinnell 1911a; Dunn 1988; MVZ 17198).

MASKED DUCK
Nomonyx dominica (Linnaeus, 1766)

A male Masked Duck reported at Wister 17 July 1977 (Luther et al. 1979) was almost certainly misidentified, as was one claimed recently from near Mecca (CBRC 2000-058). The species occurs no closer to California than southern Sinaloa (Howell and Webb 1995) and is unrecorded as a vagrant even in Sonora (Russell and Monson 1998).

NORTHERN GOSHAWK
Accipiter gentilis (Linnaeus, 1758)
The basis for the attribution of the Northern Goshawk to Salton Sea NWR (United States Fish and Wildlife Service 1993) is unknown and without warrant. There are in fact no records of this species for the region.

CRESTED CARACARA
Caracara cheriway (Jacquin, 1784)
The Crested Caracara has been reported three times in the Salton Sea region. Reports of immatures near Westmorland 14 December 1993 (Rottenborn and Morlan 2000) and 6.5 km southwest of Coachella 9 January 1994 (Howell and Pyle 1997) lack convincing documentation. An adult near Brawley 31 January 1997 (*FN* 51:801) not only lacks documentation but was felt to be an escapee (Rottenborn and Morlan 2000; R. Higson pers. comm.). To date there are no acceptable records for California. Before 1930, however, the species occupied parts of the Río Colorado delta (Patten et al. 2001), at least in winter, for which there are four records involving seven individuals from 7 December (1896, head of the Río Hardy; SDNHM 349) to 15 March (1928, 11 km west-southwest of Pilot Knob; Grinnell 1928, MVZ 52104).

APLOMADO FALCON
Falco femoralis Temminck, 1822
Contrary to claims by the California Department of Fish and Game (1979), the Aplomado Falcon has never been recorded in California, let alone in the Salton Sink.

PIED AVOCET
Recurvirostra avosetta Linnaeus, 1758
A photograph of a Pied Avocet was attributed to the Salton Sea but was likely mislabeled as this species is unknown in the Western Hemisphere (Rottenborn and Morlan 2000).

RED-NECKED STINT
Calidris ruficollis (Pallas, 1776)
A stint in worn first-alternate plumage was studied, photographed (see Roberson 1980:160), and collected near the mouth of the Alamo River 17 August 1974 (SDNHM 38887). It was published as a Red-necked Stint (McCaskie 1975) and accepted as such by the California Bird Records Committee, supported by what was considered the first (and still only) specimen for California. Some years later Richard R. Veit concluded that the specimen was in fact an adult Little Stint. During reevaluation by the California Bird Records Committee, Veit examined the specimen again and concluded that it was a Red-necked Stint after all. He published his revised conclusion, backed by a discriminant function analysis and arguments based on a tenuously small comparative sample (Veit 1988). Because of improper application of the statistics and nonindependence of the predictors used (four of the five were measures of the bill), his study is inconclusive. Attempted identification of the specimen by M. Ralph Browning and Claudia P. Wilds using the large collection at the National Museum of Natural History also proved fruitless (M. R. Browning *in litt.*). We examined the specimen in detail on several occasions, including against minimal comparative material at LACM and SDNHM. The specimen may be a Little Stint after all based on the pattern of internal markings on the outermost vestigial primary (see Cramp and Simmons 1983:309; and Hayman et al. 1986:pl. 76). Still, a firm identification may never be possible; at the least it awaits further work on this difficult group. For now the bird is best treated as an unidentified individual of this species pair (Erickson and Hamilton 2001).

An adult Red-necked Stint reported at Unit 1 on 19 July 1981 lacks convincing documentation (Heindel and Garrett 1995). One of the observers now believes the bird to have been a Sanderling in alternate plumage (B. E. Daniels pers. comm.).

SLATY-BACKED GULL
Larus schistisagus Stejneger, 1884

The Slaty-backed Gull is a Siberian species that winters in eastern Asia. There are more than 30 records for North America away from Alaska. It has been documented as far south as Brownsville, Texas, 7–22 February 1992 (Haynie 1994), and as far east as the Chicoutini River, in Quebec, 7–20 November 1993 (*AB* 48:85). We believe that a fourth-winter bird photographed at Salton City 21–28 February 1998 (*FN* 52:257; CBRC 1998-050) was a Slaty-backed Gull, California's first (and still only) record. However, the record was rejected by a vote of eight to two by the California Bird Records Committee. There was general agreement that the bird's plumage and bare parts matched this species. At issue was its late molt timing (its outermost primaries were half-grown), although timing was late for any large species of *Larus* and is plastic in gulls (S. N. G. Howell; P. Pyle pers. comm.). We predict that with the endorsement of future records for California the Salton City bird will be re-evaluated and likewise endorsed.

EURASIAN COLLARED-DOVE
Streptopelia decaocto (Frivaldskzy, 1838)

A male Eurasian Collared-Dove at Brawley 18 July 1999 (*NAB* 53:432; CBRC 1999-140) was paired with a "Ringed Turtle-Dove," the domesticated form of the African Collared-Dove, *S. roseogrisea* (Sundevall, 1957). The pair fledged two young 28 or 29 August 1999 (G. McCaskie). The female was undoubtedly a local escapee, but the provenance of the male is less clear, although it too was most likely a local escapee (guilt by association). Two Eurasian Collared-Doves were reported 3 km southeast of El Centro 7 August 2001 (G. McCaskie). Reports of potential collared-doves at Brawley 24 February 2001 (CBRC 2001-072) and at North Shore 7 July 2001 (G. McCaskie) did not exclude the Ringed Turtle-Dove. The Eurasian Collared-Dove has spread rapidly across the United States since colonizing Florida from introduced populations in the West Indies (Smith 1987; Hengeveld 1993; LeBaron 1999; Romagosa and McEneaney 1999). By fall 2001 the species had not been known to reach California from the East, but we expect that it will in the near future, as it has reached Arizona (G. H. Rosenberg pers. comm.). Inroads were made by 2002, when up to three were noted in Calipatria 26 January–14 October (*NAB* 56:357), five were noted in Niland 8 September (H. King), and 50–75 were photographed in Calipatria after late December (J. Morlan). Several birds in Calipatria had the undertail coverts white, suggesting impurity, likely as a result of interbreeding with the "Ringed Turtle-Dove," a hybrid combination documented in Florida and Illinois. Incidentally, we do not know the basis for prior claims of the Ringed Turtle-Dove at Finney and Ramer Lakes (California Department of Fish and Game 1979), but we are aware of various Imperial Valley dove breeders who keep this form (and the Eurasian Collared-Dove; M. A. Patten pers. obs.); some of these doves may escape on occasion, as at Brawley in the summer of 1999.

COMMON NIGHTHAWK
Chordelies minor (Forster, 1771)

The Common Nighthawk is a casual migrant through the California desert, with no valid records for the Sonoran Desert save a 16 October 1924 specimen from near Bard on the lower Colorado River (Rosenberg et al. 1991). We are unaware of valid records for the Salton Sink despite claims that the species is a rare spring migrant and summer visitor to Finney and Ramer Lakes (California Department of Fish and Game 1979).

BROAD-TAILED HUMMINGBIRD
Selasphorus platycercus (Swainson, 1827)

A Broad-tailed Hummingbird banded "near Mecca" 1 August 1959 (*AFN* 14:73) was not in the Salton Sink. The bird was actually captured at Cottonwood Springs, in southern Joshua Tree National Park (*fide* G. McCaskie).

GREEN KINGFISHER
Chloroceryle americana (Gmelin, 1788)

An alleged Green Kingfisher observed in a ditch near the north end 26 January 1991 would have provided the first record for California, but documentation was inadequate to establish the identification (Patten and Erickson 1994).

HAIRY WOODPECKER
Picoides villosus (Linnaeus, 1766)
Despite its extensive wanderings to various parts of the California desert, there are no valid records of the Hairy Woodpecker for the Salton Sink. We do not know the basis for attributing this species to the Imperial Valley (California Department of Fish and Game 1979), but we consider the claim erroneous.

GILDED FLICKER
Colaptes chrysoides mearnsi Ridgway, 1911
The true status of the Gilded Flicker in the Salton Sea region is difficult to determine. In the United States and northwestern Mexico this species is essentially a Sonoran Desert endemic. It was much more numerous and widespread in California a half-century ago, but it is now one of the rarest species in California, with only a few pairs remaining in Joshua Tree woodland around Cima, in the East Mojave National Preserve. The species was reported in the Imperial Valley in the 1920s (Howell 1922b) and was thought perhaps to be spreading northward into that valley even though it was not definitely attributed to the region (Grinnell and Miller 1944). Unfortunately, it was never actually documented in the Salton Sink. The closest station of record is along the Río Hardy 6 km southeast of Benito Juárez (ca. 40 km east of Mexicali) 23–27 January 1928 (Grinnell 1928; MVZ 52196–98). Subsequent records are lacking save two relatively recent sight reports from the Imperial Valley, 1 March 1969 (AFN 23:522) and 20 January–24 March 1973 (AB 27:664). These reports are best treated as tentative given the extreme difficulty of field separation of the Gilded Flicker from some hybrids between the yellow-shafted and red-shafted forms of the Northern Flicker (Kaufman 1979) and the general absence of vagrancy by the Gilded Flicker in other parts of its range.

GRAY VIREO
Vireo vicinior Coues, 1866
A singing male Gray Vireo was reportedly collected at Mecca 26 March 1911 (van Rossem 1911; Grinnell and Miller 1944; Garrett and Dunn

1981). The specimen (UCLA 10697) is in fact *V. bellii pusillus*. The original label is missing, apparently having been discarded by Donald R. Dickey when accessioned into the collection. However, van Rossem (1911) reported collecting only *V. vicinior,* and the location and date match what he reported for that species. Thus, there is no doubt that UCLA 10697 is the long accepted, but never verified, migrant "Gray" Vireo from Mecca.

BLUE-HEADED VIREO
Vireo solitarius (Wilson, 1810)
The Blue-headed Vireo is a casual visitor to California. Although there are numerous reports for the state, no more than perhaps 25 are valid, as most reports pertain to brightly colored Cassin's Vireos (see Heindel 1996). Virtually all valid records are coastal from late fall (early October and later), with a few from winter. A singing male at Westmorland 20 May 1976 (AB 30:892) would have provided one of few spring records but should be disregarded given the date and the complexity of the identification (J. L. Dunn *in litt.*).

HUTTON'S VIREO
Vireo huttoni Cassin, 1851
Hutton's Vireo is a casual vagrant to the California desert, with invalid claims far outnumbering valid records. Its attribution to the Imperial Valley (California Department of Fish and Game 1979) is erroneous.

YELLOW-GREEN VIREO
Vireo flavoviridis (Cassin, 1851)
See the Red-eyed Vireo account in the Main List.

PURPLISH-BACKED JAY
Cyanocorax beecheii (Vigors, 1829)
A first-winter Purplish-backed Jay photographed at Calexico late November–13 December 1990 was undoubtedly an escapee from captivity (Heindel and Garrett 1995).

GRAY-CHEEKED THRUSH
Catharus minimus aliciae (Baird, 1858)
The Gray-cheeked Thrush is a casual vagrant to California, where a mere 21 have been documented since the first, in 1970 (see Dunn 1988).

All but two records fall between 10 September and the end of October. The single report for the Salton Sink, of one photographed 3 km southeast of El Centro 15–16 September 2002 (K. Z. Kurland), fits nicely into this pattern. The California Bird Records Committee is currently (March 2003) reviewing the record.

Two of this thrush's three subspecies have limited breeding ranges: *C. m. minimus* (Lafresnaye, 1848) breeds only in Newfoundland, and *C. m. bicknelli* (Ridgway, 1882), only in northern New England and the Canadian Maritimes. The latter taxon is sometimes treated as a distinct species (Ouellet 1993; AOU 1998; cf. Marshall 2001). In contrast to these subspecies, *C. m. aliciae* is widespread across boreal Canada, Alaska, and eastern Siberia. As expected, the two California specimens, from Southeast Farallon Island 3 October 1970 (CAS 68501) and from Encino ca. 25 October 1997 (LACM 110224), are identifiable as the grayer (less rufous) *C. m. aliciae*, as apparently are various birds examined in hand or well photographed (see Howell and Pyle 1997). The single Arizona specimen, from Cave Creek 11 September 1932, is also *C. m. aliciae* (Monson and Phillips 1981).

RUFOUS-BACKED ROBIN

Turdus rufopalliatus (Lafresnaye, 1840)

A Rufous-backed Robin reported at the headquarters of Salton Sea NWR 21 October 1995 lacked documentation sufficient to establish the identification (McCaskie and San Miguel 1999). Furthermore, the date is slightly outside the 4 November–16 April window of the nine acceptable California records. This species has wintered near the Salton Sink on three occasions: at Imperial Dam, on the lower Colorado River, 17 December 1973–6 April 1974 (Luther et al. 1979); at Snow Creek Village, in San Gorgonio Pass, 1–20 March 1992 (Patten, Finnegan, et al. 1995); and at Borrego Springs, in the Anza-Borrego Desert, 16 March–16 April 1996 (McCaskie and San Miguel 1999).

WRENTIT

Chamaea fasciata (Gambel, 1845)

A Wrentit "seen and heard well" at the Salton Sea near Mecca 30 December 1965 (*AFN* 20:376) lacks contemporaneous documentation (J. M. Sheppard *in litt.*) and was not accepted by other authors (e.g., Garrett and Dunn 1981). Claims for Finney and Ramer Lakes (California Department of Fish and Game 1979) are baseless. This denizen of coastal sage scrub and chaparral of the California floristic province is sedentary, with no records of true vagrancy, although there are two records for the desert oasis at Morongo Valley, in southern San Bernardino County (Patten 1995a; *FN* 52:127).

WHITE-COLLARED SEEDEATER

Sporophila torqueola (Bonaparte, 1851)

There are no valid records of the White-collared Seedeater for the United States west of southern Texas, except for reports of likely escapees from southern Arizona and coastal southern California (G. H. Rosenberg pers. comm.; G. McCaskie). Its attribution to the Imperial Valley (California Department of Fish and Game 1979), particularly as an element of the native avifauna, is baseless.

BLUE BUNTING

Cyanocompsa parellina (Bonaparte, 1850)

We do not know why the Blue Bunting was attributed to Salton Sea NWR (United States Fish and Wildlife Service 1993), as this species is unrecorded in California or anywhere in the United States away from Texas and Louisiana.

VARIED BUNTING

Passerina versicolor (Bonaparte, 1838)

An adult male Varied Bunting reported at Salton Sea NWR 12 May 2001 (CBRC 2001-081) would have represented only the third record for California. The bird was seen briefly, and the identification was not corroborated.

APPENDIX: COMMON AND SCIENTIFIC NAMES OF PLANT SPECIES MENTIONED IN THE TEXT

ENGLISH NAME	SCIENTIFIC NAME (FAMILY)
Alfalfa	*Medicago sativa* (Fabaceae)
Arrowweed	*Pluchea sericea* (Asteraceae)
Asparagus	*Asparagus officinalis* (Liliaceae)
Athel	*Tamarix aphylla* (Tamaricaceae)
Barley (cultivated)	*Hordeum vulgare* (Poaceae)
Bassia	*Bassia hyssopifolia* (Chenopodiaceae)
Bottlebrush	*Melaleuca* spp. (Myrtaceae)
Bulrush	*Scirpus* spp. (Cyperaceae)
Canary Grass, Dwarf	*Phalaris minor* (Poaceae)
Catclaw	*Acacia greggii* (Fabaceae)
Cattail, Broad-leaved	*Typha latifolia* (Typhaceae)
Cattail, Southern	*Typha domingensis* (Typhaceae)
Chinaberry	*Melia azedarach* (Meliaceae)
Cholla	*Opuntia* spp. (Cactaceae)
Cottonwood, Fremont	*Populus fremontii* (Salicaceae)
Creosote	*Larrea tridentata* (Zygophyllaceae)
Crucifixion Thorn	*Castela emoryi* (Simaroubaceae)
Eucalyptus	*Eucalyptus* spp. (Myrtaceae)
Fan Palm, California	*Washingtonia filifera* (Arecaceae)
Fan Palm, Mexican	*Washingtonia robusta* (Arecaceae)
Fig	*Ficus* spp. (Moraceae)
Flax, Common	*Linum usitatissimum* (Linaceae)

ENGLISH NAME	SCIENTIFIC NAME (FAMILY)
Grass, Barnyard	*Echinochloa crus-galli* (Poaceae)
Gum, Blue	*Eucalyptus globulus* (Myrtaceae)
Gum, Red River	*Eucalyptus camadulensis* (Myrtaceae)
Honeysuckle, Cape	*Tecomaria capensis* (Bignoniaceae)
Iodine Bush	*Allenrolfea occidentalis* (Chenopodiaceae)
Ironwood	*Olneya tesota* (Fabaceae)
Joshua Tree	*Yucca brevifolia* (Liliaceae)
Lamb's Quarters	*Chenopodium album* (Chenopodiaceae)
Maple	*Acer* spp. (Aceraceae)
Mesquite, Honey	*Prosopis glandulosa* (Fabaceae)
Mesquite, Screwbean	*Prosopis pubescens* (Fabaceae)
Milo	*Sorghum bicolor* (Poaceae)
Mistletoe, Desert	*Phoradendron californicum* (Viscaceae)
Mulberry	*Morus* spp. (Moraceae)
Oleander	*Nerium oleander* (Apocynaceae)
Palm, Date	*Phoenix dactylifera* (Arecaceae)
Palo Verde, Blue	*Cercidium floridum* (Fabaceae)
Pine	*Pinus* spp. (Pineaceae)
Pomegranate	*Punica granatum* (Punicaceae)
Quail Brush	*Atriplex lentiformis* (Chenopodiaceae)
Reed, Common	*Phragmites australis* (Poaceae)
Rye, Perennial	*Lolium perenne* (Poaceae)
Sagebrush	*Artemisia* spp. (Asteraceae)
Saltbush	*Atriplex* spp. (Chenopodiaceae)
Saltcedar	*Tamarix ramosissima* (Tamaricaceae)
Sycamore	*Platanus* spp. (Platanaceae)
Sycamore, California	*Platanus racemosa* (Platanaceae)
Tamarisk	*Tamarix* spp. (Tamaricaceae)
Thistle	*Cirsium* spp. (Asteraceae)
Thistle, Russian	*Salsola tragus* (Chenopodiaceae)
Tobacco, Tree	*Nicotiana glauca* (Nyctaganaceae)
Tree, Smoke	*Dalea spinosa* (Fabaceae)
Wheat (cultivated)	*Triticum aestivum* (Poaceae)
willow	*Salix* spp. (Salicaceae)
Willow, Goodding's Black	*Salix gooddingii* (Salicaceae)
Yucca, Mojave	*Yucca schidigera* (Liliaceae)

Sources: Taxonomy and nomenclature follow Hogan (1988), Hickman (1993), and Simpson et al. (1996). English names follow a variety of sources, most notably Beauchamp (1986), Hogan (1988), Hickman (1993), and Turner et al. (1995).

LITERATURE CITED

Abbott, C. G. 1931. Four hundred Black-necked Stilts. Condor 33:38.

———. 1935. Another invasion of Wood Ibis in southern California. Condor 37:35–36.

———. 1940. Notes from Salton Sea, California. Condor 42:264–65.

Ainley, D. G. 1980. Geographic variation in Leach's Storm-Petrels. Auk 97:837–53.

Aldrich, J. W., and K. P. Baer. 1970. Status and speciation in the Mexican Duck (*Anas diazi*). Wilson Bull. 82:63–73.

Amadon, D. 1949. The seventy-five percent rule for subspecies. Condor 51:250–58.

American Ornithologists' Union (AOU). 1957. Check-list of North American birds, 5th ed. Baltimore.

———. 1973. Thirty-second supplement to the American Ornithologists' Union *Check-list of North American Birds*. Auk 90:411–19.

———. 1983. Check-list of North American birds, 6th ed. Washington, D.C.

———. 1993. Thirty-ninth supplement to the American Ornithologists' Union *Check-list of North American Birds*. Auk 110:675–82.

———. 1998. Check-list of North American birds, 7th ed. Washington, D.C.

———. 2000. Forty-second supplement to the American Ornithologists' Union *Check-list of North American Birds*. Auk 117:847–58.

Anderson, B. W., and R. D. Ohmart. 1985. Habitat use by Clapper Rails in the lower Colorado River valley. Condor 87:116–26.

Anderson, C. M., D. G. Roseneau, B. J. Walton, and P. J. Bente. 1988. New evidence of a Peregrine migration on the west coast of North America. *In* Cade, T. J., J. H. Enderson, C. G. Thelander, and C. M. White, eds., Peregrine Falcon populations: Their management and recovery, Peregrine Fund, Boise, Idaho, 507–16.

Anderson, D. W. 1983. The seabirds. *In* Case, T. J., and M. L. Cody, eds., Island biogeography in the Sea of Cortez, Univ. California Press, Berkeley, 247–65.

Anderson, D. W., L. R. Deweese, and D. V. Tiller. 1977. Passive dispersal of California Brown Pelicans. Bird-Banding 48:228–38.

Anderson, D. W., J. E. Mendoza, and J. O. Keith. 1976. Seabirds in the Gulf of California: A vulnerable, international resource. Nat. Resour. J. 16:483–505.

Andrews, R., and R. Righter. 1992. Colorado birds: A reference to their distribution and abundance. Denver Mus. Nat. Hist., Denver.

Arnold, J. R. 1935. The changing distribution of the western Mockingbird in California. Condor 37:193–99.

———. 1980. Distribution of the Mockingbird in California. W. Birds 11:97–102.

Askins, R. A. 2000. Restoring North America's birds: Lessons from landscape ecology. Yale Univ. Press, New Haven, Conn.

Bailey, A. M. 1948. Birds of arctic Alaska. Colo. Mus. Nat. Hist. Popular Ser. 8.

Bailey, A. M., and R. J. Niedrach. 1965. Birds of Colorado, vol. 1. Denver Mus. Nat. Hist., Denver.

Baldwin, S. P., H. C. Oberholser, and L. G. Worley. 1931. Measurements of birds. Cleveland Mus. Nat. Hist. Sci. Publ. 2.

Baltosser, W. H. 1989. Costa's Hummingbird: Its distribution and status. W. Birds 20:41–62.

Bancroft, G. 1920. The Harris Hawk a breeder in California. Condor 22:156.

———. 1929. A new Pacific race of Gull-billed Tern. Trans. San Diego Soc. Nat. Hist. 5:283–86.

Bangs, O., and T. E. Penard. 1921. Notes on some American birds, chiefly Neotropical. Bull. Mus. Comp. Zool. 64:365–97.

Banks, R. C. 1970. Molt and taxonomy of Red-breasted Nuthatches. Wilson Bull. 82:201–5.

———. 1986a. Subspecies of the Glaucous Gull, *Larus hyperboreus* (Aves: Charadriiformes). Proc. Biol. Soc. Washington 99:149–59.

———. 1986b. Subspecies of the Greater Scaup and their names. Wilson Bull. 98:433–44.

———. 1988. Geographic variation in the Yellow-billed Cuckoo. Condor 90:473–77.

———. 1990. Geographic variation in the Yellow-billed Cuckoo: Correction and comments. Condor 92:538.

Banks, R. C., C. Cicero, J. L. Dunn, A. W. Kratter, P. C. Rasmussen, J. V. Remsen Jr., J. D. Rising, and D. F. Stotz. 2002. Forty-third supplement to the American Ornithologists' Union *Checklist of North American Birds*. Auk 119:897–906.

Banks, R. C., and P. F. Springer. 1994. A century of population trends of waterfowl in western North America. Stud. Avian Biol. 15:134–46.

Banks, R. C., and R. E. Tomlinson. 1974. Taxonomic status of certain Clapper Rails of southwestern United States and northwestern Mexico. Wilson Bull. 86:325–35.

Barrowclough, G. F. 1980. Genetic and phenotypic differentiation in a wood warbler (genus *Dendroica*) hybrid zone. Auk 97:655–68.

Barrows, C. W. 1989. Diets of five species of desert owls. W. Birds 20:1–10.

Bartholomew, G. A., W. R. Dawson, and E. J. O'Neill. 1953. A field study of temperature regulation in young White Pelicans *Pelecanus erythrorhynchos*. Ecology 34:554–60.

Beauchamp, R. M. 1986. A flora of San Diego County. Sweetwater River Press, National City, Calif.

Bednarz, J. C. 1995. Harris' Hawk (*Parabuteo unicinctus*). The birds of North America, ed. A. F. Poole and F. B. Gill, no. 146. Acad. Nat. Sci.,

Philadelphia, and Am. Ornithol. Union, Washington, D.C.

Beezley, J. A. 1995. A coot kill site at Lake Cahuilla. Proc. Soc. Calif. Archaeol. 8:79–86.

Behle, W. H. 1942. Distribution and variation of the Horned Larks (*Otocoris alpestris*) of western North America. Univ. Calif. Publ. Zool. 46:205–316.

———. 1948. Systematic comments on some geographically variable birds occurring in Utah. Condor 50:71–80.

———. 1950. Clines in the Yellow-throats of western North America. Condor 52:193–219.

———. 1973. Clinal variation in White-throated Swifts from Utah and the Rocky Mountain region. Auk 90:299–306.

———. 1976. Systematic review, intergradation, and clinal variation in Cliff Swallows. Auk 93:66–77.

———. 1985. Utah birds: Geographic distribution and systematics. Utah Mus. Nat. Hist. Occ. Publ. 5.

Behle, W. H., E. D. Sorensen, and C. M. White. 1985. Utah birds: A revised checklist. Utah Mus. Nat. Hist. Occ. Publ. 4.

Bellrose, F. C. 1976. Ducks, geese, and swans of North America, 2nd ed. Stackpole Books, Harrisburg, Pa.

Bennett, W. W., and R. D. Ohmart. 1978. Habitat requirements and population characteristics of the Clapper Rail (*Rallus longirostris yumanensis*) in the Imperial Valley of California. Univ. Calif. Lawrence Livermore Lab., Livermore.

Berlijn, M. 1999. Blue Ross' Goose. Dutch Birding 21:161–63.

Bermingham, E., S. Rohwer, S. Freeman, and C. Wood. 1992. Vicariance biogeography in the Pleistocene and speciation in North American wood warblers: A test of Mengel's model. Proc. Natl. Acad. Sci. USA 89:6624–28.

Bevier, L. R. 1990. Eleventh report of the California Bird Records Committee. W. Birds 21:145–76.

Binford, L. C. 1971. Northern and Louisiana Waterthrushes in California. Calif. Birds 2:77–92.

———. 1979. Fall migration of diurnal raptors at Pt. Diablo, California. W. Birds 10:1–16.

———. 1983. Sixth report of the California Bird Records Committee. W. Birds 14:127–45.

———. 1985. Seventh report of the California Bird Records Committee. W. Birds 16:29–48.

———. 1989. A distribution survey of the birds of the Mexican state of Oaxaca. Ornithol. Monogr. 43.

Binford, L. C., and D. B. Johnson. 1995. Range expansion of the Glaucous-winged Gull into interior United States and Canada. W. Birds 26:169–88.

Blake, D. 1923. Sonora storms. Monthly Weather Rev. 51:585–88.

Blake, E. R. 1977. Manual of Neotropical birds, vol. 1. Univ. Chicago Press, Chicago.

Blake, W. P. 1858. Report of a geological reconnaissance in California, made in connection with the expedition to survey routes for a railroad from the Mississippi River to the Pacific Ocean, under the command of Lieutenant R. S. Williamson, Corps of Topographic Engineers, in 1853. H. Bailliere, New York.

———. 1914. The Cahuilla Basin and the desert of the Colorado. In MacDougal, D. T., ed., The Salton Sea: A study of the geography, the geology, the floristics, and the ecology of a desert basin, Carnegie Inst. Washington Publ. 193, 1–12.

———. 1915. Sketch of the region at the head of the Gulf of California: A review and history. In Cory, H. T., The Imperial Valley and the Salton Sink, John J. Newbegin, San Francisco, 1–36.

Bloom, P. H. 1994. The biology and current status of the Long-eared Owl in coastal southern California. Bull. South. Calif. Acad. Sci. 93:1–12.

Bloom, P. H., M. D. McCrary, and M. J. Gibson. 1993. Red-shouldered Hawk home-range and habitat use in southern California. J. Wildl. Manage. 57:258–65.

Bock, C. E., and L. W. Lepthien. 1976. Synchronous eruptions of boreal seed-eating birds. Am. Nat. 110:559–71.

Bolen, E. G. 1978. Notes on Blue-winged Teal × Cinnamon Teal hybrids. Southwest. Nat. 23:692–96.

Bond, R. M. 1943. Variation in western Sparrow Hawks. Condor 45:168–85.

Bourne, W. R. P., and J. R. Jehl Jr. 1982. Variation and nomenclature of Leach's Storm-Petrels. Auk 99:793–97.

Boyd, W. S., S. D. Schneider, and S. A. Cullen. 2000. Using radio telemetry to describe the fall migration of Eared Grebes. J. Field Ornithol. 71:702–8.

Boyle, R. H. 1996. Life, or death, for the Salton Sea: The plight of California's otherworldly sea. Smithsonian 27(3):86.

Bradford, D. F., L. A. Smith, D. S. Drezner, and J. D. Shoemaker. 1991. Minimizing contamination hazards to waterbirds using agricultural drainage evaporation ponds. Environ. Manage. 15:785–95.

Brandt, J. 1977. Birding at the Salton Sea. Birding 9:105–7.

Brewster, C. C., J. C. Allen, and D. D. Kopp. 1999. IPM from space: Using satellite imagery to construct regional crop maps for studying crop-insect interactions. Am. Entomol. 45:105–17.

British Ornithologists' Union (BOU). 2001. British Ornithologists' Union Records Committee: 27th report (October 2000). Ibis 143:171–75.

Brodkorb, P. 1942. Notes on some races of the Rough-winged Swallow. Condor 44:214–17.

Brown, B. T. 1993. Bell's Vireo (Vireo bellii). The birds of North America, ed. A. F. Poole and F. B. Gill, no. 35. Acad. Nat. Sci., Philadelphia, and Am. Ornithol. Union, Washington, D.C.

Brown, L., and D. Amadon. 1989. Eagles, hawks, and falcons of the world. Wellfleet Press, Secaucus, N.J.

Brown, W. H. 1976. Winter population trends in the Black and Turkey Vultures. Am. Birds 30: 909–12.

Browning, B. M. 1962. Food habits of the Mourning Dove in California. Calif. Fish Game 48:91–115.

Browning, M. R. 1974. Taxonomic remarks on recently described subspecies of birds that occur in the northwestern United States. Murrelet 55:32–38.

———. 1976. The status of Sayornis saya yukonensis Bishop. Auk 93:843–46.

———. 1977a. Geographic variation in Contopus sordidulus and C. virens north of Mexico. Great Basin Nat. 37:453–56.

———. 1977b. Geographic variation in the Dunlin, Calidris alpina, of North America. Can. Field-Nat. 91:391–93.

———. 1978. An evaluation of the new species and subspecies proposed in Oberholser's Bird Life of Texas. Proc. Biol. Soc. Washington 91:85–122.

———. 1990. Taxa of North American birds described from 1957 to 1987. Proc. Biol. Soc. Washington 103:432–51.

———. 1991. Taxonomic comments on the Dunlin Calidris alpina from northern Alaska and eastern Siberia. Bull. Brit. Ornithol. Club 111:140–45.

———. 1992. Geographic variation in Hirundo pyrrhonota (Cliff Swallow) from northern North America. W. Birds 23:21–29.

———. 1994. A taxonomic review of Dendroica petechia (Yellow Warbler) (Aves:Parulinae). Proc. Biol. Soc. Washington 107:27–51.

———. 2002. Taxonomic comments on selected species of birds from the Pacific Northwest. Ore. Birds 28:69–82.

Browning, M. R., and S. P. Cross. 1999. Specimens of birds from Jackson County, Oregon: Distribution and taxonomy of selected species. Ore. Birds 25:62–71.

Bruehler, G., and A. de Peyster. 1999. Selenium and other trace metals in pelicans dying at the Salton Sea. Bull. Environ. Contamin. Toxicol. 63:590–97.

Bryant, H. C. 1914. Occurrence of the Black-bellied Tree-Duck in California. Condor 16:94.

Burleigh, T. D. 1960. Geographic variation in the Western Wood Pewee (*Contopus sordidulus*). Proc. Biol. Soc. Washington 73:141–46.

———. 1972. Birds of Idaho. Caxton Printers, Caldwell, Idaho.

Burns, H. 1958. Salton Sea story, 6th ed. Desert Mag. Press, Palm Desert, Calif.

Butler, R. W., T. D. Williams, N. Warnock, and M. A. Bishop. 1997. Wind assistance: A requirement for migration of shorebirds? Auk 114: 456–66.

Cade, T. J. 1955. Variation of the Common Rough-legged Hawk in North America. Condor 57: 313–46.

Cade, T. J., and C. P. Woods. 1997. Changes in the distribution and abundance of the Loggerhead Shrike. Conserv. Biol. 11:21–31.

California Department of Fish and Game. 1979. Birds of the Imperial Wildlife Area Finney Ramer Unit. Sacramento.

California Department of Water Resources. 1964. Coachella Valley investigation. Calif. Dept. Water Resources Bull. 108.

Campbell, R. W., N. K. Dawe, I. McTaggart-Cowan, J. M. Cooper, G. W. Kaiser, M. C. E. McNall, and G. E. J. Smith. 1997. The Birds of British Columbia, vol. 3. Univ. Brit. Columbia Press, Vancouver.

Campbell, R. W., N. K. Dawe, I. McTaggart-Cowan, J. M. Cooper, G. W. Kaiser, A. C. Stewart, and M. C. E. McNall. 2001. Birds of British Columbia, vol. 4. Univ. Brit. Columbia Press, Vancouver.

Cardiff, E., and B. Cardiff. 1953a. Additional records for the American Redstart in the Imperial Valley of California. Condor 55:279.

———. 1953b. Records of the Coues Flycatcher and Chestnut-sided Warbler for California. Condor 55:217.

———. 1954. A winter record of Swamp Sparrow in the Imperial Valley, California. Condor 56:54.

Cardiff, E. A. 1956. Additional records for the Imperial Valley and Salton Sea area of California. Condor 58:447–48.

———. 1961. Two new species of birds for California and notes on species of the Imperial Valley and Salton Sea area of California. Condor 63:183.

Cardiff, E. A., and A. T. Driscoll. 1972. Red-headed Woodpecker in the Imperial Valley of California. Calif. Birds 3:23–24.

Cardiff, E. E. 1950. Nesting of the Black Phoebe in the Imperial Valley, California. Condor 52:166.

Cardiff, E. E., and B. E. Cardiff. 1949. The Ovenbird and the American Redstart in the Imperial Valley, California. Condor 51:44–45.

———. 1950. Late nesting record for the Abert Towhee. Condor 52:135.

———. 1951. An unusual occurrence of the Saw-whet Owl. Condor 53:154.

Carey, G. J. 1993. Hybrid male wigeon in East Asia. Hong Kong Bird Rep. 1992:160–66.

Carpelan, L. H. 1958. The Salton Sea: Physical and chemical characteristics. Limnol. Oceanogr. 3:373–86.

Carter, H. R., A. L. Sowls, M. S. Rodway, U. W. Wilson, R. W. Lowe, G. J. McChesney, F. Gress, and D. W. Anderson. 1995. Population size, trends, and conservation problems of the Double-crested Cormorant on the Pacific Coast of North America. Colonial Waterbirds Spec. Publ. 1:189–215.

Cervantes-Sanchez, J., and E. Mellink. 2001. Nesting of Brandt's Cormorants in the northern Gulf of California. W. Birds 32:134–35.

Chambers, W. L. 1921. A flight of Harris Hawks. Condor 23:65.

———. 1924. Another flight of Harris Hawks. Condor 26:75.

Chase, J. S. 1919. California desert trails. Houghton Mifflin, Boston.

Cicero, C. 1996. Sibling species of titmice in the *Parus inornatus* complex (Aves: Paridae). Univ. Calif. Publ. Zool. 128:1–217.

———. 2000. Oak Titmouse (*Baeolophus inornatus*) and Juniper Titmouse (*Baeolophus ridgwayi*). The birds of North America, ed. A. F. Poole and F. B. Gill, no. 485. Acad. Nat. Sci., Philadelphia, and Am. Ornithol. Union, Washington, D.C.

Clark, R. J. 1997. A review of the taxonomy and distribution of the Burrowing Owl (*Speotyto cunicularia*). Raptor Res. Rep. 9:14–23.

Clark, W. S., and B. K. Wheeler. 1987. A field guide to hawks. Hougton Mifflin, Boston.

———. 1998. "Dark-morph" Sharp-shinned Hawk reported from California is normal juvenile female of race *perobscurus*. Bull. Brit. Ornithol. Club 118:191–93.

Clary, B., and M. Clary. 1935. Surf Scoters on Salton Sea. Condor 37:179.

———. 1936a. Clark Nutcracker again visits the Colorado Desert. Condor 38:119.

———. 1936b. Fall and winter records from the Coachella Valley, California. Condor 38:89.

———. 1936c. Winter records of Virginia Rail and Mountain Plover in Coachella Valley, California. Condor 38:125.

Clary, B. L. 1930. Blue-footed Booby on Salton Sea. Condor 32:160–61.

Clary, B. L., and Mrs. B. L. Clary. 1935. American Golden-eye and American Merganser on Salton Sea. Condor 37:80.

Clary, B. L., Mrs. 1933. Sahuaro Screech Owl in Coachella Valley, California. Condor 35:80.

Cleere, N. 1998. Nightjars: A guide to the nightjars, nighthawks, and their relatives. Yale Univ. Press, New Haven, Conn.

Cleverly, J. R., S. D. Smith, A. Sala, and D. A. Devitt. 1997. Invasive capacity of *Tamarix ramosissima* in a Mojave Desert floodplain: The role of drought. Oecologia 111:12–18.

Cogswell, H. L. 1977. Water birds of California. Univ. California Press, Berkeley.

Cohn, J. P. 2000. Saving the Salton Sea. BioScience 50:295–301.

Collins, C. T., and K. L. Garrett. 1996. The Black Skimmer in California: An overview. W. Birds 27:127–35.

Collins, C. T., W. A. Schew, and E. Burkett. 1991. Elegant Terns breeding in Orange County, California. Am. Birds 45:393–95.

Collins, P. W., C. Drost, and G. M. Fellers. 1986. Migratory status of Flammulated Owls in California, with recent records from the Channel Islands. W. Birds 17:21–31.

Connors, P. G. 1983. Taxonomy, distribution, and evolution of Golden Plovers (*Pluvialis dominica* and *Pluvialis fulva*). Auk 100:607–20.

Connors, P. G., B. J. McCaffery, and J. L. Maron. 1993. Speciation in golden-plovers, *Pluvialis dominica* and *P. fulva*: Evidence from the breeding grounds. Auk 110:9–20.

Conover, B. 1943. The races of the Knot (*Calidris canutus*). Condor 45:226–28.

Conway, C. J., C. Sulzman, and B. E. Raulstom. 2002. Population trends, distribution, and monitoring protocols for the California Black Rail. Unpubl. rep., Coop. Fish Wildl. Res. Unit, Univ. Arizona, Tucson.

Cooke, F., and F. G. Cooch. 1968. The genetics of polymorphism in the goose *Anser caerulescens*. Evolution 22:289–300.

Cooper, J. M., and D. J. Graham. 1985. Sightings of hybrid "blue-winged" ducks (*Anas*) in British Columbia. Brit. Columbia Prov. Mus. Contr. Nat. Sci. 1.

Cory, H. T. 1915. The Imperial Valley and the Salton Sink. John J. Newbegin, San Francisco.

Coulombe, H. N. 1971. Behavior and population ecology of the Burrowing Owl, *Speotyto cunicularia*, in the Imperial Valley of California. Condor 73:162–76.

Cowan, I. M. 1938. Distribution of the races of the Williamson Sapsucker in British Columbia. Condor 40:128–29.

Craig, J. T. 1972. Two fall Yellow-throated Warblers in California. Calif. Birds 3:17–18.

Cramp, S., and K. E. L. Simmons, eds. 1977. Birds of the Western Palearctic, vol. 1. Oxford Univ. Press, Oxford, England.

———. 1983. Birds of the Western Palearctic, vol. 3. Oxford Univ. Press, Oxford, England.

Crase, F. T., and R. W. DeHaven. 1972. Current breeding status of the Yellow-headed Blackbird in California. Calif. Birds 3:39–42.

Crouch, J. E. 1943. Distribution and habitat relationships of the Phainopepla. Auk 60:319–33.

Daniels, B. E., L. Hays, D. Hays, J. Morlan, and D. Roberson. 1989. First record of the Common Black-Hawk for California. W. Birds 20:11–18.

Davis, J., and L. Williams. 1957. Irruptions of the Clark Nutcracker in California. Condor 59:297–307.

Dawson, W. L. 1921. The season of 1917. J. Mus. Comp. Ool. 2:27–36.

Deignan, H. G. 1961. Type specimens of birds in the United States National Museum. Bull. U.S. Natl. Mus. 221.

Dement'ev, G. P., N. A. Gladkov, and E. P. Spangenberg. 1951. Birds of the Soviet Union, vol. 3. Smithsonian Inst., Washington, D.C.

DeSante, D., and P. Pyle. 1986. Distributional checklist of North American Birds, vol. 1. Artemisia Press, Lee Vining, Calif.

de Stanley, M. 1966. The Salton Sea: Yesterday and today. Triumph Press, Los Angeles.

Devillers, P. 1970a. Chimney Swifts in coastal southern California. Calif. Birds 1:147–52.

———. 1970b. Identification and distribution in California of the *Sphyrapicus varius* group of sapsuckers. Calif. Birds 1:47–76.

Devillers, P., G. McCaskie, and J. R. Jehl Jr. 1971. The distribution of certain large gulls (*Larus*) in southern California and Baja California. Calif. Birds 2:11–26.

Dickerman, R. W. 1973. The Least Bittern in Mexico and Central America. Auk 90:689–91.

———. 1985. Taxonomy of the Lesser Nighthawks (*Chordeiles acutipennis*) of North and Central America. Ornithol. Monogr. 36:356–60.

———. 1986. Two hitherto unnamed populations of *Aechmophorus* (Aves: Podicipitidae). Proc. Biol. Soc. Washington 99:435–36.

Dickerman, R. W., and K. C. Parkes. 1968. Notes on the plumages and generic status of the Little Blue Heron. Auk 85:437–40.

———. 1997. Taxa described by Allan R. Phillips, 1939–1994: A critical list. *In* Dickerman, R. W., ed., The era of Allan R. Phillips: A festschrift, Horizon Commun., Albuquerque, 211–34.

Dickey, D. R. 1923. Description of a new Clapper Rail from the Colorado River valley. Auk 40: 90–94.

Dickey, D. R., and A. J. van Rossem. 1922. The occurrence of the Desert Horned Lark in southern California. Condor 24:94.

———. 1924. Notes on certain Horned Larks in California. Condor 26:110.

Dowd, M. J. 1960. Historic Salton Sea. Imperial Irrigation Dist., El Centro, Calif.

Doyle, T. J. 1997. The Timberline Sparrow, *Spizella* (*breweri*) *taverneri*, in Alaska, with notes on breeding habitat and vocalizations. W. Birds 28:1–12.

Dritschilo, W., and D. Vander Pluym. 1984. An ecotoxicological model for energy development and the Salton Sea, California. J. Environ. Manage. 19:15–30.

Dubowy, P. J. 1989. Effects of diet on selenium bioaccumulation in marsh birds. J. Wildl. Manage. 53:776–81.

Dunlap, L. 1999. Scientists respond to a Salton Sea S.O.S. Fiat Lux 9(5):26–27.

Dunn, J. 1977. The Salton Sea—a chronicle of the seasons. Birding 9:102–4.

———. 1988. Tenth report of the California Bird Records Committee. W. Birds 19:129–63.

Dunn, J., and P. Unitt. 1977. A Laysan Albatross in interior southern California. W. Birds 8:27–28.

Dunn, J. L., and K. L. Garrett. 1990. Identification of Ruddy and Common Ground-Doves. Birding 22:138–45.

———. 1997. A field guide to warblers of North America. Houghton Mifflin, Boston.

Dunn, J. L., J. Morlan, and C. P. Wilds. 1987. Field identification of the forms of Lesser Golden-Plover. Proc. 4th Internatl. Identification Meeting, Internatl. Birdwatching Center, Eilat, Israel.

Dunning, B. 1988. An unusual concentration of boobies in the northern Gulf of California. Aves Mexicanas 1(88-1):1–2.

Durham, J. W., and E. C. Allison. 1960. The geologic history of Baja California and its marine faunas. Syst. Zool. 9:47–91.

Dwight, J. 1925. The gulls (Laridae) of the world: Their plumages, moults, variations, relationships, and distribution. Bull. Am. Mus. Nat. Hist. 52:63–401.

Earnheart-Gold, S., and P. Pyle. 2001. Occurrence patterns of Peregrine Falcons on Southeast Farallon Island, California, by subspecies, age, and sex. W. Birds 32:119–26.

Eaton, S. 1957. Variation in *Seiurus noveboracensis*. Auk 77:229–39.

Eisenmann, E. 1971. Range expansion and population increase in North and Middle America of the White-tailed Kite (*Elanus leucurus*). Am. Birds 25:529–36.

Ellis, L. M. 1995. Bird use of saltcedar and cottonwood vegetation in the Middle Rio Grande Valley of New Mexico, U.S.A. J. Arid Environ. 30: 339–49.

Ely, C. R., and J. Y. Takekawa. 1996. Geographic variation in migratory behavior of Greater White-fronted Geese (*Anser albifrons*). Auk 113: 889–901.

Engelmoer, M., and C. S. Roselaar. 1998. Geographical variation in waders. Kluwer Acad. Publ., Dordrecht, Netherlands.

England, A. S., and W. F. Laudenslayer Jr. 1989. Distribution and seasonal movements of Bendire's Thrasher in California. W. Birds 20:97–123.

Erickson, R. A., and R. A. Hamilton. 2001. Report of the California Bird Records Committee: 1998 records. W. Birds 32:13–49.

Erickson, R. A., R. A. Hamilton, S. N. G. Howell, P. Pyle, and M. A. Patten. 1995. First Marbled Murrelets and third Ancient Murrelet for Mexico. W. Birds 26:39–45.

Erickson, R. A., and S. B. Terrill. 1996. Nineteenth report of the California Bird Records Committee: 1993 records. W. Birds 27:93–126.

Erickson, R. A., and T. E. Wurster. 1998. Confirmation of nesting in Mexico for four bird species from the Sierra San Pedro Mártir, Baja California. Wilson Bull. 110:118–20.

Esterly, C. O. 1920. Clarke Nutcracker on the Colorado Desert. Condor 22:40.

Evans, M. E., and W. J. L. Sladen. 1980. A comparative analysis of the bill markings of Whistling and Bewick's Swans and out-of-range occurrences of the two taxa. Auk 97:697–703.

Evens, J. G., G. W. Page, S. A. Laymon, and R. W. Stallcup. 1991. Distribution, relative abundance, and status of the California Black Rail in western North America. Condor 93:952–66.

Evermann, B. W. 1916. Fishes of the Salton Sea. Copeia 1916:61–63.

Fall, A. B. 1922. Development of the Imperial Valley. GPO, Washington, D.C.

Fialkowski, W., and W. A. Newman. 1998. A pilot study of heavy metal accumulations in a barnacle from the Salton Sea, southern California. Marine Pollution Bull. 36:138–43.

Fleischer, R. C. 2001. Taxonomic and Evolutionary Significant Unit (ESU) status of western Yellow-billed Cuckoos (Coccyzus americanus). Unpubl. rep., Natl. Zool. Park, Smithsonian Inst., Washington, D.C.

Flores, R. E., and W. R. Eddleman. 1995. California Black Rail use of habitat in southwestern Arizona. J. Wildl. Manage. 59:357–63.

Foerster, K. S., and C. T. Collins. 1990. Breeding distribution of the Black Swift in southern California. W. Birds 21:1–9.

Fogelman, R. P., J. R. Mullen, W. F. Shelton, R. G. Simpson, and D. A. Grillo. 1986. Water resources data for California, water year 1984, volume 1: Southern Great Basin from Mexican border to Mono Lake basin, and Pacific Slope basins from Tijuana River to Santa Maria River. U.S. Geol. Surv. Water-Data Rep. CA84-1.

Fortiner, J. C. 1920a. Clark Nutcracker and White-winged Dove in southern California. Condor 22:190.

———. 1920b. Winter nesting of the Ground Dove. Condor 22:154–55.

———. 1921. The doves of Imperial County, California. Condor 23:168.

Franzreb, K. E. 1987. Endangered status and strategies for conservation of the Least Bell's Vireo (Vireo bellii pusillus) in California. W. Birds 18:43–49.

———. 1989. Ecology and conservation of the endangered Least Bell's Vireo. U.S. Fish Wildl. Serv. Biol. Rep. 89.

Franzreb, K. E., and S. A. Laymon. 1993. A reassessment of the taxonomic status of the Yellow-billed Cuckoo. W. Birds 24:17–28.

Free, E. E. 1914. Sketch of the geology and soils of the Cahuilla Basin. In MacDougal, D. T., ed., The Salton Sea: A study of the geography, the geology, the floristics, and the ecology of a desert basin, Carnegie Inst. Washington Publ. 193, 21–33.

Friedmann, H. 1950. The birds of North and Middle America, pt. 11. Bull. U.S. Natl. Mus. 50.

———. 1963. Host relations of the parasitic cowbirds. Bull. U.S. Natl. Mus. 233.

Friedmann, H., L. Griscom, and R. T. Moore. 1950. Distribution check-list of the birds of Mexico, pt. 1. Pac. Coast Avifauna 29.

Friend, M. 2002. Avian disease at the Salton Sea. Hydrobiologia 473:293–306.

Garlough, F. E. 1922. Blackbirds damage grain in Imperial Valley. Calif. Fish Game 8:45.

Garrett, K., and J. Dunn. 1981. Birds of southern California: Status and distribution. Los Angeles Audubon Soc., Los Angeles.

Garrett, K. L. 1998. President's message. W. Birds 29:229–31.

Garrett, K. L., M. G. Raphael, and R. D. Dixon. 1996. White-headed Woodpecker (Picoides albolarvatus). The birds of North America, ed. A. F. Poole and F. B. Gill, no. 252. Acad. Nat. Sci., Philadelphia, and Am. Ornithol. Union, Washington, D.C.

Garrett, K. L., and D. S. Singer. 1998. Annual report of the California Bird Records Committee: 1995 records. W. Birds 29:133–56.

Garrison, B. A. 1990. Trends in winter abundance and distribution of Ferruginous Hawks in California. Trans. West. Sec. Wildl. Soc. 26: 51–56.

———. 1993. Distribution and trends in abundance of Rough-legged Hawks in California. J. Field Ornithol. 64:566–74.

Gaston, A. J., and R. Decker. 1985. Interbreeding of Thayer's Gull, Larus thayeri, and Kumlien's Gull, Larus kumlieni, on Southampton Island, Northwest Territories. Can. Field-Nat. 99:257–59.

Gauthreaux, S. A., Jr., and K. P. Able. 1970. Wind and the direction of nocturnal songbird migration. Nature 228:476–77.

Gawlik, D. E., and K. L. Bildstein. 1995. Differential habitat use by sympatric Loggerhead Shrikes and American Kestrels in South Carolina. Proc. West. Found. Vert. Zool. 6:163–66.

Gerrard, J. M. 1983. A review of the current status of Bald Eagles in North America. *In* Bird, D. M., ed., Biology and management of Bald Eagles and Ospreys. Harpell Press, Sainte-Anne-de-Bellevue, Quebec, 5–21.

Gibson, D. D., and B. Kessel. 1989. Geographic variation in the Marbled Godwit and description of an Alaska subspecies. Condor 91:436–43.

———. 1992. Seventy-four new avian taxa documented in Alaska, 1976–1991. Condor 94: 454–67.

———. 1997. Inventory of the species and subspecies of Alaska birds. W. Birds 28:45–95.

Gilligan, J., M. Smith, D. Rogers, and A. Contreras, eds. 1994. Birds of Oregon: Status and distribution. Cinclus Publ., McMinnville, Ore.

Gilman, M. F. 1903. The Phainopepla. Condor 5: 42–43.

———. 1907. Migration and nesting of the Sage Thrasher. Condor 9:42–44.

———. 1918. Minutes of Cooper Club meetings: San Bernardino chapter. Condor 20:147.

Glenn, T. J. 1983. Birds of the Coachella Valley. Coachella Valley Audubon Soc., Rancho Mirage, Calif.

Gobalet, K. W. 1992. Colorado River fishes at Lake Cahuilla, Salton Basin, southern California: A cautionary tale for zooarchaeologists. Bull. South. Calif. Acad. Sci. 91:70–83.

Godfrey, W. E. 1986. The birds of Canada, rev. ed. Natl. Mus. Nat. Sci., Ottawa.

Goldwasser, S. 1978. Distribution, reproductive success, and impact of nest parasitism by Brown-headed Cowbirds on Least Bell's Vireo. Proj. rep. W-54-R-10. Calif. Dept. Fish Game, Sacramento.

Goldwasser, S., D. Gaines, and S. Wilbur. 1980. The Least Bell's Vireo in California: A de facto endangered race. Am. Birds 34:742–45.

Grant, G. S. 1978. Foot-wetting and belly-soaking by incubating Gull-billed Terns and Black Skimmers. J. Bombay Nat. Hist. Soc. 75:148–52.

———. 1982. Avian incubation: Egg, temperature, nest humidity, and behavioral thermoregulation in a hot environment. Ornithol. Monogr. 30.

Grant, G. S., and N. Hogg. 1976. Behavior of late-nesting Black Skimmers at Salton Sea, California. W. Birds 7:73–80.

Greenway, J. C., Jr. 1958. Extinct and vanishing birds of the world. Am. Comm. Internatl. Wildl. Protection Spec. Publ. 13.

Grinnell, J. 1908. Birds of a voyage on Salton Sea. Condor 10:185–91.

———. 1911a. The Black Duck in California. Condor 13:138.

———. 1911b. Distribution of the Mockingbird in California. Auk 28:293–300.

———. 1914. An account of the mammals and birds of the lower Colorado Valley, with especial reference to the distributional problems presented. Univ. Calif. Publ. Zool. 12:51–294.

———. 1923. Observation upon the bird life of Death Valley. Proc. Calif. Acad. Sci. 8:43–109.

———. 1928. A distributional summation of the ornithology of Lower California. Univ. Calif. Publ. Zool. 32:1–300.

Grinnell, J., J. Dixon, and J. Linsdale. 1930. Vertebrate natural history of a section of northern California through the Lassen Peak region. Univ. Calif. Publ. Zool. 35:1–594.

Grinnell, J., and A. H. Miller. 1944. The distribution of the birds of California. Pac. Coast Avifauna 27.

Griscom, L. 1937. A monographic study of the Red Crossbill. Proc. Boston Soc. Nat. Hist. 41:77–210.

Grismer, M. E., and K. M. Bali. 1997. Continuous ponding and shallow aquifer pumping leaches salts in clay soils. Calif. Agric. 51(3):34–37.

Groth, J. G. 1993. Evolutionary differentiation in morphology, vocalizations, and allozymes among nomadic sibling species in the North American Red Crossbill (*Loxia curvirostra*) complex. Univ. Calif. Publ. Zool. 127.

Guers, S. L., and M. E. Flannery. 2000. Landbird migration monitoring at the Salton Sea: 1999 field survey. *In* Shuford, W. D., N. Warnock, K. C. Molina, B. Mulrooney, and A. E. Black, Avifauna of the Salton Sea: Abundance, distribution, and annual phenology, Pt. Reyes Bird Observatory, Stinson Beach, Calif., app. C.

Gullion, G. W. 1960. Ecology of the Gambel's Quail in Nevada and the arid Southwest. Ecology 41:518–36.

Gunther, J. D. 1984. Riverside County, California, place names: Their origins and their stories. Rubidoux Publ., Rubidoux, Calif.

Gurrola, L. D., and T. K. Rockwell. 1996. Timing and slip for prehistoric earthquakes on the Superstition Mountain Fault, Imperial Valley, southern California. J. Geophys. Res. 101:5977–85.

Halterman, M. D., S. A. Laymon, and M. A. Whitfield. 1989. Status and distribution of the Elf Owl in California. W. Birds 20:71–80.

Hamilton, R. A. 2001. Records of caged birds in Baja California. Monogr. Field Ornithol. 3: 254–57.

Hamilton, R. A., and N. J. Schmitt. 2000. Identification of Taiga and Black Merlins. W. Birds 31:65–67.

Hancock, J. A., and J. A. Kushlan. 1984. The herons handbook. Croom Helm, London.

Hancock, J. A., J. A. Kushlan, and M. P. Kahl. 1992. Storks, ibises, and spoonbills of the world. Academic Press, San Diego.

Hanna, W. C. 1929. Vermilion Flycatcher breeding in Coachella, California. Condor 31:75.

———. 1931. Whistling Swans on Salton Sea. Condor 33:126.

———. 1933a. Early nesting of the Le Conte Thrasher. Condor 35:74–75.

———. 1933b. Nesting of the Crissal Thrasher in Coachella Valley, California. Condor 35:79.

———. 1935. Vermilion Flycatcher increasing in Coachella Valley, California. Condor 37:173.

———. 1936. Vermilion Flycatcher a victim of Dwarf Cowbird in California. Condor 38:174.

———. 1937. A record nesting date for Le Conte's Thrasher. Oologist 54:45.

Hanna, W. C., and E. E. Cardiff. 1947. Cerulean Warbler in California. Condor 49:245.

Hardy, J. W. 1973. Feral exotic birds in southern California. Wilson Bull. 85:506–12.

Harris, S. W. 1991. Northwestern California birds. Humboldt State Univ. Press, Arcata, Calif.

Harris, S. W., and R. H. Gerstenberg. 1970. Common Teal and Tufted Duck in northwestern California. Condor 72:72:108.

Harrison, P. 1985. Seabirds: An identification guide. Houghton Mifflin, Boston.

Hart, C. M., B. Glading, and H. T. Harper. 1956. The pheasant in California. In Allen, D. L., ed., Pheasants of North America, Stackpole Books, Harrisburg, Pa., 90–158.

Hayman, P., J. Marchant, and T. Prater. 1986. Shorebirds: An identification guide to the waders of the world. Houghton Mifflin, Boston.

Haynie, C. B. 1994. Texas Bird Records Committee report for 1993. Bull. Texas Ornithol. Soc. 26:2–14.

Heindel, M. T. 1996. Field identification of the Solitary Vireo complex. Birding 28:458–71.

———. 1999. The status of vagrant Whimbrels in the United States and Canada, with notes on identification. N. Am. Birds 53:232–36.

Heindel, M. T., and K. L. Garrett. 1995. Sixteenth report of the California Bird Records Committee. W. Birds 26:1–33.

Heindel, M. T., and M. A. Patten. 1996. Eighteenth report of the California Bird Records Committee: 1992 records. W. Birds 27:1–29.

Heitmeyer, M. E., D. P. Connelly, and R. L. Pederson. 1989. The Central, Imperial, and Coachella Valleys of California. In Smith, L. M., R. L. Pederson, and R. M. Kaminski, eds., Habitat management for migrating and wintering waterfowl in North America, Texas Tech Univ. Press, Lubbock, 475–505.

Hely, A. G., G. H. Hughes, and B. Irelan. 1966. Hydrologic regimen of Salton Sea. U.S. Geol. Surv. Prof. Paper 486-C.

Hely, A. G., and E. L. Peck. 1964. Precipitation, runoff, and water loss in the lower Colorado River–Salton Sea area. U.S. Geol. Surv. Prof. Paper 486-B.

Hengeveld, R. 1993. What to do about the North American invasion by the Collared Dove? J. Field Ornithol. 64:477–89.

Henny, C. J. 1997. DDE still high in White-faced Ibis eggs from Carson Lake, Nevada. Colonial Waterbirds 20:478–84.

Henny, C. J., and D. W. Anderson. 1979. Osprey distribution, abundance, and status in western Northern America. III. The Baja California and Gulf of California population. Bull. South. Calif. Acad. Sci. 78:89–106.

Henny, C. J., and L. J. Blus. 1986. Radiotelemetry locates wintering grounds of DDE-contaminated Black-crowned Night-Herons. Wildl. Soc. Bull. 14:236–41.

Herman, S. G. 1971. The Peregrine Falcon decline in California. Am. Birds 25:818–20.

Herzog, S. K. 1996. Wintering Swainson's Hawks in California's Sacramento–San Joaquin River delta. Condor 98:876–79.

Hickey, J. J., ed. 1969. Peregrine Falcon populations: Their Biology and Decline. Univ. Wisconsin Press, Madison.

Hickman, J. C., ed. 1993. The Jepson manual: Higher plants of California. Univ. California Press, Berkeley.

Hill, G. E. 1993. Geographic variation in the carotenoid plumage pigmentation of male House Finches (Carpodacus mexicanus). Biol. J. Linnean Soc. 49:63–86.

Hill, G. E., and R. Montgomerie. 1994. Plumage colour signals nutritional condition in the House Finch. Proc. Royal Soc. London B 258:47–52.

Hill, H. M., and I. L. Wiggins. 1948. Ornithological notes from Lower California. Condor 50:155–61.

Hoffman, W., J. A. Wiens, and J. M. Scott. 1978. Hybridization between gulls (*Larus glaucescens* and *L. occidentalis*) in the Pacific Northwest. Auk 95:441–58.

Hoffmann, R. 1922. Field notes from Riverside and Imperial Counties, California. Condor 24:101.

———. 1923. Random notes from southern California. Condor 25:106–7.

———. 1927. The Gila Woodpecker at Holtville, Imperial County, California. Condor 29:162.

Hogan, E. L., ed. 1988. Western garden book. Sunset Publ., Menlo Park, Calif.

Holbrook, G. F. 1928. Probable future stages of Salton Sea. U.S. Geol. Surv. Open-File Rep.

Howell, A. B. 1920. The Wood Ibis as a winter visitant to California. Condor 22:75.

———. 1922a. Red-wings in the Imperial Valley, California. Condor 24:60–61.

———. 1922b. A winter record of Texas Nighthawk in California. Condor 24:97–98.

———. 1923. The influence of the southwestern deserts upon the avifauna of California. Auk 40:584–92.

Howell, S. N. G. 1995. Magnificent and Great Frigatebirds in the eastern Pacific. Birding 26:400–415.

———. 1999. A bird-finding guide to Mexico. Comstock Publ., Ithaca, N.Y.

Howell, S. N. G., and R. J. Cannings. 1992. Songs of two Mexican populations of the Western Flycatcher *Empidonax difficilis* complex. Condor 94:785–87.

Howell, S. N. G., and P. Pyle. 1997. Twentieth report of the California Bird Records Committee: 1994 records. W. Birds 28:117–41.

Howell, S. N. G., and S. Webb. 1995. A guide to the birds of Mexico and Northern Central America. Oxford Univ. Press, New York.

Hoyer, R. C., and S. D. Smith. 1997. Chinese Pond-Heron in Alaska. Field Notes 51:953–56.

Hubbard, J. P. 1965. Two western occurrences of the Orchard Oriole. Condor 67:265.

———. 1969. The relationship and evolution of the *Dendroica coronata* complex. Auk 86:393–432.

———. 1970. Geographic variation in the *Dendroica coronata* complex. Wilson Bull. 82:355–69.

———. 1972. The nomenclature of *Pipilo aberti* Baird (Aves: Fringillidae). Proc. Biol. Soc. Washington 85:131–38.

———. 1977. The biological and taxonomic status of the Mexican Duck. Bull. New Mex. Dept. Game and Fish 16.

———. 1978. Revised check-list of the birds of New Mexico. New Mexico Ornithol. Soc. Publ. 6.

Hubbard, J. P., and R. S. Crossin. 1974. Notes on northern Mexican birds: An expedition report. Nemouria 14.

Huels, T. R. 1984. First record of Cave Swallows breeding in Arizona. Am. Birds 38:281–83.

Huey, L. M. 1927. Observations on the spring migration of *Aphriza* and *Gavia* in the Gulf of California. Auk 44:529–31.

Hundertmark, C. A. 1978. Breeding birds of Elephant Butte Marsh. New Mexico Ornithol. Soc. Publ. 5.

Hunter, W. C., R. D. Ohmart, and B. W. Anderson. 1988. Use of exotic saltcedar (*Tamarix chinensis*) by birds in arid riparian systems. Condor 90:113–23.

Hurlbert, A. 1997. Wildlife use of agricultural drains in the Imperial Valley, California. Unpubl. rep., Salton Sea Natl. Wildl. Refuge, Calipatria, Calif.

Ingersoll, A. M. 1895. Wilson's Plover in California. Nidiologist 2:87.

Jaeger, E. C. 1947a. Stone-turning habits of some birds. Condor 49:171.

———. 1947b. White-headed Woodpecker spends winter at Palm Springs, California. Condor 49:244–45.

Jehl, J. R., Jr. 1968. The systematic position of the Surfbird, *Aphriza virgata*. Condor 70:206–10.

———. 1987a. Geographic variation and evolution in the California Gull (*Larus californicus*). Auk 104:421–28.

———. 1987b. A review of "Nelson's Gull *Larus nelsoni*." Bull. Brit. Ornithol. Club 107:86–91.

———. 1988. Biology of the Eared Grebe and Wilson's Phalarope in the nonbreeding season: A study of adaptations to saline lakes. Stud. Avian Biol. 12.

———. 1994. Changes in saline and alkaline lake avifaunas in western North America in the past 150 years. Stud. Avian Biol. 15:258–72.

———. 1996. Mass mortality events of Eared Grebes in North America. J. Field Ornithol. 67:471–76.

———. 2001. The abundance of the Eared (Black-necked) Grebe as a recent phenomenon. Waterbirds 24:245–49.

Jehl, J. R., Jr., and R. L. McKernan. 2002. Biology and migration of Eared Grebes at the Salton Sea. Hydrobiologia 473:245–53.

Jewett, S. G. 1945. The Blue Goose in California. Condor 47:167.

Johnson, D. H., D. E. Timm, and P. F. Springer. 1979. Morphological characteristics of Canada Geese in the Pacific Flyway. *In* Jarvis, R. L., and J. C. Bartonek, eds., Management and biology of Pacific Flyway geese: A symposium, Ore. State Univ., Corvallis, 56–80.

Johnson, N. K. 1965. The breeding avifaunas of the Spring and Sheep ranges in southern Nevada. Condor 76:93–114.

———. 1970. Fall migration and winter distribution of the Hammond Flycatcher. Bird-Banding 41:169–90.

———. 1980. Character variation and evolution of sibling species in the *Empidonax difficilis-flavescens* complex (Aves: Tyrannidae). Univ. Calif. Publ. Zool. 112.

———. 1994a. Old-school taxonomy versus modern biosystematics: Species-level decisions in *Stelgidopteryx* and *Empidonax*. Auk 111:773–80.

———. 1994b. Pioneering and natural expansion of breeding distributions in western North American birds. Stud. Avian Biol. 15:27–44.

———. 1995a. Seven avifaunal censuses spanning one-half century on an island of White Firs (*Abies concolor*) on the Mojave Desert. Southwest. Nat. 40:76–85.

———. 1995b. Speciation in vireos. I. Macrogeographic patterns of allozymic variation in the *Vireo solitarius* complex in the contiguous United States. Condor 97:903–19.

Johnson, N. K., and C. B. Johnson. 1985. Speciation in sapsuckers (*Sphyrapicus*). II. Sympatry, hybridization, and mate preference in *S. ruber daggetti* and *S. nuchalis*. Auk 102:1–15.

Johnson, N. K., and J. A. Marten. 1988. Evolutionary genetics of flycatchers. II. Differentiation in the *Empidonax difficilis* group. Auk 105:177–91.

———. 1992. Macrogeographic patterns of morphometric and genetic variation in the Sage Sparrow complex. Condor 94:1–19.

Johnson, N. K., J. V. Remsen Jr., and C. Cicero. 1998. Refined colorimetry validates endangered subspecies of the Least Tern. Condor 100:18–26.

Johnston, D. W. 1961. The biosystematics of American crows. Univ. Washington Press, Seattle.

Johnston, R. F., and K. L. Garrett. 1994. Population trends of introduced birds in western North America. Stud. Avian Biol. 15:221–31.

Jones, L. 1971. The Whip-poor-will in California. Calif. Birds 2:33–36.

Jones, R. M. 1999. Seabirds carried inland by tropical storm Nora. W. Birds 30:185–92.

Jurek, R. M. 1974. California shorebird surveys, 1969–1974. Proj. W-54-R Final Rep. Calif. Dept. Fish Game, Sacramento.

Kaiser, J. 1999. Battle over a dying sea. Science 284:28–30.

Kalmbach, E. R. 1934. Field observation in economic ornithology. Wilson Bull. 46:73–90.

Kaufman, K. 1976. The changing seasons: Fall 1976. Am. Birds 31:142–52.

———. 1979. Field identification of the flicker forms and their hybrids in North America. Continental Birdl. 1:4–15.

Keith, J. O., and E. J. O'Neill. 2000. Movements of juvenile American White Pelicans from breeding colonies in California and Nevada. Waterbirds 23:33–37.

Kennan, G. 1917. The Salton Sea: An account of Harriman's fight with the Colorado River. Macmillan, New York.

Kennerley, P. R., W. Hoogendoorn, and M. L. Chalmers. 1995. Identification and systematics of large white-headed gulls in Hong Kong. Hong Kong Bird Rep. 1994:127–56.

King, J. R. 2000. Field identification of adult *californicus* and *albertaensis* California Gulls. Birders J. 9:245–61.

King, W. B. 1974. Wedge-tailed Shearwater (*Puffinus pacificus*). *In* King, W. B., ed., Pelagic studies of seabirds in the central and eastern Pacific, Smithsonian Contr. Zool. 158, 53–95.

Klicka, J., A. J. Fry, R. M. Zink, and C. W. Thompson. 2001. A cytochrome-*b* perspective on *Passerina* bunting relationships. Auk 118:611–23.

Klicka, J., and R. M. Zink. 1997. The importance of recent ice ages in speciation: A failed paradigm. Science 277:1666–69.

Klicka, J., R. M. Zink, J. C. Barlow, W. B. McGillivray, and T. J. Doyle. 1999. Evidence supporting the recent origin and species status of the Timberline Sparrow. Condor 101:577–88.

Knopf, F. L. 1998. Foods of Mountain Plovers wintering in California. Condor 100:382–84.

Knopf, F. L., and B. J. Miller. 1994. *Charadrius montanus*—montane, grassland, or bare-ground plover? Auk 111:504–6.

Knopf, F. L., and J. R. Rupert. 1995. Habits and habitats of Mountain Plovers in California. Condor 97:743–51.

Koplin, J. R. 1973. Differential habitat use by sexes of American Kestrels wintering in northern California. Raptor Res. 7:39–42.

Kozlik, F. M., A. W. Miller, and W. C. Rienecker. 1959. Color-marking White Geese for determining migration routes. Calif. Fish Game 45:69–82.

Kuiken, T. 1999. Review of Newcastle disease in cormorants. Waterbirds 22:333–47.

Langham, J. M. 1991. Twelfth report of the California Bird Records Committee. W. Birds 22:97–130.

Lanyon, W. E. 1961. Specific limits and distribution of Ash-throated and Nutting Flycatchers. Condor 63:421–49.

———. 1963. Notes on a race of the Ash-throated Flycatcher *Myiarchus cinerascens* in Baja California. Am. Mus. Novitates 2129.

Laudenslayer, W. F., Jr., A. S. England, S. Fitton, and L. Saslaw. 1992. The *Toxostoma* thrashers of California: Species at risk? Trans. West. Sec. Wildl. Soc. 28:22–29.

Laughlin, J. 1947a. Baikal Teal taken in California. Condor 49:90.

———. 1947b. Black Rail at Salton Sea, California. Condor 49:132.

Laylander, D. 1997. The last days of Lake Cahuilla: The Elmore site. Pac. Coast Archaeol. Soc. Q. 33:1–138.

Laymon, S. A. 1987. Brown-headed Cowbirds in California: Historical perspectives and management opportunities in riparian habitat. W. Birds 18:63–70.

Laymon, S. A., and M. D. Halterman. 1987. Can the western subspecies of the Yellow-billed Cuckoo be saved from extinction? W. Birds 19:26.

Layton, D., and D. Ermak. 1976. A description of Imperial Valley, California, for the assessment of impacts of geothermal energy development. Univ. Calif. Lawrence Livermore Lab., Livermore.

LeBaron, G. S. 1999. Invasions, irruptions, and trends: The Christmas Bird Count database. N. Am. Birds 53:217–19.

Lee, C. A. 1995. More records of breeding Barn Swallows in Riverside, California. W. Birds 26:155–56.

Lefranc, N. 1997. Shrikes: A guide to shrikes of the world. Yale Univ. Press, New Haven, Conn.

LeGrand, H. E., Jr., P. Guris, and M. Gustafson. 1999. Bulwer's Petrel off the North Carolina coast. N. Am. Birds 53:113–15.

Lehman, P. E. 1994. The birds of Santa Barbara County, California. Univ. Calif., Santa Barbara.

Leopold, A. S. 1959. Wildlife of Mexico: The game birds and mammals. Univ. California Press, Berkeley.

Lies, M. F., and W. H. Behle. 1966. Status of the White Pelican in the United States and Canada through 1964. Condor 68:279–92.

Linsdale, J. M. 1936. The birds of Nevada. Pac. Coast Avifauna 23.

Loeltz, O. J., B. Ireland, J. H. Robison, and F. H. Olmsted. 1975. Geohydrologic reconnaissance of the Imperial Valley, California. U.S. Geol. Surv. Prof. Paper 486-K.

Luther, J. S. 1980. Fourth report of the California Bird Records Committee. W. Birds 11:161–73.

Luther, J. S., G. McCaskie, and J. Dunn. 1979. Third report of the California Bird Records Committee. W. Birds 10:169–87.

———. 1983. Fifth report of the California Bird Records Committee. W. Birds 14:1–16.

MacDougal, D. T. 1914. Movements of vegetation due to submersion and desiccation of land areas in the Salton Sink. *In* MacDougal, D. T., ed., The Salton Sea: A study of the geography, the geology, the floristics, and the ecology of a desert basin, Carnegie Inst. Washington Publ. 193, 115–72.

MacLean, S. F., Jr., and R. T. Holmes. 1971. Bill lengths, wintering areas, and taxonomy of North American Dunlins, *Calidris alpina*. Auk 88:893–901.

Madge, S., and H. Burn. 1988. Waterfowl: An identification guide to the ducks, geese, and swans of the world. Houghton Mifflin, Boston.

Manning, T. H. 1964. Geographical and sexual variation in the Long-tailed Jaeger *Stercorarius longicaudus* Vieillot. Biol. Papers Univ. Alaska 7.

Manolis, T. 1973. The Eastern Kingbird in California. W. Birds 4:33–44.

Marshall, J. T. 1967. Parallel variation in North and Middle American screech-owls. Monogr. West. Found. Vert. Zool. 1.

———. 1997. Allan Phillips and the Flammulated Owl. *In* Dickerman, R. W., ed., The era of Allan R. Phillips: A festschrift, Horizon Commun., Albuquerque, 87–92.

Marshall, J. T. 2001. The Gray-cheeked Thrush, *Catharus minimus*, and its New England subspecies, Bicknell's Thrush, *Catharus minimus bicknelli*. Nuttall Ornithol. Club, Cambridge, Mass.

Marshall, J. T., and K. G. Dedrick. 1994. Endemic Song Sparrows and yellowthroats of San Francisco Bay. Stud. Avian Biol. 15:316–27.

Massey, B. W. 1998. Guide to birds of the Anza-Borrego Desert. Anza-Borrego Desert Nat. Hist. Assoc., Borrego Springs, Calif.

Matsui, M., J. E. Hose, P. Garrahan, and G. A. Jordan. 1992. Development defects in fish embryos from Salton Sea, California. Bull. Environ. Contamin. Toxicol. 48:914–20.

Mayr, E. 1969. Principles of systematic zoology. McGraw-Hill, New York.

Mayr, E., and G. W. Cottrell, eds. 1979. Check-list of birds of the world, vol. 1, 2nd ed. Harvard Univ. Press, Cambridge, Mass.

Mayr, E., and L. L. Short. 1970. Species taxa of North American birds. Publ. Nuttall Ornithol. Club 9.

McCaskie, G. 1966. The occurrence of longspurs and Snow Buntings in California. Condor 68: 597–98.

———. 1970a. The American Redstart in California. Calif. Birds 1:41–46.

———. 1970b. The Blackpoll Warbler in California. Calif. Birds 1:95–104.

———. 1970c. The occurrences of four species of Pelecaniformes in the southwestern United States. Calif. Birds 1:117–42.

———. 1970d. Shorebird and waterbird use of the Salton Sea. Calif. Fish Game 56:87–95.

———. 1971a. A Pyrrhuloxia wanders west to California. Calif. Birds 2:99–100.

———. 1971b. Rusty Blackbirds in California and western North America. Calif. Birds 2:55–68.

———. 1973. A look at the Tree Sparrow in California. W. Birds 4:71–76.

———. 1975. A Rufous-necked Sandpiper in southern California. W. Birds 6:111–13.

———. 1983. Another look at Western and Yellow-footed Gulls. W. Birds 14:85–107.

McCaskie, G., and E. A. Cardiff. 1965. Notes on the distribution of the Parasitic Jaeger and some members of the Laridae in California. Condor 67:542–44.

McCaskie, G., and P. DeBenedictis. 1966. Notes on the distribution of certain icterids and tanagers in California. Condor 68:595–97.

McCaskie, G., P. DeBenedictis, R. Erickson, and J. Morlan. 1988. Birds of northern California: An annotated field list. 1979 Reprinted with supplement. Golden Gate Audubon Soc., San Francisco.

McCaskie, G., S. Liston, and W. A. Rapley. 1974. First nesting of Black Skimmer in California. Condor 76:337–38.

McCaskie, G., and R. R. Prather. 1965. The Curve-billed Thrasher in California. Condor 67:443–44.

McCaskie, G., and M. San Miguel. 1999. Report of the California Bird Records Committee: 1996 records. W. Birds 30:57–85.

McCaskie, G., and S. Suffel. 1971. Black Skimmers at the Salton Sea, California. Calif. Birds 2:69–71.

McCaskie, G., and R. E. Webster. 1990. A second Wedge-tailed Shearwater in California. W. Birds 21:139–40.

McCaskie, R. G. 1965. The Cattle Egret reaches the west coast of the U.S. Condor 67:89.

———. 1968. Noteworthy records of vireos for California. Condor 70:186.

McCaskie, R. G., and R. C. Banks. 1966. Supplemental list of the birds of San Diego County, California. Trans. San Diego Soc. Nat. Hist. 14: 157–68.

McCaskie, R. G., R. Stallcup, and P. DeBenedictis. 1967. The status of certain fringillids in California. Condor 69:426–29.

McKenzie, P. M., and M. B. Robbins. 1999. Identification of adult male Rufous and Allen's Hummingbirds, with specific comments on dorsal coloration. W. Birds 30:86–93.

McKernan, R. L., W. D. Wagner, R. E. Landry, and M. D. McCrary. 1984. Utilization by migrant and resident birds of the San Gorgonio Pass, Coachella Valley, and southern Mojave Desert of California, 1979–80. Rep. 84-RD-15, S. Calif. Edison, Rosemead, Calif.

McLandress, M. R., and I. McLandress. 1979. Blue-phase Ross' Geese and other blue-phase geese in western North America. Auk 96:544–50.

McLean, D. D. 1969. Some additional records of birds in California. Condor 71:433–34.

McMurray, F. B., and G. Monson. 1947. Least Grebe breeding in California. Condor 49:125–26.

McNair, D. B., and W. Post. 2001. Review of the occurrence of vagrant Cave Swallows in the United States and Canada. J. Field Ornithol. 72:485–503.

McNicholl, M. K., P. E. Lowther, and J. A. Hall. 2001. Forster's Tern (*Sterna forsteri*). The birds of North America, ed. A. F. Poole and F. B. Gill, no. 595. Acad. Nat. Sci., Philadelphia, and Am. Ornithol. Union, Washington, D.C.

Mearns, E. A. 1907. Mammals of the Mexican boundary of the United States. Bull. U.S. Natl. Mus. 56.

Mengel, R. M. 1963. The birds of Kentucky. Ornithol. Monogr. 3.

Merrifield, K. 1993. Eurasian × American Wigeons in western Oregon. W. Birds 24:105–7.

Mikuska, T., J. A. Kushlan, and S. Hartley. 1998. Key areas for wintering North American herons. Colonial Waterbirds 21:125–34.

Miles, A. K., and H. M. Ohlendorf. 1993. Environmental contaminants in Canvasbacks wintering on San Francisco Bay, California. Calif. Fish Game 79:28–38.

Miller, A. H. 1942. Differentiation of the Ovenbirds of the Rocky Mountain region. Condor 44:185–86.

Miller, A. H., H. Friedmann, L. Griscom, and R. T. Moore. 1957. Distributional check-list of the birds of Mexico, pt. 2. Pac. Coast Avifauna 33.

Miller, A. H., and R. C. Stebbins. 1964. The lives of desert animals in Joshua Tree National Monument. Univ. California Press, Berkeley.

Miller, L. 1957. Some avian flyways of western America. Wilson Bull. 69:164–69.

Miller, L., and A. J. van Rossem. 1929. Nesting of the Laughing Gull in southern California. Condor 31:141–42.

Miller, L. H. 1908. Louisiana Water-Thrush in California. Condor 10:236–37.

Mills, G. 1976. American Kestrel sex ratios and habitat separation. Auk 93:740–48.

Mindell, D. P. 1983. Harlan's Hawk (*Buteo jamaicensis harlani*): A valid subspecies. Auk 100:161–69.

———. 1985. Plumage variation and winter range of Harlan's Hawk (*Buteo jamaicensis harlani*). Am. Birds 39:127–33.

Mlodinow, S. G. 1998a. The Magnificent Frigatebird in western North America. Field Notes 52:412–19.

———. 1998b. The Tropical Kingbird north of Mexico. Field Notes 52:6–11.

———. 1999. Spotted Redshank and Common Greenshank in North America. N. Am. Birds 53:124–30.

Mlodinow, S. G., and K. T. Karlson. 1999. Anis in the United States and Canada. N. Am. Birds 53:237–45.

Mlodinow, S. G., and M. O'Brien. 1996. America's one hundred most wanted birds. Falcon Press, Helena, Mont.

Moffitt, J. 1932. Clapper Rails occur on marshes of Salton Sea, California. Condor 36:137.

Molina, K. C. 1996. Population status and breeding biology of Black Skimmers at the Salton Sea, California. W. Birds 27:143–58.

———. 2000. The recent breeding of California and Laughing Gulls at the Salton Sea, California. W. Birds 31:106–11.

Molina, K. C., and K. L. Garrett. 2001. The breeding birds of the Cerro Prieto geothermal ponds, Mexicali Valley, Baja California, Mexico. Monogr. Field Ornithol. 3:23–28.

Molina, P., H. Ouellet, and R. McNeil. 2000. Geographic variation and taxonomy of the Northern Waterthrush. Wilson Bull. 112:337–46.

Monroe, B. L., Jr. 1968. A distributional survey of the birds of Honduras. Ornithol. Monogr. 7.

Monson, G. 1954. Westward extension of the ranges of the Inca Dove and the Bronzed Cowbird. Condor 56:229–30.

———. 1958. Reddish Egret and Bronzed Cowbird in California. Condor 60:191.

Monson, G., and A. R. Phillips. 1981. Annotated checklist of the birds of Arizona, 2nd ed. Univ. Arizona Press, Tucson.

Mora, M. A. 1989. Predation by a Brown Pelican at a mixed-species heronry. Condor 91:742–43.

———. 1990. Organochlorines, reproductive success, and habitat use in birds from northwest Mexico. Ph.D. diss., Univ. Calif., Davis.

———. 1991. Organochlorines and breeding success in Cattle Egrets from the Mexicali Valley, Baja California, Mexico. Colonial Waterbirds 14:127–32.

Mora, M. A., and D. W. Anderson. 1995. Selenium, boron, and heavy metals in birds from the Mexicali Valley, Baja California, Mexico. Bull. Environ. Contamin. Toxicol. 54:198–206.

Mora, M. A., D. W. Anderson, and M. E. Mount. 1987. Seasonal variation of body condition and organochlorines in wild ducks from California and Mexico. J. Wildl. Manage. 5:132–40.

Morlan, J. 1985. Eighth report of the California Bird Records Committee. W. Birds 16:105–22.

Morrison, J., and M. Cohen. 1999. Restoring California's Salton Sea. Borderlines 7(1):1–4.

Morrison, M. L., L. S. Hall, S. K. Robinson, S. I. Rothstein, D. C. Hahn, and T. D. Rick, eds. 1999. Research and management of the Brown-headed Cowbird in western landscapes. Stud. Avian Biol. 18.

Munyer, E. A. 1965. Inland wanderings of the Ancient Murrelet. Wilson Bull. 77:235–42.

Murphy, R. C. 1917. Natural history observations from the Mexican portion of the Colorado Desert, with a note on the Lower Californian

Pronghorn and a list of the birds. Abstract Proc. Linn. Soc. New York 28:43–101.

Murray, B. W., W. B. McGillivray, J. C. Barlow, R. N. Beech, and C. Strobeck. 1994. The use of cytochrome *b* sequence variation in estimation of phylogeny in the Vireonidae. Condor 96:1037–54.

Myers, S. J. 1993. Mountain Chickadees nest in desert riparian forest. W. Birds 24:103–4.

Neff, J. A. 1947. Habits, food, and economic status of the Band-tailed Pigeon. N. Am. Fauna 58.

Nelson, E. W. 1922. Lower California and its natural resources. Mem. Natl. Acad. Sci. 16, 1st Mem.

Newcomer, M. W., and G. K. Silber. 1989. Sightings of Laysan Albatross in the northern Gulf of California. W. Birds 20:134–35.

Norris, R. M., and K. S. Norris. 1961. Algodones Dunes of southeastern California. Geol. Soc. Am. Bull. 72:605–20.

Nowak, J. H., and G. Monson. 1965. Black Brant summering at Salton Sea. Condor 67:357.

Nuechterlein, G. L., and D. Buitron. 1989. Diving differences between Western and Clark's Grebes. Auk 106:467–70.

Oberholser, H. C. 1919. A revision of the subspecies of *Passerculus rostratus* (Cassin). Ohio J. Sci. 19:344–54.

———. 1974. The Bird Life of Texas, vol. 2. Univ. Texas Press, Austin.

Ohlendorf, H. M., D. J. Hoffman, M. K. Daiki, and T. W. Aldrich. 1986. Embryonic mortality and abnormalities of aquatic birds: Apparent impacts of selenium from irrigation drainwater. Sci. Total Environ. 52:49–63.

Ohmart, R. D. 1994. The effects of human-induced changes on the avifauna of western riparian habitats. Stud. Avian Biol. 15:273–85.

Ohmart, R. D., and R. E. Tomlinson. 1977. Foods of western Clapper Rails. Wilson Bull. 89:332–36.

Olsen, K. M., and H. Larsson. 1997. Skuas and jaegers: A guide to the skuas and jaegers of the world. Yale Univ. Press, New Haven, Conn.

Olson, S. L. 1997. Towards a less imperfect understanding of the systematics and biogeography of the Clapper and King Rail complex (*Rallus longirostris* and *R. elegans*). *In* Dickerman, R. W., ed., The era of Allan R. Phillips: A festschrift, Horizon Commun., Albuquerque, 93–111.

O'Neill, E. J. 1954. Ross Goose observations. Condor 56:311.

Ouellet, H. 1993. Bicknell's Thrush: Taxonomic status and distribution. Wilson Bull. 105:545–72.

Page, G. W., F. C. Bidstrup, R. J. Ramer, and L. E. Stenzel. 1986. Distribution of wintering Snowy Plovers in California and adjacent states. W. Birds 17:145–70.

Page, G. W., and R. E. Gill Jr. 1994. Shorebirds in western North America: Late 1800s to late 1900s. Stud. Avian Biol. 15:147–60.

Page, G. W., W. D. Shuford, J. E. Kjelmyr, and L. E. Stenzel. 1992. Shorebird numbers in wetlands of the Pacific Flyway: A summary of counts form April 1988 to January 1992. Pt. Reyes Bird Observ., Stinson Beach, Calif.

Page, G. W., and L. E. Stenzel. 1981. The breeding status of the Snowy Plover in California. W. Birds 12:1–40.

Page, G. W., L. E. Stenzel, W. D. Shuford, and C. R. Bruce. 1991. Distribution and abundance of the Snowy Plover on its western North American breeding grounds. J. Field Ornithol. 62:245–55.

Palacios, E., D. W. Anderson, E. Mellink, and S. González-Guzmán. 2000. Distribution and abundance of Burrowing Owls on the peninsula and islands of Baja California. W. Birds 31:89–99.

Palacios, E., and E. Mellink. 1992. Breeding bird records from Montague Island, northern Gulf of California. W. Birds 23:41–44.

———. 1996. Status of the Least Tern in the Gulf of California. J. Field Ornithol. 67:48–58.

Palmer, R. S., ed. 1962. Handbook of North American birds, vol. 1. Yale Univ. Press, New Haven, Conn.

———. 1976a. Handbook of North American birds, vol. 2. Yale Univ. Press, New Haven, Conn.

———. 1976b. Handbook of North American birds, vol. 3. Yale Univ. Press, New Haven, Conn.

———. 1988a. Handbook of North American birds, vol. 4. Yale Univ. Press, New Haven, Conn.

———. 1988b. Handbook of North American birds, vol. 5. Yale Univ. Press, New Haven, Conn.

Parish, S. B. 1914. Plant ecology and floristics of the Salton Sink. *In* MacDougal, D. T., ed., The Salton Sea: A study of the geography, the geology, the floristics, and the ecology of a desert basin, Carnegie Inst. Washington Publ. 193, 85–114.

Parkes, K. C. 1952. Geographic variation in the Horned Grebe. Condor 54:314–15.

———. 1982. Further comments on the field identification of North American pipits. Am. Birds 36:20–22.

Parnell, J. F., R. M. Erwin, and K. C. Molina. 1995. Gull-billed Tern (*Sterna nilotica*). The birds of North America, ed. A. F. Poole and F. B. Gill, no. 140. Acad. Nat. Sci., Philadelphia, and Am. Ornithol. Union, Washington, D.C.

Patten, M. A. 1993. Notes on immature Double-crested and Neotropic Cormorants. Birding 25: 343–45.

———. 1995a. Checklist of the birds of Morongo Valley. Bureau Land Manage., Morongo Valley, Calif.

———. 1995b. Status and distribution of California birds. Condor 97:608–11.

———. 1996. Yellow-footed Gull (*Larus livens*). The birds of North America, ed. A. F. Poole and F. B. Gill, no. 243. Acad. Nat. Sci., Philadelphia, and Am. Ornithol. Union, Washington, D.C.

———. 1998. Changing seasons, fall migration, August 1–November 30, 1997: Nora, El Niño, and vagrants from far afield. Field Notes 52: 14–18.

———. 2001. The role of habitat and signalling in speciation: Evidence from a contact zone of two Song Sparrow subspecies. Ph.D. diss., Univ. Calif., Riverside.

Patten, M. A., and J. C. Burger. 1998. Spruce Budworm outbreaks and the incidence of vagrancy in eastern North American wood-warblers. Can. J. Zool. 76:433–39.

Patten, M. A., and R. A. Erickson. 1994. Fifteenth report of the California Bird Records Committee. W. Birds 25:1–34.

———. 1996. Subspecies of the Least Tern in Mexico. Condor 98:888–90.

———. 2000. Population fluctuations of the Harris' Hawk (*Parabuteo unicinctus*) and its reappearance in California. J. Raptor Res. 34:187–95.

Patten, M. A., R. A. Erickson, and P. Unitt. 2003. Population changes and biogeographic affinities of the birds of the Salton Sink, California/Baja California. Stud. Avian Biol.: in press.

Patten, M. A., S. E. Finnegan, and P. E. Lehman. 1995. Seventeenth report of the California Bird Records Committee: 1991 records. W. Birds 26: 113–43.

Patten, M. A., and M. Fugate. 1998. Systematic relationships among the Emberizid sparrows. Auk 115:412–24.

Patten, M. A., and G. W. Lasley. 2000. Range expansion of the Glossy Ibis in North America. N. Am. Birds 54:241–47.

Patten, M. A., and C. A. Marantz. 1996. Implications of vagrant southeastern vireos and warblers in California. Auk 113:911–23.

Patten, M. A., and G. McCaskie. 2003. Patterns and processes of the occurrence of pelagic and subtropical waterbirds at the Salton Sea. Stud. Avian Biol.: in press.

Patten, M. A., E. Mellink, H. Gómez de Silva, and T. E. Wurster. 2001. Status and taxonomy of the Colorado Desert avifauna of Baja California. Monogr. Field Ornithol. 3:29–63.

Patten, M. A., and R. A. Minnich. 1997. Procellariiformes occurrence at the Salton Sea and in the Sonoran Desert. Southwest. Nat. 42:303–11.

Patten, M. A., K. Radamaker, and T. E. Wurster. 1993. Noteworthy observations from northeastern Baja California. W. Birds 24:89–93.

Patten, M. A., and J. T. Rotenberry. 1998. Post-disturbance changes in a desert breeding bird community. J. Field Ornithol. 69:614–25.

Patten, M. A., and B. D. Smith-Patten. 2003. Linking the Salton Sea with its past: The history and avifauna of Lake Cahuilla. Stud. Avian Biol.: in press.

Patten, M. A., and P. Unitt. 2002. Diagnosability versus mean differences in Sage Sparrow subspecies. Auk 119:26–35.

Patten, M. A., P. Unitt, R. A. Erickson, and K. F. Campbell. 1995. Fifty years since Grinnell and Miller: Where is California ornithology headed? W. Birds 26:54–64.

Patten, M. A., and J. C. Wilson. 1996. A dark-morph Sharp-shinned Hawk in California, with comments on dichromatism in raptors. Bull. Brit. Ornithol. Club 116:266–70.

Paulson, D. R. 1993. Shorebirds of the Pacific Northwest. Univ. Washington Press, Seattle.

Paulson, D. R., and D. S. Lee. 1992. Wintering of Lesser Golden-Plovers in eastern North America. J. Field Ornithol. 63:121–28.

Payne, R. B. 1979. Family Ardeidae. *In* Mayr, E., and G. W. Cottrell, eds., Check-list of birds of the world, vol. 1, 2nd ed. Harvard Univ. Press, Cambridge, Mass., 193–244.

———. 1987. Populations and type specimens of a nomadic bird: Comments on the North American crossbills *Loxia pusilla* Gloger 1834 and *Crucirostra minor* Brehm 1845. Occ. Pap. Mus. Zool. Univ. Mich. 714.

Paynter, R. A., Jr., ed. 1968. Check-list of birds of the world, vol. 14. Harvard Univ. Press, Cambridge, Mass.

Paynter, R. A., Jr., and R. W. Storer, eds. 1970. Check-list of birds of the world, vol. 13. Harvard Univ. Press, Cambridge, Mass.

Pemberton, J. R. 1927. The American Gull-billed Tern breeding in California. Condor 29:253–58.

Peterjohn, B. G., and J. R. Sauer. 1995. Population trends of the Loggerhead Shrike from the North American Breeding Bird Survey. Proc. West. Found. Vert. Zool. 6:117–21.

Peters, J. L. 1934. Check-list of birds of the world, vol. 2. Harvard Univ. Press, Cambridge, Mass.

———. 1940. Check-list of birds of the world, vol. 4. Harvard Univ. Press, Cambridge, Mass.

———. 1960. Family Hirundinidae. *In* Mayr, E., and J. C. Greenway Jr., eds., Check-list of birds of the world, vol. 9, Harvard Univ. Press, Cambridge, Mass., 80–129.

Peterson, A. T. 1990. Birds of Eagle Mountain, Joshua Tree National Monument, California. W. Birds 21:127–35.

———. 1991. Gene flow in Scrub Jays: Frequency and direction of movement. Condor 93:926–34.

Phillips, A., J. Marshall, and G. Monson. 1964. The birds of Arizona. Univ. Arizona Press, Tucson.

Phillips, A. R. 1942. Notes on the migrations of the Elf and Flammulated Screech Owls. Wilson Bull. 54:132–37.

———. 1943. Critical notes on two southwestern sparrows. Auk 60:242–48.

———. 1950. The Great-tailed Grackles of the Southwest. Condor 52:78–81.

———. 1959. The nature of avian species. J. Ariz. Acad. Sci. 1:22–30.

———. 1961. Notas sistemáticas sobre aves Mexicanas, I. Anal. Inst. Biol. Mex. 32:333–81.

———. 1962. Notas sistemáticas sobre aves Mexicanas, II. Anal. Inst. Biol. Mex. 33:331–72.

———. 1964. Notas sistemáticas sobre aves Mexicanas, III. Rev. Soc. Mex. Hist. Nat. 25:217–42.

———. 1968. The instability of the distribution of land birds in the Southwest. Papers Archaeol. Soc. New Mexico 1:129–62.

———. 1975a. The migrations of Allen's and other hummingbirds. Condor 77:196–205.

———. 1975b. Why neglect the difficult? W. Birds 6:69–86.

———. 1986. The known birds of North and Middle America, pt. 1. Allan R. Phillips, Denver.

———. 1991. The known birds of North and Middle America, pt. 2. Allan R. Phillips, Denver.

———. 1994a. *The Known Birds of North and Middle America* versus the current AOU list. Auk 111: 770–73.

———. 1994b. A review of northern *Pheucticus* grosbeaks. Bull. Brit. Ornithol. Club 114:162–70.

———. 1994c. A tentative key to the species of kingbirds, with distributional notes. J. Field Ornithol. 65:295–306.

Phillips, J. C. 1923. A natural history of the ducks, vol. 2. Houghton Mifflin, Boston.

Phillips, J. C., and F. C. Lincoln. 1930. American waterfowl: Their present situation and the outlook for their future. Houghton Mifflin, Boston.

Pitelka, F. C. 1945. Differentiation of the Scrub Jay, *Aphelocoma coerulescens*, in the Great Basin and Arizona. Condor 47:23–26.

Pitman, R. L. 1986. Atlas of seabird distribution and relative abundance in the eastern tropical Pacific. Admin. U.S. Natl. Mar. Fish. Serv., Southwest Fish. Center Rep. LJ-86-02C.

Pittaway, R. 1999. Taxonomic history of Thayer's Gull. Ontario Birds 17:2–13.

Platter, M. F. 1976. Breeding ecology of Cattle Egrets and Snowy Egrets at the Salton Sea, southern California. M.Sc. thesis, San Diego State Univ., San Diego.

Pogson, T. H., and S. M. Lindstedt. 1991. Distribution and abundance of large Sandhill Cranes, *Grus canadensis*, wintering in California's Central Valley. Condor 93:266–78.

Portenko, L. A. 1972. Birds of the Chukchi Peninsula and Wrangel Island. Nauka Publ., Leningrad.

Porter, R. D., M. A. Jenkins, M. N. Kirven, D. W. Anderson, and J. O. Keith. 1988. Status and reproductive performance of marine Peregrines in Baja California and the Gulf of California, Mexico. *In* Cade, T. J., J. H. Enderson, C. G. Thelander, and C. M. White, eds., Peregrine Falcon populations: Their management and recovery, Peregrine Fund, Boise, Idaho, 105–14.

Post, P. W., and R. H. Lewis. 1995. The Lesser Black-backed Gull in the Americas: Occurrence and subspecific identity, part I: Taxonomy, distribution, and migration. Birding 27: 282–90.

Prater, A. J., J. H. Marchant, and J. Vuorinen. 1977. Guide to the identification and ageing of Holarctic waders. Brit. Trust Ornithol. Guide 17.

Prescott, B., B. Truesdell, and J. Zarki. 1997. Birds [of] Joshua Tree National Park: A checklist. Joshua Tree Natl. Park Assoc., Twentynine Palms, Calif.

Pugesek, B. H., K. L. Diem, and C. L. Cordes. 1999. Seasonal movements, migration, and range sizes of subadult and adult Bamforth Lake California Gulls. Waterbirds 22:29–36.

Pulich, W. M., and A. R. Phillips. 1953. A possible desert flight line of the American Redstart. Condor 55:99–100.

Putnam, D., and R. Kallenbach. 1997. Growers face critical juncture in desert forage production. Calif. Agric. 51(3):12–16.

Pyle, P. 1997. Identification guide to North American birds, pt. 1: Columbidae to Ploceidae. Slate Creek Press, Bolinas, Calif.

Pyle, P., and G. McCaskie. 1992. Thirteenth report of the California Bird Records Committee. W. Birds 23:97–132.

Quigley, R. J. 1973. First record of Sooty Shearwater for Arizona. Auk 90:677.

Rainey, D. G., S. G. Van Hoose, and J. Tramontano. 1959. Breeding of the Starling in southern California. Condor 61:57–58.

Rand, A. L., and M. A. Traylor. 1950. The amount of overlap allowable for subspecies. Auk 67:169–83.

Ratti, J. T. 1981. Identification and distribution of Clark's Grebe. W. Birds 12:41–46.

Rea, A. M. 1967. Some bird records from San Diego County, California. Condor 69:316–18.

———. 1969. Interbreeding of two subspecies of Boat-tailed Grackle, *Cassidix mexicanus nelsoni* and *Cassidix mexicanus monsoni*, in secondary contact in central Arizona. M.Sc. thesis, Ariz. State Univ., Tempe.

———. 1983. Once a river. Univ. Arizona Press, Tucson.

———. 1986a. *Corvus corax* L: Geographic variation. *In* Phillips, A. R., The known birds of North and Middle America, pt. 1, Allan R. Phillips, Denver, 65–66, 214.

———. 1986b. *Troglodytes troglodytes* (L): (2) W. races. *In* Phillips, A. R., The known birds of North and Middle America, pt. 1, Allan R. Phillips, Denver, 138–40.

Rea, A. M., and K. L. Weaver. 1990. The taxonomy, distribution, and status of coastal California Cactus Wrens. W. Birds 21:81–126.

Remsen, J. V., Jr. 1978. Bird species of special concern in California: An annotated list of declining or vulnerable bird species. Calif. Dept. Fish Game, Wildl. Manage. Branch Admin. Rep. 78-1.

Ricketts, E. D. 1928. White-winged Dove in the Imperial Valley. Calif. Fish Game 14:252.

Ridgway, R. 1901. The birds of North and Middle America, pt. 1. Bull. U.S. Natl. Mus. 50.

———. 1902. The birds of North and Middle America, pt. 2. Bull. U.S. Natl. Mus. 50.

———. 1904. The birds of North and Middle America, pt. 3. Bull. U.S. Natl. Mus. 50.

———. 1919. The birds of North and Middle America, pt. 8. Bull. U.S. Natl. Mus. 50.

Riedel, R., L. Caskey, and B. A. Costa-Pierce. 2002. Fish biology and fisheries ecology of the Salton Sea, California. Hydrobiologia 473:229–44.

Rienecker, W. C. 1965. A summary of band returns from Lesser Snow Goose of the Pacific Flyway. Calif. Fish Game 51:132–46.

———. 1968. A summary of band recoveries from Redheads (*Aythya americana*) banded in NE California. Calif. Fish Game 54:17–26.

———. 1976. Distribution, harvest, and survival of American Wigeon banded in California. Calif. Fish Game 62:141–53.

Rising, J. D. 1996. A guide to the identification and natural history of the sparrows of the United States and Canada. Academic Press, San Diego.

Robbins, C. S. 1973. Introduction, spread, and present abundance of the House Sparrow in North America. Ornithol. Monogr. 14:3–9.

Robbins, M. B., and D. A. Easterla. 1992. Birds of Missouri: Their distribution and abundance. Univ. Missouri Press, Columbia.

Roberson, D. 1980. Rare birds of the West Coast. Woodcock Publ., Pacific Grove, Calif.

———. 1985. Monterey Birds. Monterey Peninsula Audubon Soc., Carmel, Calif.

———. 1986. Ninth report of the California Bird Records Committee. W. Birds 17:49–77.

———. 1993. Fourteenth report of the California Bird Records Committee. W. Birds 24:113–66.

Roberson, D., and S. F. Bailey. 1991. *Cookilaria* petrels in the eastern Pacific Ocean: Identification and distribution. Am. Birds 45:399–403, 1067–81.

Roberson, D., and L. F. Baptista. 1988. White-shielded coots in North America: A critical evaluation. Am. Birds 42:1241–46.

Rocke, T. E., and M. D. Samuel. 1999. Water and sediment characteristics associated with avian botulism outbreaks in wetlands. J. Wildl. Manage. 63:1249–60.

Rogers, M. J. 1996. Report on rare birds in Great Britain in 1995. Brit. Birds 89:481–531.

Rogers, M. M., and A. Jaramillo. 2002. Report of the California Bird Records Committee: 1999 records. W. Birds 33:1–33.

Rohwer, S., and C. S. Wood. 1998. Three hybrid zones between Hermit and Townsend's Warblers in Washington and Oregon. Auk 115:284–310.

Romagosa, C. M., and T. McEneaney. 1999. Eurasian Collared-Dove in North America and the Caribbean. N. Am. Birds 53:348–53.

Root, T. 1988. Atlas of wintering North American birds: An analysis of Christmas Bird Count data. Univ. Chicago Press, Chicago.

Rosenberg, G. H., and J. L. Witzeman. 1998. Arizona Bird Committee report, 1974–1996: Part 1 (nonpasserines). W. Birds 29:199–224.

Rosenberg, K. V., R. D. Ohmart, W. C. Hunter, and B. W. Anderson. 1991. Birds of the lower Colorado River valley. Univ. Arizona Press, Tucson.

Rosenberg, K. V., S. B. Terrill, and G. H. Rosenberg. 1987. Value of suburban habitats to desert riparian birds. Wilson Bull. 99:642–54.

Rotenberry, J. T., M. A. Patten, and K. L. Preston. 1999. Brewer's Sparrow (*Spizella breweri*). The birds of North America, ed. A. F. Poole and F. B. Gill, no. 390. Acad. Nat. Sci., Philadelphia, and Am. Ornithol. Union, Washington, D.C.

Rothstein, S. I. 1994. The cowbird's invasion of the Far West: History, causes, and consequences experienced by host species. Stud. Avian Biol. 13:301–15.

Rottenborn, S. C., and J. Morlan. 2000. Report of the California Bird Records Committee: 1997 records. W. Birds 31:1–37.

Russell, S. M., and G. Monson. 1998. The birds of Sonora. Univ. Arizona Press, Tucson.

Ryder, R. A. 1967. Distribution, migration, and mortality of the White-faced Ibis (*Plegadis chihi*) in North America. Bird-Banding 38:257–77.

Ryser, F. A., Jr. 1985. Birds of the Great Basin: A natural history. Univ. Nevada Press, Reno.

Saiki, M. K. 1990. Elemental concentrations in fishes from the Salton Sea, southeastern California. Water Air Soil Poll. 52:41–56.

Sams, J. R. 1958. Blue Goose observed at the Salton Sea, Imperial County, California. Condor 60:191.

Sanger, G. A. 1974. Laysan Albatross (*Diomedea immutabilis*). *In* King, W. B., ed., Pelagic studies of seabirds in the central and eastern Pacific Ocean, Smithsonian Contr. Zool. 158, 129–53.

San Miguel, M. R. 1998. Extralimital breeding of the Bufflehead in California. W. Birds 29:36–40.

Saunders, G. B. 1968. Seven new White-winged Doves from Mexico, Central America, and southwestern United States. N. Am. Fauna 65.

Saunders, G. B., and D. C. Saunders. 1981. Waterfowl and their wintering grounds in Mexico, 1937–64. U.S. Fish Wildl. Serv. Resource Publ. 138.

Schmidt, R. H. 1989. The arid zones of Mexico: Climatic extremes and conceptualizationof the Sonoran Desert. J. Arid Environ. 16:241–56.

Schneider, F. B. 1928. The season, Los Angeles Region. Bird-Lore 30:208–9.

Schoenherr, A. A. 1992. A natural history of California. Univ. California Press, Berkeley.

Schram, B. 1998. A birder's guide to southern California. Am. Birding Assoc., Colorado Springs, Colo.

Sealy, S. G. 1998. The subspecies of the Northern Saw-whet Owl on the Queen Charlotte Islands: An island endemic and a nonbreeding visitant. W. Birds 29:21–28.

Sealy, S. G., A. J. Banks, and J. F. Chace. 2000. Two subspecies of Warbling Vireo differ in their responses to cowbird eggs. W. Birds 31:190–94.

Selander, R. K. 1971. Systematics and speciation in birds. *In* Farner, D. S., J. R. King, and K. C. Parkes, eds., Avian biology, vol. 1, Academic Press, New York, 57–147.

Setmire, J. G. 1998. Selenium and salinity concerns in the Salton Sea area of California. *In* Frankenberger, W. T., Jr., and R. A. Engberg, eds., Environmental chemistry of selenium, Marcel Dekker, New York, 205–21.

Setmire, J. G., R. A. Schroeder, J. N. Densmore, S. L. Goodbred, D. J. Audet, and W. R. Radke. 1993. Detailed study of water quality, bottom sediment, and biota associated with irrigation drainage in the Salton Sea area, California, 1988–90. U.S. Geol. Surv. Water-Resources Invest. Rep. 93-4014.

Setmire, J. G., J. C. Wolfe, and R. K. Stroud. 1990. Reconnaissance investigation of water quality, bottom sediment, and biota associated with irrigation drainage in the Salton Sea area, California, 1986–87. U.S. Geol. Surv. Water-Resources Invest. Rep. 89-4102.

Sheppard, J. M. 1996. Le Conte's Thrasher (*Toxostoma lecontei*). The birds of North America, ed. A. F. Poole and F. B. Gill, no. 230. Acad. Nat. Sci., Philadelphia, and Am. Ornithol. Union, Washington, D.C.

Short, L. L. 1968. Variation of Ladder-backed Wood-peckers in southwestern North America. Proc. Biol. Soc. Washington 81:1–10.

———. 1982. Woodpeckers of the world. Delaware Mus. Nat. Hist. Monogr. Ser. 4.

Shuford, W. D., C. M. Hickey, R. J. Safran, and G. W. Page. 1996. A review of the status of the White-faced Ibis in winter in California. W. Birds 27:169–96.

Shuford, W. D., N. Warnock, and R. L. McKernan. 2003. Patterns of shorebird use of the Salton Sea and adjacent Imperial Valley, California. Stud. Avian. Biol.: in press.

Shuford, W. D., N. Warnock, and K. C. Molina. 1999. The avifauna of the Salton Sea: A synthesis. Pt. Reyes Bird Observatory, Stinson Beach, Calif.

Shuford, W. D., N. Warnock, K. C. Molina, B. Mulrooney, and A. E. Black. 2000. Avifauna of the Salton Sea: Abundance, distribution, and annual phenology. Pt. Reyes Bird Observatory, Stinson Beach, Calif.

Shuford, W. D., N. Warnock, K. C. Molina, and K. Sturm. 2002. The Salton Sea as critical habitat to migratory and resident waterbirds. Hydrobiologia 473:255–74.

Sibley, C. G., and B. L. Monroe Jr. 1990. Distribution and taxonomy of birds of the world. Yale Univ. Press, New Haven.

———. 1993. A supplement to distribution and taxonomy of birds of the world. Yale Univ. Press, New Haven.

Sidle, J. G., W. H. Koonz, and K. Roney. 1985. Status of the American White Pelican: An update. Am. Birds 39:859–64.

Simpson, E. P., and S. H. Hurlbert. 1998. Salinity effects on the growth, mortality, and shell strength of *Balanus amphitrite* from the Salton Sea, California. Hydrobiologia 381:179–90.

Simpson, M. G., S. C. McMillan, B. L. Stone, J. Gibson, and J. P. Rebman. 1996. Checklist of the vascular plants of San Diego County, 2nd ed. San Diego State Univ. Herb. Press, Spec. Publ. 1.

Skrove, T. 1986. The Salton Sea: Nature's accident in the desert. Aqueduct 52(4):8–12.

Sloan, N. F. 1982. Status of breeding colonies of White Pelicans in the United States through 1979. Am. Birds 36:250–54.

Small, A. 1959. Recent occurrences of Oldsquaw in Southern California. Condor 61:302–3.

———. 1994. California birds: Their status and distribution. Ibis Publ., Vista, Calif.

Smith, B. D. 1999. Leveling the ground: Cultural investigations into precontact use of the northern shoreline of ancient Lake Cahuilla. M.A. thesis, Univ. Calif., Los Angeles.

Smith, N. G. 1966. Evolution of some Arctic gulls (*Larus*): An experimental study in isolating mechanisms. Ornithol. Monogr. 4.

Smith, P. W. 1987. The Eurasian Collared-Dove arrives in the Americas. Am. Birds 41:1371–79.

Smith, R. H., and G. H. Jensen. 1970. Black Brant on the mainland coast of Mexico. Trans. N. Am. Wildl. Conf. 35:227–41.

Snell, R. R. 1989. Status of *Larus* gulls at Home Bay, Baffin Island. Colonial Waterbirds 12:12–23.

———. 1991. Conflation of the observed and the hypothesized: Smith's 1961 research in Home Bay, Baffin Island. Colonial Waterbirds 14:196–202.

Spear, L. B., M. J. Lewis, M. T. Myres, and R. L. Pyle. 1988. The recent occurrence of Garganey in North America and the Hawaiian Islands. Am. Birds 42:385–92.

Steere, C. H. 1952. Imperial and Coachella Valleys. Stanford Univ. Press, Stanford, Calif.

Stephens, D. A., and J. E. Stephens. 1987. An American Oystercatcher in Idaho. W. Birds 18:215–16.

Stephens, F. 1919. Unusual occurrences of Bendire Thrasher, Fork-tailed Petrel, and Western Goshawk. Condor 21:87.

Stephens, T. 1997a. A river runs through desert agriculture. Calif. Agric. 51(3):6–10.

———. 1997b. Salton Sea keeps getting saltier. Calif. Agric. 51(3):8–9.

Stevenson, J. 1929. Hooded Merganser at Salton Sea, California. Condor 31:127.

———. 1932. Bird notes from southern California. Condor 34:229.

Stone, W., and S. N. Rhoads. 1905. On a collection of birds and mammals from the Colorado delta, Lower California. Proc. Acad. Nat. Sci. Philadelphia 57:676–90.

Storer, R. W., and T. Getty. 1985. Geographic variation in the Least Grebe (*Tachybaptus dominicus*). Ornithol. Monogr. 36:31–39.

Stott, K., Jr., and J. R. Sams. 1959. Distributional records of the Common Goldeneye and the Crissal Thrasher in southeastern California. Condor 61:298–99.

Streseman, E., and D. Amadon. 1979. Order Falconiformes. *In* Mayr, E., and G. W. Cottrell, eds., Check-list of birds of the world, vol. 1, 2nd ed., Harvard Univ. Press, Cambridge, Mass., 271–425.

Suffel, G. S. 1970. An Olivaceous Flycatcher in California. Calif. Birds 1:79–80.

Suffel, S. 1971. Southern California birds. Western Tanager 37(8):10.

Swarth, H. S. 1916. The Pacific Coast races of the Bewick Wren. Proc. Calif. Acad. Sci., 4th ser., 6:53–85.

———. 1917. Geographical variation in *Sphyrapicus thyroideus*. Condor 19:62–65.

———. 1933. Peale Falcon in California. Condor 35:233–34.

Swarth, H. S., and H. C. Bryant. 1917. A study of the races of the White-fronted Goose (*Anser albifrons*) occurring in California. Univ. Calif. Publ. Zool. 17:209–22.

Sykes, G. 1914. Geographical features of the Cahuilla Basin. *In* MacDougal, D. T., ed., The Salton Sea: A study of the geography, the geology, the floristics, and the ecology of a desert basin, Carnegie Inst. Washington Publ. 193, 13–20.

———. 1937. The Colorado delta. Am. Geogr. Soc. Spec. Publ. 19.

Taverner, P. A. 1936. Taxonomic comments on Red-tailed Hawks. Condor 38:66–71.

Terborgh, J. 1989. Where have all the birds gone? Princeton Univ. Press, Princeton, N.J.

Tershy, B. R., E. Van Gelder, and D. Breese. 1993. Relative abundance and seasonal distribution of seabirds in the Canal de Ballenas, Gulf of California. Condor 95:458–64.

Tetra Tech. 2000. Draft Salton Sea restoration project environmental impact statement/environmental impact report. San Bernardino, Calif.

Thompson, B. H. 1933. History and present status of the breeding colonies of the White Pelican in the U.S. Natl. Park Serv. Occ. Paper 1.

Thompson, C. W. 1991. Is the Painted Bunting actually two species? Problems determining species limits between allopatric populations. Condor 93:987–1000.

Thompson, W. L. 1964. An early specimen of the Indigo Bunting from California. Condor 66:445.

Tinkham, E. R. 1949. A record of the Scissor-tailed Flycatcher from the Colorado Desert. Condor 51:99–100.

Todd, W. E. C. 1963. Birds of the Labrador Peninsula and adjacent areas. Univ. Toronto Press, Toronto.

Tomkovich, P. S. 1992. An analysis of the geographical variability in Knots *Calidris canutus* based on museum skins. Wader Study Group Bull. 64(suppl.):17–23.

Traylor, M. A., Jr. 1979a. Check-list of birds of the world, vol. 8. Harvard Univ. Press, Cambridge, Mass.

———. 1979b. Two sibling species of *Tyrannus* (Tyrannidae). Auk 96:221–33.

Turner, R. M., J. E. Bowers, and T. L. Burgess. 1995. Sonoran Desert plants: An ecological atlas. Univ. Arizona Press, Tucson.

Twomey, A. C. 1947. Critical notes on some western Song Sparrows. Condor 49:127–28.

United States Fish and Wildlife Service. 1993. Wildlife of Salton Sea National Wildlife Refuge, California. U.S. Fish Wildl. Serv., Washington, D.C.

Unitt, P. 1977. The Little Blue Heron in California. W. Birds 8:151–54.

———. 1984. The birds of San Diego County. San Diego Soc. Nat. Hist. Mem. 13.

———. 1985. Plumage wear in *Vireo bellii*. W. Birds 16:189–90.

———. 1987. *Empidonax traillii extimus*: An endangered subspecies. W. Birds 18:137–62.

Unitt, P., R. Rodriguez E., and A. Castellanos V. 1992. Ferruginous Hawk and Pine Siskin in the Sierra de la Laguna, Baja California Sur; Subspecies of the Pine Siskin in Baja California. W. Birds 24:171–72.

Unitt, P., K. Messer, and M. Théry. 1996. Taxonomy of the Marsh Wren in southern California. Proc. San Diego Soc. Nat. Hist. 31.

Unitt, P., and A. M. Rea. 1997. Taxonomy of the Brown Creeper in California. *In* Dickerman, R. W., ed., The era of Allan R. Phillips: A festschrift, Horizon Commun., Albuquerque, 177–86.

van Rossem, A. 1911. Winter birds of the Salton Sea region. Condor 13:129–37.

van Rossem, A. J. 1926. The California forms of *Agelaius phoeniceus* (Linnaeus). Condor 28:215–30.

———. 1933a. The Gila Woodpecker in the Imperial Valley of California. Condor 35:74.

———. 1933b. Two duck records for the Imperial Valley of California. Condor 35:72.

———. 1945. A distributional survey of the birds of Sonora, Mexico. Louisiana State Univ., Occ. Pap. Mus. Zool. 21.

van Rossem, A. J., and M. Hachisuka. 1937. A further report on birds from Sonora, Mexico, with descriptions of two new races. Trans. San Diego Soc. Nat. Hist. 8:321–36.

Van Tyne, J. 1956. What constitutes scientific data for the study of bird distribution? Wilson Bull. 68:63–67.

Vaurie, C. 1965. The birds of the Palearctic fauna: Non-Passeriformes. H. F. and G. Witherby, London.

Veit, R. R. 1988. Identification of the Salton Sea Rufous-necked Sandpiper. W. Birds 19:165–69.

Verbeek, N. A. M. 1966. Wanderings of the Ancient Murrelet: Some additional comments. Condor 68:510–11.

Vidal, O., and J.-P. Gallo-Reynoso. 1996. Die-offs of marine mammals and sea birds in the Gulf of California, Mexico. Marine Mammal Science 12:627–35.

Voelker, G., and S. Rohwer. 1998. Contrasts in scheduling of molt and migration in Eastern and Western Warbling-Vireos. Auk 115:142–55.

Voous, K. H. 1988. Owls of the Northern Hemisphere. MIT Press, Cambridge, Mass.

Vorhies, C. T., R. Jenks, and A. R. Phillips. 1935. Bird records from the Tucson region, Arizona. Condor 37:243–47.

Walker, B. W., ed. 1961. The ecology of the Salton Sea, California, in relation to the sportfishery. Calif. Dept. Fish Game, Fish Bull. 113.

Walker, B. W., R. R. Whitney, and G. W. Barlow. 1961. Fishes of the Salton Sea. Calif. Dept. Fish Game, Fish Bull. 113:77–91.

Walton, B., J. Linthicum, and G. Stewart. 1988. Release and re-establishment techniques developed for Harris' Hawks—Colorado River, 1979–1986. In Glinski, R. L., B. G. Pendleton, M. B. Moss, M. N. LeFranc Jr., B. A. Millsap, and S. W. Hoffman, eds., Proceedings of the Southwest Raptor Management Symposium and Workshop, Natl. Wildl. Fed. Sci. Tech. Ser. 11, 318–20.

Warnock, N., and M. A. Bishop. 1998. Spring stopover ecology of migrant Western Sandpipers. Condor 100:456–67.

Warnock, N., W. D. Shuford, and K. C. Molina. 2003. Distribution patterns of waterbirds at the Salton Sea, California, 1999. Stud. Avian Biol.: in press.

Waters, M. R. 1983. Late Holocene lacustrine chronology and archaeology of ancient Lake Cahuilla, California. Quarternary Res. 19:373–87.

Watkins, T. J. 1976. Turkey Vulture migrations in the Mojave Desert of California. M.Sc. thesis, Calif. State Polytechnic Univ., Pomona.

Weathers, W. W. 1983. Birds of Southern California's Deep Canyon. Univ. California Press, Berkeley.

Weber, J. W. 1981. The Larus gulls of the Pacific Northwest's interior, with taxonomic comments on several forms (part 1). Continental Birdl. 2:1–10.

Weir, D. N., A. C. Kitchener, and R. Y. McGowan. 2000. Hybridization and changes in the distribution of Iceland Gulls (Larus glaucoides/ kumlieni/thayeri). J. Zool. 252:517–30.

Weitzel, N. H. 1988. Nest-site competition between European Starlings and native breeding birds in northwestern Nevada. Condor 90:515–17.

Wenink, P. W., and A. J. Baker. 1996. Mitochondrial DNA lineages in composite flocks of migratory and wintering Dunlin (Calidris alpina). Auk 113:744–56.

Wenink, P. W., A. J. Baker, and M. G. J. Tilanus. 1993. Hypervariable control-region sequences reveal global population structuring in a long-distance migrant shorebird, the Dunlin (Calidris alpina). Proc. Natl. Acad. Sci. USA 90:94–98.

Wetmore, A. 1939. Notes on the birds of Tennessee. Proc. U.S. Natl. Mus. 86:175–243.

———. 1943. The birds of southern Veracruz, Mexico. Proc. U.S. Natl. Mus. 93:215–340.

———. 1964. A revision of the American vultures of the genus Cathartes. Smithsonian Misc. Coll. 146:1–18.

Whaley, W. W., and C. M. White. 1994. Trends in geographic variation of Cooper's Hawk and Northern Goshawk in North America: A multivariate analysis. Proc. West. Found. Vert. Zool. 5:161–209.

White, C. M. 1968. Diagnosis and relationships of North American tundra-inhabiting Peregrine Falcons. Auk 85:179–91.

———. 1994. Population trends and current status of selected western raptors. Stud. Avian Biol. 15:161–72.

White, C. M., and D. A. Boyce Jr. 1988. An overview of Peregrine Falcon subspecies. In Cade, T. J., J. H. Enderson, C. G. Thelander, and C. M. White, eds., Peregrine Falcon populations: Their management and recovery, Peregrine Fund, Boise, Idaho, 789–810.

Wilbur, S. R. 1973. The Red-shouldered Hawk in the western United States. W. Birds 4:15–22.

———. 1987. Birds of Baja California. Univ. California Press, Berkeley.

Wilbur, S. R., W. D. Carrier, and G. McCaskie. 1971. The Lark Bunting in California. Calif. Birds 2: 73–76.

Wilke, P. J. 1978. Late prehistoric human ecology at Lake Cahuilla, Coachella Valley, California. Contr. Univ. Calif. Archaeol. Res. Fac. 38.

Willett, G. 1912. Birds of the Pacific slope of southern California. Pac. Coast Avifauna 7.

———. 1930. The Large-billed Sparrow at Salton Sea. Condor 32:160.

———. 1932. Sanderlings and turnstones at Salton Sea, California. Condor 34:228.

———. 1933. A revised list of the birds of southwestern California. Pac. Coast Avifauna 21.

———. 1934. The Lower California Say Phoebe in southeastern California. Condor 36:117.

Wilson, E. O., and W. L. Brown Jr. 1953. The subspecies concept and its taxonomic application. Syst. Zool. 2:97–111.

Winter, J. 1973. The California Field Ornithologists Records Committee Report, 1970–1972. W. Birds 4:101–6.

Winter, J., and G. McCaskie. 1975. Report of the California Field Ornithologists Records Committee. W. Birds 6:135–44.

Wires, L. R., and F. J. Cuthbert. 2000. Trends in Caspian Tern numbers and distribution in North America: A review. Waterbirds 23:388–404.

Woerner, J. 1989. The creation of the Salton Sea: An engineering folly. J. West 28(1):109–12.

Woodbury, D. O. 1941. The Colorado conquest. Dodd, Mead, New York.

Wooten, W. A. 1952. Roseate Spoonbill in Imperial Valley, California. Condor 54:208.

Wyman, L. E. 1918. Notes from southern California. Condor 20:192.

———. 1920. A correction concerning the European Widgeon. Condor 22:158.

———. 1922. Notes from Imperial Valley. Condor 24:181–82.

Yohe, R. M., II. 1998. Notes on the late prehistoric extension of the range for the Muskrat (*Ondatra zibethicus*) along the ancient shoreline of Lake Cahuilla, Coachella Valley, Riverside County, California. Bull. South. Calif. Acad. Sci. 97:86–88.

Yong, W., and D. M. Finch. 1997. Migration of the Willow Flycatcher along the middle Rio Grande. Wilson Bull. 109:253–68.

Yuri, T., and D. P. Mindell. 2002. Molecular phylogenetic analysis of Fringillidae, "New World nine-primaried oscines" (Aves: Passeriformes). Mol. Phylogenet. Evol. 23:229–43.

Zeleny, L. 1969. Starlings versus native cavity-nesting birds. Atl. Nat. 24:158–61.

Zimmerman, D. A. 1973. Range expansion of Anna's Hummingbird. Am. Birds 27:827–35.

Zink, R. M. 1986. Patterns and evolutionary significance of geographic variation in the *Schistacea* group of the Fox Sparrow (*Passerella iliaca*). Ornithol. Monogr. 40.

———. 1994. The geography of mitochondrial DNA variation, population structure, hybridization, and species limits in the Fox Sparrow (*Passerella iliaca*). Evolution 48:96–111.

Zink, R. M., R. C. Blackwell, and O. Rojas-Soto. 1997. Species limits in the Le Conte's Thrasher. Condor 99:132–38.

Zink, R. M., D. L. Dittmann, S. W. Cardiff, and J. D. Rising. 1991. Mitochondrial DNA variation and the taxonomic status of the Large-billed Savannah Sparrow. Condor 93:1016–19.

Zink, R. M., and A. E. Kessen. 1999. Species limits in the Fox Sparrow. Birding 31:508–17.

Zink, R. M., and M. C. McKitrick. 1995. The debate over species concepts and its implications for ornithology. Auk 112:701–19.

Zink, R. M., S. Rohwer, A. V. Andreev, and D. L. Dittmann. 1995. Trans-Beringia comparisons of mitochondrial DNA differentiation in birds. Condor 97:639–49.

INDEX

Group or species accounts are on the page(s) in **boldface** type.

PROJECT MANAGEMENT AND COMPOSITION: Princeton Editorial Associates, Inc.

TEXT: Scala

DISPLAY: Scala Sans and Scala Sans Caps

PRINTER AND BINDER: Malloy Lithographing, Inc.